Lecture Notes in Computer Science 16239

Founding Editors

Gerhard Goos
Juris Hartmanis

Editorial Board Members

Elisa Bertino, *Purdue University, West Lafayette, IN, USA*
Wen Gao, *Peking University, Beijing, China*
Bernhard Steffen, *TU Dortmund University, Dortmund, Germany*
Moti Yung, *Columbia University, New York, NY, USA*

The series Lecture Notes in Computer Science (LNCS), including its subseries Lecture Notes in Artificial Intelligence (LNAI) and Lecture Notes in Bioinformatics (LNBI), has established itself as a medium for the publication of new developments in computer science and information technology research, teaching, and education.

LNCS enjoys close cooperation with the computer science R & D community, the series counts many renowned academics among its volume editors and paper authors, and collaborates with prestigious societies. Its mission is to serve this international community by providing an invaluable service, mainly focused on the publication of conference and workshop proceedings and postproceedings. LNCS commenced publication in 1973.

Luis Martínez · David Camacho · Hujun Yin ·
Bapi Dutta · Raciel Yera ·
Rosa M. Rodríguez Domínguez ·
Antonio Tallón-Ballesteros
Editors

Intelligent Data Engineering and Automated Learning – IDEAL 2025

26th International Conference
Jaén, Spain, November 13–15, 2025
Proceedings, Part II

 Springer

Editors
Luis Martínez (iD)
University of Jaén
Jaén, Spain

David Camacho (iD)
Technical University of Madrid
Madrid, Spain

Hujun Yin (iD)
The University of Manchester
Manchester, UK

Bapi Dutta (iD)
University of Jaén
Jaén, Spain

Raciel Yera (iD)
University of Jaén
Jaén, Spain

Rosa M. Rodríguez Domínguez (iD)
University of Jaén
Jaén, Spain

Antonio Tallón-Ballesteros (iD)
University of Huelva
Huelva, Spain

ISSN 0302-9743   ISSN 1611-3349 (electronic)
Lecture Notes in Computer Science
ISBN 978-3-032-10488-5   ISBN 978-3-032-10489-2 (eBook)
https://doi.org/10.1007/978-3-032-10489-2

This Springer imprint is published by the registered company Springer Nature Switzerland AG
The registered company address is: Gewerbestrasse 11, 6330 Cham, Switzerland

If disposing of this product, please recycle the paper.

Preface

The International Conference on Intelligent Data Engineering and Automated Learning (IDEAL) is an annual international conference dedicated to emerging and challenging topics in intelligent data analytics and associated machine learning paradigms and systems.

The 26th edition, IDEAL 2025, was held in Jaén, Spain, on November 13–15. This edition represented an excellent opportunity for the scientific community to meet in person and to enhance the exchange of ideas and sharing of information. Moreover, Jaén, the olive oil capital of the world and a city steeped in Andalusian history, offered an authentic and inspiring setting for a welcoming and memorable event.

The IDEAL conference is designed as a global meeting point for researchers, fostering collaboration and the exchange of innovative ideas and insights. It provides an opportunity not only to present ground-breaking research but also to engage in meaningful discussions and share experiences that drive progress in these challenging scientific areas. The core themes of IDEAL 2025 included Trustworthy Artificial Intelligence and Generative AI, Agentic AI, LLMs, Federated Learning, Machine Learning & Deep Learning for Real-World Applications, Data Mining and Pattern Recognition, Cybersecurity, Information Retrieval and Management, and Hybrid Intelligent Systems and Agents.

In total 146 submissions were received. Through a single-blind peer review process, 95 papers were accepted for publication in the main track and special sessions. The Programme Committee provided fundamental feedback and peer reviews to all the submissions. Submissions received 2.2 reviews each on average.

In addition to the IDEAL 2025 main track, there were 10 very interesting and relevant special sessions:

- Trustworthy and Intelligent Data Infrastructures: Distributed Ledger Technologies and Cybersecurity for Automated Learning Systems;
- Trustworthy and Explainable Artificial Intelligence: An Interdisciplinary Perspective;
- 2nd International Workshop on Anomaly Detection with Machine Learning;
- Federated Learning, Intelligent Fusion and Applications (FLIFA);
- Workshop on Intelligent Decision and Recommender Systems (WIDRS);
- Artificial Intelligence Applied to Image Analysis;
- Recent Advances in Generative Artificial Intelligence;
- AI-Driven Innovations: LLMs in Lifelong and Self-Supervised Learning;
- Machine Learning Under Concept Drift in Highly Dynamic Environments: Advances and Challenges (MLCD 2025); and
- Hybrid Intelligent Decision-Making: Integrating AI and MCDM for Real-World Applications.

We would like to thank the Special Sessions Chairs for their invaluable efforts in organizing these sessions. We would also like to extend a special thank you to our esteemed keynote speaker – Francisco Herrera, for delivering his exceptional lecture on Next Frontier in Explainable AI: Human-AI Synergy.

We would like to thank all reviewers for their time and effort in reviewing the papers. The published papers have gone through a refinement process to accommodate discussions during the conference as well as comments from the reviewers who directed any necessary improvements. Also, we would like to thank all the committee members and organizers who participated in any way in the conference and especially the sponsors: Universidad de Jaén, IEEE CIS UK and Ireland Chapter, and Springer LNCS.

September 2025

Luis Martínez
David Camacho
Bapi Dutta
Raciel Yera
Rosa M. Rodríguez Domínguez
Antonio J. Tallón-Ballesteros
Hujun Yin

Organization

General Chairs

Luis Martínez	University of Jaén, Spain
David Camacho	Technical University of Madrid, Spain
Hujun Yin	University of Manchester, UK

Program Chairs

Rosa M. Rodríguez Domínguez	University of Jaén, Spain
Juan M. Alberola	Universitat Politècnica de València, Spain
Antonio J. Tallón-Ballesteros	Universidad de Huelva, Spain

Special Session Chairs

Antonio J. Tallón Ballesteros	Universidad de Huelva, Spain
Diego García Zamora	University of Jaén, Spain

Local Chairs

Bapi Dutta	University of Jaén, Spain
Raciel Yera	University of Jaén, Spain

Publicity Chairs

Álvaro Labella	University of Jaén, Spain
Francisco José Quesada Real	University of Jaén, Spain

Web Chairs

Francisco J. Martínez	University of Jaén, Spain
Bruno Ramos Cruz	University of Jaén, Spain

Local Organizing Committee

Álvaro Labella	University of Jaén, Spain
Bruno Ramos Cruz	University of Jaén, Spain
Bapi Dutta	University of Jaén, Spain
Diego García Zamora	University of Jaén, Spain
Jessica Zaqueros Martínez	University of Jaén, Spain
Francisco J. Martínez	University of Jaén, Spain
Francisco José Quesada Real	University of Jaén, Spain
Luis Martínez	University of Jaén, Spain
Manuel J. Barranco García	University of Jaén, Spain
Pedro J. Sánchez Sánchez	University of Jaén, Spain
Raciel Yera	University of Jaén, Spain
Rosa M. Rodríguez Domínguez	University of Jaén, Spain

Steering Committee

Hujun Yin	University of Manchester, UK
Colin Fyfe	University of the West of Scotland, UK
Guilherme Barreto	Federal University of Ceará, Brazil
Jimmy Lee	Chinese University of Hong Kong, China
John Keane	University of Manchester, UK
Jose A. Costa	Federal University of Rio Grande do Norte, Brazil
Juan Manuel Corchado	University of Salamanca, Spain
Laiwan Chan	Chinese University of Hong Kong, China
Malik Magdon-Ismail	Rensselaer Polytechnic Institute, USA
Marc van Hulle	KU Leuven, Belgium
Ning Zhong	Maebashi Institute of Technology, Japan
Paulo Novais	University of Minho, Portugal
Peter Tino	University of Birmingham, UK
Samuel Kaski	Aalto University, Finland
Vic Rayward-Smith	University of East Anglia, UK
Yiu-ming Cheung	Hong Kong Baptist University, China
Zheng Rong Yang	University of Exeter, UK

Publicity and Liaison Chairs

Bin Li	University of Science and Technology of China, China
Guilherme Barreto	Federal University of Ceará, Brazil

Jose A. Costa	Federal University of Rio Grande do Norte, Brazil
Yimin Wen	Guilin University of Electronic Technology, China

Programme Committee

Juan M. Alberola	Universitat Politècnica de València, Spain
Ricardo Aler	Universidad Carlos III de Madrid, Spain
Cesar Analide	University of Minho, Portugal
Francisco Salas-Molina	Universitat Politècnica de València, Spain
Jarvin Anton	Universidad de Sancti Spiritus José Martí Pérez, Cuba
Romis Attux	FEEC/UNICAMP, Brazil
Reyhan Aydoğan	Özyeğin Üniversitesi, Turkey
Clara Baltasar	Universidad Politécnica de Madrid, Spain
Bruno Baruque	Universidad de Burgos, Spain
Carmelo J. A. Bastos Filho	University of Pernambuco, Brazil
Riza Batista-Navarro	University of Manchester, UK
Vitor Beires Nogueira	University of Évora, Portugal
Fran Bellas	University of A Coruña, Spain
Holger Billhardt	Universidad Rey Juan Carlos, Spain
Eduardo Borges	Universidade Federal do Rio Grande, Brazil
Lourdes Borrajo	University of Vigo, Spain
Juan A. Botia	University of Murcia, Spain
Vicent Botti	Universitat Politècnica de València, Spain
Jose Luis Calvo Rolle	University of A Coruña, Spain
David Camacho	Universidad Politécnica de Madrid, Spain
Hung Cao	University of New Brunswick, Canada
Roberto Carballedo	University of Deusto, Spain
Edianny Carballo	University of Jaén, Spain
Carlos Carrascosa	Universitat Politècnica de València, Spain
Luis Cavique	Universidade Aberta, Portugal
Songcan Chen	Nanjing University of Aeronautics and Astronautics, China
Ning Chen	Beijing Academy of Science and Technology, China
Stelvio Cimato	University of Milan, Italy
Manuel Cobo	University of Cádiz, Spain
Luís Conceição	Instituto Politécnico do Porto, Portugal
Roberto Confalonieri	Free University of Bozen-Bolzano, Italy
Juan M. Corchado	University of Salamanca, Spain
Paulo Cortez	University of Minho, Portugal

Raúl Cruz-Barbosa	Universidad Tecnológica de la Mixteca, Mexico
Sergio D'Antonio	Universidad Politécnica de Madrid, Spain
Andre de Carvalho	University of São Paulo, Brazil
Fernando De la Prieta	University of Salamanca, Spain
Javier Díez-González	Universidad de León, Spain
Gracaliz Dimuro	Universidade Federal do Rio Grande, Brazil
Dinu Dragan	University of Novi Sad, Serbia
Nestor Duque Méndez	Universidad Nacional de Colombia, Colombia
Dalila Durães	Universidade do Minho, Portugal
Bapi Dutta	University of Jaén, Spain
Ziga Emersic	University of Ljubljana, Slovenia
Francisco Enguix	Universidad Politécnica de Valencia, Spain
Florentino Fdez-Riverola	University of Vigo, Spain
Alberto Fernandez	Universidad Rey Juan Carlos, Spain
Antonio Fernández-Caballero	Universidad de Castilla-La Mancha, Spain
Joao Carlos A. Ferreira	ISCTE-IUL, Portugal
Ruben Ferrero Guillen	Universidad de León, Spain
Cèsar Ferri	Universitat Politècnica de València, Spain
Felipe França	Federal University of Rio de Janeiro, Brazil
Pedro Freitas	Universidade Católica Portuguesa, Portugal
Marcus Gallagher	University of Queensland, Australia
Diego Garces	Universidad Complutense de Madrid, Spain
Manuel Garcia-Piqueras	Universidad de Castilla-La Mancha, Spain
Isaías García-Rodríguez	Universidad de León, Spain
Diego Garcia-Zamora	University of Jaén, Spain
Javier Huertas	Universidad Politécnica de Madrid, Spain
María José Ginzo Villamayor	University of Santiago de Compostela, Spain
Adrián Girón Jiménez	Universidad Politécnica de Madrid, Spain
Luis Gómez	Universidad de Las Palmas de Gran Canaria, Spain
Teresa Gonçalves	University of Évora, Portugal
Cesar Gonzalez	Universidad Politécnica de Valencia, Spain
Maciej Grzenda	Warsaw University of Technology, Poland
Mohamed Ali Hadj Taieb	University of Sfax, Tunisia
Stella Heras	Universitat Politècnica de València, Spain
Emilcy Hernández-Leal	Universidad de Medellín, Colombia
J. Michael Herrmann	University of Edinburgh, UK
Wei-Chiang Hong	Harbin Engineering University, China
Dariusz Jankowski	Wrocław University of Science and Technology, Poland
Jaume Jordán	Universitat Politècnica de València, Spain
Angel Juan	Universitat Politècnica de València, Spain

Vicente Julian	Universitat Politècnica de València, Spain
Alexandros Konios	Nottingham Trent University, UK
Helena Liz	Universidad Politécnica de Madrid, Spain
Giancarlo Lucca	Universidade Católica de Pelotas, Brazil
Wenjian Luo	Harbin Institute of Technology, China
Jose Machado	University of Minho, Portugal
Rui Neves	Universidade Nova de Lisboa, Portugal
Rafael Magdalena	University of Valencia, Spain
Antón Makarov	CSIC, Spain
Cedric Marco-Detchart	Universidad Pública de Navarra, Spain
Pasqual Martí	Universitat Politècnica de València, Spain
Sergio Marti Sanz	Universitat Politècnica de València, Spain
Alejandro Martín	Universidad Politécnica de Madrid, Spain
Jose Francisco Martínez Trinidad	INAOE, Mexico
Diogo Martinho	Instituto Politécnico do Porto, Portugal
Pablo Miralles-González	Universidad Politécnica de Madrid, Spain
Pablo Miranda Rodríguez	Tecnalia Research & Innovation, Spain
Jose Molina	Universidad Carlos III de Madrid, Spain
Carlos Monserrat	Universitat Politècnica de València, Spain
Joan Ciprià Moreno	Universitat Politècnica de València, Spain
Paulo Moura Oliveira	University of Trás-os-Montes and Alto Douro, Portugal
Silvia Nadal	Universitat Politècnica de València, Spain
Tatsuo Nakajima	Waseda University, Japan
Susana Nascimento	NOVA University Lisbon, Portugal
Debashis Naskar	University of Salamanca, Spain
Antonio Neme	Universidad Autónoma de México, Mexico
Paulo Novais	Universidade do Minho, Portugal
Bogdan Okreša Đurić	University of Zagreb, Croatia
José Valente de Oliveira	University of Algarve, Portugal
Eva Onaindia	Universitat Politècnica de València, Spain
Izaskun Oregi	TECNALIA, Basque Research and Technology Alliance, Spain
Eneko Osaba	TECNALIA, Spain
Cristina Padro-Ferragut	Universitat Politècnica de València, Spain
Javi Palanca	Universitat Politècnica de València, Spain
Jose Tomas Palma	University of Murcia, Spain
Marco Antonio Peña Cubillos	International University of Andalucia, Spain
Carlos Pereira	ISEC, Portugal
Yilena Pérez	Universidad de Holguín, Cuba
Iñigo Perez Delgado	Ayesa, Spain
Laura Po	University of Modena and Reggio Emilia, Italy

Alessandro Sebastian Podda	University of Cagliari, Italy
Radu-Emil Precup	Politehnica University of Timişoara, Romania
Abraham Prieto García	Universidade da Coruña, Spain
Akash Punhani	SRM Institute of Science and Technology, India
Paulo Quaresma	University of Évora, Portugal
Francisco Quesada Real	University of Jaén, Spain
Julio Quintana	Universidad de Ciego de Ávila, Cuba
Bruno Ramos Cruz	University of Jaén, Spain
Izabela Rejer	West Pomeranian University of Technology in Szczecin, Poland
Jaime Andres Rincon Arango	Universidad de Burgos, Spain
Elsa Rodrigues	Instituto Politécnico de Beja, Portugal
Sara Rodríguez González	University of Salamanca, Spain
Alejandro Romero	Universidade da Coruña, Spain
Ruben Ruiz-Gonzalez	Universidad de Burgos, Spain
Victor Sanchez-Anguix	Universitat Politècnica de València, Spain
María Inmaculada Santamaría	Universidad Politécnica de Madrid, Spain
Helida Santos	Federal University of Rio Grande, Brazil
Ricardo Santos	Polytechnic Institute of Porto, Portugal
Matilde Santos Peñas	Universidad Complutense de Madrid, Spain
José Santos Reyes	University of A Coruña, Spain
Yuya Sasaki	Osaka University, Japan
Markus Schatten	University of Zagreb, Croatia
Marcin Szpyrka	AGH University of Science and Technology, Poland
Antonio J. Tallón-Ballesteros	University of Huelva, Spain
Joaquin Taverner	Universitat Politècnica de València, Spain
Nikolay Tcholtchev	Fraunhofer FOKUS, Germany
Jan Telle	University of Bergen, Norway
Murat Testik	Hacettepe University, Turkey
Darian Tomašević	University of Ljubljana, Slovenia
Stefania Tomasiello	University of Salerno, Italy
Alexandros Tzanetos	Jönköping University, Sweden
Eiji Uchino	Yamaguchi University, Japan
Eva Vallada	Universitat Politècnica de València, Spain
Alfredo Vellido	Universitat Politècnica de Catalunya, Spain
Paula Verde	Universidad de León, Spain
María Fulgencia Villa	Universitat Politècnica de València, Spain
Matej Vitek	University of Ljubljana, Slovenia
Tzai-Der Wang	Cheng Shiu University, Taiwan
Agnieszka Wendland	Warsaw University of Technology, Poland
Michal Wozniak	University of Wrocław, Poland

Xin-She Yang	Middlesex University London, UK
Hua Yang	Zhongyuan University of Technology, China
Raciel Yera	University of Jaén, Spain
Hujun Yin	University of Manchester, UK
Avik Adhikary	Southwest Jiaotong University, China
Pragya Gupta	IIT Kharagpur, India
Jessica Zaqueros-Martinez	University of Jaén, Spain
Jose M. Molina	Universidad Carlos III de Madrid, Spain
Tugce Arican	University of Twente, Netherlands
Álvaro Labella Romero	University of Jaén, Spain
Juan Moreno Garcia	University of Castilla-La Mancha, Spain
David Muñoz-Valero	University of Castilla-La Mancha, Spain
N. Gowtham Reddy	IIT Kharagpur, India
Alejandro Antonio Torres García	INAOE, Mexico
Dikshit Chauhan	National University of Singapore, Singapore
Antoine Marmonie	École Nationale Supérieure des Mines de Saint-Étienne, France
Soumita Guria	IIT Kharagpur, India
César Guevara	CUNEF, Spain
Luis Zhinin-Vera	University of Castilla-La Mancha, Spain
Francisco Jesús Martínez Mimbrera	University of Jaén, Spain
Tanmay Kayal	IIT Mandi, India
Luís Matos	Universidade do Minho, Portugal
Sung-Bae Cho	Yonsei University, South Korea
Dmitriy Fedorov	ITMO, Russia
Manuel Germán Morales	University of Jaén, Spain
Mateus Toledo	University of Jaén, Spain
Sanchita Mondal	IIT Kharagpur, India
Weslley Moura	Universidade do Minho, Portugal
Danilo Valdes-Ramirez	University of Salamanca, Spain
Rubén Rodríguez Fernández	Rey Juan Carlos University, Spain
Pedro José de Oliveira	Universidade do Minho, Portugal
Alejandro Bellogin	Autonomous University of Madríd, Spain
Maria Salamó	University of Barcelona, Spain
Antonio Moreno	Rovira i Virgili University, Spain
Alejandro Ariza-Casabona	University of Barcelona, Spain
Iván Cantador	Autonomous University of Madríd, Spain
Pablo Pérez-Núñez	University of Oviedo, Spain
Pablo Sánchez	Comillas Pontifical University, Spain

Special Session on Trustworthy and Intelligent Data Infrastructures: Distributed Ledger Technologies and Cybersecurity for Automated Learning Systems

Organizers

Ana Isabel Azevedo	CEOS.PP/ISCAP/Polytechnic Institute of Porto, Portugal
Antonio Santos-Olmo	University of Castilla-La Mancha, Spain
Óscar Carlos Medina	Universidad Tecnológica Nacional – Facultad Regional Córdoba, Argentina
Francisco José Quesada Real	University of Jaén, Spain

Special Session on Hybrid Intelligent Decision-Making: Integrating AI and MCDM for Real-World Applications

Organizers

Pavel Anselmo Álvarez-Carrillo	Universidad Autónoma de Occidente, Mexico
Luis Pérez-Domínguez	Universidad Autónoma de Ciudad Juárez, Mexico
Ernesto León-Castro	Universidad Católica de la Santísima Concepción, Chile

Special Session on Trustworthy and Explainable Artificial Intelligence: an Interdisciplinary Perspective

Organizers

Manuel Germán Morales	University of Jaén, Spain
Ángel M. García Vico	University of Jaén, Spain
Cristóbal J. Carmona	University of Jaén, Spain

Workshop on Intelligent Decision and Recommender Systems

Organizers

Luis Martínez   University of Jaén, Spain
Bapi Dutta   University of Jaén, Spain
Raciel Yera   University of Jaén, Spain
Dikshit Chauhan   National University of Singapore, Singapore

Special Session on Machine Learning Under Concept Drift in Highly Dynamic Environments: Advances and Challenges

Organizers

José Tomás Palma Méndez   University of Murcia, Spain
Juan A. Botía   University of Murcia, Spain
Grzegorz J. Nalepa   Halmstad University, ELLIIT, Sweden and
                            Jagiellonian University, Poland
Martin Atzmueller   Osnabrück University and German Research
                            Center for Artificial Intelligence (DFKI),
                            Germany

The 2nd International Workshop on Anomaly Detection with Machine Learning

Organizers

Jason J. Jung   Chung-Ang University, South Korea
Gen Li   Chengdu University, China
Szymon Bobek   Jagiellonian University, Poland
Grzegorz Nalepa   Jagiellonian University, Poland

Special Session on Artificial Intelligence Applied to Image Analysis

Organizers

Gema Parra Cabrera   University of Jaén, Spain
Francisco Daniel Pérez Cano   University of Jaén, Spain

Special Session on AI-Driven Innovations: LLMs in Lifelong and Self-supervised Learning

Organizers

Youcef Djenouri   Norwegian Research Center, Norway
Gautam Srivastava   Brandon University, Canada
Tomasz Michalak   Warsaw University, Poland

Special Session on Recent Advances in Generative Artificial Intelligence

Organizers

Alberto Fernández-Isabel   Rey Juan Carlos University, Spain
Rubén Rodríguez Fernández   Rey Juan Carlos University, Spain

Special Session on Federated Learning, Intelligent Fusion and Applications (FLIFA)

Organizers

Cedric Marco-Detchart   Universidad Pública de Navarra, Spain
Jaime A. Rincon Arango   Universidad de Burgos, Spain
Vicente Julian   Universitat Politècnica de València, Spain
Carlos Carrascosa Casamayor   Universitat Politècnica de València, Spain
Paulo Novais   Universidade do Minho, Portugal
Carlos Lopez-Molina   Universidad Pública de Navarra, Spain
Laura de Miguel Turullols   Universidad Pública de Navarra, Spain

Giancarlo Lucca   Universidade Católica de Pelotas, Brazil
Graçaliz P. Dimuro   Universidade Federal do Rio Grande, Brazil
Jose Guillermo Guarnizo Marin   Universidad Distrital Francisco José de Caldas,
                                       Colombia
Humberto Bustince   Universidad Pública de Navarra, Spain

Contents – Part II

Workshop on Intelligent Decision and Recommender Systems

Special Session on Machine Learning Under Concept Drift in Highly Dynamic Environments: Advances and Challenges

Special Session on Trustworthy and Explainable Artificial Intelligence: an Interdisciplinary Perspective

Special Session on Trustworthy and Intelligent Data Infrastructures: Distributed Ledger Technologies and Cybersecurity for automated Learning Systems

Special Session on Hybrid Intelligent Decision-Making: Integrating AI and MCDM for Real-World Applications

Contents – Part I

Main Track

Special Session on Federated Learning, Intelligent Fusion and Applications (FLIFA)

**AI-Driven Innovations: LLMs in Lifelong and Self-Supervised
Learning**

Workshop on Intelligent Decision and Recommender Systems

The Impact of Contexts in Contextual Matrix Factorization Models

Alejandro Bellogín[✉]

Universidad Autónoma de Madrid, Madrid, Spain
alejandro.bellogin@uam.es

Abstract. While matrix factorisation and neural network techniques have become increasingly popular for incorporating contextual information in recommendation, a gap exists in understanding the impact of multidimensional contexts on the performance of these methods. Our study highlights this gap, noting the absence of analysis determining which types of multidimensional contexts yield the greatest benefits from the application of these approaches. In our experiments with real data from the tourism domain, where several contexts are available at the same time, we conclude that inherent venue features – such as their categories or whether a parking is available – impact more in the results than other contexts that can be captured with other data signals – such as venue popularity.

Keywords: Recommender Systems · Deep Learning · Context

1 Introduction

In today's dynamic and personalized online environment, context plays an increasingly vital role in shaping user behaviour and preferences. Whether it is the time of day influencing purchasing decisions, the location affecting restaurant choices, or the presence of social companions impacting entertainment preferences, context is inextricably linked to user needs and desires [2]. Recognising this importance, Context-Aware Recommender Systems (CARS) have gained significant traction [3], since they allow to capture the complexity of real-world decision-making, where context plays a crucial role. By explicitly incorporating contextual information into the recommendation process, CARS lead to more personalised and relevant recommendations [3], where Matrix Factorisation (MF) and Neural Networks (NNs) have become dominant paradigms to achieve this goal [9].

As presented in [10, 12], incorporating contexts in deep learning architectures allows for better contextual representations, which in turn help learning more accurately the user preferences. Indeed, those two works propose deep context-aware recommendation models supporting explicit, latent, or structured context embeddings. To do that, the authors from [10] extend the neural collaborative filtering models previously proposed in [8] (NCF and NeuMF) by adding a new

L. Martínez et al. (Eds.): IDEAL 2025, LNCS 16239, pp. 3–9, 2026.
https://doi.org/10.1007/978-3-032-10489-2_1

component of contextual information. In this way, a generalisation of NeuMF is defined that consists of two major components which incorporate contextual information in different ways: Generalised Matrix Factorisation (GMF) and MultiLayer Perceptron (MLP). In that model, these GMF and MLP towers allows to obtain better performance, either in terms of error prediction or ranking [10].

Following that line of work, in [12] the authors extended that method by exploiting in which components the context vectors can be incorporated and how they should be learned. By doing this, authors found that there was always a variant of their approach (named NeuCMF) which obtained better results than DeepFM [7], which can take rich information into consideration and was not included in the experiments from [10]. Hence, NeuCMF seems like a promising state-of-the-art algorithm for CARS, however, no analyses have been made to understand how this approach depends on context variability.

In this work, we analyse the impact of different contexts on the performance of NeuCMF. In particular, we consider how the model is affected by being trained with or without some contexts in the tourism domain. By doing this, we conclude that inherent venue features – such as their categories or whether a parking is available – impact more in the results than other contexts that can be captured with other signals of the data – such as how popular a venue is. Our main contribution is, thus, a deeper understanding on how contextual recommendation approaches based on MF and NNs depend on the type of contexts being exploited. Even though this is an ongoing work, we aim to bring the current gap in the literature about this issue and motivate future research to explore the interplay between multidimensional context characteristics and complex recommendation techniques such as NeuCMF or others based on MF and NNs.

2   Contextual Matrix Factorization Models

NeuCMF models, as defined in [12], are a family of approaches comprising the 6 variations summarised in Table 1 and implemented in the DeepCARSKit library [11]. These models may use up to 3 GMF towers (called simply MF towers in that work) and an MLP tower. The MF towers execute the elementwise product to simulate a Matrix Factorisation method either between the user and item embeddings (UI tower), user and context embeddings (UC tower), or item and context embeddings (IC). At the same time, the MLP tower may concatenate embeddings from users (MLP_u), from items (MLP_i), or from the contexts (MLP_c) as the first input to several layers of the multilayer perceptron.

Moreover, [12] proposes 3 options to incorporate the context in the model, based on the components previously defined:

- Option 1, where the contexts are only considered in the MLP tower, ignoring the UC and IC towers.
- Option 2, where contexts are only incorporated in the MF towers, hence using the UC and IC towers, and discarding MLP_c from the MLP tower.
- Option 3, where contexts are used in both the MF towers and MLP towers.

Table 1. Variations of NeuCMF model as defined in [12].

Variation	Option	Towers in NeuCMF	Mode for Context Embedding
$NeuCMF_{0i}$	1	UI + MLP Towers	i mode
$NeuCMF_{0w}$		with MLP_c	w mode
$NeuCMF_{i0}$	2	All 4 towers	i mode
$NeuCMF_{w0}$		without MLP_c	w mode
$NeuCMF_{ii}$	3	All 4 towers	i mode
$NeuCMF_{ww}$		with MLP_c	w mode

Furthermore, the authors propose 2 methods to represent the context as the embeddings: the w mode, where the embedding is created for the whole context situation; or the i mode, where embeddings are created for each context condition, and the representation of a context situation is obtained by concatenating each individual embedding vector for the conditions in the context situation. This method, in principle, is expected to alleviate the sparsity issue, especially when there are many context dimensions in the data set, even though it assumes an independent relationship among context conditions, which may not always hold.

3   Experiments and Results

In this section, we first present the settings followed in the reported experiments (Sect. 3.1) and how the contexts were obtained from the data (Sect. 3.2). Then, in Sects. 3.3 and 3.4 we present the results found to answer the main research question we consider in this work: *what is the impact of contexts in NeuCMF models?* We first consider only the effect of individual contexts (Sect. 3.3), consequently, some preliminary tests for pairs of contexts are included and analysed (Sect. 3.4).

3.1   Experimental Settings

We consider the Yelp dataset[1], which was not included as a tested domain in [12]. This dataset includes user reviews and ratings to different tourism venues, especially restaurants, of several cities in the United States. Because of limitations with the library, we selected a city large enough to perform the experiments while having a decent amount of items. After filtering out interactions with incomplete information, or users with few reviews (to have enough facformation to train and evaluate the algorithm), we selected the city of *Tampa* with 13K reviews, a size comparable to the second-largest dataset used in [12].

We performed a 5-fold cross validation split, and report Precision and Recall metrics at various cutoffs [6]. Even though in [12] the authors reported error

---

[1] https://business.yelp.com/data/resources/open-dataset/.

Table 2. Precision and Recall values of NeuCMF when all the contexts are considered (row 'all'), when one context is removed (middle block), or two contexts are removed (last block). Best improvement with respect to 'all' in bold, worst result in italic. Arrows denote whether results with respect to 'all' are improved ($\uparrow$) or not ($\downarrow$) when removing those contexts.

Contexts		P@10	P@20	R@10	R@20
all		0.0176	0.0111	0.1760	0.2213
no categories	$\downarrow$	*0.0094*	*0.0084*	*0.0646*	*0.1112*
no day	$\uparrow$	0.0220	0.0149	0.2187	0.2964
no parking	$\downarrow$	0.0139	0.0099	0.1386	0.1964
no meal	$\downarrow$	0.0170	0.0111	0.1689	0.2208
no popularity	$\uparrow$	**0.0270**	**0.0207**	**0.2686**	**0.4117**
no pop, no day	$\downarrow$	0.0154	0.0095	0.1531	0.1876
no cat, no par	$\downarrow$	*0.0085*	*0.0084*	*0.0520*	*0.0977*

metrics (i.e., MAE and RMSE), in this work we decided to focus on ranking metrics, as they are more appropriate for the top-N recommendation problem [6].

3.2   Considered Contextual Dimensions

One of the motivations to use this dataset is that several contexts become available. In this section, we present which ones we considered. In every case, they were extracted from information derived from data, not inferred from textual reviews as in recent work [4].

First of all, taking the timestamp of the review into account, we extracted two contextual dimensions: based on the day of the week, whether it was a weekday (WD) or weekend day (WE); and based on the hour, if the meal was dinner (Din, after 5PM) or lunch (Lch). These two dimensions represent the temporal context attached to the user interaction.

Secondly, we understand the category of a venue can be interpreted as a contextual dimension representing the type of item a user is interested in at that moment. Since items in this dataset may have more than one category assigned, we use all the categories combined and use their value after hashing to obtain unique representations.

Finally, we also consider two more contexts related to characteristics of the items (item features). Whether the venue had its own parking lot (Gar) or not (NoGar) and whether it could be considered a popular place. For this, we determined that it was popular (Pop) if its number of reviews was within the top 20% of all the venues in that city (for Tampa, this meant that it had to be greater than 418), otherwise the venue was considered not popular (NoPop).

We understand the last three contextual dimensions (category, parking, popularity), while being strictly item features, they can be associated to contextual preferences of users at a specific time, such as whether they need a restaurant with parking or from a specific category.

3.3 Performance Analysis When One Contextual Dimension Is Removed

Table 2 shows the performance obtained when the selected NeuCMF model[2] exploits all the dimensions explained in Sect. 3.2, and how the performance is affected by removing one context at a time.

We observe that performance improves whenever the popularity or the day contexts are removed. On the other hand, if the other dimensions (categories, parking, or meal) are removed, performance decreases. This behaviour is consistent for both metrics and cutoffs.

Based on these results, we conclude that the most impactful contexts are those that are not easy to be represented already by other signals of the data while, at the same time, embed enough information by themselves. For example, *popularity* can already be captured by rating patterns, and several methodologies have been proposed to model (and uncover) this signal [1]. The case of the contextual dimension *day* is related to the granularity of this context; whereas in some domains it might be enough to discriminate between weekends and week days (such as restaurant or movie recommendation [5]), for the overall domain of tourism, and in particular, for the Yelp dataset, this is not enough, as interactions are spread across the entire week.

3.4 Performance Analysis When Removing Pairs of Contextual Dimensions

The last rows of Table 2 show the performance of NeuCMF when a pair of contextual dimensions was removed. For the sake of space, only two pairs are included in the table. The first one is a combination *(no pop, no day)* where both contexts have demonstrated to be bad for the model (i.e., their removal increased performance); the second one *(no cat, no par)* illustrates a situation where both are individually good for the model.

According to the obtained results, for the two analysed combinations performance decrease. This may be attributed, in the first case, to the fact that too much information is removed, since even though both dimensions (individually) improve the model when removed, it seems the method is not able to learn user preferences and discriminate the corresponding contextual situations well enough without this pair of dimensions. For the second case, since the performance decreases when each context was removed, it is no surprise that when removing both, performance decreases too. However, it should be noted that, in

---

[2] For this preliminary study, we conducted some tests with our data and the best performing method was NeuCMF_{ii}, so this is the NeuCMF variation used here. We leave for future work an exhaustive analysis of the dependency of these contextual dimensions for the other variations described in Sect. 2.

terms of obtained values, this second combination produces the worst results, evidencing that these contextual dimensions are more useful than other dimensions, and it is not only a matter of reducing too much the amount of information available for the model.

4 Discussion and Limitations

In the study presented herein, we have discovered the impact of multidimensional contexts in the performance of NeuCMF; we now dig further in this analysis. We note that the analysed contextual dimensions have a larger impact on Recall than Precision, and this impact is more positive in higher cutoffs (20), both in terms of actual improvement (from 53.4% at cutoff 10 to 86.5% for *no popularity*, for example), but also in lower losses (from -46.6% at cutoff 10 to -24.3% for *no categories*, for example).

We have also observed that removing two dimensions at the same time is not convenient, at least for the combinations tested in our study. But the fact that removing one contextual dimension may improve the performance, raises the importance of properly understanding the research question of this work (*what is the impact of contexts in NeuCMF models?*), since it could bring advantages in terms of efficiency, but also in terms of the accountability of the algorithms, as the less information an algorithm takes into account, the less exposed users will be and less privacy risks they would face. Hence, when evidencing comparable performances, models using (and requiring) less data from users should be preferred.

One important limitation of this work is that more contextual dimensions should be tested. In this work, we extracted dimensions where some contextual meaning could be attached to, but according to recent works, many others could be extracted directly from the user reviews [10], so it would be worthwhile to analyse if either types of contexts impact in the same way. Other NeuCMF models should be considered, as we focused on the one that performed the best in some preliminary experiments. By comparing other variations, it would also make it possible to analyse the impact of the model complexity on how capable they are to continue working at decent performance levels with a varying number of contexts.

5 Conclusions

Better exploiting and integrating the contexts of user interactions is key to understand the user behaviour and improve their recommendations. However, for such purpose it is necessary to know which context information is more valuable for a given contextual recommendation algorithm. The work reported in this paper presents a preliminary study to analyse this issue. We observed that some dimensions have a larger impact on the performance than others, in particular, those encoding signals that could be derived from interactions (such as popularity) are good candidates to not be considered as contextual dimensions. No conclusive answer was obtained regarding the impact of pairs of contexts, as the ones analysed always reduced the performance, so further work is deserved to fully understand how and which contexts should be exploited by contextual MF neural models.

Acknowledgments. This work was supported by Grant PID2022-139131NB-I00 funded by MCIN/AEI/10.13039/501100011033 and by "ERDF, a way of making Europe." We thank Fernando Huidobro for running the experiments.

References

1. Abdollahpouri, H., Burke, R., Mobasher, B.: Managing popularity bias in recommender systems with personalized re-ranking. In: Barták, R., Brawner, K.W. (eds.) Proceedings of the 32nd FLAIRS, pp. 413–418. AAAI Press (2019)
2. Abowd, G.D., Dey, A.K., Brown, P.J., Davies, N., Smith, M., Steggles, P.: Towards a better understanding of context and context-awareness. In: Gellersen, H.-W. (ed.) HUC 1999. LNCS, vol. 1707, pp. 304–307. Springer, Heidelberg (1999). https://doi.org/10.1007/3-540-48157-5_29
3. Adomavicius, G., Bauman, K., Tuzhilin, A., Unger, M.: Context-aware recommender systems: From foundations to recent developments. In: Ricci, F., Rokach, L., Shapira, B. (eds.) Recommender Systems Handbook, pp. 211–250. Springer, New York (2022). https://doi.org/10.1007/978-1-0716-2197-4_6
4. Bauman, K., Tuzhilin, A.: Know thy context: parsing contextual information from user reviews for recommendation purposes. Inf. Syst. Res. **33**(1), 179–202 (2022)
5. Campos, P.G., Díez, F., Cantador, I.: Time-aware recommender systems: a comprehensive survey and analysis of existing evaluation protocols. User Model. User Adapt. Interact. **24**(1–2), 67–119 (2014)
6. Gunawardana, A., Shani, G., Yogev, S.: Evaluating recommender systems. In: Ricci, F., Rokach, L., Shapira, B. (eds.) Recommender Systems Handbook, pp. 547–601. Springer, US (2022)
7. Guo, H., Tang, R., Ye, Y., Li, Z., He, X.: DeepFM: a factorization-machine based neural network for CTR prediction. In: Sierra, C. (ed.) Proceedings of the 26th IJCAI, pp. 1725–1731 (2017)
8. He, X., Liao, L., Zhang, H., Nie, L., Hu, X., Chua, T.: Neural collaborative filtering. In: Barrett, R., Cummings, R., Agichtein, E., Gabrilovich, E. (eds.) Proceedings of the 26th WWW, pp. 173–182. ACM (2017)
9. Mateos, P., Bellogín, A.: A systematic literature review of recent advances on context-aware recommender systems. Artif. Intell. Rev. **58**(1), 20 (2025)
10. Unger, M., Tuzhilin, A., Livne, A.: Context-aware recommendations based on deep learning frameworks. ACM Trans. Manag. Inf. Syst. **11**(2), 8:1–8:15 (2020)
11. Zheng, Y.: DeepCARSKit: a demo and user guide. In: Proceedings of 30th UMAP, pp. 18–21. ACM (2022)
12. Zheng, Y., Arias, G.F.: A family of neural contextual matrix factorization models for context-aware recommendations. In: Proceedings of 30th UMAP, pp. 1–6. ACM (2022)

A Comparative Fairness Study in News Recommendation Systems

Marta Salcedo and Alejandro Bellogín[(⊠)]

Universidad Autónoma de Madrid, Madrid, Spain
`marta.salcedos@estudiante.uam.es, alejandro.bellogin@uam.es`

Abstract. Recommender systems have transformed how users access information, particularly in the news domain. While these systems enhance personalization, they also introduce fairness concerns, potentially limiting users' exposure to diverse perspectives. This experimental study aims to evaluate fairness across various recommendation algorithms by comparing collaborative and neural network-based approaches. To assess performance, both classical evaluation metrics, together with bias and fairness-aware metrics, such as Bias Disparity, Ranking-based Equal Opportunity, and Average Recommendation Popularity, are considered. The Adressa dataset is used to analyze recommendation behavior and their impact when using various user and item attributes. The results highlight important variations in fairness distribution across different algorithms, underscoring the necessity of incorporating fairness considerations into recommender system design.

Keywords: News recommendation · Fairness · Algorithmic bias

1 Introduction

News articles are continuously published, while less-recent ones rapidly become not relevant or interesting for users. With a vast volume of news emerging daily, users often struggle to efficiently find content that aligns with their interests. At the same time, Recommender Systems (RS), while designed to enhance user experience by personalizing content, often inadvertently propagate media bias [21]. This occurs as these systems tend to favor content that aligns with users' existing beliefs, leading to an echo chamber effect that restricts exposure to diverse viewpoints [6]. Empirical studies indicate that advanced news RS are more likely to suggest biased news articles to users who exhibit a preference for such content, thereby amplifying the issue of media bias in digital news consumption [7].

Thus, developing fair RS in general, but especially in the news domain, is a critical step towards improving equal opportunities among users, items, and other stakeholders. However, there are several definitions of what fairness means [9], and many proposals of fair RSs use various fairness metrics focused on specific criteria [29]. Because of these issues, there is a lack of a standardized evaluation

© The Author(s), under exclusive license to Springer Nature Switzerland AG 2026
L. Martínez et al. (Eds.): IDEAL 2025, LNCS 16239, pp. 10–22, 2026.
https://doi.org/10.1007/978-3-032-10489-2_2

framework and benchmarking results across metrics [16]. Moreover, while it is important to improve (or guarantee some minimum) fairness, it might be possible that the overall recommendation performance may be negatively affected when introducing mitigation strategies with that goal. Hence, a balance between fairness and recommendation accuracy is recognized as necessary for these systems to be practical and useful in the real world [5, 17].

In this context, we study the performance of various state-of-the-art collaborative filtering and neural network recommendation algorithms in the news domain, where these challenges are critical. Indeed, as stated in a recent paper [26], news RS should inform users about the existence of opinions they are not familiar with (diversity), without favoring some of them (fairness); however, no previous study has considered several fairness and bias metrics while comparing their results. Even though our main focus is on bias and fairness, we also measure classical accuracy, to understand the overall impact on all dimensions. For this, we use a real-world benchmark dataset, Adressa, where various user and item attributes can be extracted and analyzed. The results demonstrate the well-known tradeoff between fairness and accuracy, but they also evidence a less obvious outcome: that it might be more difficult to guarantee fairness for some attributes, whereas for others (probably because they are linked to other signals, such as popularity), classical and simple techniques already provide *fair* results.

2 Fairness and Bias Evaluation

When discussing fairness within RS, it is important to recognize the distinction between fairness and bias, two concepts that usually converge as synonyms or closely related [9]. **Bias** is usually associated (especially when referring to *item bias*) to the propensity of RS to disproportionately favor popular items, thereby creating a feedback loop that limits exposure to a diverse range of content [19]. Conversely, **fairness** concerns both users and items, and addresses disparities of recommendations received by different user groups while impacting on items of different characteristics [10]. Here we consider the following metrics to measure bias and fairness. First, regarding bias and, hence, mostly focused on popularity:

Average Coverage of Long Tail Items (ACLT) measures the extent to which items in the long tail (less popular items) are recommended. Higher ACLT indicates better representation of niche or less popular items [2].

Average Percentage of Long Tail Items (APLT). This metric reflects the recommender system's ability to diversify its suggestions beyond mainstream, popular items [1].

Average Recommendation Popularity (ARP). Captures the average popularity of recommended items, with lower ARP indicating a preference for recommending less popular items [31].

Popularity-based Ranking-based Equal Opportunity (PopREO). Focuses on equalizing the exposure of popular and less popular items in the top ranks, ensuring balanced representation regardless of item popularity [33]. Lower values are better.

Popularity-based Ranking-based Statistical Parity (PopRSP).
Ensures that the proportion of popular and less popular items in recommendation lists is statistically similar across different user groups, promoting parity and mitigating popularity bias [33].

Expanding on the notions of item and user fairness, it becomes evident that fairness in RS extends beyond the interactions between users and items to encompass a more comprehensive ecosystem. This ecosystem includes both consumers–the users who interact with recommendations–and providers–the creators or organizations whose content is being recommended [22]. Understanding fairness through these dual lenses of consumer fairness and provider fairness provides a more holistic approach to addressing biases and disparities in recommendation outcomes. Consumer fairness ensures equitable user experiences by safeguarding against biases that disproportionately affect specific user groups, while provider fairness focuses on creating balanced opportunities for content creators to reach diverse audiences. Together, these dimensions underscore the interconnectedness of fairness concerns, where neglecting one can amplify inequities in the other, ultimately shaping the broader fairness landscape of recommender systems [10].

With growing concerns regarding fairness in recommendations, specific metrics have been developed to evaluate how equitably algorithms serve different user or item groups. The following will be used for the evaluation of the methods tested in this work:

Bias Disparity (BD) measures the difference in recommendation performance between user or item groups, highlighting systemic biases in the model [27]. A value close to 0 indicates the recommendation model mirrors the bias found in the data, meaning it neither amplifies nor mitigates existing biases, whereas a positive (negative) value suggests the model introduces a higher (lower) bias than what is present in the original data.
Item Mean Absolute Deviation Ranking-based (MADR). This metric evaluates how evenly ranking performance is distributed in recommendation lists among the different (item) groups, helping to reduce item unfairness [8].
Ranking-based Equal Opportunity (REO) ensures that items from different groups have an equal probability of appearing at the top ranks of recommendation lists [33].

3   Experiments and Results

In this section, we address our main research question: **how do state-of-the-art RS based on collaborative filtering and neural networks perform in terms of fairness and accuracy in the news domain?** For this, in Sect. 3.1 we present the data used in our study and how the user and item groups were identified; later, Sects. 3.2, 3.3, and 3.4 show the results obtained when considering accuracy, bias, and fairness metrics. In Sects. 3.5 and 3.6 we discuss the main outcomes and limitations derived from our study.

3.1 Experimental Settings

The dataset used in this study is the Adressa Dataset (Light Version 1), a publicly available dataset designed for research in news RS[1]. Provided by Adresseavisen, a Norwegian newspaper based in Trondheim, the dataset was collected using Cxense APIs and has been widely used in news recommendation research to analyze user behavior and content personalization strategies [12]. The dataset consists of user interactions with news articles recorded over one week (January 1–7, 2017), while providing insights into user demographics, device usage, and content distribution. Specifically, the dataset comprises 640,502 users and 20,428 items, with a total of 3,101,991 recorded transactions. The high sparsity indicates that user interactions with items are relatively infrequent, highlighting the challenges in generating personalized recommendations effectively.

To analyze user and item fairness, we identified and clustered in groups user and item attributes from the dataset. Item attributes selected included `access` type (distinguishing subscriber-only from open-access content), `author` name (which was further processed to infer gender through the use of Genderize API[2]), `category` (defining the topic of the article), `classification` (originally containing multiple values related to automatically assigned topics but later concatenated for consistency), and `sentiment` (derived from article metadata). User attributes selected included `country` (obtained from metatada and estimated from IP address), `device` type (such as mobile, desktop, or tablet), and operating system `OS` (e.g., Windows, iOS, Android).

The following recommendation algorithms were considered in the experiments, all of them available in the Elliot library [3]: Random produces random suggestions and serves as a non-personalized baseline; MostPop is also non-personalized but recommends according to how popular items are; ItemKNN is a nearest-neighbor method based on items [20]; BPRMF is a Matrix Factorization (MF) method that uses Bayesian Personalized Ranking optimization to handle pairwise preferences [18]; EASE^R is a linear autoencoder-based algorithm [25] popular to perform well across datasets [4]; NeuMF is a neural approach that extends traditional MF [14]. Finally, experiments were executed at different ranking cutoffs, but for the sake of space, only results at cutoff 10 are reported.

3.2 Accuracy Evaluation

Table 1 presents the evaluation results in terms of accuracy metrics for different recommendation models at cutoff 10. The models were assessed using multiple evaluation metrics, including Normalized Discounted Cumulative Gain (nDCG), Recall, Hit Rate (HR), Precision, Mean Average Precision (MAP), and Mean Reciprocal Rank (MRR) [13]. These results provide insights into the overall performance of each model in terms of ranking accuracy effectiveness.

---

[1] https://reclab.idi.ntnu.no/dataset/.
[2] https://genderize.io/documentation.

Table 1. Accuracy evaluation, higher values, better. Subindices denote relative rank of the algorithm with respect to that column (metric).

Model	nDCG	Recall	HR	Precision	MAP	MRR
Random	0.0011_6	0.0017_6	0.0054_5	0.0005_5	0.0006_5	0.0017_5
MostPop	0.0065_4	0.0111_4	0.0216_4	0.0022_4	0.0027_4	0.0075_4
ItemKNN	0.0396_2	0.0701_2	0.0943_2	0.0099_2	0.0125_2	0.0360_2
BPRMF	0.0085_3	0.0139_3	0.0278_3	0.0029_3	0.0034_3	0.0099_3
EASE^R	$\mathbf{0.0478_1}$	$\mathbf{0.0888_1}$	$\mathbf{0.1234_1}$	$\mathbf{0.0129_1}$	$\mathbf{0.0147_1}$	$\mathbf{0.0419_1}$
NeuMF	0.0014_5	0.0028_5	0.0052_6	0.0005_5	0.0005_6	0.0015_6

The Random model, as expected, performs the worst across all metrics, reinforcing the importance of personalized over non-personalized approaches. The MostPop model, despite being a simple popularity-based approach, maintains moderate performance and consistently outperforms the Random baseline. However, it remains weaker compared to ItemKNN and BPRMF in capturing personalized recommendations, although it outperforms NeuMF.

ItemKNN demonstrates significantly better performance in Recall and Hit Rate. EASE^R emerges as one of the strongest models, outperforming both ItemKNN and BPRMF in terms of nDCG and Recall. It exhibits a particularly strong ability to retrieve relevant items efficiently, making it a competitive alternative to traditional collaborative filtering models. Interestingly, NeuMF, despite being a deep learning-based approach, performs worse than traditional collaborative filtering techniques in Recall and Hit Rate. This may be due to dataset sparsity or suboptimal hyperparameter tuning, which could affect its ability to generalize effectively.

3.3 Bias Evaluation

Table 2 presents the bias evaluation measurements according to the metrics presented in Sect. 2. These metrics provide insight into how recommendation algorithms handle popularity bias and content exposure when suggesting news articles. First, the Random model exhibits the lowest level of popularity bias, based on its high ACLT and APLT values and relatively low ARP values, evidencing it does not prioritize recommending popular content. Additionally, its PopREO and PopRSP values are significantly lower than those of personalized models. While this model minimizes bias, it lacks personalization, making it less suitable for real-world recommendation scenarios. In the other extreme, the MostPop model, as expected, strongly favors popular content, reflected in its consistently high ARP values, and maximum PopREO and PopRSP values. The BPRMF model exhibits a strong popularity bias – an observation already known in the area for other domains [11, 23, 32] –, as evidenced by all the values that mimic those from the MostPop algorithm. The ItemKNN model, on the other hand, demonstrates lower popularity bias compared to MostPop and BPRMF,

Table 2. Bias evaluation, better (less biased) values indicated by ↑ (higher) and ↓ (lower) arrows, depending on the metric.

Model	ACLT ↑	APLT ↑	ARP ↓	PopREO ↓	PopRSP ↓
Random	**8.689**$_1$	**0.955**$_1$	**84.056**$_1$	**0.020**$_1$	**0.004**$_1$
MostPop	0.000$_5$	0.000$_5$	4604.950$_5$	1.000$_6$	1.000$_6$
ItemKNN	1.396$_3$	0.131$_3$	2722.296$_3$	0.739$_3$	0.985$_3$
BPRMF	0.000$_5$	0.000$_5$	4671.479$_6$	1.000$_6$	1.000$_6$
EASE^R	0.498$_4$	0.050$_4$	3351.067$_4$	0.860$_4$	0.995$_4$
NeuMF	7.274$_2$	0.727$_2$	423.002$_2$	0.181$_2$	0.784$_2$

as indicated by its higher ACLT and APLT values. This suggests that ItemKNN distributes recommendations across a broader range of items, mitigating bias to some extent. Similarly, the EASE^R model introduces a slight reduction in popularity bias compared to BPRMF, as reflected in its nonzero ACLT and APLT values, which indicate some level of long-tail recommendation. However, its ARP value remains relatively high, suggesting that it continues to prioritize popular items, something that should be taken into account when considering its high accuracy performance (see Table 1). The NeuMF model shows moderate bias compared to the other models; while it achieves slightly better long-tail coverage than MostPop and BPRMF, its PopREO and PopRSP values suggest a strong preference for popularity, though to a lesser extent than the others.

These results highlight how different recommendation models handle popularity bias in news recommendation. While some models reduce popularity bias by promoting diverse content, they often do so at the cost of engagement and relevance. Among the evaluated approaches, ItemKNN appears to offer the most balanced tradeoff, effectively reducing bias while maintaining competitive recommendation performance.

3.4 Fairness Evaluation

The Bias Disparity evaluates disparities in exposure and representation by analyzing combinations of user clusters (`device type`, `OS`, `country`) and item clusters (`access` type, `author` gender, `category`, `classification`, `sentiment`). Due to the large number of possible user-item attribute combinations, individual values for each pair cannot be analyzed in detail. Instead, an averaged Bias Disparity value has been computed for each combination to provide a clearer understanding of the general trends present in the data. As an example, Table 3 presents specific Bias Disparity values for different `device` and `sentiment` clusters[3], whereas Table 4 presents the aggregated Bias Disparity values for all user

---

[3] Note that for this example, 6 columns are obtained, due to the combination of a user cluster with 3 values (`device`: mobile, desktop, or tablet) and an item cluster with 2 values (`sentiment`: no sentiment or negative). For other clusters or combinations, this way of presenting the results is unfeasible.

and item attributes. It can be noted how the aggregated values do not capture all the nuances that appear in Table 3; for example, for MostPop, even though 3 of the values (those related to S0) are negative, the aggregated value in Table 4 (0.456) is positive.

Table 3. BD for device (D) and sentiment (S) clusters.

Model	D0-S0	D0-S1	D1-S0	D1-S1	D2-S0	D2-S1
Random	0.147	−0.426	0.085	−0.312	0.133	−0.405
MostPop	−0.387	1.122	−0.443	1.634	−0.396	1.207
ItemKNN	−0.128	0.372	−0.088	0.325	−0.119	0.361
BPRMF	−0.255	0.740	−0.315	1.160	−0.266	0.811
EASE^{R}	−0.052	0.151	−0.037	0.137	−0.052	0.157
NeuMF	0.151	−0.437	0.087	−0.322	0.135	−0.411

Table 4. Aggregated BD for user (device, OS, country) and item (access type, author gender, category, classification, sentiment) clusters.

Model	Device					OS					Country				
	Access	Author	Category	Class.	Sent.	Access	Author	Category	Class.	Sent.	Access	Author	Category	Class.	Sent.
Random	5.129	0.672	15.577	4.755	−0.130	2.522	0.532	4.697	1.741	−0.068	0.627	0.259	0.556	0.367	0.085
MostPop	−0.621	−0.251	−0.439	−0.240	0.456	−0.419	−0.059	−0.166	−0.032	0.407	−0.104	0.143	0.032	0.193	0.690
ItemKNN	−0.097	−0.096	0.410	0.185	0.121	−0.120	0.001	0.121	0.156	0.090	0.071	0.123	0.088	0.308	0.160
BPRMF	−0.374	−0.205	−0.414	−0.261	0.313	−0.218	−0.023	−0.141	−0.054	0.289	0.078	0.178	0.064	0.172	0.539
EASE^{R}	−0.375	−0.162	−0.249	−0.159	0.051	−0.247	−0.048	−0.088	0.040	0.037	−0.006	0.118	0.055	0.218	0.123
NeuMF	5.071	0.211	0.132	0.081	−0.133	2.460	0.159	0.428	0.132	−0.072	0.648	0.183	0.747	0.094	0.033

These results offer a broader perspective on how each model interacts with different demographic and content groups. By averaging the results, it is possible to compare the overall fairness of recommendation models, highlighting which models distribute recommendations more equitably across diverse user and item characteristics, providing insights into how recommendation models favor certain user groups or content types, indicating disparities in exposure. In particular, the Random model consistently exhibits the highest absolute BD values across most attributes (all except **sentiment**). Since its values are usually positive, this shows that it tends to increase the bias (for that pair of attributes) with respect to the original data. The rest of the models, except for NeuMF, seem to simply replicate the biases that already exist in the training data (with values very close to zero). NeuMF, on the other hand, produces very high values for the **access** attribute, comparable to the Random model; however, for the rest of the attributes, even though it increases the bias with respect to the data, its personalization component, produces more limited disparities than Random.

Table 5 presents the MADR evaluation metric, assessing how different recommendation models distribute ranking performance across various item attributes, hence, highlighting disparities in ranking fairness. One key observation is that the Random model exhibits the lowest values across all attributes excepts for `category`, where the MostPop model obtains slightly better performance. This indicates that the performance of the Random recommender is the same independently of the item attribute, however, since its performance is, by definition, very low, this good result in fairness does not translate in the model being practical for real-world recommendation tasks. At the other side of the spectrum we find BPRMF and ItemKNN, which seem to perform quite differently depending on the item attribute (in particular, `access`, `author` gender, and `sentiment` for BPRMF and `access`, `category`, and `classification` for ItemKNN). NeuMF and EASE^{R}, as a consequence, obtain medium values of this metric for most of the attributes; considering the high performance achieved by EASE^{R} in terms of accuracy (see Table 1), this model could be a good option to address the accuracy-fairness tradeoff.

Table 5. MADR values for item attributes.

Model	Access	Author	Category	Classification	Sentiment
Random	**0.0009**$_1$	**0.0027**$_1$	0.0011$_2$	**0.0012**$_1$	**0.0009**$_1$
MostPop	0.0026$_2$	0.0033$_2$	**0.0006**$_1$	0.0016$_2$	0.0019$_2$
ItemKNN	0.0270$_5$	0.0162$_4$	0.0212$_6$	0.0135$_6$	0.0024$_3$
BPRMF	0.0293$_6$	0.0271$_6$	0.0027$_4$	0.0055$_4$	0.0543$_6$
EASE^{R}	0.0189$_4$	0.0263$_5$	0.0167$_5$	0.0117$_5$	0.0030$_4$
NeuMF	0.0036$_3$	0.0038$_3$	0.0011$_2$	0.0025$_3$	0.0054$_5$

Table 6. REO values for item attributes.

Model	Access	Author	Category	Classification	Sentiment
Random	**0.2726**$_1$	**0.1384**$_1$	**2.9657**$_1$	**2.1699**$_1$	0.0193$_2$
MostPop	1.4142$_6$	0.9365$_5$	11.2957$_6$	5.7810$_6$	0.5178$_5$
ItemKNN	0.4775$_2$	0.4050$_2$	3.4245$_2$	2.6967$_2$	0.1675$_3$
BPRMF	1.3251$_5$	0.8653$_4$	9.4444$_4$	4.2546$_5$	0.3425$_4$
EASE^{R}	1.0804$_3$	0.5441$_3$	4.1486$_3$	2.9322$_3$	**0.0030**$_1$
NeuMF	1.3001$_4$	0.9882$_6$	9.5606$_5$	4.1440$_4$	0.7668$_6$

Finally, Table 6 presents the REO performance values, assessing whether items from different groups have an equal probability of appearing at the top ranks of recommendation lists, promoting fairness according to item visibility across various item attributes. Thus, this metric helps determine whether certain content types receive disproportionate exposure. Again, we observe the Random model exhibits the lowest values for this metric except, in this case, for the `sentiment` attribute. The worst-performing models according to this metric correspond to MostPop (`access`, `category`, `classification`, `author` gender, and `sentiment`) and NeuMF (`author` gender, `sentiment`, and `category`).

3.5   Discussion

In this section we analyze how various personalized and non-personalized algorithms perform with respect to different accuracy, bias, and fairness metrics. The first obvious observation is that, as expected, there is not a clear winner with respect to the *accuracy-fairness* (or *accuracy-bias*) tradeoff. However, what also stems from these results is the **bias-fairness** tradeoff. Since the selected bias metrics are mainly related to popularity measurements, any fairness metric not aligned with this attribute will result in incompatible conclusions. For example, MADR ranks MostPop as the first or second best algorithm for any item attribute, where this method is, by definition, the most biased one when considering popularity-biased metrics. A similar observation can be made between REO and bias metrics.

However, the fact that fairness metrics may behave very differently depending on the user or item attribute being considered, makes this analysis richer and with far more implications for this dimension. According to the relative ranking of the recommendation algorithms for MADR, we may group the item attributes in 3 groups: `access` and `author` gender (although `category` behaves in a very similar way as these two attributes), `classification` is more similar to `category` than to `access` or `author` gender, and `sentiment`. At the same time, the performance of the attributes according to the REO metric, although being slightly different (for example, MostPop is the worst performing one in REO for `access` attribute, but the second-best in MADR), in terms of the derived groups according to the relative performance of the algorithms remains the same: `category` and `classification` are very similar to each other, `sentiment` is the least similar to the rest, and `access` and `author` gender correlate to some extent to each other and with `category`. This highlights that, even though the recommendation algorithms were not trained with any information about these attributes whatsoever, the inherent signals available in the data allow to produce recommendations that can be interpreted as more or less fair with respect to these attributes, depending on the considered fairness definition.

A little bit more complex to analyze is the bias disparity (BD) metric, since it takes into account the original bias in the data and, as a consequence, it is not bounded and the sign is important. However, this is the only metric, among the ones considered in this study, that can integrate both user and item attributes at the same time. In this case, the `sentiment` attribute again stands out with a distinctive pattern. Also `country`, as user attribute, performs differently than `device` and `OS`, obtaining almost in every case positive BD values, meaning the algorithms are increasing the bias with respect to the original data.

3.6   Limitations

While this study provides valuable insights into fairness in news recommender systems, several limitations must be acknowledged. First, the choice of the Adressa dataset, a local Norwegian news outlet, constrains the generalizability of the findings. As a regional newspaper, Adressa does not reflect the diversity or editorial plurality of a global news aggregator, where balancing across multiple news sources and ideological perspectives is more critical. Datasets like MIND[4], which include a broader range of publishers and user demographics, may be more appropriate for evaluating fairness in such contexts. Additionally, the use of 'country' as a fairness-relevant user attribute is limited in this setting, as most users are likely Norwegian, and international interactions may still represent Norwegian expatriates, thus reducing the attribute's discriminatory power.

Moreover, the evaluation relies on standard recommendation metrics that may not fully capture the normative and democratic roles of news recommenders. Metrics such as nDCG and Recall, while useful for general performance assessment, overlook domain-specific concerns like viewpoint diversity, civic engagement, and stakeholder equity. As highlighted in recent work [15,24,28], fairness in news recommendation requires more context-aware evaluation approaches. Future work should better align methodological choices with the unique ethical and societal demands of the news domain.

Additionally, this work is focused on classical collaborative filtering approaches or neural network-based recommendation algorithms. Other algorithmic families, such as those based on content or knowledge, are not considered and the effects on fairness and bias metrics by those methods remain unexplored.

4   Conclusions

This study conducted a comparative assessment of fairness in news RS by evaluating multiple algorithms using classical performance metrics, bias measurements, and fairness indicators. The findings reveal that recommendation algorithms inherently introduce bias, with certain models disproportionately favoring specific item categories, user demographics, or sentiment-driven content. Traditional algorithms like MostPop and BPRMF exhibited a strong tendency to prioritize popular articles, reinforcing the visibility of widely consumed content. EASE^R demonstrated a more balanced item distribution, though biases remained, influenced by both user attributes and content characteristics. Fairness-specific metrics further revealed systematic biases in content ranking. For example, REO and BD metrics indicated that certain models disproportionately favored sentiment-driven news, whereas long-tail exposure metrics (ACLT and APLT) confirmed that less popular items were consistently underrepresented, underscoring the ongoing challenge of achieving equitable content distribution. Despite incorporating fairness-aware evaluation, no model successfully

---

[4] https://msnews.github.io/.

eliminated bias. Instead, this study highlights the *accuracy-fairness tradeoff*, where efforts to mitigate bias often come at the cost of reduced recommendation performance. This finding suggests that multi-objective optimization approaches to balance accuracy, diversity, and equity are essential for fairness-aware recommender systems [30], and more work, specifically tailored to the news domain, is missing to measure and mitigate these aspects.

Acknowledgments. This work was supported by Grant PID2022-139131NB-I00 funded by MCIN/AEI/10.13039/501100011033 and by "ERDF, a way of making Europe."

References

1. Abdollahpouri, H., Burke, R., Mobasher, B.: Controlling popularity bias in learning-to-rank recommendation. In: Proceedings of RecSys, Como, Italy, 27–31 August 2017, pp. 42–46. ACM (2017)
2. Abdollahpouri, H., Burke, R., Mobasher, B.: Managing popularity bias in recommender systems with personalized re-ranking. In: Proceedings of FLAIRS, Sarasota, Florida, USA, 19–22 May 2019, pp. 413–418. AAAI Press (2019)
3. Anelli, V.W., et al.: Elliot: a comprehensive and rigorous framework for reproducible recommender systems evaluation. In: Proceedings of SIGIR, Virtual Event, Canada, 11–15 July 2021, pp. 2405–2414. ACM (2021)
4. Anelli, V.W., Bellogín, A., Noia, T.D., Jannach, D., Pomo, C.: Top-N recommendation algorithms: a quest for the state-of-the-art. In: Proceedings of UMAP, Barcelona, Spain, 4–7 July 2022, pp. 121–131. ACM (2022)
5. Buijsman, S.: Navigating fairness measures and trade-offs. AI Ethics **4**(4), 1323–1334 (2024)
6. Cinelli, M., Morales, G.D.F., Galeazzi, A., Quattrociocchi, W., Starnini, M.: The echo chamber effect on social media. Proc. Natl. Acad. Sci. USA **118**(9), e2023301118 (2021). Accessed Nov 2024
7. Cinus, F., Minici, M., Monti, C., Bonchi, F.: The effect of people recommenders on echo chambers and polarization. In: Proceedings of ICWSM, Atlanta, Georgia, USA, 6–9 June 2022, pp. 90–101. AAAI Press (2022)
8. Deldjoo, Y., Anelli, V.W., Zamani, H., Bellogín, A., Di Noia, T.: A flexible framework for evaluating user and item fairness in recommender systems. User Model. User-Adap. Inter. **31**(3), 457–511 (2021). https://doi.org/10.1007/s11257-020-09285-1
9. Deldjoo, Y., Jannach, D., Bellogín, A., Difonzo, A., Zanzonelli, D.: Fairness in recommender systems: research landscape and future directions. User Model. User Adapt. Interact. **34**(1), 59–108 (2024)
10. Ekstrand, M.D., Das, A., Burke, R., Diaz, F.: Fairness in recommender systems. In: Ricci, F., Rokach, L., Shapira, B. (eds.) Recommender Systems Handbook, pp. 679–707. Springer, US (2022). https://doi.org/10.1007/978-1-0716-2197-4_18
11. Guíñez, F., Ruiz, J., Sánchez, M.I.: Quantification of the impact of popularity bias in multi-stakeholder and time-aware environments. In: Boratto, L., Faralli, S., Marras, M., Stilo, G. (eds.) BIAS 2021. CCIS, vol. 1418, pp. 78–91. Springer, Cham (2021). https://doi.org/10.1007/978-3-030-78818-6_8

12. Gulla, J.A., Zhang, L., Liu, P., Özgöbek, Ö., Su, X.: The Adressa dataset for news recommendation. In: Proceedings of ICWI, Leipzig, Germany, 23–26 August 2017, pp. 1042–1048. ACM (2017)

13. Gunawardana, A., Shani, G.: Evaluating recommender systems. In: Ricci, F., Rokach, L., Shapira, B. (eds.) Recommender Systems Handbook, pp. 265–308. Springer, Boston, MA (2015). https://doi.org/10.1007/978-1-4899-7637-6_8

14. He, X., Liao, L., Zhang, H., Nie, L., Hu, X., Chua, T.: Neural collaborative filtering. In: Proceedings of WWW, Perth, Australia, 3–7 April 2017, pp. 173–182. ACM (2017)

15. Helberger, N.: On the democratic role of news recommenders. Digit. J. **7**(8), 993–1012 (2019)

16. Jin, D., et al.: A survey on fairness-aware recommender systems. Inf. Fusion **100**, 101906 (2023)

17. Karimi, S., Rahmani, H.A., Naghiaei, M., Safari, L.: Provider fairness and beyond-accuracy trade-offs in recommender systems. CoRR abs/2309.04250 (2023)

18. Koren, Y., Rendle, S., Bell, R.M.: Advances in collaborative filtering. In: Ricci, F., Rokach, L., Shapira, B. (eds.) Recommender Systems Handbook, pp. 91–142. Springer, US (2022). https://doi.org/10.1007/978-1-0716-2197-4_3

19. Mansoury, M., Abdollahpouri, H., Pechenizkiy, M., Mobasher, B., Burke, R.: Feedback loop and bias amplification in recommender systems. In: Proceedings of CIKM, Virtual Event, Ireland, 19–23 October 2020, pp. 2145–2148. ACM (2020)

20. Nikolakopoulos, A.N., Ning, X., Desrosiers, C., Karypis, G.: Trust your neighbors: a comprehensive survey of neighborhood-based methods for recommender systems. In: Ricci, F., Rokach, L., Shapira, B. (eds.) Recommender Systems Handbook, pp. 39–89. Springer, US (2022). https://doi.org/10.1007/978-1-0716-2197-4_2

21. Ruan, Q., Xu, J., Leavy, S., Namee, B.M., Dong, R.: Rewriting bias: mitigating media bias in news recommender systems through automated rewriting. In: Proceedings of UMAP, Cagliari, Italy, 1–4 July 2024, pp. 67–77. ACM (2024)

22. Sacharidis, D., Mouratidis, K., Kleftogiannis, D.: A common approach for consumer and provider fairness in recommendations. In: Proceedings of ACM RecSys Late-Breaking Results, Copenhagen, Denmark, 16–20 September 2019. CEUR Workshop Proceedings, vol. 2431, pp. 1–5. CEUR-WS.org (2019)

23. Sánchez, P., Bellogín, A., Boratto, L.: Bias characterization, assessment, and mitigation in location-based recommender systems. Data Min. Knowl. Discov. **37**(5), 1885–1929 (2023)

24. Smets, A., Hendrickx, J., Ballon, P.: We're in this together: a multi-stakeholder approach for news recommenders. Digit. Journal. **10**(10), 1813–1831 (2022)

25. Steck, H.: Embarrassingly shallow autoencoders for sparse data. In: Proceedings of WWW, San Francisco, CA, USA, 13–17 May 2019, pp. 3251–3257. ACM (2019)

26. Treuillier, C., Castagnos, S., Özgöbek, Ö., Brun, A.: Beyond trade-offs: unveiling fairness-constrained diversity in news recommender systems. In: Proceedings of UMAP, Cagliari, Italy, 1–4 July 2024, pp. 143–148. ACM (2024)

27. Tsintzou, V., Pitoura, E., Tsaparas, P.: Bias disparity in recommendation systems. In: Proceedings of the Workshop on Recommendation in Multi-stakeholder Environments, Copenhagen, Denmark, 20 September 2019. CEUR Workshop Proceedings, vol. 2440. CEUR-WS.org (2019)

28. Vrijenhoek, S., Bénédict, G., Granada, M.G., Odijk, D., de Rijke, M.: Radio - rank-aware divergence metrics to measure normative diversity in news recommendations. In: Proceedings of RecSys, Seattle, WA, USA, 18–23 September 2022, pp. 208–219. ACM (2022)

29. Wang, Y., Ma, W., Zhang, M., Liu, Y., Ma, S.: A survey on the fairness of recommender systems. ACM Trans. Inf. Syst. **41**(3), 52:1–52:43 (2023)
30. Wu, H., Ma, C., Mitra, B., Diaz, F., Liu, X.: A multi-objective optimization framework for multi-stakeholder fairness-aware recommendation. ACM Trans. Inf. Syst. **41**(2), 47:1–47:29 (2023)
31. Yin, H., Cui, B., Li, J., Yao, J., Chen, C.: Challenging the long tail recommendation. Proc. VLDB Endow. **5**(9), 896–907 (2012)
32. Zhu, Z., He, Y., Zhao, X., Zhang, Y., Wang, J., Caverlee, J.: Popularity-opportunity bias in collaborative filtering. In: Proceedings of WSDM, Virtual Event, Israel, 8–12 March 2021, pp. 85–93. ACM (2021)
33. Zhu, Z., Wang, J., Caverlee, J.: Measuring and mitigating item under-recommendation bias in personalized ranking systems. In: Proceedings of SIGIR, Virtual Event, China, 25–30 July 2020, pp. 449–458. ACM (2020). Accessed Nov 2024

Empirical Comparison of Four Recommendation Models for Course Selection

Diego Cheuquepán-Maldonado[1(✉)] [iD] and Jordi Escayola Mansilla[2]

[1] Departamento de Ingeniería Informática, Universidad de Santiago de Chile,
Santiago, Chile
`diego.cheuquepan@usach.cl`
[2] Universitat Oberta de Catalunya, Barcelona, Spain
`jescayolam@uoc.edu`

Abstract. Higher education institutions design faculty development programs consisting of a set of training courses. These programs are organized into training pathways that are offered during the academic year. The pathways allow participants to develop skills on various educational topics. Therefore, faculty members are expected to continually update their skills to improve their teaching practices. However, selecting a training course that is part of a pathway with multiple options is a difficult task for instructors because it is a time-consuming task. To provide instructors with training courses tailored to their interests and needs, we evaluated four collaborative filtering models based on matrix factorization using implicit feedback on a dataset of 2,113 instructors course enrollments from a faculty development program. The results are encouraging; a Weighted Approximate-Rank Pairwise loss model achieves 0.86 area under the curve score and a top-3 precision of 0.24, while an Alternating Least Squares model achieves 4.62 novelty metric. We discuss how the two models with the best performance operate in a real-world situation using their metrics results in order to understand the recommendations offered by each model to the user.

Keywords: Course Recommendation · Collaborative Filtering · Matrix Factorization · Implicit Feedback · Recommender System

1 Introduction

In higher education, the quality of teaching depends on the pedagogical work of the instructors who teach, who often do not have pedagogical skills because they have not been formally trained as teachers. To support them, higher education institutions design faculty training and development programs consisting of a series of training courses [21]. These courses are organized into training pathways that are typically offered throughout the academic year. These pathways allow participants to develop skills in various educational fields such as teaching,

L. Martínez et al. (Eds.): IDEAL 2025, LNCS 16239, pp. 23–34, 2026.
https://doi.org/10.1007/978-3-032-10489-2_3

assessment, and educational technology, among others. Therefore, faculty members are expected to continually update their teaching skills in order to improve their teaching practices. However, selecting a training course from a pathway with many options is time-consuming for instructors. Moreover, offering personalized recommendations is complex given the large number of courses available. Consequently, there is a recognized need for tools that personalize training course recommendations and guide individual instructors toward courses that promote their professional growth as educators, considering that the effectiveness of professional development improves significantly when it is personalized and tailored to the characteristics of individual teachers [14]. In the educational context, recommender systems have proven useful on multiple fronts, from guiding students in course selection to recommending learning objects on online platforms [6]. Although most of the literature related to recommender systems in education focuses on supporting learners, a smaller fraction addresses recommendations aimed at educators [18]. For example, a study of open educational resource recommender systems showed that approximately 64% of the research related to learning objectives is student-oriented, compared to only 36% focused on teachers [16]. This suggests that course recommendation for professional development is a relatively underexplored area, despite its strategic importance in fostering continuous instructor improvement. In higher education, where university instructors must balance teaching, research, and training duties, a training course recommendation system could alleviate information overload and facilitate informed decision-making about which courses to take for their professional development.

In this work, the gap is addressed by evaluating and comparing four collaborative filtering algorithms on a faculty training dataset. Furthermore, their effectiveness in instructor course recommendations is analyzed. A collaborative filtering approach was adopted because recommender systems and collaborative filtering in particular have proven effective in education for guiding learners, suggesting their potential to assist instructors as well. Therefore, this study investigates whether collaborative filtering recommender algorithms (Implicit Alternating Least Squares (ALS), Bayesian Personalized Ranking (BPR), Weighted Approximate-Rank Pairwise loss (WARP), and Logistic Matrix Factorization (LMF)) can effectively personalize faculty course recommendations, and which of these approaches best balances recommendation precision and novelty.

To our knowledge, this study is the first to empirically compare multiple recommendation models for faculty development courses using different evaluation metrics. Therefore, four matrix factorization-based models were evaluated on instructors' course enrollment data set and their trade-offs in precision and novelty were identified. Given the success of collaborative filtering in related domains, for instance, student course recommendations, its application for instructor training recommendations was explored.

In summary, the contributions of this work are: (1) an empirical comparison of four recommendation algorithms (ALS, BPR, WARP, LMF) on an instructor training course dataset; (2) an evaluation across multiple metrics (AUC, Precision@K, Novelty) to assess accuracynovelty trade-offs; (3) identification of the best performing models for different objectives and discussion of their practical implications for faculty development.

This paper is organized as follows. In Sect. 2 main concepts and the problem setting are presented. Section 3 related works on collaborative filtering based on matrix factorization models are presented. Section 4 the methods used in this study are explained. The experimental and evaluation results are presented in Sect. 5. In Sect. 6 the main findings are discussed. Finally, Sect. 7 concludes the paper and gives information about our future work.

2 Background

2.1 Collaborative Filtering and Implicit Feedback

In the Recommender Systems (RS) field, the Collaborative Filtering (CF) method uses information from a user's past behavior to predict which items another user will like or be interested in. This approach has become very popular for designing a recommender system. Among them, the Matrix Factorization (MF) model that jointly learns the user and item latent factors for CF is tremendously successful and has become one of the solution approaches. Recommender systems methods rely on the ratings, also called scores, from users to items indicating an agree or disagree level. In our training course recommender problem, the feedback is a unary scale [13]. In this kind of rating, a user can only indicate a like for an item but cannot specify a dislike for that item. Therefore, this scale allows one to indicate the presence or absence of an attribute. For that reason, in our problem the rating only indicates whether an instructor approved a training course as follows, each element is either "?" (unknown) or "1" (approved). This source of knowledge refers to the actions that a user performs on items that cannot be directly interpreted as being of explicit interest. That is, the user does not explicitly declare their preference for an item through a rating, but rather their preference for that item is interpreted through their behavior. Such unary feedback indicates a positive interaction but not a graded preference score, so it falls under implicit feedback. To model this scenario, the data was represented as a binary matrix $R \in \{0, 1\}^{m \times n}$, where $R_{ui} = 1$ indicates that the instructor u has approved the training course i, and all other entries are considered unknown or unobserved.

2.2 Matrix Factorization Models

In this work, four models based on matrix factorization and using implicit feedback were evaluated. These models capture the nuances of one-class data by either re-weighting unobserved entries or directly optimizing ranking criteria, rather than treating missing interactions as zeros. These four algorithms were selected as they cover both pointwise (ALS, LMF) and pairwise (BPR, WARP) learning strategies for implicit feedback, allowing us to compare how each objective function influences the recommendation outcomes. The focus is on four representative matrix factorization algorithms, as they are proven strong baselines in recommender systems, often matching or surpassing more complex models in accuracy [4].

Implicit Alternating Least Squares (ALS) is a matrix factorization algorithm for collaborative filtering problems [7]. This algorithm uses the alternating least squares optimization method. Starting from a sparse matrix of user-item interactions, ALS factors it into two smaller matrices, one representing the user and the other representing items. It also incorporates an alpha parameter to weight observed interactions, treating known interactions with higher confidence.

Bayesian Personalized Ranking (BPR) seeks to directly optimize the relative ranking between items rather than predicting absolute values [15]. This algorithm uses a log-likelihood function to transform a probability expression into a sum of logarithms. For each user, observed items are ranked above unobserved ones, using stochastic gradient descent on a sum-of-logs loss.

Weighted Approximate Rank Pairwise loss (WARP) is a latent embedding algorithm for collaborative filtering problems [20]. This algorithm uses pairwise loss as an optimization method. WARP weights the loss according to a function of the approximate rank of the positive item.

Logistic Matrix Factorization (LMF) models user-item interactions as probabilities using a logistic function applied to the inner product of latent factors [8]. This algorithm employs a loss function derived from the log-likelihood used in logistic regression.

2.3   Recommender Systems Evaluation

Evaluating recommender systems demands a balance between measuring how accurately we predict user preferences and how engaging or surprising those suggestions feel [17].

Metrics like the Area Under the ROC Curve (AUC) and Precision@K (p@k) assess ranking quality, but from different angles: AUC captures the model's global ability to place truly interacted-with items above unobserved ones, offering a threshold-agnostic view that is robust to the extreme class imbalance inherent in implicit feedback. Precision@K, in contrast, zeros in on the user's experience of the top-K recommendations, if the few items presented at the top of their list are not relevant, the overall utility suffers, regardless of the broader ranking performance. However, Precision@K focus purely on relevance, which can lead models to favor overly popular content. That is where a novelty metric becomes essential, by penalizing recommendations of already well-known or universally popular items, novelty encourages the system to surface less obvious courses, fostering discovery and personalization.

In practice, one rarely optimizes for a single metric in isolation. High AUC and p@k scores indicate a model that effectively ranks known positives, but without considering novelty, these same metrics may mask a popularity bias that stifles long-term engagement. Conversely, emphasizing novelty too strongly can erode immediate relevance. Novelty is defined as an average popularity-based metric, where a lower popularity yields higher novelty. In practice, this was calculated as the average inverse logarithmic popularity of the recommended items, following the established definitions of novelty [22].

3   Related Works

It is difficult to compare the results of this study with previous work, as most of these focused on course recommendations for students over instructors. A recent review notes that even when teachers are the target users, evaluation is often done indirectly through student outcomes [18]. In addition, the models used in this study were evaluated using different evaluation metrics than those used in previous works [1]. Finally, most of the previous approach used a continuous rating score, while this study used implicit feedback [7].

Systematic reviews of the literature on recommender systems applied to education [1,6,16] show the efforts that have been made regarding the problem of course recommendation. One of the findings is the prevalence of course recommendations for students over instructors [18]. For this reason, although this section will present course recommendation systems for students, the analysis will focus on the collaborative filtering models based on matrix factorization that have been used in these studies.

A course recommender using the Bayesian Personal Ranking Matrix Factorization (BPR-MF) model was proposed by [10]. The purpose was to recommend advanced courses to students who are entering their senior year. The parameters used in the experiment were as follows. The number of latent features $k = 12$. The regularization $\lambda = 0.05$. The learning rate $\gamma = 0.7$. The best result was 0.97 using the Area Under the Receiver Operating Characteristic Curve (AUC-ROC). The work does not report other evaluation metrics.

A course recommender using the Alternating Least Squares (ALS) algorithm was proposed by [2]. The aim was to recommend courses based on the similarity between the course templates of the students. The parameters used in the algorithm were not informed. The best result was 0.86 using the threshold-based accuracy metric.

As noted in [12], different methods were evaluated to recommend elective courses to students based on their academic records and personal preferences. One of them was Non-Negative Matrix Factorization (NMF). This model was used to factorize the student-elective matrix into two matrices, student attributes, and elective attributes. The best result was 0.85 using the accuracy metric.

A course recommender was proposed by [11] to assist students in selecting the most appropriate elective courses in higher education using Matrix Factorization with Stochastic Gradient Descent (MF-SGD) model. In the best model, the parameters were set as follows. The number of latent features $k = 2$, and the regularization factor $\lambda = 0.001$. The best result was 6.97 using the mean absolute error (MAE) metric.

A course recommender for students based on the courses previously preferred by them using the Low-Rank Matrix Factorization (LRMF) model with similarity-based regularization was proposed by [19]. In the best model, the regularization parameter was set as 0.00001. The best result was 0.0023 using the normalized mean absolute error (NMAE) metric.

Because each of these works optimizes different metrics in different data sets, their numerical results are not directly comparable with the present study.

In summary, prior course recommenders have been geared towards students and often report a single accuracy metric. In contrast, our work focuses on instructors and evaluates multiple algorithms across accuracy and novelty metrics using implicit feedback. This dual focus on a new user group and multi-metric evaluation distinguishes our contribution.

4   Method

4.1   Data Source

The source used in this work corresponds to the database of trained instructor records in the Institutional Faculty Development Program provided by Santo Tomás University in 2021. This had information on instructors who participated in at least two training courses of the Faculty Development Program.

4.2   Data Set

The data set contained 8,150 total records, corresponding to 34 training courses and 2,113 instructors, who were part of the Institutional Faculty Development Program in 2021. From these data, an interaction matrix was constructed that exhibits a sparsity of 88.66%. The data set was randomly divided into two disjoint subsets in a 70–30 proportion, safeguarding that every user was in both subsets. Then, two interaction matrices were created using these subsets. Our data set lacked temporal information that would allow us to divide the courses based on the chronological sequence in which they were passed by the instructor. For this reason, that sequential patterns are not explicitly modeled in the study.

4.3   Preprocessing

A usercourse interaction matrix with $1\,$s (indicating that the instructor completed the course) and $0\,$s (for unknown interactions) was constructed. Then, this matrix was split into train/test. Each model was trained on the training matrix and evaluated for how well it ranks held-out interactions among unknowns.

4.4   Training

Recommender system models were built in a Colab Notebook using Python's LightFM [9] and Implicit [5] libraries.

A grid search strategy was used in order to search for the best parameters for each model. The grid search was performed using the following parameters: number of latent factors (k), learning rate (γ), regularization (λ), and weight factor (α). Table 1 describes the values for the parameters α, γ, λ, and k.

Table 1. Values for the grid of the parameters α, γ, λ and k.

Parameter	Values
α	1, 10, 50, 100, 200
γ	0.002, 0.005, 0.01, 0.05, 0.1
λ	0.002, 0.005, 0.01, 0.05, 0.1
k	20, 30, 40, 50, 80, 100

4.5 Evaluation

In order to evaluate the performance for each fit model, metrics were used to capture different aspects of recommendation quality.

To ensure that course recommendations are not only accurate, but also diverse and capable of presenting instructors with new learning opportunities, we have adopted an evaluation strategy that combines precision-oriented measures (Precision@k) with objectives that go beyond precision (Novelty).

Finally, the performance of each model was measured using the Area Under the ROC Curve (AUC-ROC). Additionally, the DeLong test was used to compare the AUC-ROC values in order to determine whether the best models were statistically different. In all tests, a 1% nominal level ($p < 0.01$) was considered significant.

5 Results

5.1 Quantitative Performance

The results of the performance of the four models for all the experiments carried out are shown in Table 2. In these, the second column shows the parameters used in the best model for each case.

Table 2. Results of each experiment in terms of area under curve, precision, and novelty.

Model	Parameters	AUC	p@3	Novelty
ALS	$k = 100$ $\lambda = 0.1$ $\alpha = 1$	0.8441	0.2115	**4.6201**
LMF	$k = 80$ $\lambda = 0.002$ $\gamma = 0.1$	0.8365	0.2162	3.3392
BPR-MF	$k = 50$ $\lambda = 0.005$ $\gamma = 0.1$	0.7635	0.2104	3.7048
WARP	$k = 50$ $\lambda = 0.1$ $\gamma = 0.005$	**0.8578**	**0.2394**	3.7369

The main observation derived from Table 2 is that the best models can generate a robust recommendation because three of them obtain an AUC value greater than 0.8, suggesting a considerable result. In addition, this result is validated by the DeLong test, which indicates that there is a significant difference between the WARP model with respect to the other three models ($p < 0.01$ in all cases).

Therefore, the WARP model performs better than the others according to the AUC and p@3 metrics. The ALS model performs better than the other three models according Novelty metric. This suggests that the pairwise ranking optimization of WARP is highly effective in placing courses an instructor has taken above those they have not, thus improving AUC and top-3 precision. In contrast, ALS, which factors in all data with regularization, tends to recommend a broader range of courses, as reflected in its higher novelty score.

Figure 1 shows the performance of the best model for every recommendation model using the ROC curve. The AUC metric is shown for each model and the highest AUC is for the WARP model. The WARP curve lies above the others, indicating better ranking performance.

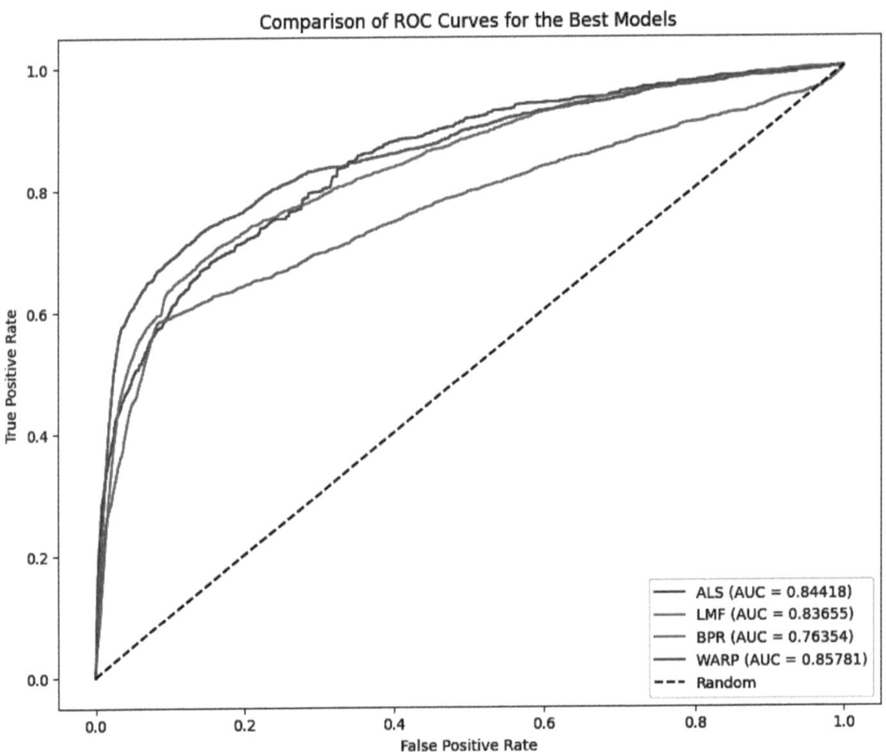

Fig. 1. ROC curves for each of the four models on the test set.

5.2   Application in Real World

User 1002 had completed the following training courses: ST0192, ST0391, ST0433, ST0439, ST0369, and ST0438, as you can see in Fig. 2. Furthermore, while WARP recommended courses ST0382, ST0375, and ST0437, ALS recommended ST0382, ST0435, and ST0413. Although there is overlap with ST0382 between the models, there are also two recommendations that differentiate both.

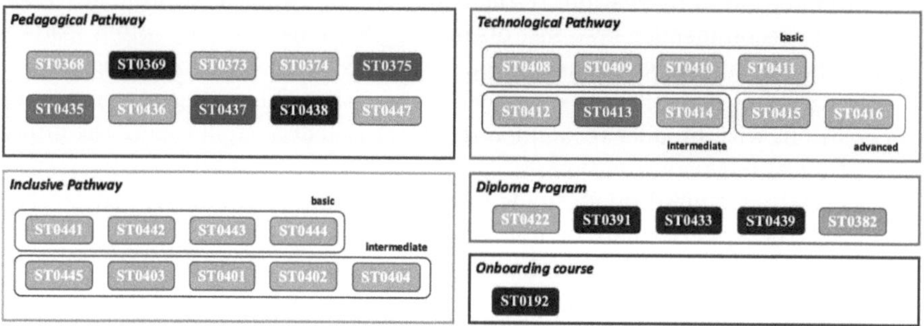

Fig. 2. Recommendations for user 1002. Blue represents courses completed by the user. Yellow represents overlapping course recommendations. Red represents courses recommended by WARP model. Green represents courses recommended by ALS model. (Color figure online)

When analyzing the results of each model comparatively, we first notice that both suggest the same course to the user (ST0382) from the Diploma program. Then, there is a second recommendation that is different, but from the same Pedagogical pathway. However, there is a third recommendation that differentiates the two models. We note that ST0413 (from ALS) belongs to the Technology pathway and was not taken by many similar users, reflecting a novel suggestion. Whereas ST0437 (from WARP) is another Pedagogical pathway course, closely aligned with the past interests of the user.

Therefore, while WARP suggests a course in the same Pedagogical pathway, ALS recommends a course in the Technology pathway. This is explained by the metrics obtained by each model. Although WARP only guarantees that the recommendations are correct (AUC, p@3), ALS also offers diverse suggestions (Novelty).

6   Discussion

In general, WARP emerged as the top performer in accuracy-related measures, slightly ahead of BPR and notably above ALS and LMF in AUC and precision. This aligns with prior findings that pairwise ranking losses can yield better top-N recommendation performance in implicit settings [15]. However, ALS's strength in novelty indicates that it surfaces less popular courses. In practice, this means that ALS might help instructors discover courses they would not normally consider, an aspect not captured by only precision metrics. Thus, there is a trade-off. If an institution prioritizes recommending highly relevant courses, perhaps those most similar to what an instructor has already taken, then WARP is preferable. Conversely, if the goal is to broaden the development of an instructor with novel suggestions, ALS might be more suitable. A combination of both approaches could also be considered to balance relevance and novelty.

Therefore, our observation resonates with the known accuracy-diversity dilemma in recommender systems, optimizing for pure accuracy often leads to popular, homogeneous recommendations, whereas explicitly accounting for novelty can increase diversity at some cost to precision. This suggests our results are consistent with broader recommender system theory, emphasizing the importance of multi-metric evaluation [3].

This study is limited by its single institution dataset from one year. For that reason, instructors elsewhere or in other years might have a different behavior, so results may vary. In addition, another limitation of our approach is that the model cannot provide personalized recommendations for completely new instructors without prior course data. This is the cold start problem in recommender systems and could be improved by providing the most popular courses as a recommendation. A different strategy could be to use content-based models, using instructor profiles or course descriptions, which could alleviate, but this approach is left for future work. Finally, a last limitation is that our evaluation is off-line, based on historical enrollments as implicit feedback, and actual deployment might reveal additional considerations like instructor satisfaction or course availability, which are not captured here.

Therefore, we caution that our results may not directly generalize to other institutions, distinct instructors or different time spans.

7   Conclusion and Future Work

In this study, we addressed the underexplored area of instructor course recommendations by evaluating four recommendation models on real training course participation data.

The results are promising; three of the four evaluated models achieved an AUC greater than 0.8, which is a strong performance in this context. In particular, the best model was WARP, with an AUC of 0.8578. This result was validated by the DeLong test, showing that there is a significant difference between the best WARP model and the other models ($p < 0.01$ in all cases). Additionally, the ALS model obtained a Novelty score greater than the one obtained by the other models.

Therefore, we found that such models can indeed personalize faculty development by offering a list with course recommendations, and even different models strike different balances between precision and novelty of recommendations.

This finding suggest that a hybrid recommender could be beneficial, using a two-stage approach where ALS-derived predictions ensure baseline accuracy and a second stage re-ranks or filters results to inject novelty from a WARP model.

This is an important contribution; the best model could constitute the core of a course-recommender system, which would likely reduce the time invested in the recommendation task.

In summary, our findings highlight a fundamental trade-off; recommending courses closely aligned with an instructor's past for immediate relevance versus suggesting new, unfamiliar courses for exploratory learning.

Future work will explore hybrid models and incorporate course content to handle new instructors and further improve the relevance and diversity of the recommendations. Furthermore, it could explore methods to achieve a better balance between precision and novelty, for example, by blending recommendations from WARP and ALS or even incorporating a weighted objective that includes both relevance and diversity. In addition, incorporating content (course descriptions, learning outcomes) might help provide explanations for recommendations and handle new courses or instructors. Another promising direction is to conduct an online study or pilot program in which instructors receive recommendations to evaluate actual uptake and satisfaction, as offline metrics may not tell the whole story.

Acknowledgments. This work is funded by the National Agency for Research and Development of Chile. ANID BECAS/DOCTORADO NACIONAL 21250717.

References

1. Algarni, S., Sheldon, F.: Systematic review of recommendation systems for course selection. Mach. Learn. Knowl. Extr. **5**(2), 560–596 (2023). https://doi.org/10.3390/make5020033
2. Bhumichitr, K., Channarukul, S., Saejiem, N., Jiamthapthaksin, R., Nongpong, K.: Recommender systems for university elective course recommendation. In: 2017 14th International Joint Conference on Computer Science and Software Engineering (JCSSE), pp. 1–5. IEEE, Nakhon Si Thammarat, Thailand (2017). https://doi.org/10.1109/JCSSE.2017.8025933
3. Castells, P., Hurley, N.J., Vargas, S.: Novelty and diversity in recommender systems. In: Ricci, F., Rokach, L., Shapira, B. (eds.) Recommender Systems Handbook, pp. 881–918. Springer, Boston, MA (2015). https://doi.org/10.1007/978-1-4899-7637-6_26
4. Ferrari Dacrema, M., Boglio, S., Cremonesi, P., Jannach, D.: A troubling analysis of reproducibility and progress in recommender systems research. ACM Trans. Inf. Syst. **39**(2), 1–49 (2021). https://doi.org/10.1145/3434185
5. Implicit: Fast Python collaborative filtering for implicit feedback datasets. https://pypi.org/project/implicit/. Accessed 30 June 2025
6. Guruge, D.B., Kadel, R., Halder, S.J.: The state of the art in methodologies of course recommender systems-a review of recent research. Data **6**(2), 18 (2021). https://doi.org/10.3390/data6020018
7. Hu, Y., Koren, Y., Volinsky, C.: Collaborative filtering for implicit feedback datasets. In: Eighth IEEE International Conference on Data Mining, Pisa, Italy, pp. 263–272. IEEE (2008). https://doi.org/10.1109/ICDM.2008.22
8. Johnson, C.C.: Logistic matrix factorization for implicit feedback data. In: Workshop on Distributed Machine Learning and Matrix Computations, pp. 1–9. [n. p.], Montreal, Canada (2014)

9. Kula, M.: Metadata embeddings for user and item cold-start recommendations. In: Proceedings of the 2nd Workshop on New Trends on Content-Based Recommender Systems, pp. 14–21. CEUR-WS.org, Vienna, Austria (2015)
10. Lee, E.L., Kuo, T.T., Lin, S.D.: A collaborative filtering-based two stage model with item dependency for course recommendation. In: 2017 IEEE International Conference on Data Science and Advanced Analytics (DSAA), Tokyo, Japan, pp. 496–503. IEEE (2017). https://doi.org/10.1109/DSAA.2017.18
11. Muzdybayeva, G., Khashimova, D., Amirzhanov, A., Kadyrov, S.: A matrix factorization-based collaborative filtering framework for course recommendations in higher education. In: 2023 17th International Conference on Electronics Computer and Computation (ICECCO), Kaskelen, Kazakhstan, pp. 1–4. IEEE (2023). https://doi.org/10.1109/ICECCO58239.2023.10147152
12. Naik, A., Kumar, A.B., Virupaksha, D., Priyanka, H.: Elective recommendation system. In: 2023 IEEE Technology and Engineering Management Conference-Asia Pacific (TEMSCON-ASPAC), Bengaluru, India, pp. 1–5. IEEE (2023). https://doi.org/10.1109/TEMSCON-ASPAC59527.2023.10531403
13. Pan, R., et al.: One-class collaborative filtering. In: Eighth IEEE International Conference on Data Mining, Pisa, Italy, pp. 502–511. IEEE (2008). https://doi.org/10.1109/ICDM.2008.16
14. Qazi, A.G., Pachler, N.: Conceptualising a data analytics framework to support targeted teacher professional development. Prof. Dev. Educ. **51**(3), 495–518 (2024). https://doi.org/10.1080/19415257.2024.2422066
15. Rendle, S., Freudenthaler, C., Gantner, Z., Schmidt-Thieme, L.: BPR: Bayesian personalized ranking from implicit feedback. In: Twenty-Fifth Conference on Uncertainty in Artificial Intelligence, Montreal, Canada, pp. 452–461. AUAI Press (2009)
16. Rima, S., Meriem, H., Najima, D., Rachida, A.: Systematic literature review on open educational resources recommender systems. Int. J. Interact. Mob. Technol. **16**(8), 44–77 (2022). https://doi.org/10.3991/ijim.v16i18.32197
17. Shani, G., Gunawardana, A.: Evaluating recommendation systems. In: Ricci, F., Rokach, L., Shapira, B., Kantor, P.B. (eds.) Recommender Systems Handbook, pp. 257–297. Springer, Boston, MA (2011). https://doi.org/10.1007/978-0-387-85820-3_8
18. Siafis, V., Rangoussi, M., Psaromiligkos, Y.: Recommender systems for teachers: a systematic literature review of recent (2011–2023) research. Educ. Sci. **14**(7), 723 (2024). https://doi.org/10.3390/educsci14070723
19. Shah, D., Shah, P., Banerjee, A.: Similarity based regularization for online matrix-factorization problem: an application to course recommender systems. In: 2017 IEEE Region 10 Conference, Penang, Malaysia, pp. 1874–1879. IEEE (2017). https://doi.org/10.1109/TENCON.2017.8228164
20. Weston, J., Bengio, S., Usunier, N.: WSABIE: scaling up to large vocabulary image annotation. In: Proceedings of the Twenty-Second International Joint Conference on Artificial Intelligence, Barcelona, Spain, pp. 2764–2770. AAAI Press (2011)
21. Yao, X., Li, X.: Optimizing university teachers learning progress: evolving study efficiency with artificial intelligence agent. In: 7th International Conference on Big Data and Artificial Intelligence, Beijing, China, pp. 241–246. IEEE (2024). https://doi.org/10.1109/BDAI62182.2024.10692441
22. Zhang, L.: The definition of novelty in recommendation system. J. Eng. Sci. Technol. Rev. **6**(3), 141–145 (2013). https://doi.org/10.25103/jestr.063.25

Fuzzy Logic for Layer-Wise Explainability in Deep Neural Networks: A First Approach

Arturo Fernández-Mora[1], Juan Moreno-Garcia[2]([✉]), David Muñoz-Valero[2], and Luis Martínez[3]

[1] Escuela de Doctorados, Universidad de Jaén, 23071 Jaén, Andalucía, Spain
afm00108@red.ujaen.es
[2] E. Ingeniería Industrial y Aeroespacial Toledo, Universidad de Castilla-La Mancha, 45004 Toledo, Castilla-La Mancha, Spain
{Juan.Moreno,David.Munoz}@uclm.es
[3] Escuela Politécnica Superior de Jaén, Universidad de Jaén, 23071 Jaén, Andalucía, Spain
martin@ujaen.es

Abstract. This work presents an initial approach to studying the explainability of deep learning models at the layer level using fuzzy logic techniques. The aim is to link neural activations to input text concepts, analyzing their coherence and relationships. For this, a dataset of movie reviews is used, where each text is classified into one of seven emotions via a sentiment analysis model. The resulting activation vectors are clustered, allowing each vector to belong to multiple clusters, thus reflecting the ambiguity of textual data.

The resulting clusters are analyzed through coherence measures to assess their internal consistency and uncertainty. These analyses help reveal how similar concepts are grouped in the neural activations. Preliminary results are promising, motivating further exploration of this approach.

As an additional contribution, this work proposes the application of the same methodology to a domain with more objective concepts, such as wine recommendation. In this case, fuzzy profiles are defined based on organoleptic attributes of wine, and affinity scores between wines and user profiles are computed using TSK FIS models. These affinity values are used as neural network training targets, allowing subsequent analysis of how the network organizes these clearer and more structured concepts in its hidden activations.

Keywords: Explainability · Deep Learning · Fuzzy Logic

1 Introduction

The interpretability of deep learning models is a rapidly expanding field of research, driven by the growing need to understand and justify the decisions made by these systems in critical applications. In contexts where transparency and trust are essential—such as medicine, industrial automation, or decision-making in the social domain—the ability to explain a model's internal behavior is key to its acceptance and responsible use. However, despite their high performance, deep neural networks continue to be perceived as black boxes due to the inherent difficulty in understanding the internal processes that lead to their predictions.

This work presents an initial approach to the extraction of concepts learned by a neural network, with the aim of providing understandable explanations of its behavior. To this end, the study builds on the approach proposed by Templeton et al. [1], who employ autoencoders as a mechanism for identifying meaningful internal representations. Autoencoders are neural networks trained to reconstruct their input from a compressed latent space, which forces the network to learn an intermediate representation that captures the most relevant features or concepts from the original data.

These intermediate representations can be interpreted as a compressed form of the essential information present in the input data, acting as a filter that separates noise from the most relevant patterns [2]. The analysis of these intermediate layers therefore enables the identification of the aspects of the data captured by the network during learning, opening the door to more transparent interpretations of the model's behavior.

This work is structured as follows: Sect. 2 provides a detailed description of the process followed throughout the study, including dataset selection, neural network design and architecture, the procedure used to extract emotions from texts, and the methodology for concept analysis using clustering techniques. Section 3 presents and analyzes the results obtained after applying the proposed methods, evaluating the model's ability to segment neural activations into coherent clusters that reflect emotional patterns. Then, Sect. 4 addresses a critical analysis of the challenges encountered during the study and introduces an initial proposal to address these limitations in future work. Methodological improvements and new approaches are suggested to help overcome the identified issues. Finally, Sect. 5 summarizes the main conclusions drawn from the study, presents new contributions, and proposes future research directions aimed at improving the results and extending the proposed approach.

2   Architecture

This section describes the architecture used to extract emotions as learned concepts by a neural network when performing text polarity classification, following the general workflow shown in Fig. 1. The workflow can be summarized in the following steps:

1. First, the IMDB dataset [3] was selected for the experimentation. This dataset contains 50,000 reviews published by users about various movies, providing a suitable basis for the analysis of textual opinions.

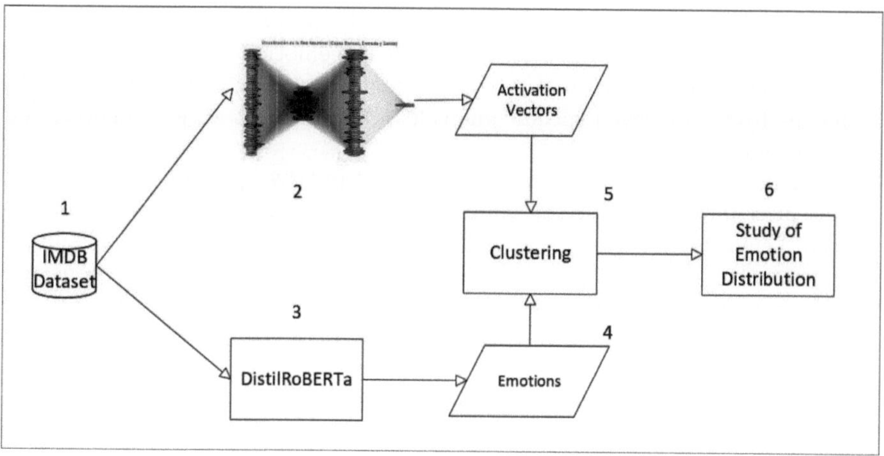

Fig. 1. Schematic representation of the complete process carried out for text analysis.

2. Next, a neural network was designed with the goal of classifying the polarity (positive or negative) of the texts in the IMDB dataset. The architecture of this network is shown in Fig. 2. The network includes an embedding layer that transforms the texts into 32-dimensional vectors, followed by a flattening layer that produces a one-dimensional vector. On top of this representation, an autoencoder is incorporated, composed of three dense layers (with 64, 15, and 64 neurons, respectively), where the intermediate layer is intended to capture the most relevant features for the task. To enhance generalization, a 50% dropout rate is applied to the connections between the input layer and the latent layer of the autoencoder, as well as between the latent layer and the output layer. The activations extracted from the intermediate layer are normalized and interpreted as concepts learned by the neural network.

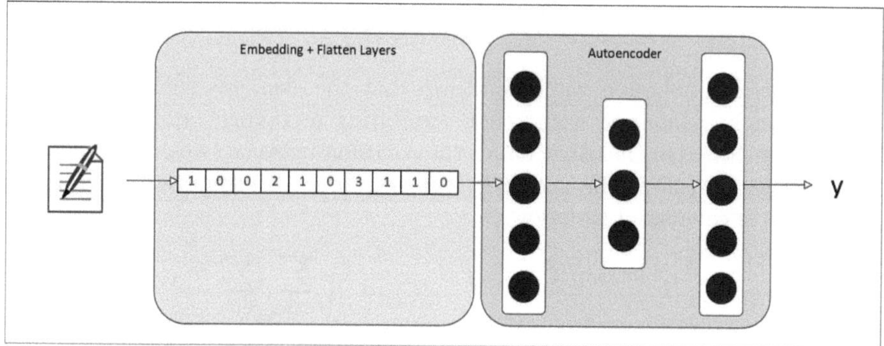

Fig. 2. Visual diagram of the neural network for polarity classification

3. Complementarily, the DistilRoBERTa model, an optimized and efficient version of RoBERTa [4], was used to extract the emotions present in each text. This model, widely used in natural language processing tasks, allows classifying the texts into seven specific emotions: disgust, neutral, joy, surprise, fear, anger and sadness.

4. The emotion detection process was carried out by splitting each text into sentences and identifying the dominant emotion in each one. Afterwards, the overall emotion of the text was calculated by weighting the emotions sentence by sentence, assigning greater importance to the first and last sentences, as well as to those containing exclamation or question marks. This strategy is based on the observed fact that in opinion texts, the main feelings are often expressed at the beginning or end, or emphasized through emphatic punctuation, as commonly occurs in movie reviews.

5. Finally, using the activation vectors and the emotions associated with each text, several clustering methods were applied to group the neural representations: Fuzzy C-means [5], K-Means [6], Agglomerative Clustering [7], and Gaussian Mixture Model (GMM) [11]. In the particular case of Fuzzy C-means, where a vector may partially belong to multiple clusters, the percentage of each emotion present in each cluster was calculated using the following expression:

$$C_{C,e} = 100 \times \sum_{i \in I_C} u_{C,i} \tag{1}$$

where u represents the degree of membership of text i, which expresses emotion e, in cluster C.

6. Finally, to evaluate the cohesion of the clusters, cosine similarity was used:

$$S_{j,i} = \frac{v_i \cdot v_j}{|v_i||v_j|} \tag{2}$$

where v_i and v_j are activation vectors. Therefore, the internal cohesion of each cluster C_k, which is a metric that reflects the average degree of similarity among the elements belonging to the same cluster, is calculated as follows:

$$Cohesion(C_k) = \frac{1}{|C_k|(|C_k| - 1)} \sum_{i \neq j \in C_k} S_{i,j} \tag{3}$$

High internal cohesion values indicate that the elements within the cluster are strongly related to each other, exhibiting a compact and homogeneous grouping [8,9]. On the other hand, the cohesion between two clusters C_m and C_n, which measures the average similarity between elements from different clusters, is calculated as follows:

$$Cohesion(C_m, C_n) = \frac{1}{|C_m||C_n|} \sum_{i \in C_m} \sum_{j \in C_n} S_{i,j} \tag{4}$$

Low values in this metric are desirable, as they indicate that the clusters are well differentiated from each other and exhibit low similarity, which suggests a clearer and more effective segmentation [10].

3 Results Analysis

In this section, the results obtained after applying the different clustering methods are presented, and the cohesion of the formed clusters is analyzed, with the aim of evaluating whether neural activations allow for a structured segmentation of emotional concepts.

The clustering methods used are:

- **Fuzzy C-means:** This algorithm is a fuzzy extension of the classic K-means. Instead of assigning each element to a single cluster, it allows each element to partially belong to several clusters, according to a membership degree that is iteratively optimized.
- **K-means:** One of the most well-known and widely used algorithms, which seeks to divide the dataset into k clusters by minimizing the intra-cluster distance. Each element is exclusively assigned to a cluster, and the centroid of each group is recalculated until convergence is reached.
- **Agglomerative Clustering:** A hierarchical method that starts by considering each element as an independent cluster and, in each iteration, merges the most similar clusters according to a distance metric (such as average or complete linkage), thus building a hierarchy of groups.
- **Gaussian Mixture Model (GMM):** This probabilistic model assumes that the data have been generated by a combination of several Gaussian distributions. Each cluster is represented by a Gaussian, and the membership of elements to clusters is calculated based on probabilities.

The corresponding tables show the results obtained, where the left part, a), of the table represents the distribution of emotions and the right part, b), represents the values of internal cohesion (Eq. 3) and inter-cluster cohesion (Eq. 4).

The analysis of emotional distributions shows that there is no clear assignment of emotions to specific clusters. Additionally, in K-means, Agglomerative Clustering, and GMM, one cluster appears to be practically empty.

Table 1. Combined matrix of emotion distribution by cluster and cohesion obtained through Fuzzy C-means.

	disgust	neutral	joy	surprise	fear	anger	sadness	0	1	2	3	4
0	21.995	20.709	20.938	18.058	19.557	18.378	23.142	0.915	0.798	−0.832	−0.821	−0.143
1	21.366	21.779	22.911	19.515	23.221	25.361	23.638	0.798	0.763	−0.783	−0.703	−0.051
2	16.546	18.214	17.384	17.004	16.761	17.581	14.533	−0.832	−0.783	0.811	0.742	0.069
3	20.829	21.519	20.499	21.400	19.577	24.068	19.023	−0.821	−0.703	0.742	0.755	0.151
4	19.265	17.780	18.268	24.023	20.884	14.612	19.664	−0.143	−0.051	0.069	0.151	0.112

Regarding cohesion:

- **Fuzzy C-means (Table** 1): Clusters 0, 1, 2, and 3 present high internal cohesion, with cluster 0 standing out. Cluster 4 shows low cohesion. Between clusters, relevant positive cohesion is observed between 0 and 1, and between 2 and 3, as well as marked negative cohesion between these pairs, indicating opposite activations.
- **K-means (Table** 2): Clusters 1, 2, 3, and 4 show good internal cohesion, especially clusters 3 and 4. Cluster 0 exhibits low cohesion. Regarding inter-cluster cohesion, strong negative values are observed between cluster 2 and clusters 1 and 3.
- **Agglomerative Clustering (Table** 3): High internal cohesion is observed in clusters 0, 1, 3, and 4, with clusters 3 and 4 standing out. Cluster 2 shows low cohesion. Negative inter-cluster relationships are observed between cluster 1 and clusters 0 and 3.
- **GMM** (Table 4): Only clusters 3 and 4 show high internal cohesion, while the rest present low cohesion. No significant inter-cluster relationships are observed.

Table 2. Combined matrix of emotion distribution by cluster and cohesion obtained through K-Means.

	disgust	neutral	joy	surprise	fear	anger	sadness	0	1	2	3	4
0	33.670	30.897	32.184	23.563	27.402	25.843	35.983	0.094	−0.063	0.079	−0.149	−0.039
1	33.165	36.766	33.972	38.506	34.520	40.449	28.452	−0.063	0.795	−0.835	0.479	−0.250
2	30.013	29.913	31.929	36.207	37.011	29.214	33.891	0.079	−0.835	0.890	−0.558	0.219
3	2.774	1.552	1.533	1.149	1.068	4.494	1.255	−0.150	0.479	−0.558	0.941	−0.160
4	0.378	0.871	0.383	0.575	0.000	0.000	0.418	−0.039	−0.250	0.219	−0.160	0.959

Table 3. Combined matrix of emotion distribution by cluster and cohesion obtained through Agglomerative Clustering.

	disgust	neutral	joy	surprise	fear	anger	sadness	0	1	2	3	4
0	40.479	42.863	40.485	41.954	38.078	43.820	35.146	0.678	−0.763	−0.173	0.439	−0.245
1	30.895	29.610	31.290	35.632	36.655	29.214	33.054	−0.763	0.869	0.208	−0.562	0.233
2	25.473	24.915	26.054	20.690	23.843	22.472	30.544	−0.173	0.208	0.190	−0.232	−0.011
3	2.774	1.666	1.660	1.149	1.068	4.494	0.837	0.439	−0.562	−0.232	0.943	−0.175
4	0.378	0.947	0.511	0.575	0.356	0.000	0.418	−0.245	0.233	−0.011	−0.175	0.941

In summary, the neural activations do not exhibit a clear structure that enables the effective segmentation of emotional concepts into well-defined groups, which seems to be due to the difficulty of capturing the emotions conveyed by

Table 4. Combined matrix of emotion distribution by cluster and cohesion obtained through GMM.

	disgust	neutral	joy	surprise	fear	anger	sadness	0	1	2	3	4
0	47.037	49.716	48.148	42.529	46.619	48.315	52.720	0.106	−0.106	−0.059	0.037	−0.018
1	24.969	21.242	22.733	24.713	22.064	22.472	20.502	−0.106	0.243	0.012	−0.057	−0.054
2	1.513	2.007	1.533	1.149	1.424	0.000	0.837	−0.059	0.012	0.242	0.087	−0.112
3	12.736	14.237	13.665	10.345	12.811	19.101	8.787	0.037	−0.057	0.087	0.861	−0.868
4	13.745	12.798	13.921	21.264	17.082	10.112	17.155	−0.018	−0.054	−0.112	−0.868	0.953

the texts. However, some clusters show a tendency to represent certain emotions more strongly than others, and, on the other hand, good internal cohesion is observed within the clusters themselves. This motivates continuing research in this direction with the aim of obtaining concept-based clusters from neural activations, using datasets whose concepts are less ambiguous and more interpretable. For this reason, an alternative approach is proposed based on the oenological domain, specifically wine and food pairing, as it is considered to have a less subjective and ambiguous interpretation than that associated with emotions.

4   Framework for the Study on Wine Pairing

It is proposed to apply the same procedure used for emotion analysis, but adapted to the domain of wine pairing. The new framework, detailed throughout this section, is illustrated in Fig. 3. In this figure, it can be seen that the emotion extraction stage has been replaced by a profiling system based on fuzzy

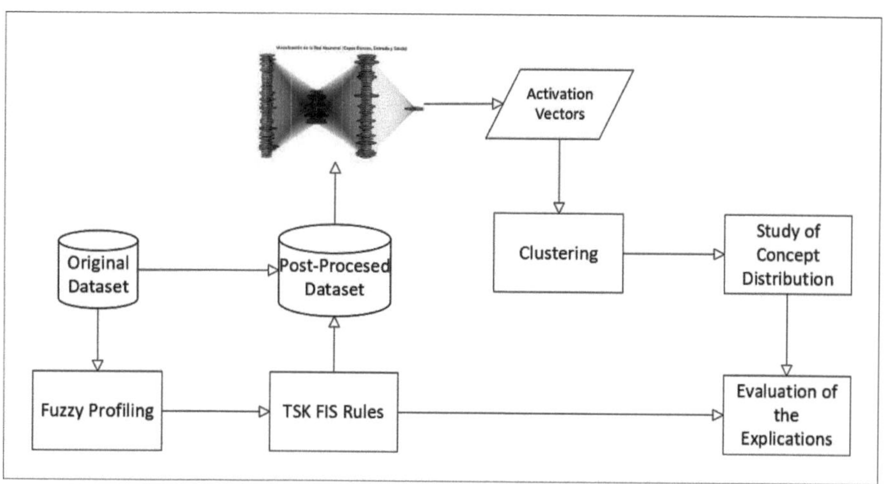

Fig. 3. Visual scheme of the new complete process for extracting wine concepts.

rules. This system is used both to generate the outputs that train the neural network and to facilitate the comparison of the final explanations derived from the model.

The original dataset to be used is the wine dataset [12], which includes the following organoleptic characteristics:

- **Sweetness (D):** Indicates the residual sugar in the wine. It can take the values: [Very Dry, Dry, Semi-Dry, Sweet and Very Sweet].
- **Acidity (AC):** Provides freshness and balance, and can be perceived as a tingling sensation in the mouth. It can take the values: [Very Low, Low, Medium, High and Very High].
- **Temperature (T):** Indicates the optimal temperature for serving the wine. The fuzzy sets defined for the serving temperature are shown in Fig. 4.
- **Tannins (TA):** Compounds derived from the grape skin that provide astringency, a drying sensation, and structure to the wine. It can take the values: [Very Low, Low, Medium, High and Very High].
- **Color (C):** Indicates the type of wine. Possible values include [Red, White, Rosé, Sparkling].
- **Alcohol (AL):** Contributes warmth and body to the wine. The fuzzy sets defined for alcohol content in wine are shown in Fig. 5.
- **Body (CU):** Refers to the sensation of weight and density in the mouth, influenced by alcohol, tannins, and residual sugar. It can take the values: [Very Light, Light, Medium, Full-bodied and Very Full-bodied].

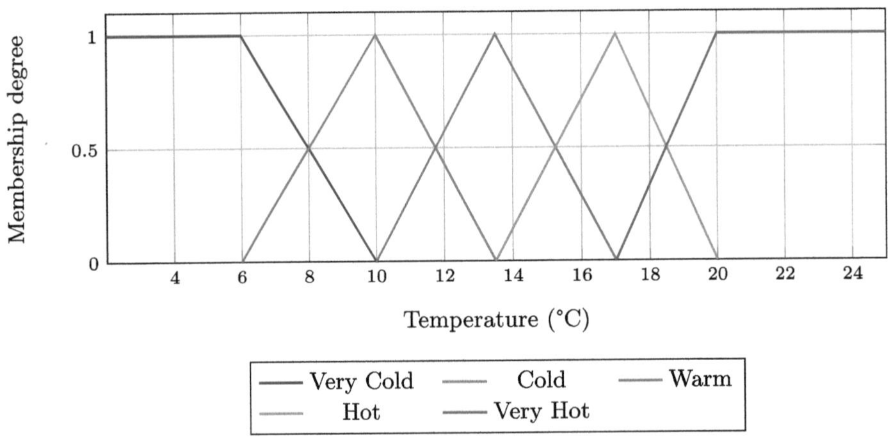

Fig. 4. Fuzzy sets for the wine serving temperature

Based on this data, the goal is to classify each wine according to the pairing that best matches it, facilitating the extraction of latent concepts that reflect meaningful relationships between the wine's characteristics and its possible gastronomic combinations. To achieve this, different pairings are modeled as profiles

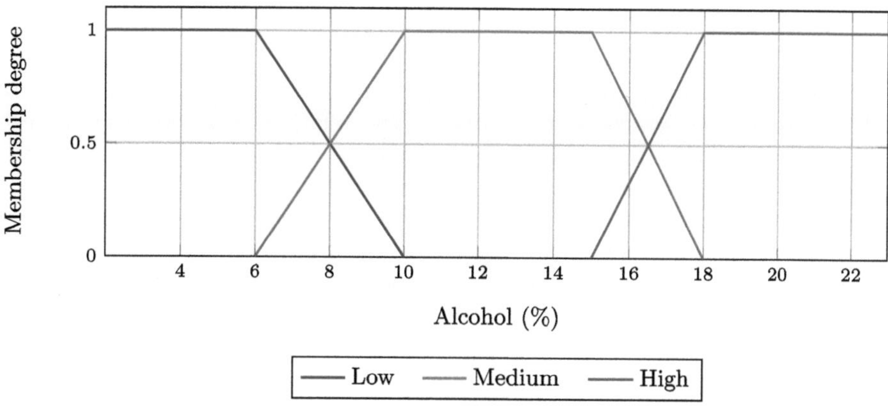

Fig. 5. Fuzzy sets for the alcohol percentage present in the wine

that capture expert knowledge through a fuzzy model [13], allowing the evaluation of the affinity degree between each wine and the defined profiles. These models, built from relevant combinations of attributes, enable the recommendation of suitable pairings for each wine according to its organoleptic characteristics.

For example, Table 5 shows the model created for the pairing profile that includes Asian food, seafood, salads, and fried dishes. This type of food pairs better with light-bodied wines, that is, with light or very light body (CU – Light or Very Light), served at low temperatures (T – Very Cold), exhibiting relatively high acidity (AC – High or Very High), and having low tannins (TA – Very Low). Since these foods contain less fat, alcohol is absorbed less effectively, so it is advisable to choose wines with low alcohol content (AL – Low). White or sparkling wines are most suitable for this type of cuisine (C – White or

Table 5. Representative values for Asian food, salads, fried dishes and seafood

Body (CU)	Very Light
	Light
Temperature (T)	Very Cold
Acidity (AC)	Very High
	High
Tannins (TA)	Very Low
Alcohol (AL)	Low
Sweetness (D)	Very Dry
	Dry
Color (C)	White
	Sparkling

Sparkling). Regarding sweetness, drier wines tend to pair better (D – Very Dry or Dry).

To determine which pairings work best with each wine, it is necessary to define a preference vector that represents the relative importance of the different co-occurrences between attributes. This vector allows for differentiated weighting of combinations involving different numbers of attributes, enabling the representation of varying levels of complexity in the profiles. Additionally, weight vectors are defined for each order of specific combinations of the wine's organoleptic characteristics—such as the relationship between acidity, serving temperature, and alcohol content, or between sweetness and tannins—based on their relevance within a given profile. In this way, a weighting mechanism is introduced to more accurately reflect the implicit preferences in each user profile, enabling a more flexible and expressive modeling through fuzzy rules. Following the example of Asian food, salads, fried dishes, and seafood, the following fuzzy rules are obtained for this profile:

- R1: If (CU Medium) AND ((TA Low) OR (TA Medium)) Then $100 * 0.55 * 0.5 = 27.5$
- R2: If ((C White) OR (C Rosé)) AND (AC Medium) AND (T Medium) Then $100 * 0.55 * 0.5 = 27.5$
- R3: If ((C Red) OR (C Rosé)) AND (AC Medium) AND (AL Medium) Then $100 * 0.45 * 0.5 = 22.5$
- R4: If ((D Dry) OR (D Semi-dry)) AND (TA Medium) Then $100 * 0.45 * 0.5 = 22.5$

The weight vectors and utility degree vectors used to determine the rules are as follows:

- $w_2 = \{w^{CU-TA}, w^{D-TA}\} = \{0.55, 0.45\}$
- $w_3 = \{w^{C-AC-T}, w^{C-AC-AL}\} = \{0.55, 0.45\}$
- $w_1, w_4, w_5, w_6, w_7 = \text{Not applicable}$
- $DU = \{0, 0.5, 0.5, 0, 0, 0, 0\}$

In this way, these profiles will be used to assign each wine a membership value for each pairing profile using the TSK FIS approach [14]. These membership values will be used as outputs to train the neural network from which the activations are extracted in order to study the learned concepts. Furthermore, since the fuzzy rules encapsulate expert knowledge for each profile, they will also serve to compare the reliability of the explanations obtained through the network activations.

5   Conclusions and Future Works

In conclusion, it has been shown that using emotions as base concepts for analysis presents a high level of complexity, due both to their abstract nature and the inherent ambiguity in their interpretation. This difficulty makes concept segmentation and analysis imprecise and challenging to interpret accurately. For this reason, an alternative approach is proposed: to work with a dataset whose concepts are clearly defined and more objective and quantifiable, thereby enabling more interpretable and robust results.

Despite the difficulties encountered in the case of emotions, we remain confident in the proposed framework thanks to the levels of internal cohesion observed in the resulting clusters. These results indicate that, although the conceptual segmentation was not entirely satisfactory due to the complexity of the emotional domain, the architecture used shows potential for grouping data of a more objective and structured nature. Therefore, it is proposed to continue exploring this approach, adapting it to contexts where concepts are more easily identifiable, such as in wine pairing profiling.

An interesting line of research involves modeling the uncertainty of neural weights using fuzzy logic. Unlike traditional approaches, which consider weights as fixed values optimized during training, this approach would allow their variability to be represented as ranges or membership functions, thus capturing factors such as dataset size, problem complexity, or noise. In this way, each neural connection would include an associated degree of confidence, providing a more flexible and robust model in the face of data variability. This representation would facilitate more interpretable and resilient networks, and would enable a deeper analysis of how uncertainty affects model behavior and tasks such as clustering or classification.

Acknowledgments. This work was supported by grant PID2020-112967GBC32 funded by MCIN/AEI/10.13039/501100011033, by ERDF A Way of Making Europe and by grant ProyExcel_00257.

References

1. Templeton, A.: Scaling monosemanticity: extracting interpretable features from claude 3 sonnet. Anthropic (2024)
2. Scherlis, A., Sachan, K., Jermyn, A.S., Benton, J., Shlegeris, B.: Polysemanticity and capacity in neural networks (2022). arXiv preprint arXiv:2210.01892
3. Maas, A., Daly, R.E., Pham, P.T., Huang, D., Ng, A.Y., Potts, C.: Learning word vectors for sentiment analysis. In: Proceedings of the 49th Annual Meeting of the Association for Computational Linguistics: Human Language Technologies, pp. 142–150, June 2011
4. Liu, Y., et al.: RoBERTa: a robustly optimized BERT pretraining approach (2019). arXiv preprint arXiv:1907.11692
5. Bezdek, J.C., Ehrlich, R., Full, W.: FCM: the fuzzy c-means clustering algorithm. Comput. Geosci. **10**(2–3), 191–203 (1984)
6. MacQueen, J.: Some methods for classification and analysis of multivariate observations. In: Proceedings of the Fifth Berkeley Symposium on Mathematical Statistics and Probability, vol. 1, pp. 281–297. University of California Press (1967)
7. Johnson, S.C.: Hierarchical clustering schemes. Psychometrika **32**(3), 241–254 (1967)
8. Han, J., Kamber, M., Pei, J.: Data Mining: Concepts and Techniques, 3rd edn. Elsevier (2011)

9. Jain, A.K., Murty, M.N., Flynn, P.J.: Data clustering: a review. ACM Comput. Surv. **31**(3), 264–323 (1999)
10. Xu, R., Wunsch, D.: Clustering and Density Estimation. Wiley-IEEE Press (2005)
11. McLachlan, G.J., Peel, D.: Finite Mixture Models. John Wiley & Sons (2004)
12. Kim, G.-R.: Wine information Version 7. Kaggle. https://www.kaggle.com/datasets/dev7halo/wine-information. Accessed 28 May 2025
13. Muñoz-Valero, D., Moreno-García, J., López-Gómez, J.A., Villarrubia-Martín, E.A., Jiménez-Linares, L.: A knowledge-driven fuzzy logic framework for supporting decision-making entities. Appl. Soft Comput. **139**, 113415 (2025). https://doi.org/10.1016/j.asoc.2025.113415
14. Takagi, T., Sugeno, M.: Fuzzy identification of systems and its applications to modeling and control. IEEE Trans. Syst. Man Cybern. **1**, 116–132 (1985)

Monte Carlo-Based Interval TOPSIS for Navigating Decision Support Under Uncertainty

Jingda Ying[1(✉)], Christian Wagner[1], Isaac Triguero[1,2,3], and Shaily Kabir[1]

[1] Lab for Uncertainty in Data and Decision Making (LUCID), School of Computer Science, University of Nottingham, Nottingham, UK
{psxjy6,christian.wagner,isaac.triguero1,shaily.kabir5}@nottingham.ac.uk
[2] Department of Computer Science and Artificial Intelligence, University of Granada, Granada, Spain
[3] DaSCI Andalusian Institute in Data Science and Computational Intelligence, Granada, Spain

Abstract. Multi-criteria decision-making (MCDM) techniques are widely used to facilitate systematic and informed decision making. However, real-world problems in diverse fields such as business, cybersecurity, and environmental management often involve significant levels of uncertainty. While progress has been made in developing MCDM methods capable of addressing uncertainty, many challenges remain–particularly in capturing uncertainty, communicating it effectively to stakeholders, and selecting the most suitable MCDM technique for a given context. This paper focuses on interval extensions of the widely used Technique for Order Preference by Similarity to Ideal Solution (TOPSIS), highlighting the need to (1) effectively communicate uncertainty to decision makers and (2) establish a way to compare these extensions correctly. To address these, we propose a Monte Carlo-based approach for interval TOPSIS, offering a new interval extension of the original TOPSIS to handle uncertainty independently. Meanwhile, we leverage a data set from the literature and use the proposed extension as a reference point to compare some of the most recent TOPSIS extensions and analyse their underlying differences in uncertainty handling.

Keywords: MCDM · Uncertainty · Interval · TOPSIS · Monte Carlo

1 Introduction

Uncertainty is often inevitable and a critical aspect in decision-making scenarios. In many cases, an expert's assessment of specific criteria is subject to uncertainty due to incomplete data or the subjectivity of the given context. For example, if a cybersecurity expert is to assess the vulnerability of a network without having access to the firmware versions used by some of the components, they may not be able to provide a precise (and consistent) assessment, even if they are highly trained and experienced. When data is collected through questionnaires, the use of intervals helps to express the uncertainty in their responses [8]. Unlike

L. Martínez et al. (Eds.): IDEAL 2025, LNCS 16239, pp. 47–58, 2026.
https://doi.org/10.1007/978-3-032-10489-2_5

numerical values, intervals enable direct modelling of uncertainty through their range/width.

Multi-criteria decision making (MCDM) is the domain of structured decision-support, primarily concerned with techniques designed to help decision makers (DMs) choose alternatives by considering an array of potentially conflicting criteria [2,21]. One of the most popular MCDM methods is the Technique for Order Preference by Similarity to Ideal Solution (TOPSIS) [9,22]. TOPSIS employs a geometric approach to evaluate alternatives based on their proximity to the ideal solution and the distance from the worst solution, generally deriving a ranking of the available alternatives as its output. Despite its success in numerous applications [18], the original version of TOPSIS is not designed to handle uncertainty in experts' assessments. This has led to the development of a wide variety of TOPSIS extensions that allow different input formats such as intervals [1,11] or fuzzy sets [4,16].

While these extensions have advanced the handling of uncertain criteria assessments, there remains substantial room for further development with respect to the effective and efficient handling of uncertainty for better–informed decision support. To illustrate this, Fig. 1 shows a conceptual, yet typical supplier selection MCDM example addressed with TOPSIS where the aim is to choose between two alternatives/suppliers A_1 and A_2. For each alternative, the same two criteria, i.e. Cost and Quality, have been assessed using an interval, reflecting uncertainty in the given assessments. In an ideal case, extensions of TOPSIS designed to handle interval inputs eventually produce an interval-valued Relative Closeness (RC)[1]. Generally, even when the RC is provided as an interval, the final step is to adopt a ranking approach to generate the final ranking of the alternatives/suppliers, as shown in Fig. 1. However, note how such a ranking may be misleading, even in this simplistic case: a conservative DM may well prefer A_2 over A_1: while it may not be as good, it also does not risk to be as bad. A DM more open to risk may prefer A_1. Critically, this decision lies with the DM, and the role of an MCDM system is to put them in a better informed position to make this choice, *not* to make the decision for them.

The above example demonstrates some of the high-level challenges associated with handling uncertainty with TOPSIS, including existing extensions. As part of an ongoing body of work, we are focusing on three specific aspects of this gap in the literature.

1. How can uncertain, e.g. interval-valued criteria assessments be practically elicited in real-world contexts?
2. How do we effectively map uncertainty affecting the criteria assessments to the decision support outputs (such as the RC in the example above) and communicate them meaningfully to DMs?
3. How do we compare and contrast different (interval) TOPSIS extensions with a view to selecting the most appropriate techniques given a specific context?

---

[1] Where extensions do not generate interval-valued outputs, they cannot communicate uncertainty arising from the inputs.

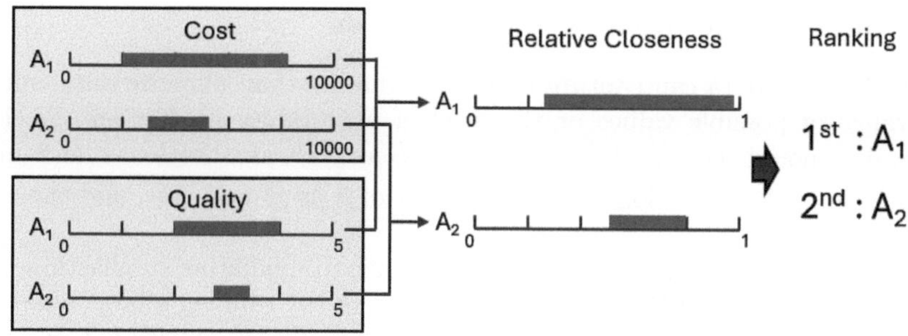

Fig. 1. Is ranking of A_1 really (always) better than A_2?

This paper primarily addresses the third point above–seeking to develop the means to compare various TOPSIS techniques. While substantial literature on TOPSIS extensions exists, there are quasi-no standard benchmark problems or other means to compare techniques, making it very challenging to establish the advantages/disadvantages of one approach vs another.

As an initial step towards enabling a more direct comparison of techniques, we propose a new extension of the original TOPSIS [22] for interval-valued criteria assessments using Monte Carlo (MC) simulation [14]. The latter is a widely adopted approach to handling uncertain, such as interval [3,14,23] inputs, while using the original model developed for numeric data. It thus offers a simple but principled extension which may serve as a reference point facilitating the comparison of the behaviour of other TOPSIS extensions–leveraging other techniques such as interval arithmetic [10], or additional components, such as an auxiliary decision matrix [13], or weight entropy [12]–to the original TOPSIS approach.

Beyond introducing and evaluating the Monte Carlo TOPSIS extension, we also compare the results to Jahanshahloo's 2009 [10] and Kaczynska's 2023 [13] TOPSIS extensions as the most widely used and most recent extensions, respectively. We do not suggest that the proposed extension is in any way ideal–but we discuss how it may serve as a reference point for supporting mutual comparisons of interval TOPSIS extensions, and thus may provide better guidance for DMs regarding the behaviours associated with a specific extension selected for use.

This paper is organised as follows: Sect. 2 briefly introduces key concepts. Section 3 proposes the MC TOPSIS extension for intervals and demonstrates its use. Section 4 compares the MC TOPSIS with two existing extensions. Lastly, Sect. 5 concludes the paper and highlights future work.

2   Background

In this section, we briefly review key concepts, including interval uncertainty modelling, MC simulation, the original as well as two interval TOPSIS extensions.

2.1 Intervals and Monte Carlo Simulation

Interval-valued data can capture richer information than numeric data, such as the range of possible values or the level of uncertainty in data [6]. A closed interval is noted as $\bar{x} = [x^l, x^u]$, where x^l and x^u are the lower and upper bounds of $\bar{x}$ $(x^l \leq x^u)$ along with the center of $\bar{x}$ as $x^c = \frac{x^u + x^l}{2}$, and the range as $x^r = x^u - x^l$.

MC simulation provides an effective approach to simulating stochastic scenarios within an interval-valued data set [14]. It works by running a large number of simulations on numeric samples drawn from the a priori interval-valued inputs to approximate the behaviour of a system under uncertainty. MC simulation has already been used to evaluate linear/non-linear interval regression models [3,14,23] and to solve complex MCDM problems under uncertainty [15,17]. However, existing studies have not yet attempted to use MC simulation to compare and evaluate TOPSIS extensions.

2.2 The Original TOPSIS

An MCDM problem D can be concisely expressed in a matrix format as defined in (1).

$$
D =
\begin{array}{c}
\\
A_1 \\
A_2 \\
\vdots \\
A_m
\end{array}
\begin{array}{c}
\begin{array}{cccc}
C1 & C2 & \cdots & Cn
\end{array} \\
\left(
\begin{array}{cccc}
x_{11} & x_{12} & \cdots & x_{1n} \\
x_{21} & x_{22} & \cdots & x_{2n} \\
\vdots & \vdots & \vdots & \vdots \\
x_{m1} & x_{m2} & \cdots & x_{mn}
\end{array}
\right)
\end{array}
\tag{1}
$$

$$
W = \left[w_1, w_2, \ldots, w_n \right]^T
$$

where $A_1, A_2 \cdots A_m$, are possible alternatives among which DMs have to choose, $C_1, C_2 \cdots C_n$ are criteria with which each alternative's performance is measured, x_{ij} is the assessment of alternative A_i with respect to criterion C_j. A weight vector W measures the importance of each criterion, and w_j is the weight corresponding to the criterion C_j.

In TOPSIS, the positive ideal solution (PIS) and negative ideal solution (NIS) are defined as all the best/worst achievable values of each criterion, respectively. TOPSIS is based on the idea that the best alternative should have the relative shortest distance from the PIS and the relative farthest distance from the NIS.

TOPSIS addresses the MCDM problem (1) via the following six steps:

Step 1: normalise D column-wise for each assessment as $N = (n_{ij})_{m \times n}$ with:

$$
n_{ij} = \frac{x_{ij}}{\sqrt{\sum_{k=1}^{m} x_{kj}^2}}, i = 1, ..., m, j = 1, ..., n.
\tag{2}
$$

Step 2: Calculate the weighted decision matrix $V = (v_{ij})_{m \times n}$:

$$v_{ij} = w_j n_{ij}, i = 1, ..., m, j = 1, ..., n. \tag{3}$$

Step 3: Determine PIS and NIS for each criterion as A^+ and A^-:

$$A^+ = [v_1^+, v_2^+, \ldots, v_n^+] = \{\max_i v_{ij} | j \in K_b, \min_i v_{ij} | j \in K_c\},$$
$$A^- = [v_1^-, v_2^-, \ldots, v_n^-] = \{\min_i v_{ij} | j \in K_b, \max_i v_{ij} | j \in K_c\}. \tag{4}$$

where K_b is the set of benefit criteria, K_c is the set of cost criteria.

Step 4: Calculate the Euclidean distances between each alternative and PIS, respectively NIS as d_i^+ and d_i^-:

$$d_i^+ = \sqrt{\sum_{j \in K_b} (v_{ij} - v_j^+)^2 + \sum_{j \in K_c} (v_{ij} - v_j^-)^2}, i = 1, \ldots, m,$$
$$d_i^- = \sqrt{\sum_{j \in K_b} (v_{ij} - v_j^-)^2 + \sum_{j \in K_c} (v_{ij} - v_j^+)^2}, i = 1, \ldots, m. \tag{5}$$

Step 5: Calculate the relative closeness (RC) of each alternative:

$$RC_i = \frac{d_i^-}{d_i^- + d_i^+}, i = 1, 2, ..., m \tag{6}$$

Step 6: Rank the alternatives such that the bigger the RC_i, the better ranked alternative A_i.

2.3   Interval TOPSIS Extensions

Interval TOPSIS extensions have been developed to address and manage uncertainty in decision-making processes. To handle uncertainty in interval-valued criteria assessments, Jahanshahloo et al. [11] first extended the TOPSIS to intervals in 2006 and later proposed a new model in 2009 that used interval arithmetic to generate an interval-valued RC [10]. Tsaur [20] modified the normalisation method to adapt to DMs with different risk preferences. Dymova et al. [7] argued that the determined ideal solutions should exist in the data set, and proposed Direct TOPSIS. Furthermore, Jiang et al. [12] introduced the use of entropy weights to leverage richer information from the data set and defined a flexible parameter to apply entropy weights more comprehensively. Kaczynska et al. [13] proposed another method that calculates RC as intervals by building auxiliary decision matrices for each alternative, incorporating combinations of the best and worst cases of each assessment.

Jahanshahloo et al. 2009 [10] and Kaczynska et al. [13] consider RC to be calculated as intervals to capture additional information. Both methods proceed to provide the final rankings, providing an indication in terms of which alternative is 'ideal'.

In other words, the approaches consider uncertainty but do not engage with the broader challenge of how this should be communicated, for example, in the sense discussed in respect to Fig. 1.

In the following, we briefly review the above two interval TOPSIS extensions [10,13] with standardised notations.

Jahanshahloo's Approach [10] **Step 1:** Generate the normalised decision matrix $\bar{N}$. The normalisation method for each interval-valued assessment $\bar{x}_{ij}$ in decision matrix $\bar{D}$ is:

$$n_{ij}^l = \frac{x_{ij}^l}{\sqrt{\sum_{k=1}^m (x_{kj}^l)^2 + (x_{kj}^u)^2}}, i = 1, ..., m, j = 1, ..., n,$$

$$n_{ij}^u = \frac{x_{ij}^u}{\sqrt{\sum_{k=1}^m (x_{kj}^l)^2 + (x_{kj}^u)^2}}, i = 1, ..., m, j = 1, ..., n. \tag{7}$$

Step 2: Determining the weighted decision matrix $\bar{V} = (\bar{v}_{ij})_{m \times n}$:

$$\bar{v}_{ij} = w_j \bar{n}_{ij} = [w_j n_{ij}^l, w_j n_{ij}^u], i = 1, ..., m, j = 1, ..., n. \tag{8}$$

Step 3: Determining the PIS and NIS respectively.

The authors consider that the definition of ideal solutions should depend on the specific conditions of each interval-valued assessment. To address this, the interval-valued PIS ($\bar{A}_k^+ = [A_k^{+l}, A_k^{+u}]$) and NIS ($\bar{A}_k^- = [A_k^{-l}, A_k^{-u}]$) of alternative k are determined separately based on each assessment $\bar{v}_{kj}$ using (9) where K_b is the set of benefit criteria, K_c is the set of cost criteria.

$$A_k^{+l} = [v_{k1}^{+l}, v_{k2}^{+l}, \ldots, v_{kn}^{+l}] = \{\max_{i \neq k}\{v_{ij}^u, v_{kj}^l\} | j \in K_b, \min_{i \neq k}\{v_{ij}^l, v_{kj}^u\} | j \in K_c\},$$

$$A_k^{+u} = [v_{k1}^{+u}, v_{k2}^{+u}, \ldots, v_{kn}^{+u}] = \{\max_i v_{ij}^u | j \in K_b, \min_i v_{ij}^l | j \in K_c\},$$

$$A_k^{-l} = [v_{k1}^{-l}, v_{k2}^{-l}, \ldots, v_{kn}^{-l}] = \{\min_i v_{ij}^l | j \in K_b, \max_i v_{ij}^u | j \in K_c\},$$

$$A_k^{-u} = [v_{k1}^{-u}, v_{k2}^{-u}, \ldots, v_{kn}^{-u}] = \{\min_{i \neq k}\{v_{ij}^l, v_{kj}^u\} | j \in K_b, \max_{i \neq k}\{v_{ij}^u, v_{kj}^l\} | j \in K_c\}. \tag{9}$$

Step 4: Calculate the distances between each alternative to PIS and NIS. Four Euclidean distances d_i^{+u}, d_i^{+l}, d_i^{-u}, and d_i^{-l} are computed, where d_i^{+u} represents the distance from the worst case (lower bounds for benefit criteria, and upper bounds for cost criteria) of the alternative $\bar{V}_i$ to the best case of positive ideal solutions A_i^+ (upper bounds for benefit criteria, and lower bounds for cost criteria); d_i^{+l} is the distance from the best case of the alternative $\bar{V}_i$ to the worst case of positive ideal solutions A_i^+; d_i^{-u} represents the distance from the best case of the alternative $\bar{V}_i$ to the worst case of negative ideal solutions A_i^-; d_i^{-l} represents the distance from the worst case of the alternative $\bar{V}_i$ to the best case of negative ideal solutions A_i^-.

Step 5: Derive the interval-valued $\overline{RC}$ for each alternative with the above four distances:

$$\overline{RC_i} = [\frac{d_i^{-l}}{d_i^{-u} + d_i^{+u}}, \frac{d_i^{-u}}{d_i^{-l} + d_i^{+l}}], i = 1, 2, ..., m. \tag{10}$$

Step 6: Rank the alternatives using the $\overline{RC}$, adopting interval comparison approaches from [5, 19] to compare intervals.

Kaczynska's Approach [13]. In this method, the first two steps are the same as in Jahanshahloo's approach. Then, the authors introduce numeric auxiliary decision matrices $M_j^{'}$, generated by conducting a Cartesian product of the lower and upper bounds for each alternative against each criterion:

$$M_j^{'} = \{v_{1j}^l, v_{1j}^u\} \times \{v_{2j}^l v_{2j}^u\} \times \cdots \times \{v_{nj}^l, v_{nj}^u\}. \tag{11}$$

The authors applied the original TOPSIS and recorded the RC for every auxiliary decision matrix. The interval-valued $\overline{RC}$ $[min(RC_i), max(RC_i)]$ is defined as the combination of the minimum and maximum values obtained from all the auxiliary matrices. Since each auxiliary matrix represents a combination of extreme cases of assessment, the maximum and minimum values of the auxiliary matrices to some extent reveal the uncertainty of the alternative. The authors proceed to rank the alternatives in three ways, specifically the lower bounds, the centres, and the upper bounds of the $\overline{RC}$s.

They compare their proposed model in terms of ranking results with the initial Jahanshahloo approach [11] that calculates RC as a numeric value, rather than an interval. The authors do not further discuss the potential for interpretation or communication of the range of $\overline{RC}$s with respect to decision support.

3   An MC-Based TOPSIS Extension for Intervals

As mentioned in Sect. 1, there are no standard evaluation frameworks or data sets to evaluate or compare interval TOPSIS extensions. In this paper, we propose the Monte Carlo (MC) TOPSIS as a potential reference point for comparing interval TOPSIS extensions by adopting uniformly sampled MC simulations on the interval-valued data set. Here, the original TOPSIS is adopted to calculate the RC and associated alternative ranking for each sample before combining sample results. MC TOPSIS not only provides an interval- or distribution-valued RC to communicate the uncertainty, but also provides probable rankings along with their corresponding probabilities. The method consists of the following steps:

1. Uniformly sample a point in every interval-valued assessment, then record these crisp assessments as a possible (now numeric) assessment scenario D_k. Repeat the process k times, where we use $k = 100,000$ in this paper.
2. Apply the original TOPSIS to determine the numeric RCs for each assessment scenario D_k.

3. Record and summarise the ranking of alternatives for each scenario D_k.
4. Use the minimum and maximum values of the set of resulting RCs as an interval-valued RC ($\overline{RC}$) for each alternative.[2]

To illustrate the proposed MC TOPSIS, the experiment data set from [10], shown in Table 1 is used. In this data set, six alternatives and four criteria are involved, where Criteria 1 and 2 are cost criteria and Criteria 3 and 4 are benefit criteria. The weight for every criterion is set to 1.

Table 1. MC Simulation data set, adopted from [10].

Alternatives	Criterion 1	Criterion 2	Criterion 3	Criterion 4
Alternative 1	1451	[2551, 3118]	[40, 50]	[153, 187]
Alternative 2	843	[3742, 4573]	[63, 77]	[459, 561]
Alternative 3	1125	[3312, 4049]	[48, 58]	[153, 187]
Alternative 4	55	[5309, 6488]	[72, 88]	[347, 426]
Alternative 5	356	[3709, 4534]	[59, 71]	[151, 189]
Alternative 6	391	[4884, 5969]	[72, 88]	[388, 474]

As an example, Fig. 2 shows the histogram of the RCs for each alternative with the proposed MC TOPSIS. The histogram shows a roughly symmetrical bell-shaped distribution, indicating that the RC result from MC TOPSIS is approximately normally distributed. In this paper, we select the minimum and maximum values, to form RC intervals for subsequent comparisons, as we aim to compare it with other extensions where the output RC is interval-valued. A more detailed discussion of the richer information within the RC interval will be addressed in future research.

As ranking interval-valued RC is challenging, the proposed MC TOPSIS method offers a new approach to assist DMs in identifying the most appropriate alternative in the sense discussed in Sect. 1. While the original TOPSIS provides precise rankings for individual scenarios, MC TOPSIS offers rankings along with their corresponding probabilities, derived from simulation experiments. In this case, a summary of the ranking results shows that, out of 100,000 simulations, 99,417 instances (99.417% probability) resulted in the ranking $A_4 > A_6 > A_5 > A_2 > A_3 > A_1$, while 583 instances (0.583% probability) resulted in the ranking $A_6 > A_4 > A_5 > A_2 > A_3 > A_1$.

Notably, the extent of overlapping between the $\overline{RC}$s does not determine the likelihood of one alternative being superior to the other. As Fig. 3 shows, the $\overline{RC}$s of MC TOPSIS corresponding to Alternative 1 and Alternative 3 overlap significantly. However, no cases are observed in the simulation results where alternative 1 is superior to alternative 3. The reason for this lies at the heart of how RC is computed within TOPSIS, with the individual alternatives' criteria

---

[2] As discussed in the remainder of this section, the RCs will not be uniformly distributed. In this paper we do not make further use of this distribution, focusing on the minimums and maximum.

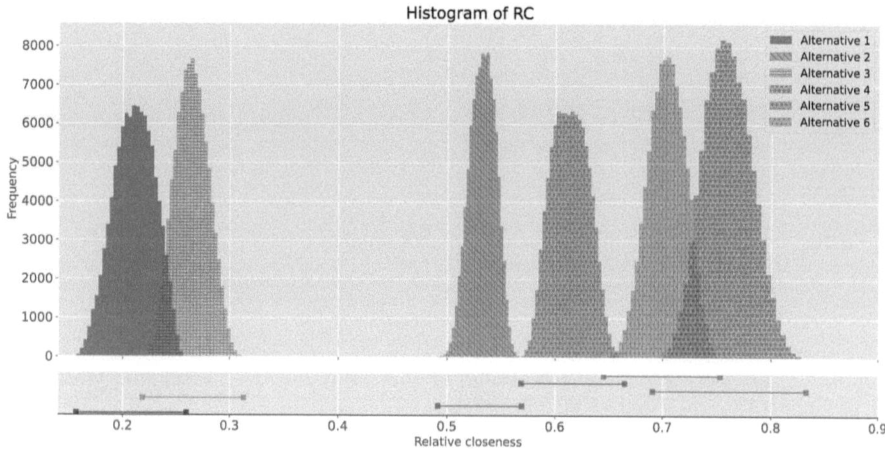

Fig. 2. Histogram of the MC TOPSIS RC. From left to right, RC for Alternatives: 1, 3, 2, 5, 6, 4.

assessments in one iteration interacting via the normalisation process. However, the range or size of $\overline{RC}$ is of value, reflecting a measure of uncertainty in the performance of the given alternative. For example, when the probabilities of one alternative being superior to or inferior to another are roughly the same, then the range of the corresponding intervals for both alternatives can serve as an important reference for selection, as the example shown in Fig. 1 shows. We will further discuss the interpretation of the uncertainty captured in the MC TOPSIS output distributions and associated intervals in a future journal article.

4   Comparing Extensions with MC TOPSIS

We proceed to compare the results of the proposed MC TOPSIS to Jahanshahloo's [10] and Kaczynska's [13] methods as the most popular and most recent interval extensions of TOPSIS, respectively.

The RC outputs of [10, 13], and MC TOPSIS for the data set in Table 1 are shown in Table 2. The rankings of the alternatives are obtained from Sengupta's approach [19] introduced in [10], and rankings in brackets are obtained from the method of comparing the centre of RC intervals used in [13]. Overall, the two existing extensions produce the same ranking specifically $A_4 > A_6 > A_5 > A_2 > A_3 > A_1$. This ranking also aligns with the most probable ranking which of MC TOPSIS at 99.417% probability.

Figure 3 demonstrates the results of $\overline{RC}$s of Jahanshahloo's [10] and Kaczynska's [13] approaches compared to the proposed MC TOPSIS at the interval level as discussed.

Considering the results, and as highlighted in the introduction, the articulation and communication of uncertainty is essential for meaningful MCDM decision support. In terms of the interval TOPSIS extensions, the interval-valued

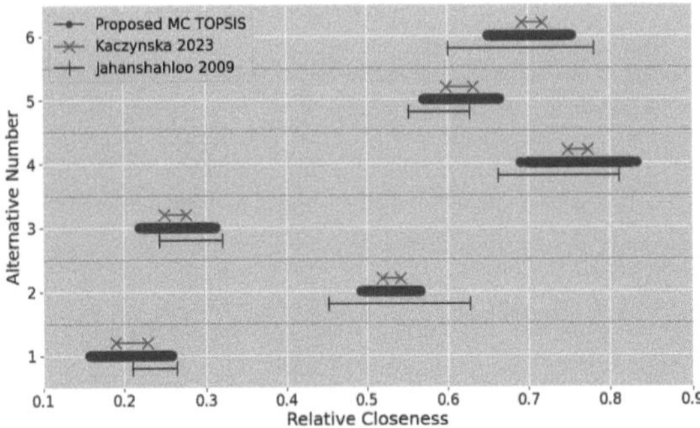

Fig. 3. Comparison of interval RC results of Jahanshahloo et al. 2009 [10], Kaczynska et al. 2023 [13] and proposed MC TOPSIS.

Table 2. Alternative Relative Closeness ($\overline{RC}$) Results and Rankings based on [19], and interval midpoints in brackets respectively.

Model	Jahanshahloo's approach $\overline{RC}$	Ranking	Kaczynska's approach $\overline{RC}$	Ranking	Proposed MC TOPSIS $\overline{RC}$	Ranking
Alternative 1	[0.211, 0.264]	6 (6)	[0.189, 0.229]	6 (6)	[0.157, 0.260]	6 (6)
Alternative 2	[0.452, 0.623]	4 (4)	[0.520, 0.542]	4 (4)	[0.492, 0.569]	4 (4)
Alternative 3	[0.243, 0.320]	5 (5)	[0.249, 0.276]	5 (5)	[0.219, 0.313]	5 (5)
Alternative 4	[0.663, 0.810]	1 (1)	[0.748, 0.773]	1 (1)	[0.691, 0.833]	1 (1)
Alternative 5	[0.551, 0.628]	3 (3)	[0.598, 0.632]	3 (3)	[0.569, 0.665]	3 (3)
Alternative 6	[0.601, 0.780]	2 (2)	[0.692, 0.717]	2 (2)	[0.645, 0.753]	2 (2)

RC provides insights into the degree of uncertainty associated with each alternative. We assume for the discussion that the MC technique provides results which are the most faithful to the original TOPSIS method. Thus, the range of $\overline{RC}$ obtained from MC TOPSIS can be a potential reference when comparing existing approaches.

From this point of view, we note that, for example, the $\overline{RC}$ of Alternative 2 in Jahanshahloo's approach [10] has the widest range of 0.171, while the $\overline{RC}$ of MC TOPSIS has the narrowest range of 0.067. In addition, the $\overline{RC}$ for Alternative 2 completely covers the $\overline{RC}$ of Alternative 5, which implies that in some situations, Alternative 2 could be superior to Alternative 5. However, this contradicts the simulation results. Kaczynska's method [13] on the other hand generates very narrow $\overline{RC}$, for example, implying that Alternative 4 is always superior to Alternative 6, again contradicting simulation results.

It is well known that one challenge of interval arithmetic as leveraged in [10] is that it may over-estimate uncertainty which in turn here may lead to potential for misinterpretation. While MC simulation provides no guarantees on bounds either, the results indicate that current techniques such as [10,13] may over respectively under-estimate uncertainty in interval TOPSIS outputs.

5   Conclusions

In this paper, we discuss challenges, research gaps, and opportunities within the MCDM and specifically TOPSIS literature in terms of handling uncertainty. We highlight how such techniques have the potential to ensure that DMs are better informed and thus can be empowered to have more agency in making the final decisions. We proceed to focus on interval extensions of TOPSIS as one of the most popular MCDM techniques, exploring how to establish which extension is the most suitable for a given context. As an initial step toward this goal, we propose a relatively simple interval TOPSIS extension based on MC simulation, intended as a reference point for comparing other interval TOPSIS extensions and gaining insights into their underlying differences in uncertainty handling and communication.

We show how the proposed MC TOPSIS technique provides valuable insight in respect to uncertainty in potential rankings, allowing for the association of probabilities with the given possible rankings. We also discuss one approach to distilling the results into interval-valued $\overline{RC}$ coefficients which reflect the quasi-final output of TOPSIS techniques–prior to ranking. Based on the latter, and an example data set from the TOPSIS literature, we discuss potential challenges in the RCs generated by current extensions, particularly when these are used as a primary source of information for DMs.

Having said this, the visualisation of the proposed MC TOPSIS $\overline{RC}$s on their own, also exhibit challenges, for example suggesting the possibility that Alternative 3 is better than 1, even though the actual simulation indicates otherwise. This behaviour is linked to the underlying numeric TOPSIS model, specifically its normalisation stage. In future, we will conduct further research into this behaviour and, more broadly, into how to generate TOPSIS interval outputs where the interval-valued $\overline{RC}$ independently provides meaningful and informative insights to DMs.

Acknowledgments. This work was supported by the School of Computer Science, University of Nottingham. Triguero was also funded by a Maria Zambrano Senior Fellowship at the University of Granada.

References

1. Akram, M., Kahraman, C., Zahid, K.: Extension of TOPSIS model to the decision-making under complex spherical fuzzy information. Soft. Comput. **25**(16), 10771–10795 (2021). https://doi.org/10.1007/s00500-021-05945-5
2. Basílio, M.P., Pereira, V., Costa, H.G., Santos, M., Ghosh, A.: A systematic review of the applications of multi-criteria decision aid methods (1977–2022). Electronics **11**(11), 1720 (2022)
3. Beyaztas, U., Shang, H.L., Abdel-Salam, A.S.G.: Functional linear models for interval-valued data. Commun. Stat.-Simul. Comput. **51**(7), 3513–3532 (2022)
4. Chen, C.T.: Extensions of the TOPSIS for group decision-making under fuzzy environment. Fuzzy Sets Syst. **114**(1), 1–9 (2000)

58   J. Ying et al.

5. Delgado, M., Vila, M.A., Voxman, W.: On a canonical representation of fuzzy numbers. Fuzzy Sets Syst. **93**(1), 125–135 (1998)
6. Diday, E., Noirhomme-Fraiture, M.: Symbolic Data Analysis and the SODAS Software. Wiley (2008)
7. Dymova, L., Sevastjanov, P., Tikhonenko, A.: A direct interval extension of TOPSIS method. Expert Syst. Appl. **40**(12), 4841–4847 (2013)
8. Ellerby, Z., Wagner, C.: Do people prefer to give interval-valued or point estimates and why? In: 2021 IEEE International Conference on Fuzzy Systems (FUZZ-IEEE), pp. 1–6. IEEE (2021)
9. Hwang, C.L., Yoon, K., Hwang, C.L., Yoon, K.: Methods for multiple attribute decision making. In: Multiple Attribute Decision Making: Methods and Applications a State-of-the-Art Survey, pp. 58–191 (1981)
10. Jahanshahloo, G.R., Lotfi, F.H., Davoodi, A.: Extension of TOPSIS for decision-making problems with interval data: interval efficiency. Math. Comput. Model. **49**(5–6), 1137–1142 (2009)
11. Jahanshahloo, G.R., Lotfi, F.H., Izadikhah, M.: An algorithmic method to extend TOPSIS for decision-making problems with interval data. Appl. Math. Comput. **175**(2), 1375–1384 (2006)
12. Jiang, J., Ren, M., Wang, J.: Interval number multi-attribute decision-making method based on TOPSIS. Alex. Eng. J. **61**(7), 5059–5064 (2022)
13. Kaczyńska, A., Sulikowski, P., Watróbski, J., Sałabun, W.: Enhancing sustainable assessment of electric vehicles: a comparative study of the TOPSIS technique with interval numbers for uncertainty management. Energies **16**(18), 6652 (2023)
14. Lima Neto, E.D.A., de Carvalho, F.D.A.: Nonlinear regression applied to interval-valued data. Pattern Anal. Appl. **20**, 809–824 (2017)
15. Madani, K., Lund, J.R.: A Monte-Carlo game theoretic approach for multi-criteria decision making under uncertainty. Adv. Water Resour. **34**(5), 607–616 (2011)
16. Madi, E.N., Garibaldi, J.M., Wagner, C.: Exploring the use of type-2 fuzzy sets in multi-criteria decision making based on TOPSIS. In: 2017 IEEE International Conference on Fuzzy Systems (FUZZ-IEEE), pp. 1–6. IEEE (2017)
17. Marcondes, G.A.B., Vilela, M.D.S.: Project selection with uncertainty using Monte Carlo simulation and multi-criteria decision methods. In: Parlier, G.H., Liberatore, F., Demange, M. (eds.) ICORES ICORES 2020 2021. CCIS, vol. 1623, pp. 152–170. Springer, Cham (2020). https://doi.org/10.1007/978-3-031-10725-2_8
18. Ortiz-Barrios, M., Miranda-De la Hoz, C., López-Meza, P., Petrillo, A., De Felice, F.: A case of food supply chain management with AHP, DEMATEL, and TOPSIS. J. Multi-Criteria Decis. Anal. **27**(1-2), 104–128 (2020)
19. Sengupta, A., Pal, T.K.: On comparing interval numbers. Eur. J. Oper. Res. **127**(1), 28–43 (2000)
20. Tsaur, R.C.: Decision risk analysis for an interval TOPSIS method. Appl. Math. Comput. **218**(8), 4295–4304 (2011)
21. Tzeng, G.H., Huang, J.J.: Multiple Attribute Decision Making: Methods and Applications. CRC Press (2011)
22. Yoon, K.: A reconciliation among discrete compromise solutions. J. Oper. Res. Soc. **38**, 277–286 (1987)
23. Zhao, Q., Wang, H., Wang, S.: Robust regression for interval-valued data based on midpoints and log-ranges. Adv. Data Anal. Classif. **17**(3), 583–621 (2023)

Advancing Sustainable Tourism Recommendations Through Innovative Hierarchical Evaluation and Data Acquisition Methods

Monir Yahya Salmony[(✉)] [ID], Antonio Moreno [ID], and Aida Valls [ID]

Dept. Enginyeria Informàtica i Matemàtiques, Intelligent Technologies for Advanced Knowledge Acquisition (ITAKA), Universitat Rovira i Virgili, Tarragona, Catalonia, Spain
`moniryahyaali.salmony@urv.cat`

Abstract. The uncontrolled growth of global tourism requires a new generation of recommendation systems that guide users toward more sustainable choices. Existing systems, however, often prioritize popularity, failing to account for the complex sustainability impacts of tourism. Addressing this gap, this paper introduces a novel, comprehensive hierarchical framework for evaluating Points of Interest based on both user-centric and sustainability criteria. We further propose a detailed, multifaceted data acquisition and estimation strategy designed to overcome the critical challenge of data scarcity and populate this framework with robust values. This hybrid approach integrates primary data from platforms and open repositories with advanced inference methods. In particular, we propose to employ proxy variables and leverage Artificial Intelligence methods including Large Language Models to estimate hard-to-measure criteria. Building upon this approach, we present a systematic guide for implementing each criterion, specifying its relevant data source, acquisition method and evaluation scale. This work provides a foundational methodology that spans from conceptual structure to data implementation, enabling the development of next-generation recommendation systems that guide tourists toward genuinely sustainable choices.

Keywords: Tourism sustainability · Recommendation systems · Hierarchical modeling · Data acquisition

1 Introduction

Global tourism is a double-edged sword; while it increases economic growth, it also poses significant threats to environmental stability, local cultures, and social fairness in host destinations [1,2]. Traditional Point of Interest (POI) recommendation systems [3], which typically focus on popular sites or on the users' main preferences [4], often fail to consider these increasing sustainability challenges [1,5]. These systems can worsen these issues by guiding tourists to

L. Martínez et al. (Eds.): IDEAL 2025, LNCS 16239, pp. 59–70, 2026.
https://doi.org/10.1007/978-3-032-10489-2_6

crowded sites, increasing environmental damage and putting pressure on local resources. They also tend to promote a standardized rather than authentic travel experience, which can harm the unique culture and identity of each destination.

Consequently, there is a notable and increasing need for decision-support tools capable of guiding tourists towards choices that align with both their individual preferences and established sustainability objectives [6]. Recognizing these limitations, the research community has begun exploring more sustainable approaches, such as promoting outdoor (environmentally friendly) POIs using re-ranking [7], or recommending destinations (rather than individual POIs) based on aggregating sustainability indicators [8]. Recent studies have highlighted the significant environmental impact of recommender systems research itself, revealing that deep learning-based methods, particularly those employing Graph Neural Networks, can generate substantially higher CO_2 emissions compared to traditional algorithms [9,10], highlighting the need to balance performance with sustainability.

However, effective sustainable tourism requires a more holistic and multi-dimensional perspective that assesses environmental management, cultural preservation, social well-being, and local economic benefit, all while ensuring traveler satisfaction [6]. To manage this complexity, a structured, hierarchical approach is essential. Yet, a significant barrier remains as there is a lack of comprehensive, reliable data for the very criteria that define sustainability.

To address these challenges, this paper introduces a novel, comprehensive solution in two integrated parts. First, we propose a hierarchical multi-criteria evaluation framework that systematically organizes evaluation criteria. The framework evaluates both user preference alignment and sustainability, with sustainability further divided into environmental, cultural, social and economic aspects. Second, recognizing that the effectiveness of this framework relies on trustable data, we present a multi-faceted data acquisition and estimation strategy. This strategy details how to populate the framework by combining direct data sourcing with advanced estimation techniques, utilizing proxy variables and Artificial Intelligence (AI) technologies including Natural Language Processing (NLP) and Large Language Models (LLMs) to approximate values for criteria that lack direct measurement.

The resulting dataset will support robust multi-criteria sustainability assessment and enable the application of advanced decision-making methods such as ELECTRE [12], AHP, or PROMETHEE [13]. Among these methods, ELECTRE-H [12] stands out as an ideal choice for our hierarchical framework, as it can operate directly on original and diverse data values, reducing the need for transformation and preserving data interpretability. It also allows weight adjustments across hierarchy levels to reflect user preferences or policy priorities.

This paper makes the following contributions:

- We introduce a hierarchical multi-criteria evaluation framework that organizes both user-centric and sustainability criteria for POI recommendations.
- We propose a novel data acquisition strategy that combines direct data sourcing with proxy variables, AI methodologies and LLMs.

– We provide a practical guide for implementing each criterion, detailing its potential data sources, acquisition method, scale, and optimization goal.

This paper is structured as follows. Section 2 provides an overview of the hierarchical evaluation framework. Section 3 details our proposed multi-faceted data acquisition strategies. Finally, Sect. 4 concludes the paper and discusses future work.

2 Overview of a Hierarchical Multi-criteria Evaluation Framework for POIs

To address the need for a more holistic assessment of a POI from different perspectives, we propose a novel, comprehensive hierarchical multi-criteria evaluation framework. This framework is designed to systematically organize the complex factors that balance user-centric and sustainability-focused criteria, moving beyond simplistic metrics like popularity. It is structured into multiple levels to ensure an in-depth evaluation, as shown in Fig. 1, and it is organized as follows:

– **Level 1:** At the highest level, we establish a fundamental distinction between two primary dimensions: (a) Adequacy to User Preferences & Trip Characteristics, which captures the personalized fit of a POI for a specific traveler and trip, and (b) Sustainability, which assesses the POI's broader sustainability impacts.
– **Level 2:** These primary dimensions are then decomposed into more granular sub-dimensions. Criterion (a) is split into General Preferences and Trip Characteristics. Sustainability, critically, is broken down into four universally recognized pillars (Environmental, Cultural, Social, and Economic), ensuring a balanced and comprehensive assessment.
– **Levels 3 & 4:** These lower levels define the specific and measurable criteria that populate and operationalize the level 2 dimensions of the framework. Level 3 introduces key evaluation areas under each pillar in level 2. For example, *General Preferences* captures broad user preferences including favoured POI categories, popularity and indoor/outdoor activity preferences. *Trip Characteristics* considers specific aspects of the trip such as the composition of the travel group, travel budget, and seasonal considerations. *Environmental Sustainability* is assessed through water consumption, CO_2 emissions, biodiversity protection, pollution control, and noise level, focusing on environmental impact and conservation efforts. *Cultural Sustainability* measures the preservation of cultural heritage, traditions, and authenticity. *Social Aspects* considers the social impact of the POI, including overcrowding levels, local cultural authenticity, and community impact. *Economic Sustainability* evaluates the local economic benefits from different views, including local ownership, local businesses, and local employment opportunities. Level 4 then further breaks down the *Cultural Heritage* criterion into detailed and measurable indicators such as historical significance, architectural value, cultural and social impact, touristic interest, official recognition, etc. These indicators summarize the POI's overall cultural and community importance.

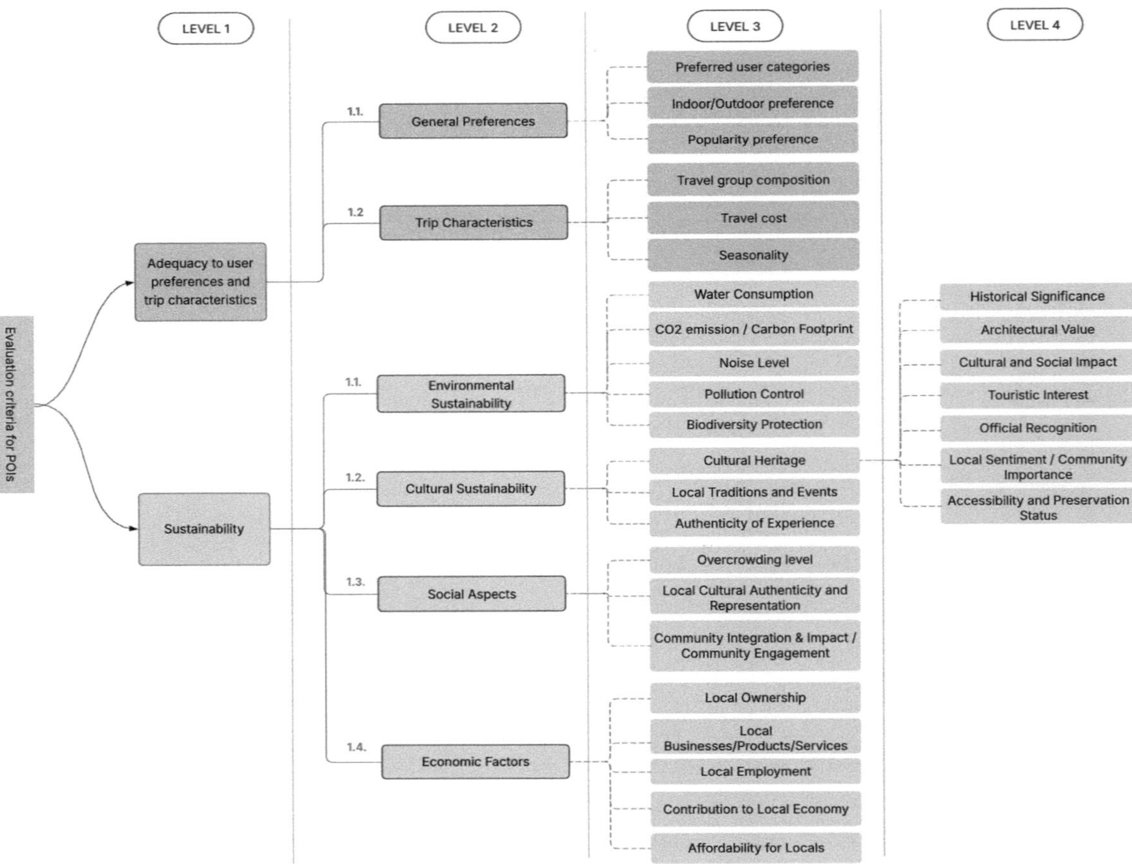

Fig. 1. Hierarchical Framework for Sustainable Tourism POI Evaluation.

While the above hierarchical structure provides a high level conceptual overview of the framework, its practical value depends entirely on the ability to populate it with reliable data. The complete operationalization of this framework, detailing how each criterion can be sourced and measured, is the second core contribution of this paper, which we detail in the subsequent section.

3 Multi-faceted Data Acquisition Strategies

To operationalize the framework proposed in Sect. 2, we introduce a comprehensive data acquisition and estimation strategy. It systematically integrates diverse data sources with direct and advanced inference information techniques to ensure the hierarchical framework can be populated with robust, actionable, and personalized values. The strategy is organized into two principal components: Primary Data Sources and Estimation & Inference Methods.

3.1 Primary Data Sources

This component comprises foundational data sources that provide explicit, factual, or systematically extracted information. The related data for each POI is retrieved automatically via APIs or controlled web scraping from the relevant sources, thereby minimizing the need for manual collection.

Platform-Derived Data: Structured data from large-scale platforms including both general (e.g., Google Maps (GM)) and touristic (e.g., TripAdvisor (TA), Booking (BK)) ones serves as the backbone for evaluating our criteria. These platforms provide direct access to key attributes such as POI categories, popularity metrics, and basic cost indicators, ensuring depth in data coverage.

Official POI Websites: To supplement platform data and address gaps, POI specific information may be gathered directly from official POI websites (OW) (e.g., a museum's official site). This is crucial for acquiring specific data points, particularly for criteria such as local ownership, explicit sustainability policies (e.g., pollution control, biodiversity protection), or unique operational details not fully covered by tourism platforms. This method ensures the inclusion of up-to-date and context-specific information where available.

Open and Public Data Repositories: Publicly available datasets from governmental agencies, research institutions, tourism public reports and maintained knowledge bases such as Wikipedia (WP) may be systematically integrated to enrich the dataset. These may contain environmental quality indicators, public transport accessibility, cultural heritage value, and comprehensive information about POIs. These sources provide objective, often regionally aggregated or community-curated data that can be linked to specific POIs through geocoding and advanced data integration techniques.

User-Contributed Input: Personalization is driven by direct input from the users of the recommender system we intend to design and develop. These users would share their preferences for categories, group composition, and subjective criteria such as popularity or indoors/outdoors settings. This information would then be turned into standard scores used by the system.

3.2 Estimation and Inference Methods

When direct or explicit data are unavailable or incomplete, the framework proposes the use of advanced techniques to estimate or infer the required values for our criteria as follows:

Proxy and Derived Variables: For criteria that are difficult to measure directly, in some cases it is possible to identify and utilize proxy variables based on related, more accessible data that can serve as reliable indicators. For example, review counts may serve as a proxy for popularity, and normalized visitor traffic data relative to capacity can proxy the overcrowding level. Derived variables may also be constructed by aggregating multiple indicators, such as creating a composite cultural heritage score.

AI and LLM-Based Inference: AI methods and LLMs can serve as supplemental strategies to extract qualitative and quantitative insights from unstructured text (e.g., reviews, descriptions, blogs). AI techniques such as machine learning algorithms and NLP techniques can be employed for systematic scoring and categorization. For example, sentiment analysis can assess visitor satisfaction, while topic modeling can help in identifying frequent sustainability themes within textual data. In addition, LLMs are well-suited for these tasks due to their extensive open-world knowledge, logical and commonsense reasoning abilities, and cultural awareness acquired from large-scale corpora [11]. By leveraging their embedded internal knowledge, we can extract structured information such as estimating CO_2 emissions, classifying noise levels, assessing authenticity, or estimating community integration. To ensure reliability, LLM outputs will be generated with deterministic settings, constrained to predefined formats, and validated against trusted sources or manually labeled samples to mitigate non-determinism, hallucinations, and bias.

Several specific methods are proposed across different evaluation criteria:

- **Named Entity Recognition (NER):** It extracts entities from text. For example, in the pollution control criterion shown in Table 2, NER identifies eco-certification entities (e.g., "ISO 14001") and sustainability-related terms within POI descriptions, with the total count forming the score. It is also proposed to detect official recognition (e.g., "UNESCO") as shown in Table 4.
- **Sentiment Analysis (SA):** It analyzes user reviews to quantify subjective opinions. For touristic interest and local sentiment, SA is applied to visitor reviews to estimate the overall appeal of a POI in Table 4, providing a score that reflects community satisfaction.

- **Topic Modeling (TM):** It identifies recurring themes in large text corpora. For the local traditions and events criterion shown in Table 4, TA is proposed to detect reviews focused on traditional activities, with the number or proportion of such reviews forming the score.
- **Text Similarity (TS):** It compares POI descriptions to reference definitions. For local cultural authenticity and representation see (Table 5), TS measures the semantic alignment between POI content and local cultural themes, producing a similarity score.
- **Rule-Based (RB) Logic:** It is suggested for criteria with clear textual indicators. For local ownership see (Table 3), RB checks for chain indicators versus local ownership terms in descriptions, assigning binary scores accordingly.

Tables 1, 2, 3, 4 and 5 outline how each criterion is sourced and acquired- whether directly (codified with a green background, ●) or via proxy variables through AI and LLM-based methods (codified with a blue background, ●). They also specify the user involvement, the type and scale of each criterion (e.g., categorical (C), ordinal (O), binary (B), or numeric (N)), and the goal (e.g., maximize ⇑, minimize ⇓, or match with user preferences). For example, "Travel Cost" is a numeric criterion that can be matched to the user's budget, while "Contribution to Local Economy", with ordinal Low to High [L-H] values, should be maximized for higher sustainability. Table 6 presents a sample of sustainability-related criteria for the ten most popular POIs in Barcelona using LLM.

Table 1. General Preferences Criteria for POI Evaluation with User Preferences Goal.

Criterion	Data Source(s)	AI/LLM Approach	Type	Goal and Methodology
POI Category	TA, BK	-	C	👤: Extract category (e.g., museum, park) directly from source listings
Indoor/ Outdoor	TA	LLM	B/O	👤: Assign 0 if POI is indoor, 1 otherwise, based on direct source listings or LLM inference: [1 = fully indoor, 5 = fully outdoor].
Popularity	TA, BK, GM	-	N	👤: Extract POI position directly from source listing
Travel Group Composition	TA	-	B	👤: Extract POI group tags from TA (e.g., good for kids, family). Assign 1 if tag present, 0 otherwise, for each tag.
Travel Cost	TA, BK, OW	-	N	👤: Extract ticket prices, entry fees from TA/BK listings or official websites
Seasonality	TA	-	N	👤: Extract TA review timestamps, calculate seasonal visitor distribution, i.e., [0-1] per season

Table 2. Environmental Criteria for POI Evaluation.

Criterion	Data Source(s)	AI/LLM Approach	Type	Goal and Methodology
Water Consumption		LLM	N	⇓ : LLM-inferred
CO_2 emissions		LLM	N	⇓ : LLM-inferred
Noise Level		LLM	O/N	⇓ : LLM-inferred:[v low - v high]
Pollution Control	TA, BK, OW, GM, WP	NER	N	⇑ : NER extracts eco-certification entities (e.g., LEED) and eco-related keywords (e.g., sustainable, green energy) from POI descriptions; an aggregated score can be computed from their appearance.
Biodiversity Protection	TA, BK, OW, WP	LLM	N	⇑ : LLM analyzes POI description for biodiversity protection measures; assigns score [0-1]. Ex.: 0.8 (POI maintains native gardens and supports local wildlife).

Table 3. Economic Criteria for POI Evaluation

Criterion	Data Source(s)	AI/LLM Approach	Type	Goal and Methodology
Local Ownership	TA, OW, WP	RB	B	⇑ : RB to check POI descriptions for chain indicators (e.g., known brand names) vs local ownership terms (e.g., family-owned, municipal). Assign 1 for local, 0 for chain.
Local Businesses/ Products/ Services		LLM	O	⇑ : LLM inferred [L-H].
Local Employment	OW, WP	NER + RB	B	⇑ : NER to extract ownership and employment-related entities from descriptions, then RB logic to assign 1 (local employment) or 0 (not local/unknown).
Contribution to Local Economy		LLM	O	⇑ : LLM inferred [L-H].
Affordability for Locals	TA, BK, OW	NER+ RB	B	⇓ : Compare ticket price to local income or check for free/discounted entry; assign 1 (affordable) if criteria are met, 0 otherwise.

Table 4. Cultural Criteria for POI Evaluation

Criterion	Data Source(s)	AI/LLM Approach	Type	Goal and Methodology
Historical Significance	TA, OW, WP	-	N	⇑ : Proxied by calculating the age of the POI: (current year - construction year) using data from sources. Ex: 200 (high historical value).
Architectural Value	Internal LLM knowledge	LLM	N	⇑ : LLM-inferred:[0-1].
Cultural and Social Impact	TA, OW, WP	LLM	N	⇑ : LLM analyzes POI descriptions for cultural or social role indicators; assigns a score in [0-1]. Ex: 0.9 (POI hosts major cultural festivals)
Touristic Interest	TA, BK, GM	SA	N	⇑ : SA on visitor reviews to estimate interest; score within [0-1]. Ex: 0.8 (Mostly positive reviews, high tourist appeal).
Official Recognition	TA, OW, WP	NER	N	⇑ : NER extracts official recognition entities (e.g., UNESCO, Local Heritage Site) from POI descriptions or listings. Assign a score using a rule-based lookup table: None = 0.0, Local = 0.5, National/Regional = 0.75, International = 1.0. Ex: 1.0 (UNESCO World Heritage Site).
Local Sentiment	TA, BK, OW, GM	SA	N	⇑ : SA on local feedback and reviews; score between 0 and 1. Ex: 0.7 (good local opinions).
Accessibility and Preservation Status	TA, BK, GM	TM/NER + SA	N	⇑ : NER to filter certain criterion-related review terms (e.g., wheelchair, restored), then SA on these reviews and average the scores. Ex: 0.7 (Mostly positive feedback on accessibility and maintenance).
Local Traditions and Events	TA, BK, OW, GM	TM/NER + SA	N	⇑ : TM to identify reviews focused on local traditions and events; score is the number or proportion of reviews assigned to this topic. Ex: 10 (reviews discuss annual traditional festival).
Authenticity of Experience	TA, BK, GM, WP	TM/NER + SA	N	⇑ : TM to filter reviews/descriptions for authenticity-related terms (e.g., authentic, local traditions), then SA and average the scores. Ex: 0.8 (Frequent positive mentions of authentic local culture).

Table 5. Social Criteria for POI Evaluation

Criterion	Data Source(s)	AI/LLM Approach	Type	Goal and Methodology
Overcrowding Level	TA, OW, BK, GM	-	N	$\Downarrow$: Proxied using waiting time or visitor numbers from reviews to capacity; resulting ratio is the score. Ex: 1.5 (indicating significant overcrowding).
Local Cultural Authenticity and Repre- sentation	TA, BK, OW, WP	TS	N	$\Uparrow$: Prioritize POIs (e.g., historical sites and local art museums); then TS (e.g., embedding similarity) between POI descriptions and local cultural definitions; the similarity value is the score. Ex: 0.9 (High similarity to local cultural themes).
Community Impact	TA, GMR, WP; Internal LLM knowledge	LLM	O	$\Downarrow$: LLM analyzes reviews for signs of conflict or limited resident access; assigns value within [very low - very high] based on detected impact. Ex: M (Some reviews mention resident complaints about crowding).

Table 6. Key Attributes of 10 Most Popular POIs in Barcelona

POI Name	Indoor/ Outdoor	Noise Level	Water Consumption (m^3/yr)	Historical Significance	Official Recognition
Sagrada Familia	Mostly Indoor	Medium	30	0.8	UNESCO, BCIN
Park Güell	Mostly Outdoor	Medium	40	0.7	UNESCO, BCIN
Gothic Quarter	Mostly Outdoor	High	100	1.0	BCIN
La Rambla	Outdoor	Very High	50	0.7	BCIN
Camp Nou	Both	Very High	150	0.3	BCIN
Casa Batlló	Mostly Indoor	Medium	8	0.7	UNESCO, BCIN
Casa Milà	Both	Medium	7.5	0.7	UNESCO, BCIN
Magic Fountain of Montjuc	Outdoor	High	5,000	0.4	BCIL
Picasso Museum	Indoor	Low	6	0.8	BCIN
Barcelona Cathedral	Mostly Indoor	Medium	15	1.0	BCIN

4   Conclusion and Future Work

In this paper, we introduced a novel, comprehensive hierarchical framework for the holistic evaluation of POI, combining user preferences with the four pillars of sustainability. Recognizing that this framework is only valuable if it can be populated with data, we then presented a practical and robust multi-faceted data acquisition strategy. This strategy systematically combines direct data sourcing with advanced estimation techniques to overcome the critical challenge of data scarcity in sustainability assessments. These two contributions provide a complete blueprint, from conceptual structure to practical data implementation, for building the next generation of recommendation systems that can guide tourists toward genuinely responsible and sustainable choices and consequently contribute positively to the visited destination. While our framework and data strategy provide a comprehensive approach to sustainable POI evaluation, limitations remain. Reliable data for many sustainability criteria are often scarce or inconsistently formatted among data sources. Although estimation techniques can help address data gaps, they also have limitations. In particular, LLM-based inference may suffer from potential bias, and the need for careful prompt engineering and validation against ground truth.

Looking ahead, the immediate future work involves the full implementation and empirical validation of the proposed frameworks. It includes the development of a data acquisition pipeline, validating proxy indicators against ground truth, refining the proposed AI techniques, and developing specific LLM prompts for each inferred criterion. A key next step will be to integrate this enriched dataset with a MCDM method like ELECTRE-H and test the end-to-end system on real-world destinations. ELECTRE-H is suitable as it can integrate user preferences with sustainability dimensions, allows weight adjustments across hierarchy levels and can rank POIs with respect to a single dimension (e.g., environmental or social). This will be crucial for assessing the framework's real-world effectiveness and demonstrating its potential to foster more sustainable tourism practices.

Acknowledgments. The first author acknowledges the support of the URV predoctoral grant 2023PMF-PIPF-18. This work was also partially funded by Universitat Rovira i Virgili (URV) [projects number 2023PFR-URV-114 and 2022PFR-URV-41], and by the Generalitat de Catalunya (AGAUR) through the funding of the activities of the ITAKA consolidated research group [2021-SGR-00114]. The support of the Spanish Research Network on Recommender Systems (ELIGE-IA, RED2022-134302-T) is also acknowledged.

Disclosure of Interests. The authors have no competing interests to declare that are relevant to the content of this article.

References

1. Baena, V., Cerviño, J.: Tourism in the era of social responsibility and sustainability: understanding international tourists' destination choices. Sustainability **16**(19), 8509 (2024). https://doi.org/10.3390/su16198509

2. Said, A.: Recommender systems for social good: the role of accountability and sustainability. International Workshop on Recommender Systems for Sustainability and Social Good, pp. 1–4. Springer, Cham (2024). https://doi.org/10.1007/978-3-031-87654-7_1

3. Zhang, Q., et al.: A survey on point-of-interest recommendation: models, architectures, and security. IEEE Trans. Knowl. Data Eng. (2025). https://doi.org/10.1109/TKDE.2025.3551292

4. Merinov, P., Ricci, F.: Simulating the impact of recommendation salience on tourists experienced utility. In: International Workshop on Recommender Systems for Sustainability and Social Good, pp. 35–51. Springer, Cham (2024). https://doi.org/10.1007/978-3-031-87654-7_4

5. Mauro, N., Scarpinati, L., Ferrero, F., Geninatti Cossatin, A., Mattutino, C.: Point-of-interest recommender systems: nudging towards sustainable tourism. In: Adjunct Proceedings of the 32nd ACM Conference on User Modeling, Adaptation and Personalization, pp. 491–495 (2024). https://doi.org/10.1145/3631700.3664904.

6. Felfernig, A., et al.: Recommender systems for sustainability: overview and research issues. Front. Big Data **6**, 1284511 (2023). https://doi.org/10.3389/fdata.2023.1284511

7. Sánchez Pérez, P., Bellogín, A.: Towards sustainability optimization in touristic route recommendation. In: Intelligent Management of Data and Information in Decision Making: Proceedings of the 16th FLINS Conference on Computational Intelligence in Decision and Control & the 19th ISKE Conference on Intelligence Systems and Knowledge Engineering (FLINS-ISKE 2024), pp. 9–16 (2024). https://doi.org/10.1142/9789811294631_0002

8. Banerjee, A., Satish, A., Wörndl, W.: Enhancing tourism recommender systems for sustainable city trips using retrieval-augmented generation. In: International Workshop on Recommender Systems for Sustainability and Social Good, pp. 19–34. Springer, Cham (2024). https://doi.org/10.1007/978-3-031-87654-7_3

9. Vente, T., Wegmeth, L., Said, A., Beel, J.: From clicks to carbon: the environmental toll of recommender systems. In: Proceedings of the 18th ACM Conference on Recommender Systems, pp. 580–590 (2024). https://doi.org/10.1145/3640457.3688074.

10. Purificato, A., Silvestri, F.: Eco-aware graph neural networks for sustainable recommendations. In: International Workshop on Recommender Systems for Sustainability and Social Good, pp. 111–122. Springer, Cham (2024). https://doi.org/10.1007/978-3-031-87654-7_11

11. Lin, J., et al.: How can recommender systems benefit from large language models: a survey. ACM Trans. Inf. Syst. **43**(2), 1–47 (2025). https://doi.org/10.1145/3678004

12. Del Vasto-Terrientes, L., Fernández-Cavia, J., Huertas, A., Moreno, A., Valls, A.: Official tourist destination websites: hierarchical analysis and assessment with ELECTRE-III-H. Tour. Manag. Perspect. **1**(15), 16–28 (2015). https://doi.org/10.1016/j.tmp.2015.03.004

13. Greco, S., Ehrgott, M., Figueira, J.R. (eds.): Multiple Criteria Decision Analysis. ISORMS, vol. 233. Springer, New York (2016). https://doi.org/10.1007/978-1-4939-3094-4

Mitigating Overfitting in Recommender Systems via Intra-domain Transfer Learning

Eva Blanco-Mallo[1], Pablo Pérez-Núñez[2(✉)], Verónica Bolón-Canedo[1], and Beatriz Remeseiro[2]

[1] Department of Computer Science and Information Technologies, Universidade da Coruña, CITIC, Campus de Elviña s/n, 15071 A Coruña, Spain
`{eva.blanco,vbolon}@udc.es`
[2] Department of Computer Science, Universidad de Oviedo, Campus de Gijón s/n, 33203 Gijón, Spain
`{pabloperez,bremeseiro}@uniovi.es`

Abstract. Real-world recommender systems often suffer from extreme imbalance in user feedback, as most users provide only positive opinions. This paper demonstrates that such imbalance has substantial negative effects on the system's ability to generalize, particularly in predicting negative preferences. To address this issue, the study investigates whether intra-domain transfer learning, leveraging users with more complete profiles (those who provide both positive and negative feedback) can help mitigate these limitations. A synthetic evaluation framework is introduced based on false-positive and false-negative user profiles, allowing for a detailed assessment of model behavior under incomplete user data. Experiments on image-based datasets reveal consistent improvements in true negative rates for underrepresented user groups. Although the mitigation is partial, the results suggest that selecting appropriate user subsets as a source domain may offer a promising direction, and that identifying which users generalize best could be key to future progress.

Keywords: Recommender systems · Intra-domain transfer learning · Feedback imbalance · Evaluation framework · Image-based recommendation

1 Introduction

The ability to connect people across the globe in real-time is one of the most notable achievements of the digital era. This development promotes universal

This work was supported by MICIU/AEI (Grants PID2023-147404OB-I00, PID2021-128045OA-I00, and FPI PRE2020-092608), Xunta de Galicia (Grants ED431C 2022/44 and ED431G 2023/01), the Agency for Science, Business Competitiveness, and Innovation of the Principality of Asturias (Grant GRU-GIC-24-018), and EU-FEDER Galicia 2021-27 operational program.

access to millions of products and services, radically transforming consumer behavior. The sheer number of available options is, in fact, overwhelming. Grounded in social theories, the *paradox of choice* [15] highlights how excessive alternatives can complicate decision-making processes and diminish user satisfaction. This phenomenon is closely linked to the concept of *information overload* [1], where the overwhelming volume of available data leads to anxiety and stress in individuals. These insights underscore the critical need for personalized solutions to help users manage and select from the wide range of options available.

Recommender systems (RS) have emerged as a crucial tool for filtering and personalizing content. The goal of a personalized RS is to predict whether a user will like a specific item based on their personal preferences, thereby facilitating more efficient decision-making processes. RS typically operate through methodologies such as content-based filtering, which analyzes item attributes and user preferences; and collaborative filtering, which leverages patterns derived from user interactions. These systems often rely on data collected from recommendation platforms such as TripAdvisor or Amazon, where users submit reviews for a wide range of products and services. These reviews typically include a rating, often accompanied by an image and/or a text comment. Images offer valuable visual insights into the items being reviewed, capturing aspects such as design, appearance, and specific features that can significantly influence user preferences. This makes such user-generated data invaluable for learning and predicting personalized preferences, enabling RS to suggest items that align closely with individual tastes.

Real-world datasets gathered from the aforementioned platforms present significant challenges that hinder the optimal performance of RS. The first major issue is data sparsity, since most users contribute only a limited number of reviews, constraining the system's ability to capture their unique preferences [16]. The second critical problem is class imbalance, which stems from the overwhelming predominance of positive opinions [20]. This imbalance arises from the dominance of users providing only positive reviews, while those contributing negative or mixed reviews account for a much smaller proportion of the dataset. Consequently, the model leans toward overly optimistic predictions due to limited exposure to diverse feedback, struggling to discriminate between items that genuinely match user preferences and those that do not. Additionally, the scarcity of counterexamples for most users complicates evaluation, making it difficult to determine whether the system is truly effective or simply overgeneralizes its recommendations.

To investigate this issue, this work examines whether intra-domain transfer learning can help mitigate the effects of feedback imbalance by leveraging users with more complete profiles, specifically those who provide both positive and negative opinions. These mixed users offer richer information by indicating not only what they appreciate but also what they avoid, and they often contribute a greater volume of data overall. The approach considered involves first learning from these users and then transferring the acquired knowledge to improve recommendation performance for users with less complete profiles, such as those who provide only positive or only negative feedback.

The motivation for this approach stems from insights provided by social science research, including social identity theory [17] and self-categorization theory [18]. These theories demonstrate how individual preferences are often shaped by alignment with broader groups sharing similar traits. Numerous studies have shown that individual preferences are influenced by factors such as social identity, culture, or age, among others [6]. This suggests that the underlying patterns found in the preferences of users with richer, more balanced profiles are likely to partially reflect the preferences of users with less diverse feedback.

While transfer learning is often used across domains [5], we propose the first approach to knowledge transfer between different types of users to enhance system performance. We apply this idea within an image-based RS architecture, reflecting the growing importance of visual information in user decision-making. Visual cues often carry valuable context that complements textual reviews or ratings, making image-based RS an increasingly important area of research. The approach aims to bridge gaps in user representation and improve generalization under sparse or imbalanced conditions. To evaluate the proposed method, we introduce a synthetic evaluation framework designed to assess the model's ability to infer missing preferences. Specifically, we construct edge cases where users lack explicit feedback on dislikes (e.g., users with only positive reviews) or likes (e.g., users with only negative reviews). By assessing the system's performance on these synthetic profiles, we evaluate whether it can generalize patterns learned from richer, mixed-user profiles to predict missing preferences. This methodology ensures that the model is not only learning to replicate the patterns seen during training but can generalize to scenarios where explicit data is missing.

In summary, this work makes three key contributions:

1. An in-depth study of the class imbalance caused by the predominance of users providing only positive feedback, analyzing its impact on both system performance and evaluation.
2. An intra-domain transfer learning strategy that leverages mixed-user patterns to improve recommendation quality across all user types, with a particular focus on addressing the needs of underrepresented groups.
3. A novel evaluation framework that incorporates synthetic user profiles to rigorously assess system performance under conditions of severe data imbalance.

Together, these contributions focus on a characteristic problem in these contexts and propose a way to mitigate it through the use of transfer learning, providing a solid foundation for future research.

2 Methodology

A training method is proposed to provide recommendations based on users' preferences inferred from the images they have posted. In general terms, the problem is defined as a binary classification task:

$$(u, i) \rightsquigarrow 0|1, \tag{1}$$

where u is a user who posted the image i, which represents a particular item. As for the two labels, 0 indicates that user u does not like the item, while 1 indicates the opposite.

2.1 User Typology

As discussed above, a critical challenge in real-world datasets used for recommender systems lies in the significant imbalance in the type and quantity of reviews provided by users. Depending on the nature of the reviews in their history, user profiles in online review datasets can generally be grouped into three categories:

- **Mixed users (MU).** Users with both positive and negative reviews. These users provide the most complete information. This dual perspective enables the identification of features they value and those they avoid in items, which allows for a richer overall understanding of the characteristics that influence their choices. Additionally, they typically have the highest average number of images per user.
- **Negative users (NU).** Users with only negative reviews. These users are the worst represented, as they tend to provide the least amount of information. The main issue lies in the fact that, based on the available information, it is impossible to identify the features they seek in an item, only the ones they avoid. Therefore, it remains necessary to know more about the features they favor to make accurate recommendations based on their preferences.
- **Positive users (PU).** Users with only positive reviews. This group overwhelmingly dominates compared to the others. For these users, the features they seek in an item are well-represented, allowing recommendations to be more data-driven and tailored to their specific input.

Given the significant imbalance in sample sizes according to these user types, the system is expected to prioritize aligning with the preferences of positive-type users. However, the ideal case would involve primarily inferring knowledge from users whose preferences are more comprehensively represented (i.e., mixed users). Notably, since positive-type users provide no negative samples, it becomes difficult to determine whether the system is overly biased toward these cases or genuinely capable of making discriminative recommendations. As detailed later, our proposal aims to address both of these challenges.

2.2 Model Architecture

To address this binary classification task, we propose the model depicted in Fig. 1: a network that learns from user-photo dyads (u, i) to provide personalized image-based recommendations. The encoding of each input dyad is defined as follows: users are codified using an embedding layer, and images are represented using a convolutional neural network (CNN). For user representation, encodings were obtained using an embedding layer based on their identifiers.

Their usefulness in finding representations of similar objects has been demonstrated in the field of NLP [19], although their use is increasingly widespread in the context of user representation [4,14]. The embedding layer generates a 2048-dimensional feature vector for each user. Regarding the representation of images, feature vectors of the same size were extracted using ResNet50 [9], a CNN selected due to its effectiveness in numerous image-based recommender systems [11]. These models can process continuous dense vectors and infer the complex, intricate patterns that input images contain.

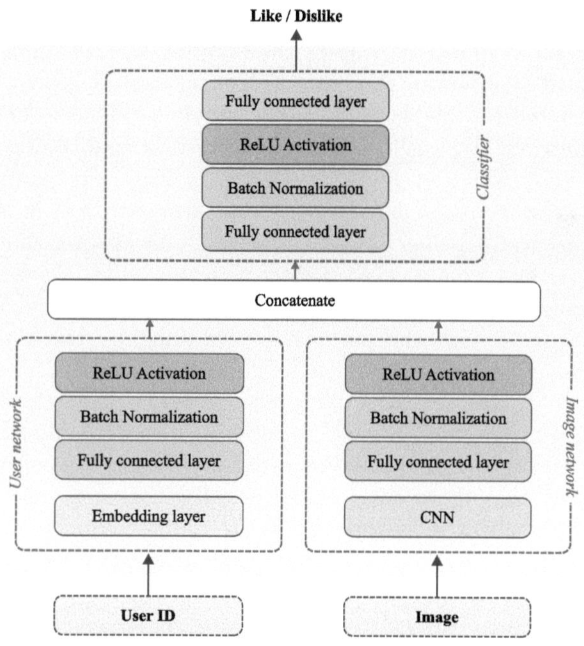

Fig. 1. Proposed architecture.

Next, the two encodings (vectors of size 2048) are mapped to their corresponding feature representations, each composed of a 256-dimensional vector. To achieve this, a fully connected (FC) layer is used, followed by a batch normalization (BN) layer [10], and a rectified linear unit (ReLU) [12] as the activation function. These two feature vectors are concatenated, resulting in a 512-dimensional vector, which includes an individual representation of each element in the dyad (u, i). This vector feeds into an FC layer of 256 units, followed by BN and ReLU activation layers. Finally, an FC layer, with a size equal to the number of classes, and the softmax activation function generate a probability output for class membership in the range $[0, 1]$. The value of this output will determine the system's confidence level (ψ) when providing a recommendation.

2.3   Training Process

This work explores whether intra-domain transfer learning can mitigate the effects of feedback imbalance by leveraging mixed-user profiles. To this end, a two-stage training process is carried out. The goal is to determine whether the preferences learned from mixed users (source domain), those who provide both positive and negative feedback, can be used to improve recommendations for all users (target domain), including those with only partial feedback.

To achieve this, a two-stage training process is conducted. The first stage is designed to learn the features that define the preferences of mixed users, so that the RS is trained using only their reviews. In the second stage, the RS is retrained using all available data from the training set. The goal is to identify the preferences of all users in the dataset (i.e., across all types) leveraging the knowledge extracted by the classifier in the previous stage. In other words, the parts of the network responsible for extracting and fine-tuning features from users and images (*User* and *Image Networks* in Fig. 1) are adjusted, but the weights of the network responsible for deciding whether certain features appeal to a particular user (*Classifier* in Fig. 1) are not. This ensures that, regardless of the features extracted from a user's preferences, they will be classified according to the patterns learned from mixed users. As for the loss function, the categorical cross-entropy is used with the predictions obtained by the classifier in both training phases.

The idea is that, after learning to represent a user's preferences in the second phase, the model attempts to classify them according to the decision boundaries learned from more complete users. For instance, If certain positive features are shared, it is likely that the corresponding negative features will also be similar. This transfer may help the system reject irrelevant recommendations, even in cases where the target user's profile lacks explicit negative feedback.

3   Evaluation Procedure

Due to the significant imbalance characterizing the datasets used in RS, both in terms of class (*like* or *dislike*) and user type (mixed, negative, or positive), defining an appropriate evaluation methodology becomes challenging. Given that the majority of users are of the positive type, it is expected that the system's performance for these users will dominate over others. This affects both metrics and behavior, as demonstrated later in the experimentation. One of the main issues stemming from this is that it is not possible to determine if the system is truly discriminative for these users, as there are no negative examples available. At least, this limitation persists under current evaluation methodologies. Most metrics commonly used in these systems, such as accuracy, precision, F1-score, or AUC [11], do not accurately reflect this type of imbalance. Some approaches employ the *top@n* metric, which involves checking whether the items a user likes are among the top n recommended [8]. However, considering that most datasets of this kind contain only a few examples among thousands, it is difficult to evaluate system performance using these metrics. Additionally, when evaluating, the RS may not recommend exactly the items that appear in the test set, but this may not always have to be an error.

Addressing these complexities requires a detailed evaluation of the system. The following insights offer a clearer understanding of its classification behavior:

1. According to the *the paradox of choice* [15], a large number of options can overwhelm users and be counterproductive. Consequently, it is logical to prioritize the quality, not quantity, of recommendations. This means aiming for high precision in the *like* class, regardless of overall accuracy. Incorrect recommendations can lead to user frustration, diminished trust in the system, and decreased overall satisfaction, which may result in reduced user engagement and retention. Therefore, the accuracy of the *like* class, referred to as the true positive rate (TPR), is not as critical as the precision.

2. It is also essential to ensure that the system is discriminative with regard to the recommended items. Given the extreme class imbalance, the system could achieve high precision in the *like* class even if all the examples in the *dislike* class were misclassified. Thus, the simplest way to monitor this behavior is by tracking the accuracy of the *dislike* class, known as the true negative rate (TNR).

In addition, when analyzing system behavior by user type, the following considerations should also be taken into account:

3. Most samples come from users with only positive reviews, meaning the system can only identify features they like, but not those they dislike. As a result, the system can rank items by preferences but cannot effectively discriminate between them. It is reasonable to assume that all items could potentially appeal to these users, though some more than others. To address this, recommendations should be evaluated using a confidence threshold (ψ), representing the likelihood that an item belongs to the *like* class. While some items may align more closely with the user's preferences, there must be high confidence in recommending items the user has explicitly shown interest in. This ensures that items lacking relevant features are not recommended.

4. In contrast to the previous case, users with only negative samples provide the necessary information for the system to be discriminative. However, information on what features they actually seek in an item is not available. Nonetheless, it can be simulated. Our approach involves extracting a small subset of mixed users and converting them into negative users during training, referred here as false negative users (FNU). The process involves removing all positive ratings from these users in the training and validation sets and later assigning them to the test set. This allows us to evaluate whether the model can provide accurate positive recommendations, even when it lacks this information during training. The same procedure is applied to a subset of positive users, referred to as false positive users (FPU), allowing us to analyze how the system behaves with all user types.

5. Every user with only positive or only negative samples will ideally tend to become a mixed user as they provide more information about their

preferences. Achieving high performance with mixed users is essential so that the system can adapt to all user types in the future.

In summary, to properly analyze a recommender system's performance, we propose evaluating reported accuracy and precision, by class, for users categorized as mixed, false negative users, and false positive users.

4 Experimentation

This section discusses the selected datasets, the procedure used to generate the training partitions, and the system implementation specifications.

4.1 Datasets

The two datasets used in this study, both publicly available [13], were collected on the TripAdvisor platform in 2023. They contain reviews posted by TripAdvisor users for all restaurants in Madrid and Barcelona up to the download date. Each selected review includes the identifier of the user who posted it, one or more images, and a star rating (from one to five). In line with standard procedures in recommender systems, reviews with one to three stars (< 4) are labeled as 0 (*dislike*), while those with four or five stars (≥ 4) are labeled as 1 (*like*).

Each dataset was divided into three subsets: training, validation, and test. Given that this is a user-dependent problem, certain restrictions must be considered. For each user u, their images were split into two groups based on the assigned rating, positive or negative. For each group, one image was allocated to the test set, and the remaining images were assigned to the training set. If only one image is available, it is assigned to the training set. This ensures that all users evaluated in the test set were considered during training. The initial training set was further divided following the same procedure, ultimately producing three partitions: training, validation, and test.

Table 1 reports the overall statistics of the datasets and partitions for each of the two cities considered. Due to space constraints, only the values for the training and test sets are decomposed by user type. It is important to note that, in the first training phase, only samples belonging to mixed-type users were used. Similarly, for evaluation purposes, 100 mixed-type users were selected from the training set. For the first 50, samples from positive reviews in the training and validation sets were moved to the test set. For the remaining 50, the same procedure was followed, selecting samples from negative reviews. As shown in Table 1, the imbalance is extreme, both by user type and class. Positive-type users account for more than 80% of the total, and the ratio between positive and negative classes exceeds 6:1 in both cities. Moreover, excluding mixed-type users, the average number of images per user is remarkably low. Note that only one image (per class) per user was selected for the test set to maximize the number of images in the training set. Nonetheless, positive and negative users have an average of less than two images.

Table 1. Training and test partitions used in the experimentation. The analysis includes the number of users (total and percentage), reviews, and images (total and by class), grouping the data based on different user types.

Partitions →	All	Train			Validation	Test				
User type →	All	Mixed	Negative	Positive	All	Mixed	Negative	Positive	FNU	FPU
Madrid Users (%)	140474	12770 (9.09)	13969 (9.94)	113735 (80.96)	59709	11797 (13.88)	5176 (6.09)	68010 (80.03)	50	50
Reviews	255889	78565	14340	141237	65251	20178	5176	68010	478	150
Images ($\bar{x}$)	588218 (4.19)	192498 (15.07)	16440 (1.18)	219668 (1.93)	65251 (1.09)	21175 (1.79)	5176 (1.00)	68010 (1.00)	1295 (25.90)	330 (6.60)
✓ ($\bar{x}$)	502633 (3.58)	146495 (11.47)	0	219668 (1.93)	56215 (0.94)	12245 (1.04)	0	68010 (1.00)	1245 (24.90)	50 (1.00)
✗ ($\bar{x}$)	85585 (0.61)	46003 (3.60)	16440 (1.18)	0	9036 (0.15)	8930 (0.76)	5176 (1.00)	0	50 (1.00)	280 (6.60)
Barcelona Users (%)	182148	12699 (6.97)	14462 (7.94)	154987 (85.09)	73394	11554 (10.70)	5340 (4.95)	91044 (84.34)	50	50
Reviews	287403	62504	14744	184901	78331	19244	5340	91044	251	149
Images ($\bar{x}$)	627769 (3.45)	142991 (11.26)	16568 (1.14)	273698 (1.76)	78331 (1.06)	19797 (1.71)	5340 (1.00)	91044 (1.0)	641 (12.82)	312 (6.24)
✓ ($\bar{x}$)	553217 (3.04)	107259 (8.45)	0	273698 (1.76)	69917 (0.95)	11299 (0.98)	0	91044 (1.0)	591 (11.82)	50 (1.0)
✗ ($\bar{x}$)	74552 (0.41)	35732 (2.81)	16568 (1.14)	0	8414 (0.11)	8498 (0.73)	5340 (1.00)	0	50 (1.00)	262 (5.24)

FNU: False Negative Users. FPU: False Positive Users. ✗ : *dislike.* ✓ : *like.* $\bar{x}$: Average per user of the accompanying value.

4.2   Implementation

The model architecture and the training approach for both phases were implemented using PyTorch. For the CNN, ResNet50, pre-trained weights from ImageNet [3] were utilized. The proposed system was trained in each phase using a batch size of 16 for 20 epochs, with early stopping applied based on validation accuracy and a patience of 5 epochs. Regarding the learning rate, the procedure outlined in [7] was followed, with the rate progressively adjusted during stochastic gradient descent, starting from an initial value of 0.01. The experiments were conducted on a machine with an Intel Core i7-11700K processor and a GeForce RTX3090 GPU.

5   Results

The results obtained from training the system using the proposed two-phase approach (*mix*) are reported in Table 2. Performance is compared with that achieved through conventional training (*base*), i.e., single-phase training using all user types without distinction. As detailed in Sect. 3, to evaluate the system's behavior properly, we analyze the accuracy and precision of both classes (*like* and *dislike*) by user type (mixed, negative, and positive). Additionally, the performance achieved with false users, both negative (FNU) and positive (FPU), is analyzed separately. In the case of false positive users, various confidence thresholds (ψ) were considered when predicting the *like* class. As mentioned above, confidence refers to the probability that a sample belongs to a specific class, determined by the output of the system.

The system's performance under the intra-domain transfer strategy (*mix*) is reported in Table 2, alongside results from a conventional single-phase baseline (*base*) trained on the full dataset without user-type distinction. As outlined in Sect. 3, we report class-wise accuracy and precision for like and dislike across different user groups: mixed, positive-only, and negative-only users. Additionally, we evaluate the system on synthetic profiles—false negative users (FNU) and false positive users (FPU)—designed to probe generalization under data asymmetry. For FPU, we also assess prediction reliability at increasing confidence thresholds (ψ), based on the model's output probabilities.

Table 2. Results in terms of accuracy (acc.) and precision (prec.) by class (✗ : *dislike*, ✓ : *like*) based on different user types: all users, mixed, negative, positive, false negative users (FNU), and false positive users (FPU). For the latter, various confidence levels (ψ) were also evaluated.

User type →	All		Mixed		Negative	Positive	FNU		FPU		FPU $(\psi \geq 0.90)$		FPU $(\psi \geq 0.95)$		FPU $(\psi \geq 0.99)$	
Class →	✗	✓	✗	✓	✗	✓	✗	✓	✗	✓	✗	✓	✗	✓	✗	✓
base acc.	0.3816	0.9958	0.0334	0.9779	**0.9824**	0.9990	0.0400	0.9855	**0.0321**	1.0000	0.0750	0.9200	0.1107	0.9000	0.2214	0.8400
base prec.	0.9404	0.9016	0.5237	0.5811	-	-	0.1000	0.9624	1.0000	**0.1558**	0.8400	0.1508	0.8611	0.1531	0.8857	0.1615
mix acc.	**0.4597**	0.9582	**0.3585**	0.7767	0.6345	0.9909	**0.8200**	0.1880	0.0107	1.0000	**0.1929**	0.9000	**0.4857**	0.5600	**0.9893**	0.0200
mix prec.	0.6593	**0.9098**	0.5393	**0.6241**	-	-	0.0390	**0.9630**	1.0000	0.1529	0.9153	**0.1661**	0.8608	**0.1628**	0.8497	**0.2500**

(a) Madrid

User type →	All		Mixed		Negative	Positive	FNU		FPU		FPU $(\psi \geq 0.90)$		FPU $(\psi \geq 0.95)$		FPU	
Class →	✗	✓	✗	✓	✗	✓	✗	✓	✗	✓	✗	✓	✗	✓	✗	✓
base acc.	**0.3929**	0.9977	0.0233	0.9850	**0.9811**	0.9993	0.0200	0.9814	**0.0649**	1.0000	0.1489	0.9400	0.1603	0.9000	0.2099	0.8200
base prec.	0.9591	**0.9240**	0.5395	0.5728	-	-	0.0833	0.9221	1.0000	**0.1695**	0.9286	**0.1741**	0.8936	**0.1698**	0.8594	0.1653
mix acc.	0.3797	0.9791	**0.3030**	0.8480	0.5017	0.9953	**0.7600**	0.3384	0.0153	1.0000	**0.2405**	0.8200	**0.4046**	0.6200	**0.9351**	0.0800
mix prec.	0.7104	0.9211	0.5998	**0.6180**	-	-	0.0886	**0.9434**	1.0000	0.1623	0.8750	0.1708	0.8480	0.1658	0.8419	**0.1905**

(b) Barcelona

acc. (✗ : TNR, ✓ : TPR)

When aggregating all users, overall performance differences between *mix* and *base* are relatively small. In the Madrid dataset, *mix* yields a higher true negative rate (*TNR*) and slightly higher precision for the *like* class. In Barcelona, the baseline performs marginally better. These aggregate results, however, obscure key differences across user groups, particularly in cases of feedback imbalance.

The most evident improvements appear in mixed users. The *mix* approach reaches a true negative rate of 0.36 in Madrid and 0.30 in Barcelona, compared to just 0.03 and 0.02 with the baseline. This suggests that learning from users with both positive and negative feedback improves the model's ability to recognize non-relevant items. In contrast, for negative-only and positive-only users, the baseline achieves higher accuracy, particularly for negative users. These results are expected, since the baseline is directly optimized for the overall distribution, which is dominated by homogeneous profiles.

A deeper understanding comes from analyzing the behavior of the system on synthetic users. For false negative users (FNU), the baseline fails to discriminate and classifies nearly all items as positive, resulting in extremely low true negative rates. In contrast, the *mix* approach achieves substantially higher *TNR*, with 0.82 in Madrid and 0.76 in Barcelona. Even though these users lack positive training examples, the model still manages to correctly classify a notable portion of like examples (18% in Madrid, 34% in Barcelona), suggesting it generalizes knowledge from mixed users. For false positive users (FPU), the difference is even more pronounced. At the highest confidence threshold ($\psi \geq 0.99$), the *mix* approach reaches *TNR* values of 0.98 in Madrid and 0.93 in Barcelona, while the baseline remains below 0.23. This indicates that *mix* produces more conservative and better-calibrated predictions when faced with ambiguous or underspecified profiles.

To complement this evaluation, a multi-criteria decision analysis (MCDA) was conducted using the weighted sum model (WSM) [2]. This approach aggregates performance across user types and class metrics to assess robustness and balance. Table 3 shows that *mix* consistently outperforms *base* in global evaluations and in most individual categories, particularly those requiring generalization across unseen user behavior. Although *base* performs slightly better for some isolated user types, these gains are offset by poor performance with mixed and synthetic profiles, an indication of limited generalization capacity.

Table 3. Results of the multi-criteria decision analysis considering true negative and true positive rates across different user groups according to the weighted sum model. The table presents the ranking for each approach (*base* and *mix*) in each scenario, with the corresponding aggregated score shown in parentheses.

	Approach	Mixed	Negative	Positive	FNU	FPU	All Criteria
MAD	*base*	2 (0.3213)	**1 (0.6076)**	**1 (0.5020)**	2 (0.4431)	2 (0.2732)	2 (0.4740)
	mix	**1 (0.6787)**	2 (0.3924)	2 (0.4980)	**1 (0.5568)**	**1 (0.7267)**	**1 (0.5260)**
BCN	*base*	2 (0.3044)	**1 (0.6616)**	**1 (0.5010)**	2 (0.3846)	2 (0.2628)	2 (0.4691)
	mix	**1 (0.6956)**	2 (0.3383)	2 (0.4990)	**1 (0.6154)**	**1 (0.7372)**	**1 (0.5309)**

In summary, these results provide initial empirical support for the hypothesis that intra-domain transfer from mixed users can help mitigate the effects of user-level imbalance in recommender systems. While not universally superior, the *mix* strategy appears to provide more balanced predictions across user groups and to better handle extreme or underspecified cases. These findings highlight the potential of using targeted user subsets as a source of transferable knowledge, and suggest that further exploration of user-level transfer is warranted.

6   Conclusions

This work shows that training recommender systems on majority user groups, such as positive-only profiles, can lead to overfitting and poor generalization. Despite being the most represented in the data, these profiles provide limited signals for distinguishing between relevant and irrelevant content.

To investigate whether this effect can be mitigated, we explore intra-domain transfer learning as a means to improve personalization under feedback imbalance. By learning from users who provide both positive and negative feedback, the system gains access to richer preference signals that may generalize to users with less complete profiles. The results suggest that this strategy improves the model's ability to identify negative preferences and handle edge cases where explicit counterexamples are absent.

The study also highlights limitations in standard evaluation procedures, which often overlook class imbalance and fail to capture generalization gaps. To address this, we propose an evaluation framework based on synthetic user profiles, allowing for a more nuanced assessment of model behavior under asymmetrical feedback. This approach offers insights into how systems perform in scenarios that resemble real-world limitations, such as users with only one type of feedback.

While the findings are promising, they also point to clear limitations. The effectiveness of the proposed strategy depends on the availability of informative user profiles, which may be scarce in many practical settings. Future work should explore alternative sources of transferable knowledge, including profile clustering, cross-domain signals, or the controlled generation of synthetic user behavior. In addition, analyzing the feature space covered by mixed users could offer a principled way to identify which user profiles are most representative and informative for transfer, improving the selection of source users and reducing reliance on dataset-specific heuristics.

References

1. Bawden, D., Robinson, L.: The dark side of information: overload, anxiety and other paradoxes and pathologies. J. Inf. Sci. **35**(2), 180–191 (2009)
2. Cabral, J.B., Luczywo, N.A., Zanazzi, J.L.: Scikit-criteria: Colección de métodos de análisis multi-criterio integrado al stack científico de Python. In: XLV Jornadas Argentinas de Informática e Investigación Operativa (45JAIIO)- XIV Simposio Argentino de Investigación Operativa (SIO), pp. 59–66 (2016)
3. Deng, J., Dong, W., Socher, R., Li, L.J., Li, K., Fei-Fei, L.: ImageNet: a large-scale hierarchical image database. In: IEEE Conference on Computer Vision and Pattern Recognition, pp. 248–255 (2009)
4. Díez, J., Pérez-Núnez, P., Luaces, O., Remeseiro, B., Bahamonde, A.: Towards explainable personalized recommendations by learning from users' photos. Inf. Sci. **520**, 416–430 (2020)
5. Fang, X.: Making recommendations using transfer learning. Neural Comput. Appl. **33**(15), 9663–9676 (2021)
6. Froehlich, L., Dorrough, A.R., Glöckner, A., Stürmer, S.: Similarity predicts cross-national social preferences. Soc. Psychol. Pers. Sci. **12**(8), 1486–1498 (2021)
7. Ganin, Y., et al.: Domain-adversarial training of neural networks. J. Mach. Learn. Res. **17**(59), 1–35 (2016)
8. Harte, J., Zorgdrager, W., Louridas, P., Katsifodimos, A., Jannach, D., Fragkoulis, M.: Leveraging large language models for sequential recommendation. In: ACM Conference on Recommender Systems, pp. 1096–1102 (2023)
9. He, K., Zhang, X., Ren, S., Sun, J.: Deep residual learning for image recognition. In: IEEE Conference on Computer Vision and Pattern Recognition, pp. 770–778 (2016)

10. Ioffe, S., Szegedy, C.: Batch normalization: accelerating deep network training by reducing internal covariate shift. In: International Conference on Machine Learning, pp. 448–456 (2015)
11. Lops, P., Musacchio, E., Musto, C., Polignano, M., Silletti, A., Semeraro, G.: Reproducibility analysis of recommender systems relying on visual features: traps, pitfalls, and countermeasures. In: ACM Conference on Recommender Systems, pp. 554–564 (2023)
12. Nair, V., Hinton, G.E.: Rectified linear units improve restricted Boltzmann machines. In: 27th International Conference on Machine Learning, pp. 807–814 (2010)
13. Pérez-Núñez, P., Blanco-Mallo, E., Bolón-Canedo, V., Remeseiro, B.: Tripadvisor restaurant reviews (2025). https://doi.org/10.5281/zenodo.14622324
14. Rashed, A., Elsayed, S., Schmidt-Thieme, L.: Context and attribute-aware sequential recommendation via cross-attention. In: ACM Conference on Recommender Systems, pp. 71–80 (2022)
15. Schwartz, B.: The Paradox of Choice: Why More is Less, New York (2004)
16. Shafqat, W., Byun, Y.C.: A hybrid GAN-based approach to solve imbalanced data problem in recommendation systems. IEEE Access **10**, 11036–11047 (2022)
17. Tajfel, H., Billig, M.G., Bundy, R.P., Flament, C.: Social categorization and intergroup behaviour. Eur. J. Soc. Psychol. **1**(2), 149–178 (1971)
18. Turner, J.C., Oakes, P.J., Haslam, S.A., McGarty, C.: Self and collective: cognition and social context. Pers. Soc. Psychol. Bull. **20**(5), 454–463 (1994)
19. Wang, S., Zhou, W., Jiang, C.: A survey of word embeddings based on deep learning. Computing **102**(3), 717–740 (2020)
20. Zhang, X., Li, R., Zhang, B., Yang, Y., Guo, J., Ji, X.: An instance-based learning recommendation algorithm of imbalance handling methods. Appl. Math. Comput. **351**, 204–218 (2019)

Post-hoc Recommendation Explanations in E-Learning: An Empirical Study Using Fuzzy Tools

Ouahiba Remadnia[1], Raciel Yera[2(✉)], Rosa M. Rodríguez[2],
and Faiz Maazouzi[1]

[1] Department of Computer Science, University of Souk Ahras, Souk Ahras, Algeria
{w.remadnia,f.maazouzi}@univ-soukahras.dz
[2] Department of Computer Science, University of Jaén, Jaén, Spain
{ryera,rmrodrig}@ujaen.es

Abstract. The increasing reliance on artificial intelligence (AI) in education has highlighted the necessity for transparent and trustworthy recommender systems. This paper presents a framework that combines collaborative filtering, specifically Singular Value Decomposition (SVD), with post-hoc, model-agnostic explanation techniques to generate interpretable recommendations in e-learning environments. To achieve this, we implement both crisp and fuzzy association rule mining approaches for explaining recommendations based on learners' historical behaviors. The fuzzy model leverages membership degrees to account for interaction intensity, while the crisp model uses binarized user–item transactions. We generate explanations by mapping recommended items to association rules, and explanation quality is assessed using the Average Fidelity metric. Experimental evaluations on a real-world MOOC dataset demonstrate the effectiveness of the framework in delivering recommendations with interpretable justifications. The results validate the feasibility of integrating fuzzy-based explainability into recommendation pipelines, as the proposed approach enhances the crisp-based explanation approach, taking into account fidelity-based measures.

Keywords: Explainable Recommender Systems · E-Learning · Fuzzy Association Rules · Crisp explanation approach · Collaborative Filtering · Singular Value Decomposition (SVD)

1 Introduction

AI has grown in popularity in recent years, leading to widespread adoption across a range of services. One of its key applications is in education [13], where recommender systems (RS) have significantly enhanced its overall impact. These systems have become an essential part of daily life, serving as effective tools to help people access online information that best fits their preferences and needs. Typically, RSs provide a list of suggestions based on the user's profile and

© The Author(s), under exclusive license to Springer Nature Switzerland AG 2026
L. Martínez et al. (Eds.): IDEAL 2025, LNCS 16239, pp. 84–95, 2026.
https://doi.org/10.1007/978-3-032-10489-2_8

behavior, which may include interactions with available options, item attributes, and other relevant data.

Recently, successful recommendation approaches have focused on computational intelligence techniques such as matrix factorization and deep learning-based methods to predict users' unknown preferences and generate recommendation lists [16,23]. However, these strategies suffer from a major drawback: their black-box nature reduces transparency and undermines credibility. Research in recommender systems (RS) has primarily emphasized improving the accuracy of recommendation algorithms [18]. At the same time, governments, society, and several studies—including [20] stress the importance of explainability in recommendation outcomes, alongside accuracy improvements.

Explainable recommendation has emerged as an important dimension across various domains, including e-health and e-business, where it enhances decision-making by providing transparency and trust. Its relevance has also extended to highly active sectors such as e-commerce and e-learning [26], underscoring its role in supporting informed and confident user decisions.

This contribution is motivated by the need to enhance trust and transparency in RSs for the e-learning domain. As learning recommendation continues to emerge as a primary research area [10], it is increasingly evident that current educational recommendation frameworks often rely on complex, black-box algorithms [8]. Although prior studies have explored explainability in recommendation systems [26], few have focused specifically on the e-learning context [26]. In addition, evaluating the impact of fuzzy tools in this task remains necessary [23]. Our work addresses these gaps by making the following contributions:

- The developing of a post-hoc recommendation explanation framework to be used on an e-learning domain.
- The evaluation of the impact of uncertainty management using fuzzy tools, as opposed to information processing through crisp models.

The paper is organized as follows. Section 2 introduces a brief overview on recommendation explanation in e-learning. Section 3 details the proposed framework, respectively considering fuzzy and crisp information processing. Section 4 develops a set of experiments for measuring the discussed approaches, including dataset analysis, evaluation protocol, and experimental results. Section 5 concludes the paper, pointing out future directions.

2   The Explainable E-Learning Recommender System Domain

Explainable e-learning RS represent a growing yet relatively underexplored area within the broader landscape of RS. Traditionally, e-learning RS focused on improving personalization and learner engagement through tailored content. However, the dimension of explainability has only recently gained relevance, driven by the need to increase transparency, user trust, and pedagogical alignment within educational environments [26].

In this context, surveys such as that of Adadi and Berrada [1] mapped the landscape of explainable AI in RS, highlighting both the potential and the challenges of incorporating explainability into e-learning. Researchers have integrated traditional collaborative filtering (CF) and content-based (CB) approaches with explainability techniques, including memory-based methods [12], matrix factorization approaches with explicit factors [27], and post-hoc model-agnostic methods such as LIME and SHAP to complement black-box recommendations [17].

Moreover, explainable e-learning RS increasingly leverage heterogeneous data sources, such as learner performance metrics, course attributes, semantic content features, and engagement signals extracted from interactions. These diverse data sources offer rich contexts to generate recommendations that are not only personalized but also interpretable and aligned with learner goals.

Recent studies have focused on enhancing transparency, trust, and engagement in personalized learning environments. For example, researchers have examined how explanations affect trust in e-learning platforms [14], user satisfaction and participation in learning projects [4], and student engagement through course recommendations [25]. Other works proposed ontology-driven frameworks for generating structured, learner-centered explanations using knowledge graphs [2] and applied Bayesian models to improve transparency in quiz recommendations [19].

These approaches rely on diverse datasets, ranging from small-scale user studies to large-scale educational repositories, and employ methods including hybrid CF, matrix factorization, CB filtering, knowledge graphs, and Bayesian Knowledge Tracing. However, the literature still shows limited exploration of post-hoc and model-agnostic explanation methods within e-learning recommender systems. Our contribution aims to fill this gap.

3   Exploring Post-hoc Explanation Generation in E-Learning Recommendations

The current section presents the framework used to generate post-hoc explanations for the recommendations provided by an e-learning platform, specifically a MOOC platform.

Figure 1 illustrates the overall framework, which builds on ExplanationMining [15], one of the pioneering approaches for post-hoc recommendation generation. This method has recently been applied in domains such as e-health [22] and group recommendation [24].

The framework first preprocesses the raw interaction data of the system, converting it into a format suitable for subsequent stages. It also filters out users or items with very limited activity on the platform, since these could introduce noise and negatively affect later steps. Next, the framework develops two parallel processes. On one side, it trains a recommendation model and then generates the top-n recommendations. On the other side, it discovers frequent items from the filtered dataset, leading to rule generation and pruning to avoid redundancy. Finally, the system uses these rules to generate recommendation explanations, which it categorizes according to model fidelity. This framework produces explanations of the form:

"We recommend i because you engaged with X."

The next subsections will present two implementations of this framework, respectively using fuzzy (Sect. 3.1) and crisp approaches (Sect. 3.2).

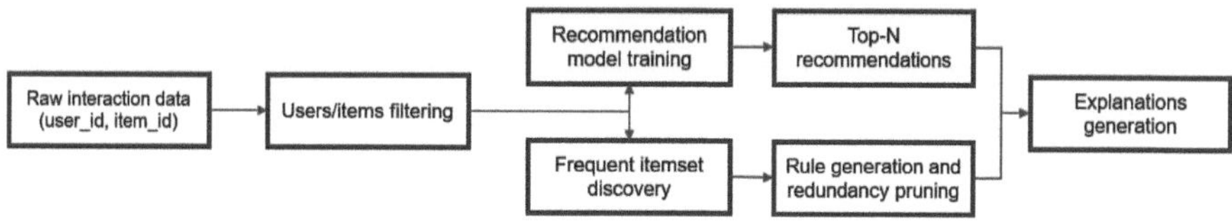

Fig. 1. The general framework used for generating post-hoc explanations.

3.1   Explanations Generation Through a Fuzzy Approach

As pointed out in the Introduction section, the use of fuzzy approaches has not been sufficiently explored in the explanation generation tasks in RS. Nevertheless, it is important to highlight that in some scenarios, such as e-learning related, it could have a remarkable role given the continuous nature of some of the underlying data (e.g. watching time, as will be discussed below), where the use of techniques for managing the uncertainty could lead to better performance. Based on these premises, our proposal is driven by the following stages:

Interaction Processing and Filtering. We begin by filtering the raw dataset of user-item interactions to retain only the most active users and the most popular items. The goal of this step is to increase the density of the interaction matrix and improving the statistical reliability of the discovered knowledge. Alternatively, we could opt for keeping the full dataset to be used in the next stages of the proposal. The experimental stage of the current contribution details how this filtering process was specifically done at this scenario.

In any case, the filtered interactions are sorted chronologically per user.

Recommendation Model Training and Top-n Recommendations. We train a CF model based on matrix factorization (SVD) using the rating values from the dataset. Particularly, we will use the popular FunkSVD approach [7], a matrix factorization technique used for rating prediction in RS. Rather than performing a full singular value decomposition, it learns low-dimensional latent factors for users and items by minimizing the regularized squared error between predicted and actual ratings using stochastic gradient descent (Eq. 1).

$$\min_{b_*,\mathbf{p}_*,\mathbf{q}_*} \sum_{(u,i)\in\mathcal{K}} (r_{ui} - \hat{r}_{ui})^2 + \lambda \left(b_u^2 + b_i^2 + \|\mathbf{p}_u\|^2 + \|\mathbf{q}_i\|^2 \right) \tag{1}$$

The predicted rating $\hat{r}_{ui}$ (Eq. 2) for a user-item pair is computed as the sum of the global average rating (μ), user and item bias terms (b_u and b_i), and the dot product of the user and item latent vectors ($\mathbf{q}_i^\top \mathbf{p}_u$).

$$\hat{r}_{ui} = \mu + b_u + b_i + \mathbf{q}_i^\top \mathbf{p}_u \tag{2}$$

This allows the model to capture both general tendencies (via biases) and nuanced interactions (via latent factors), making it well-suited for collaborative filtering tasks.

We use this model to predict unknown ratings for user-item pairs, and generate top-N recommendations for each user.

Frequent Itemset Discovery, Rule Generation, and Redundancy Pruning. In parallel, from the obtained dataset we construct fuzzy transactions, where each item is associated with a degree of membership corresponding to the preference degree of the item (between 0 and 1). These fuzzy transactions are used to mine frequent itemsets using a fuzzy version of the Apriori algorithm [9]. In this version, the support of an itemset is computed as the average of the mean membership degrees across all transactions in which the itemset appears.

Let $\mathcal{T} = \{T_1, T_2, \ldots, T_N\}$ be a set of fuzzy transactions, where each transaction T_i is a mapping $T_i : \mathcal{I} \to [0,1]$ assigning to each item $j \in \mathcal{I}$ a degree of membership $T_i(j)$. For an itemset $X \subseteq \mathcal{I}$, the fuzzy support with mean aggregation is defined as:

$$\mathrm{supp}_{\mathrm{mean}}(X) = \frac{1}{|\mathcal{T}|} \sum_{T \in \mathcal{T}} \left(\frac{1}{|X|} \sum_{j \in X} T(j) \right) \tag{3}$$

Herein, we have defined the fuzzy support based on the arithmetic mean, taken into account that it is less restrictive than the largely-used minimum function, and therefore could be more adequate to e-learning scenarios where interaction levels varies across the different users.

Based on these itemsets, we extract fuzzy association rules of the form $A \implies B$, where A and B are item subsets. Rules are filtered by a minimum confidence threshold, and then pruned to eliminate redundant rules—those whose antecedents and consequents are subsets of stronger rules with similar confidence.

Explanation Generation. In this phase, a recommended item is considered explainable if there exists a fuzzy rule $A \implies B$ such that the user's training history contains all elements of A and the recommended item belongs to B. Finally, we compute the model fidelity as the ratio of recommended items that are explainable in this way. Fidelity can be calculated globally or at the user level, allowing for fine-grained evaluation of explainability effectiveness. Algorithm 1 illustrates this approach. Herein, for each recommendation list (Line 4), first it is extracted the user preference history (Line 5). Subsequently, for each item in the recommendation list (Line 6), it is verified whether for some previously extracted rule, the items in the antecedent are in the user preference history, and the consequent matches with the current (Line 9). In such case, the rule is registered as the explanation (Line 10). Finally, the model fidelity is calculated as the proportion of explainable items (Line 13).

Algorithm 1. Post-hoc Explanation and Fidelity Calculation

Require: Top-N recommendations $\mathcal{R}$, user histories $\mathcal{H}$, fuzzy rules $\mathcal{F}$
Ensure: Explanations $\mathcal{E}$, global fidelity F

1: $\mathcal{E} \leftarrow \emptyset$
2: $explained \leftarrow 0$
3: $total \leftarrow 0$
4: **for all** $(u, rec\_list) \in \mathcal{R}$ **do**
5:      $H_u \leftarrow \mathcal{H}[u]$
6:      **for all** $i \in rec\_list$ **do**
7:          $total \leftarrow total + 1$
8:          **for all** $(A \rightarrow B) \in \mathcal{F}$ **do**
9:             **if** $A \subseteq H_u$ **and** $i \in B$ **then**
10:               $\mathcal{E}[u] \leftarrow \mathcal{E}[u] \cup \{(i, A \rightarrow B)\}$
11:               $explained \leftarrow explained + 1$
12:               **break**
13: $F \leftarrow explained/total$
14: **return** $(\mathcal{E}, F)$

3.2 Explanations Generation Through a Crisp Approach

This section is focused on formalizing an alternative crisp approach for implementing the framework screened at Fig. 1. With this aim in mind, we will provide formalization and implementation details at the distinctive stage in relation to the previous fuzzy approach.

In this alternative, transactions are built from the training set by binarizing interactions. An item is included in a user's transaction if the user has either rated it or watched at least some portion of it. These binary transactions are also processed with the standard Apriori algorithm [3] to discover frequent itemsets above a given minimum support. Crisp association rules $A \implies B$ are then extracted from these itemsets, using standard confidence thresholds. To reduce redundancy and in a similar way to the former approach, we apply a pruning step that removes rules whose antecedents and consequents are subsets of higher-confidence rules.

Herein, the classic support is defined as follows [3]. Let $\mathcal{T} = \{T_1, T_2, \ldots, T_N\}$ be a set of transactions, where each transaction $T_i \subseteq \mathcal{I}$ is a subset of items. Given an itemset $X \subseteq \mathcal{I}$, its crisp support is defined as:

$$\text{supp}_{\text{crisp}}(X) = \frac{1}{|\mathcal{T}|} \sum_{T \in \mathcal{T}} \delta(X \subseteq T) \tag{4}$$

where $\delta(X \subseteq T) = 1$ if $X \subseteq T$, and 0 otherwise.

The remaining stages of this approach are performed in a similar way to those presented in the previous section.

4   Experiments

In this section, we present the experimental validation of our proposed approach. We detail the dataset used, the evaluation protocol followed, and discuss the obtained results.

4.1   Dataset

For our work, we used the *E-learning Recommender System Dataset* [5], a real-world dataset provided by Mandarine Academy, an Ed-Tech company specializing in corporate training through MOOCs, web conferences, and other digital learning methods. The dataset is publicly available and is designed to facilitate research in RS, particularly within the e-learning domain.

The dataset contains user interaction data collected from the MOOC platform `mooc.office365-training.com`, spanning from early 2016 to late 2021. It captures user behavior in both explicit and implicit formats. Explicit ratings include interactions such as likes, social shares, and bookmarks, whereas implicit feedback is derived from watch time and page views. Given the low frequency of explicit interactions, the dataset primarily emphasizes implicit feedback, enabling the analysis of user engagement without requiring direct user input.

In total, the dataset comprises a large number of explicit and implicit ratings provided by anonymized users over learning items that include associated metadata such as content descriptions and duration. Specifically, the ratings reflect user engagement with learning materials via explicit feedback (e.g., ratings) and implicit feedback (e.g., views or clicks).

For the purpose of our study, we focus on the rating data. Two language-based versions of the dataset are available: (1) a French content dataset comprising 85,339 ratings, and (2) an English content dataset comprising 3,659 ratings. Each interaction in the dataset includes the following fields: `user_id`, `item_id`, `watch_percentage`, `created_at`, and `rating`. Here, the user ratings are derived by normalizing the watch percentage to a 1–10 scale. Further information on these dataset can be checked at Hafsa et al. [5].

4.2   Evaluation Protocol

This subsection is focused on discussing the evaluation protocol and some implementation details associated to the developed experiments.

- *Interaction processing and filtering.* At this stage we will consider two experimental scenarios, by using the whole dataset (full), or just a dataset composed of the top 500 more active users, and their evaluations over the top 500 popular items (top-500). 90% of each user data were using from training, leaving the remaining 10% for building the test set in each case.

- *Recommendation model training and top-n recommendations.* We will use the SVD approach implemented inside the Surprise library [6], which is a popular toolkit for deploying experiments in RS. The default parameters for this approach, defined in Surprise, will be used across our experiments. Top 10 recommendations will be only considered in the current contribution, due to the limitation in space.
- *Frequent itemsets discovery, rule generation, and pruning.* In this stage, it will be used a fixed value of $minconf = 0.2$ and several values of minimum support, tailored to each experimental scenario and dataset. Furthermore, for the case of the fuzzy support, it is necessary the definition of the function $T()$. Figure 2 illustrates the membership function that will be considered in this work, that receives as input the rating values in the range $[1; 10]$.
- *Explanation generation.* The explanation generation is performed across Algorithm 1, which receives as input the top n recommendations, user profiles and extracted rules, and does not depend on any specific parameter in the version provided in the current contribution. As pointed out, to quantify the alignment between the recommendations and the generated explanations, we employ the Average Fidelity metric, defined as:

$$Fidelity = \frac{Number\ of\ explained\ recommendations}{Total\ number\ of\ recommendations} \tag{5}$$

Higher fidelity indicates better interpretability of the recommendations without sacrificing personalization.

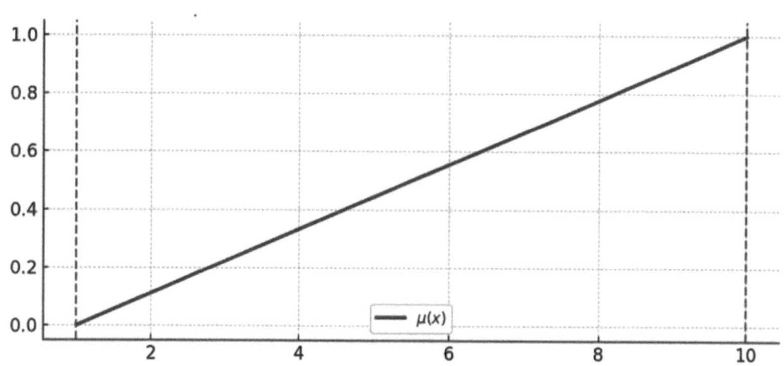

Fig. 2. Membership function T() used for calculating the fuzzy support.

4.3   Results and Discussion

We present a comparative evaluation of our fuzzy and crisp explanation approaches across multiple configurations, using model fidelity as the main evaluation metric. Fidelity captures the proportion of recommended items that can be post-hoc explained by mined association rules. Across all experimental conditions, the fuzzy explanation approach consistently outperforms the crisp alternative. This difference is especially notable at lower support thresholds, where the use of fuzzy logic enables partial evidence accumulation, resulting in higher explanatory coverage. For example, in the English dataset limited to the top 500 users and items, the fuzzy approach achieves a fidelity of 0.183 at a support of 0.03, which approximately duplicated the value obtained by the crisp explanation method under the same conditions.

This regularity is repeated across the French dataset and the full datasets, suggesting that fuzzy rules are more effective at capturing latent, approximate relationships among items that crisp rules might overlook. The flexibility associated to the management of the fuzzy membership grades seems to be an important component for mitigating the effects of data sparsity and interaction variability, especially common in e-learning contexts (Tables 1 and 2).

Table 1. English dataset. Top 500 active users and items. Minconf$=0.2$

Approach-Support values	0.03	0.04	0.05	0.06	0.07	0.08
Fuzzy explanation approach	**0.183**	**0.091**	**0.082**	**0.082**	**0.064**	**0.008**
Crisp explanation approach	0.097	0.071	0.046	0.046	–	–

Table 2. French dataset. Top 500 active users and items. Minconf$=0.2$

Approach-Support values	0.2	0.3	0.4
Fuzzy explanation approach	**0.094**	**0.013**	**0.004**
Crisp explanation approach	0.027	0.008	–

As expected, increasing the minimum support threshold has a negative impact on fidelity, as fewer rules are mined and fewer recommendations can be explained. This decline is more relevant in the crisp setting, which quickly reaches zero fidelity at moderate support values, such as 0.07 or 0.08. In contrast, the fuzzy approach maintains non-zero fidelity across a wider range of thresholds, although it also declines as support increases. For instance, in the English full dataset, fuzzy fidelity decreases from 0.057 to 0.005 as the support increases from 0.04 to 0.08.

The differences between experiments using the full dataset and those filtered to include only the top 500 most active users and most popular items further highlight the impact of data sparsity. Higher fidelity scores are obtained in the filtered case, as the denser interaction patterns lead to more frequent item co-occurrences and better rule coverage. In contrast, the full datasets include many users and items with few interactions, which reduce the chances of identifying statistically relevant patterns and contribute to lower fidelity.

In summary, fuzzy rules provide better coverage for recommendation explanation, particularly in sparse or noisy environments and under low support thresholds. Crisp rules tend to produce low fidelity results in such conditions, often failing to explain any recommendations when moderate support is required. These results underline the importance of balancing rule quality and explanation coverage when designing explainability modules for RS (Tables 3 and 4).

Table 3. English dataset. Full dataset. Minconf $= 0.2$

Approach-Support values	0.04	0.05	0.06	0.07	0.08
Fuzzy explanation approach	**0.057**	**0.051**	**0.051**	**0.04**	**0.005**
Crisp explanation approach	0.044	0.028	0.025	–	–

Table 4. French dataset. Full dataset. Minconf $= 0.2$

Approach-Support values	0.06	0.07	0.08	0.09	0.1
Fuzzy explanation approach	**0.029**	**0.027**	**0.016**	**0.016**	**0.001**
Crisp explanation approach	0.014	0.014	0.014	–	–

5   Conclusions

This work presented an explainable recommendation approach for e-learning, combining CF via Singular Value Decomposition (SVD) with post-hoc, model-agnostic explanation techniques using both fuzzy and crisp association rule mining. Our framework addresses the increasing necessity for trust and transparency in AI-driven educational systems, working towards the balance between explainability and accuracy.

The fuzzy rule-based approach leverages membership functions to capture the intensity of learner interactions, while the crisp model operates on binary interactions, both contributing to interpretable justifications for recommended learning resources. Experimental results on a real-world MOOC dataset validate the effectiveness of the proposed system in providing accurate recommendations with explainable support, as quantified by the Average Fidelity metric. Furthermore, it is relevant to highlight the positive effect of the management of the underlying uncertainty through fuzzy tools.

Although the framework shows promising performance, challenges remain in improving the fidelity of explanations, particularly in sparse datasets. Future work will explore integrating deep learning models with explainability components, conducting user-centric evaluations to assess perceived trust and usefulness [21], and aligning generated explanations with pedagogical objectives.

We also plan to include other computational intelligence-based approaches with this goal in mind [11,23]. Ultimately, our goal is to develop robust, user-aware, and pedagogically aligned recommendation systems that foster meaningful and transparent learning experiences.

Acknowledgments. This work is supported by the Spanish Thematic Network on Recommender Systems (RED2022-134302-T) and by the Spanish Ministry of Science, Innovation, and Universities, through the Knowledge Generation Projects grant (PID2024-161073NB-I00). Raciel Yera is supported by the European Union's Horizon Europe research and innovation program under the Marie Sklodowska-Curie Grant Agreement No.101106164. Rosa M. Rodríguez is supported by the Andalusian PAIDI 2020 Program under the project PROYEXCEL-00257.Ouahiba Remadnia thanks her supervisors, collaborators, Mohamed-Cherif Messaadia University (Algeria), and the Laboratory of Computer Science and Mathematics (LIM).

References

1. Adadi, A., Berrada, M.: Peeking inside the black-box: a survey on explainable artificial intelligence (XAI). IEEE Access **6**, 52138–52160 (2018)
2. Afreen, N., et al.: Learner-centered ontology for explainable educational recommendation. In: Adjunct Proceedings of the 32nd ACM Conference on User Modeling, Adaptation and Personalization, pp. 567–575 (2024)
3. Agrawal, R., Imieliński, T., Swami, A.: Mining association rules between sets of items in large databases. In: Proceedings of the 1993 ACM SIGMOD International Conference on Management of Data, pp. 207–216 (1993)
4. Ben Zaken, D., Segal, A., Cavalier, D., Shani, G., Gal, K.: Generating recommendations with post-hoc explanations for citizen science. In: Proceedings of the 30th ACM Conference on User Modeling, Adaptation and Personalization, pp. 69–78 (2022)
5. Hafsa, M., Wattebled, P., Jacques, J., Jourdan, L.: E-learning recommender system dataset. Data Brief **47**, 108942 (2023)
6. Hug, N.: Surprise: a Python library for recommender systems. J. Open Source Softw. **5**(52), 2174 (2020)
7. Koren, Y., Bell, R., Volinsky, C.: Matrix factorization techniques for recommender systems. IEEE Comput. **42**(8), 30–37 (2009)
8. Kulkarni, P.V., Rai, S., Kale, R.: Recommender system in elearning: a survey. In: Bhalla, S., Kwan, P., Bedekar, M., Phalnikar, R., Sirsikar, S. (eds.) Proceeding of International Conference on Computational Science and Applications. AIS, pp. 119–126. Springer, Singapore (2020). https://doi.org/10.1007/978-981-15-0790-8_13
9. Kuok, C.M., Fu, A., Wong, M.H.: Mining fuzzy association rules in databases. ACM SIGMOD Rec. **27**(1), 41–46 (1998)
10. Lu, J., Wu, D., Mao, M., Wang, W., Zhang, G.: Recommender system application developments: a survey. Decis. Support Syst. **74**, 12–32 (2015)
11. Martínez, L., Castro, J., Yera, R.: Managing natural noise in recommender systems. In: Martín-Vide, C., Mizuki, T., Vega-Rodríguez, M.A. (eds.) TPNC 2016. LNCS, vol. 10071, pp. 3–17. Springer, Cham (2016). https://doi.org/10.1007/978-3-319-49001-4_1

12. Mu, R.: A survey of recommender systems based on deep learning. IEEE Access **6**, 69009–69022 (2018)
13. Ni, B., Xie, X.: Distributed distribution and scheduling of teaching resources based on a random matrix educational leadership model. Informatica **48**(8) (2024)
14. Ooge, J., Kato, S., Verbert, K.: Explaining recommendations in e-learning: effects on adolescents' trust. In: Proceedings of the 27th International Conference on Intelligent User Interfaces, pp. 93–105 (2022)
15. Peake, G., Wang, J.: Explanation mining: post hoc interpretability of latent factor models for recommendation systems. In: Proceedings of the 24th ACM SIGKDD International Conference on Knowledge Discovery & Data Mining, pp. 2060–2069 (2018)
16. Remadnia, O., Maazouzi, F., Chefrour, D.: Hybrid book recommendation system using collaborative filtering and embedding based deep learning. Informatica **49**(8) (2025)
17. Ribeiro, M.T., Singh, S., Guestrin, C.: "why should i trust you?" explaining the predictions of any classifier. In: Proceedings of the 22nd ACM SIGKDD International Conference on Knowledge Discovery and Data Mining, pp. 1135–1144 (2016)
18. Ricci, F., Rokach, L., Shapira, B.: Introduction to recommender systems handbook. In: Ricci, F., Rokach, L., Shapira, B., Kantor, P.B. (eds.) Recommender Systems Handbook, pp. 1–35. Springer, Boston, MA (2011). https://doi.org/10.1007/978-0-387-85820-3_1
19. Takami, K., Flanagan, B., et al.: Toward educational explainable recommender system: explanation generation based on Bayesian knowledge tracing parameters. In: International Conference on Computers in Education (2021)
20. Tintarev, N., Masthoff, J.: Explaining recommendations: design and evaluation. In: Ricci, F., Rokach, L., Shapira, B. (eds.) Recommender Systems Handbook, pp. 353–382. Springer, Boston (2015). https://doi.org/10.1007/978-1-4899-7637-6_10
21. Waterschoot, C., Yera Toledo, R., Tintarev, N., Barile, F.: With friends like these, who needs explanations? Evaluating user understanding of group recommendations. In: Proceedings of the 33rd ACM Conference on User Modeling, Adaptation and Personalization, pp. 253–262 (2025)
22. Yera, R., Alzahrani, A.A., Martínez, L.: Exploring post-hoc agnostic models for explainable cooking recipe recommendations. Knowl.-Based Syst. **251**, 109216 (2022)
23. Yera, R., Martinez, L.: Fuzzy tools in recommender systems: a survey. Int. J. Comput. Intell. Syst. **10**(1), 776–803 (2017)
24. Yera, R., Martínez, L.: LORE4GroupRS: explaining group recommendations supported by a local rule-based approach. In: International Conference on Intelligent Data Engineering and Automated Learning, pp. 300–312. Springer (2024)
25. Yu, R., Pardos, Z., Chau, H., Brusilovsky, P.: Orienting students to course recommendations using three types of explanation. In: Adjunct proceedings of the 29th ACM Conference on User Modeling, Adaptation and Personalization, pp. 238–245 (2021)
26. Zhang, Y., Chen, X., et al.: Explainable recommendation: a survey and new perspectives. Found. Trends® Inf. Retrieval **14**(1), 1–101 (2020)
27. Zhang, Y., Lai, G., Zhang, M., Zhang, Y., Liu, Y., Ma, S.: Explicit factor models for explainable recommendation based on phrase-level sentiment analysis. In: Proceedings of the 37th International ACM SIGIR Conference on Research & Development in Information Retrieval, pp. 83–92 (2014)

GREX: A Platform for Supporting Explanations in Group Recommender Systems

Mateus Toledo[(✉)] [iD], Raciel Yera [iD], Manuel J. Barranco [iD], and Bapi Dutta [iD]

University of Jaén, Campus Las Lagunillas s/n, 23007 Jaén, Spain
mtleao@ujaen.es
https://sinbad2.ujaen.es/

Abstract. Although explainability is critical for trustworthy AI, Group Recommender Systems (GRS) lack dedicated, flexible software tools. To address this, we present GREX, a novel, open-source Python library designed to facilitate the development and evaluation of explanations in group settings. GREX is built on a modular and extensible architecture, providing implementations of three distinct explanation paradigms: counterfactual (Sliding-Window-Weighted), rule-based (EXPGRS), and local model-agnostic (LORE4GROUPS). We conducted a comprehensive description of GREX stages, overall including data preparation, training, recommendation, and explanation. GREX is focused on empowering researchers and practitioners to systematically develop and compare Explainable AI (XAI) methods, fostering progress in transparent and user-centric GRS.

Keywords: Group Recommender Systems (GRS) · Explainable AI (XAI) · Software library

1 Introduction

Recommender systems have become an integral part of many online platforms, guiding users to discover relevant products, services, or content amidst overwhelming choices [4]. Traditionally, these systems focus on individual users by analyzing their preferences and behaviors to generate personalized suggestions. However, in many real-world scenarios, decisions are made collectively by groups rather than individuals. Group recommender systems address this need by aggregating the preferences and characteristics of multiple users to provide recommendations that satisfy the group as a whole [6]. This emerging area extends the challenges of traditional recommender systems, introducing unique complexities related to preference aggregation, fairness, and group dynamics.

Artificial intelligence (AI) technologies influence various aspects of daily life; therefore, the importance of explainability has grown significantly. Explainability in AI refers to the ability to provide clear, understandable reasons behind

© The Author(s), under exclusive license to Springer Nature Switzerland AG 2026
L. Martínez et al. (Eds.): IDEAL 2025, LNCS 16239, pp. 96–108, 2026.
https://doi.org/10.1007/978-3-032-10489-2_9

the outputs or decisions of an algorithm. In recommender systems, explanations can enhance user trust and satisfaction by helping users understand why certain items are suggested [20]. Furthermore, explainability supports transparency and accountability, which are essential for addressing ethical concerns such as bias and fairness. In the context of group recommendations, providing explanations becomes even more challenging, as it involves communicating the rationale behind suggestions made to diverse users with potentially conflicting preferences [21].

Despite increasing research interest in explainable recommender systems (XRS), there is still an important gap in the availability of dedicated software libraries that support explanation generation, especially for group recommendation. Although there are several frameworks and tools for building and evaluating recommendation algorithms [5,6,8–10,17], few offer modular and extensible components explicitly designed to produce interpretable explanations for groups. This lack limits the ability of researchers and practitioners to systematically develop, benchmark, and deploy explainability methods in group recommendation contexts, thereby hindering progress in the field.

In this work, we present a novel open-source library aimed at bridging this gap by providing a comprehensive set of explanation approaches tailored for group recommender systems. Our library offers modular implementations of state-of-the-art techniques, flexible integration with existing recommendation models, and evaluation tools for explainability quality. We aim to facilitate research and practical adoption of explainable group recommendations, ultimately fostering more transparent and trustworthy recommender systems.

2   Related Works

2.1   Explanations in Group Recommender Systems

Several studies have explored ways to justify or explain group recommendations from diverse perspectives. Early works have incorporated visual tools: Wang et al. [17] used hierarchical graphs and pie charts to depict group influence, while Quijano et al. [14] emphasized the social dynamics behind group decisions to improve acceptance. Similarly, Marquez et al. [11] focused on negotiation-based recommendation, offering visual explanations of evolving group preferences. More recently, Yera et al. [21,22] have been focused on discovering relationships based on previous preferences over item and item features, for justifying item recommendations. The latest research has explored novel interfaces and explanation types; Al-Hazwani et al. [1] combine conversational AI with bivariate map visualizations to foster transparent consensus-building, while Stratigi et al. [15] have proposed sliding windows-based approaches for identifying counterfactual explanations for item recommendations.

Explanations based on social choice and aggregation strategies have also attracted the attention of the research community. Najafian et al. [12], Trang Tran et al. [16], and Barile et al. [3] examined how preference distributions influence the effectiveness of different aggregation-based explanations, showing that group configuration affects interpretability. Recent work has turned to innovative paradigms: Waterschoot et al. [18,19] explored large language models for generating and assessing group explanations. Still, as noted by Alvarado et al. [2], the field calls for deeper investigation beyond user studies.

2.2 Previous Libraries for Supporting Explanations in Recommender Systems

Several libraries have supported research in recommender systems across the last decade [8–10]. However, there is a limited availability on libraries for supporting recommendation explainability.

A remarkable exception to this shortcoming is RecoXplainer [5], which is a software library designed to facilitate the development and offline evaluation of explainable recommender systems. The authors highlight a common challenge in the field: while there is growing interest in explainability due to its potential to improve user trust, understanding, and acceptance, researchers often lack access to standardized tools and must re-implement existing methods from scratch. RecoXplainer addresses this gap by offering a unified and extensible framework that brings together a variety of explainability techniques and recommendation algorithms in a single platform.

The library supports both model-based and post-hoc explanation approaches and includes several black-box recommender models. In addition to its modular design, it provides a set of offline evaluation metrics specifically aimed at assessing the quality of explanations. This makes it possible not only to prototype new explainability strategies but also to compare them systematically.

Inspired by RecoXplainer, the goal of our contribution is the development of a library for supporting the explanation generation process in group recommender systems.

3 GREX: A Novel Library for Supporting Explanations in Group Recommendation

In this section, we introduce GREX, a software library developed to facilitate research and experiments on explanations for group recommendation. We begin by detailing its foundational architecture, which is designed for modularity and flexibility. Subsequently, we describe the specific components implemented within the library, covering a range of techniques for recommendation, aggregation, and explanation.

3.1 General Architecture

The library operates within a highly defined pipeline workflow, which consists of Data Preparation, Group Handling, Model Selection and Training, Recommendation, Explanation, and Evaluation. A key feature of our library is that these stages are modular. This means that the choice of a specific machine learning

(ML) model, for instance, does not restrict the selection of a group strategy or an explanation method. This design allows for a flexible combination of components, enabling users to configure them as required. The entire architecture is illustrated in Fig. 1.

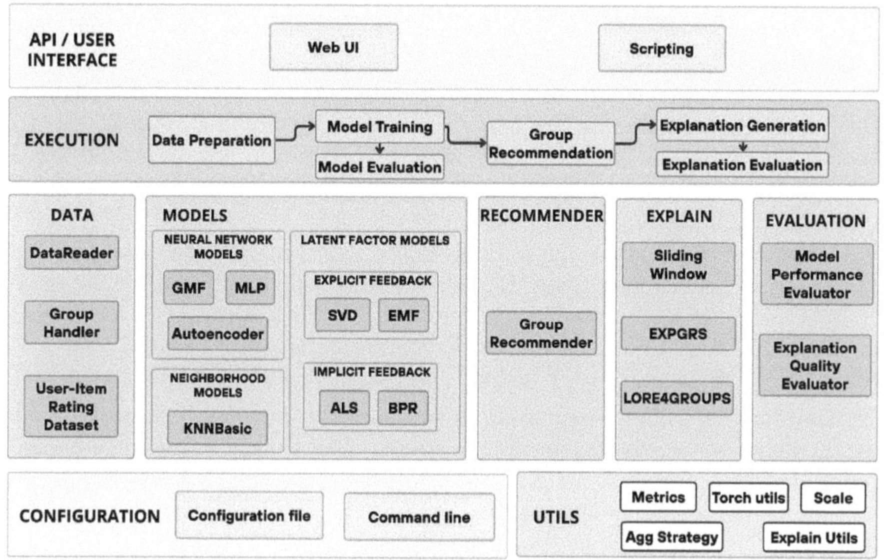

Fig. 1. GREX diagram.

3.2 Components of the Library: Group Recommender and Explanation Approaches

Recommendation Generation. Once a model has been trained (ML models available, taken as reference Coba et al. [5]: ALS, BPR, Autoencoder, EMF, GMF, MLP, KNNBasic, and SVD), the Recommendation Generation stage uses it to produce ranked lists of items. The library provides distinct components for this process, reflecting the different requirements for recommending to individual users versus groups, both following a Recommender abstract component. Although the library supports standard individual recommendations through the Individual Recommender component, its main contribution lies in the specialized process for generating group recommendations, which is handled by the GroupRecommender component. The workflow is as follows:

– Setup and Candidate Selection: The process is initiated by calling the setup recommendation method. This crucial first step takes the list of group members, the trained model, and the aggregation strategy it will use as input. It then prepares a pool of candidate items for recommendation, which filters out any items that have already been seen by at least one member of the group. This ensures that the recommendations are novel for all members.

- Individual Preference Collection: After establishing the candidate items, the preferences of each individual member are gathered for these items. It iterates through each user in the group and utilizes the trained model's prediction to obtain a set of prediction scores for every item in the candidate pool. This step results in a collection of individual prediction lists, one for each group member.
- Score Aggregation: With the individual predictions collected, the next step is to aggregate them into a single score that represents the group's collective preference for each item by applying an aggregation strategy. The library currently supports strategies for the aggregation of individual predictions, which have been detailed in Felfernig et al. [7].
- Final Recommendation List: Finally, the aggregated scores are sorted in descending order to produce a top-N recommended list of items for the group.

Explanation Generation. Following the generation of recommendations, the Explanation Generation stage aims to provide transparency by clarifying why certain items were recommended to a group. The library is designed to be a comprehensive framework for XAI in group recommender systems, supporting multiple paradigms to generate explanations. The current approaches are:

- Counterfactual Explanations (Sliding-Window-Weighted (SWW)): Inspired by the work of Stratigis et al. [15], the Sliding-Window-Weighted (SWW) approach generates counterfactual explanations by identifying the minimal set of items that, if removed from a group's interaction history, would alter the recommended item. We extend their method by introducing a novel metric and a weighted final score. The full set of scores used to rank candidate explanations is:
  - Item Popularity Score: $\mathrm{pop}(U, i) = \sum_{\forall u \in U} \rho(u, i)$.
  - Rating Score: $\mathrm{rate}(G, i) = \frac{\sum_{\forall u \in G} \rho(u,i)}{|G|}$.
  - Item Intensity Score: $\mathrm{int}(G, i) = \frac{\sum_{\forall u \in G} r(u,i)}{|G|}$.
  - Relevance Score: $\mathrm{rel}(G, i, t) = \sum_{\forall u \in G} h(u, i, t)$.
  - Trending Score: Our novel metric analyzes interaction timestamps to identify "hype periods" and scores items based on whether group members engaged with them during these trends.
    $\mathrm{trend}(G, i) = \frac{1}{|G|} \sum_{\forall u \in G} \mathrm{trend}(u, i)$

  Furthermore, unlike the original approach, we combine these metrics into a weighted composite normalized score. This allows for adjusting each component's influence for a more balanced and configurable ranking. $\mathrm{composite}(G, i) = \sum_{k \in K} w_k \cdot \mathrm{score}_k(G, i)$ The final ranking is determined by sorting items by this composite score: $\mathrm{rank}_{\mathrm{items}} = \mathrm{argsort}_{\mathrm{desc}}\{\mathrm{composite}(G, i) : \forall i \in I\}$

- Rule-Based Explanations (EXPGRS): Based on the framework by Yera et al. [22], this method uses pre-computed association rules to explain recommendations. An explanation is generated if enough group members satisfy a rule's antecedent for the recommended item. The method proposes two types of rule explanations:
  - Type I: At least 'n' members of the group share the same preferences ($I = I_u$).
  - Type II: The group's combined preferences satisfy the rule ($I = \bigcup_{u \in G} I_u$).
- Local Explanations (LORE4GROUPS): Being an implementation of the LORE4GROUPS method by Yera et al. [21], this approach first generates local explanations for each member. A decision tree is built based on the neighborhood of similar items. It is used to extract a factual rule (r) justifying the recommendation and counterfactuals (ϕ) representing the minimal changes that would prevent it.

The group explanation is then created by merging the factual rules from all members into a single rule r_G, removing any contradictions. This combined rule is presented alongside the collection of all individual counterfactuals.

$$e(G) = <r_G, \phi_{u_1}, \phi_{u_2}, \ldots, \phi_{u_k}> \quad \text{where} \quad r_G = \bigcup_{u \in G} r_u$$

Evaluation. We adopt a comprehensive evaluation protocol following recent XAI research [5,13,15,23]. The `Evaluator` component implements metrics for both recommendation accuracy and explanation quality.

For the case of recommendation accuracy, we will use two main metrics [5]:

- Hit Ratio (HR): Fraction of users with at least one relevant item in top-N recommendations [5].
- Normalized Discounted Cumulative Gain (nDCG): Evaluates ranking quality, giving higher scores to relevant items appearing earlier in the list [5].

Additionally, three explanation quality metrics are considered in GREX:

- Model Fidelity (MF): The proportion of recommendations for which an explanation was successfully generated. This was applied to EXPGRS and LORE4GROUPS. [23]
- Gaussian Intra-List Distance and Dispersion (GILD): Measures the diversity of explanations within a group using the average kernel distance between them. Higher values indicate greater diversity. This was applied to EXPGRS and LORE4GROUPS. [13]
- Item Metrics: Refers to the individual component scores (e.g., Popularity, Rating, Trending) and the final Composite Value calculated for the Sliding Window method. [15]

4   Scenario of Use and Discussion

This section presents our framework and discusses it within practical scenarios of use. The primary goal is to understand the performance and inherent trade-offs of the different explanation methods when applied to group recommendations. The source code and our results could be explored in our GitHub repository[1].

4.1   Scenario of Use-Web Application

Here, a scenario of the use of GREX is presented, which is implemented through a web application that goes through the different components of the library.

Initially, Fig. 2 illustrates how the different data can be loaded, including the input of the rating data, the groups composed, and the binarization threshold for prediction models based on implicit feedback.

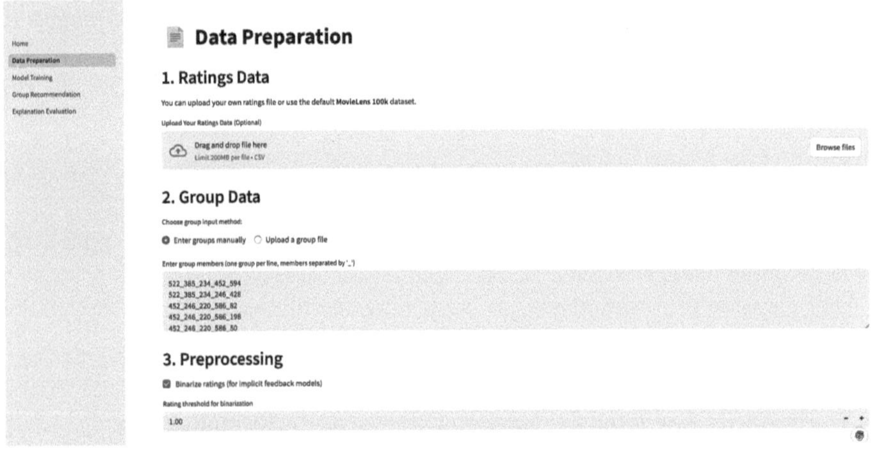

Fig. 2. Data preparation stage in GREX.

Subsequently, the model selection and training stage allows the choice of the model that will be used for the rating prediction (Fig. 3), among the several well-known approaches such as Alternative Least Squares, Bayesian Personalized Ranking, etc. [5]. For each approach, the values of each associated parameter could be also assigned. After training, the user can get their offline evaluation.

Once the individual rating prediction is configured, the next stage (Fig. 4) is centered on selecting the group recommendation aggregation strategy to use from those pointed out in Felfernig et al. [7], and the group for which the recommendation will be delivered. The size of the top n recommendation list can be also fixed as parameter.

---

[1] https://github.com/toledomateus/pygrex.

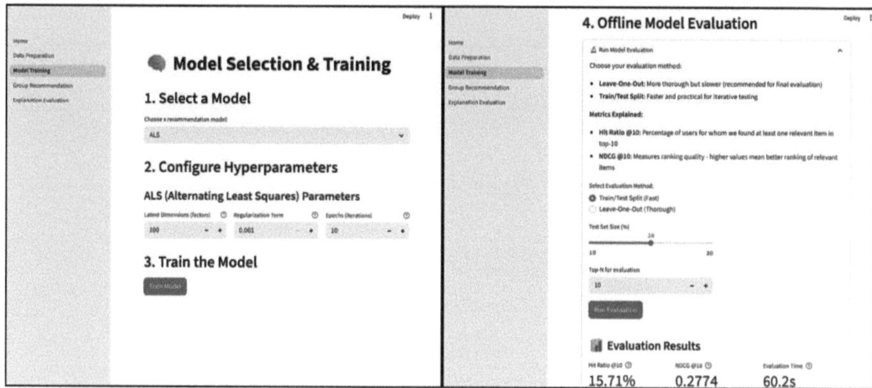

Fig. 3. Training stage in GREX (a). Getting the offline model evaluation (b).

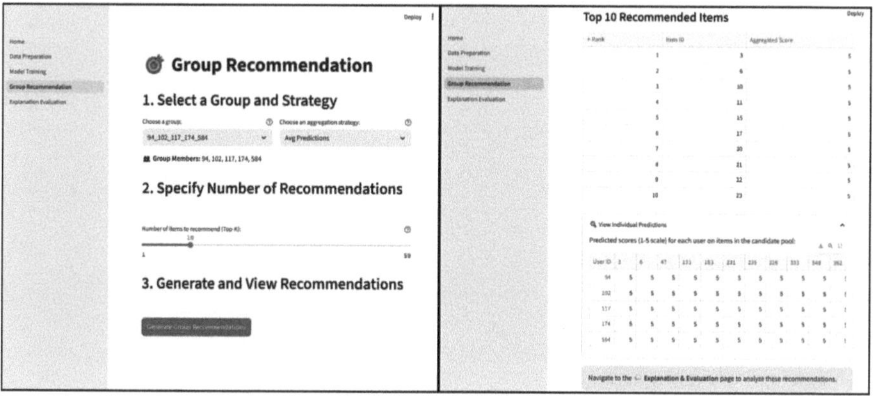

Fig. 4. Recommendation stage in GREX.

Finally, Fig. 5 illustrates the explanation stage, which settings depend on the used explanation approach. Particularly, this figure illustrates the interface associated to the approach Yera et al. [21]. In Fig. 6 is possible to visualize the explanation in 4 tabs: 1) Decision Tree, shows the complete local model generated for the item; 2) Decision Path, highlights the specific rules that led to the recommendation; 3) Alternatives, presents individual members counterfactuals explaining what changes would lead to a negative recommendation; and 4) Group Analysis, summarizes the group's consensus and individual members' concerns.

For the remaining two approaches Sliding Window [15] and EXPGRS [22], tailored interfaces are also developed.

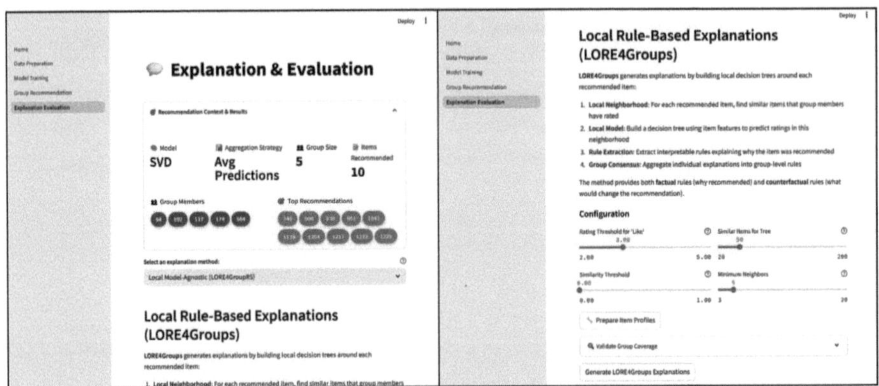

Fig. 5. Explanation stage in GREX.

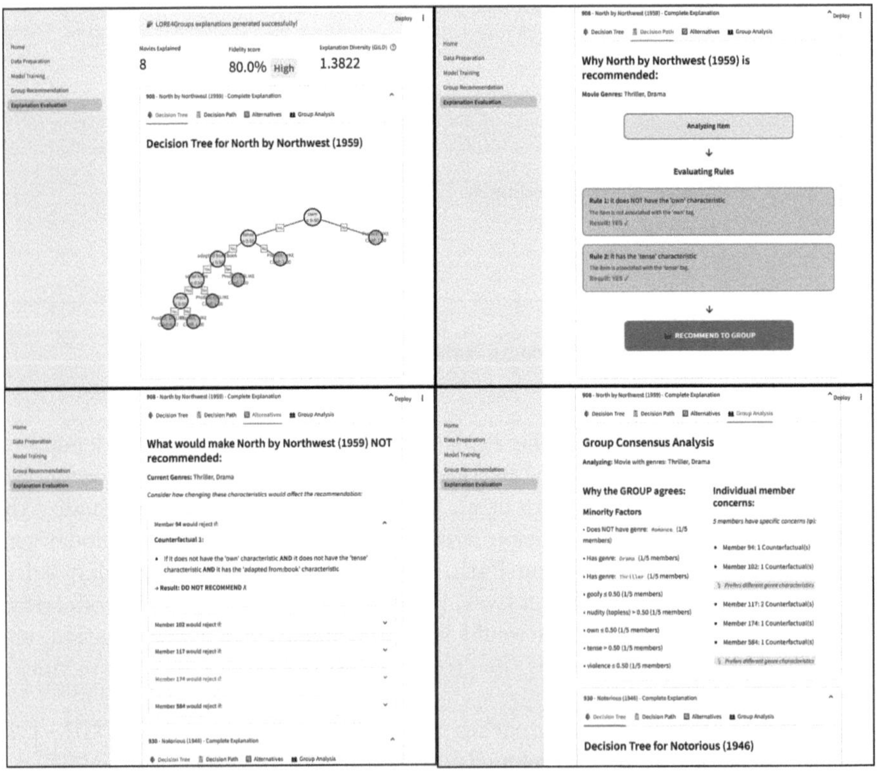

Fig. 6. Details of LORE4GROUPS explanation. (a) Decision Three, (b) Decision Path, (c) Alternatives, (d) Group consensus.

4.2   Scenario of Use - Scripting

In addition to web application, the library is designed for programmatic use through API calls, making it suitable for integration into production systems or for research purposes within environments like Jupyter notebooks. Installation is straightforward via PyPI ('pip install pygrex') or by building the artifact from the source repository.

Below we provide a minimal end-to-end workflow, from data loading to explanation. This demonstrates how the core components of the library can be scripted to perform the main tasks of the recommendation pipeline.

Firstly, it is necessary to perform all the data preparation step including, the binarization for implicit models, and prepare the dataset with consecutive IDs for model training.

```python
data_reader = DataReader(**cfg.data.ml1m)
data_reader.binarize(binary_threshold=1.0) # for implicit
    models
data_reader.make_consecutive_ids_in_dataset()
```

Secondly, we instantiate the model and train it using its fit method.

```python
model = ALS(**cfg.model.als)
model.fit(data_reader)
```

For the Group Recommendation step, we build the group handling and generate the group recommendation by its aggregation method.

```python
group_handler = GroupInteractionHandler(**cfg.data.groups
    )
recommender = GroupRecommender(data=data_reader)
recommender.setup_recommendation(
    model=model,
    members=group_members,
    data=data_reader,
    aggregation_strategy=AggregationStrategy.AVERAGE,
)
top_10_recommendations = recommender.
    get_group_recommendations(top_k=10)
```

Once the group recommendation is generated, the subsequent step is to get its explanation. Below, we provide an example of the Sliding Window approach, even so the other approaches introduced before can also be used.

```python
target_item_to_explain = top_10_recs[0]
items_rated_by_group = group_handler.
    get_rated_items_by_all_group_members(
    group=group_members, original_data=data_reader
)
sliding_window = SlidingWindow(sequence=
    items_rated_by_group, window_size=3)
explainer = SlidingWindowExplainer(
```

```
    cfg=cfg,
    data=data_reader,
    group_handler=group_handler,
    members=group_members,
    target_item=target_item_to_explain,
    candidate_items=top_10_recs,
    aggregation_strategy=aggregation_strat,
    sliding_window=sliding_window,
    model=model,
)
explanation = explainer.find_explanation()
```

A comprehensive Jupyter notebook demonstrating advanced usage, including the implementation of all available explanation methods, is available in the project's official repository[2].

4.3 Discussion and Implications from the Scenarios of Use

The GREX framework is built on a modular architecture, where components such as data handlers, prediction models, group aggregation methods, and explanation techniques can be operated and tested independently. This design allows researchers to interchange and evaluate different algorithms and datasets without rebuilding the entire system, facilitating experimentation across various group recommendation scenarios.

This inherent flexibility makes the framework a suitable platform for benchmarking explainable group recommendation approaches. It enables the systematic comparison of different strategies under controlled conditions. The evaluation can extend beyond standard accuracy or ranking metrics to include factors related to explainability, such as the clarity and diversity of the recommendations.

Furthermore, GREX is designed to support user studies that examine the human-computer interaction aspects of group recommendations. The framework offers flexibility for researchers with diverse technical backgrounds. From the developer viewpoint the researcher could extend and customize core components, while from the UI viewpoint, they can readily explore the implemented methods using their own datasets and use cases. This accessibility allows a broader range of studies into how different aggregation methods and explanation styles impact user satisfaction and acceptance, effectively bridging the gap between system performance and user experience.

5   Conclusions and Future Works

The current contribution is motivated by the lack of a platform for supporting explanations for group recommender systems. Inspired by the RecoX-

---

[2] https://github.com/toledomateus/pygrex.

plainer library [5] built for individual recommendations, the current contribution presents GREX, which is a library focused on covering the mentioned shortcoming.

The development of GREX opens up several avenues for both industrial and educational applications. In industry, the framework can be leveraged to build more transparent and trustworthy recommendation features for groups in domains such as media streaming, platforms for reviewing, travel planning, and e-commerce. In an educational context, GREX serves as a practical tool for students and researchers, allowing them to experiment with and understand the interplay between different datasets, models, aggregation strategies, and explanation methods in a controlled environment.

For future work, we have identified several promising directions and invite the open-source community to contribute. Key areas for expansion include: 1) we plan to leverage the framework's extensibility by integrating a wider variety of recommendation algorithms, particularly more advanced deep learning models, and evaluating a broader set of group preference aggregation strategies, and 2) we will explore the integration of other popular model-agnostic explanation methods, such as LIME and SHAP, adapting them to the group recommendation context to provide a more comprehensive comparison of the state-of-the-art.

Acknowledgments. Raciel Yera is supported by the European Union's Horizon Europe research and innovation program under the Marie Sklodowska-Curie Grant Agreement No. 101106164. This work is also supported by the Andalusian PAIDI 2020 Program under the project PROYEXCEL-00257, and by the Spanish Thematic Network on Recommender Systems (RED2022-134302-T).

References

1. Al-Hazwani, I., Aschwanden, O., Inel, O., Bernard, J., Boratto, L.: Prism: from individual preferences to group consensus through conversational ai-mediated and visual explanations. In: Proceedings of the Nineteenth ACM Conference on Recommender Systems, RecSys 2025. pp. 1343–1345. Association for Computing Machinery, New York (2025). https://doi.org/10.1145/3705328.3759340
2. Alvarado, O., Htun, N.N., Jin, Y., Verbert, K.: A systematic review of interaction design strategies for group recommendation systems. Proc. ACM Hum.-Comput. Interact. **6**(CSCW2), 1–51 (2022)
3. Barile, F., et al.: Evaluating explainable social choice-based aggregation strategies for group recommendation. User Model. User-Adapted Interact., 1–58 (2023)
4. Bobadilla, J., Ortega, F., Hernando, A., Gutiérrez, A.: Recommender systems survey. Knowl.-Based Syst. **46**, 109–132 (2013)
5. Coba, L., Confalonieri, R., Zanker, M.: Recoxplainer: a library for development and offline evaluation of explainable recommender systems. IEEE Comput. Intell. Mag. **17**, 46–58 (2022)
6. De Pessemier, T., Dooms, S., Martens, L.: Comparison of group recommendation algorithms. Multimedia Tools Appli. **72**, 2497–2541 (2014)
7. Felfernig, A., Boratto, L., Stettinger, M., Tkali, M.: Group Recommender Systems: An Introduction, 1st edn. Springer Publishing Company, Incorporated (2018)

8. Gantner, Z., Rendle, S., Freudenthaler, C., Schmidt-Thieme, L.: Mymedialite: a free recommender system library. In: Proceedings of the fifth ACM Conference on Recommender Systems, pp. 305–308 (2011)
9. Guo, G., Zhang, J., Sun, Z., Yorke-Smith, N.: LibRec: a java library for recommender systems. In: Posters, Demos, Late-breaking Results and Workshop Proceedings of the 23rd Conference on User Modelling, Adaptation and Personalization (UMAP) (2015)
10. Hug, N.: Surprise: a python library for recommender systems. J. Open Source Softw. **5**(52), 2174 (2020)
11. Márquez, J.O.Á., Ziegler, J.: Negotiation and reconciliation of preferences in a group recommender system. J. Inform. Process. **26**, 186–200 (2018)
12. Najafian, S., Tintarev, N.: Generating consensus explanations for group recommendations: an exploratory study. In: Adjunct Publication of the 26th Conference on User Modeling, Adaptation and Personalization, pp. 245–250 (2018)
13. Ohsaka, N., Togashi, R.: A critical reexamination of intra-list distance and dispersion. In: Proceedings of the 46th International ACM SIGIR Conference on Research and Development in Information Retrieval, SIGIR 2023, pp. 1619–1628. Association for Computing Machinery, New York (2023). 10.1145/3539618.3591623
14. Quijano-Sanchez, L., Sauer, C., Recio-Garcia, J.A., Diaz-Agudo, B.: Make it personal: a social explanation system applied to group recommendations. Expert Syst. Appl. **76**, 36–48 (2017)
15. Stratigi, M., Bikakis, N., Stefanidis, K.: Counterfactual explanations for group recommendations. In: Proceedings of the 27th International Workshop on Design, Optimization, Languages and Analytical Processing of Big Data (DOLAP 2025) (Feb 2025)
16. Tran, T.N.T., Atas, M., Felfernig, A., Le, V.M., Samer, R., Stettinger, M.: Towards social choice-based explanations in group recommender systems. In: Proceedings of the 27th ACM Conference on User Modeling, Adaptation and Personalization, pp. 13–21 (2019)
17. Wang, W., Zhang, G., Lu, J.: Hierarchy visualization for group recommender systems. IEEE Trans. Syst. Man Cybernet. Syst. **49**(6), 1152–1163 (2017)
18. Waterschoot, C., Tintarev, N., Barile, F.: The pitfalls of growing group complexity: Llms and social choice-based aggregation for group recommendations. In: Adjunct Proceedings of the 33rd ACM Conference on User Modeling, Adaptation and Personalization, pp. 322–330 (2025)
19. Waterschoot, C., Yera, R., Tintarev, N., Barile, F.: With friends like these, who needs explanations? evaluating user understanding of group recommendations. In: Proceedings of the 33rd ACM Conference on User Modeling, Adaptation and Personalization, pp. 253–262 (2025)
20. Yera, R., Alzahrani, A.A., Martínez, L.: Exploring post-hoc agnostic models for explainable cooking recipe recommendations. Knowl.-Based Syst. **251**, 109216 (2022)
21. Yera, R., Martínez, L.: LORE4GroupRS: explaining group recommendations supported by a local rule-based approach. In: Julian, V., et al. Intelligent Data Engineering and Automated Learning – IDEAL 2024. IDEAL 2024. LNCS, vol. 15347. Springer, Cham (2025). https://doi.org/10.1007/978-3-031-77738-7_25
22. Yera, R., Martinez, L.: Towards post-hoc explanation strategies in group recommender systems. In: Proceedings of the Spanish National Conference on Artificial Intelligence (CAEPIA 2024), pp. 621–623. AEPIA, La Coruña, Spain (Jun 2024)
23. Zhang, Y., Chen, X., et al.: Explainable recommendation: A survey and new perspectives. Foundat. Trends® Inform. Retrieval **14**(1), 1–101 (2020)

Feature-Importance Aware Deep Neural Network Model for Explainable Recommender Systems

Pragya Gupta[1], Aishwaryaprajna[2(✉)], Debashree Guha[3],
and Debjani Chakraborty[1]

[1] Department of Mathematics, Indian Institute of Technology Kharagpur, Kharagpur
721302, West Bengal, India
`g.pragya@kgpian.iitkgp.ac.in`, `debjani@maths.iitkgp.ac.in`
[2] Department of Computer Science, University of Exeter, Exeter, UK
`aishwaryaprajna@exeter.ac.uk`
[3] School of Medical Science and Technology, Indian Institute of Technology
Kharagpur, Kharagpur 721302, West Bengal, India
`debashree_smst@smst.iitkgp.ac.in`

Abstract. An interpretable decision-making process is vital for developing recommender systems. It should identify key items and relevant information to improve the recommendations. Deep learning models have shown promising performance, but simultaneously achieving explainability and efficiency is challenging. Existing explainable recommender systems interpret feature importance after training, which may not be able to determine crucial feature importance for enhancing performance. Additionally, the model's learning paradigm depends on a single-objective function that ignores the captured information of user-item interaction. This study aims to develop a feature importance aware deep neural network model for explainable recommender systems, which facilitates improved predictive performance by considering the feature importance within the training procedure and generates a recommendation list based on the user's preference. The proposed method minimizes the combination of model losses and introduces a penalty term that specifically targets and diminishes the detrimental impact of a particular feature on the solution's efficacy. This strategy ensures the determination of the optimal solution by satisfying the explainable conditions imposed during the training framework. Extensive experiments on publicly accessible FilmTrust and MovieLens-100K datasets show notable recommendation performance (The source code is available at: https://github.com/PS297/ExplainableRS.git).

Keywords: Recommender system · Shapley value · DNN · Stochastic gradient method

L. Martínez et al. (Eds.): IDEAL 2025, LNCS 16239, pp. 109–120, 2026.
https://doi.org/10.1007/978-3-032-10489-2_10

1   Introduction

In recent years, the exponential growth of digital data has opened the essential requirement for new users' personalized services to navigate the world. Recommender systems (RS) are widely applied solutions in a spectrum of applications that expedite the function of identifying items such as movies, music, and videos that align with users' interests and facilitate this with high precision. RS determines the most favored items for each user, and it can be associated with a machine learning problem where the model aims to learns the feature interactions using the pre-defined information for ratings prediction. Popular RS like the Amazon and Netflix provide the numeric rating [17], which defines the user preference. In general, RS collects and processes a high volume of user information to apprehend the user's favored items and improve the accuracy of recommendations. The development of transparent, interpretable RS models has evolved into a popular research topic [15]. Interpretable models can provide detailed explanations for their recommendations. For example, they can explain how the user's input affects the predicted outcome for non-interacted users. This emphasises the importance of aligning the interpretation with the model's predictions.

1.1   Related Work

Several methods have been introduced and benchmarked on publicly accessible datasets. In essence, the developed methods can be categorised into sequential and explainable recommendations.

Sequential Recommendations: Early development of recommender models such as collaborative filtering [9] follows a simple architecture design and enables to provide of decent performance with a certain degree of interpretation. Lack of sequential information restricts them to static user profiles despite having satisfactory performance. Due to that, sequential models [7] have been introduced, for instance, GRU4Rec [5], and have shown impressive recommendation performance. However, the improved performance of the developed models lacks transparency and explainability.

Explainable Recommendations: Compared to existing sequential recommendation models, explainable RS improves user trust by providing rationality for their recommendation while preserving the high predictive performance [15]. The interpretation of these models is presented by demonstrating the preference weight for the items that users have recommended or interacted with. Restricted Boltzmann Machines (RBMs) are employed as an explainable model to generate a selected recommendation list [1]. Jendal et al. [6] introduced a hypergraph-based model to apprehend the high-order interaction among users, items, and other features. The explainability mechanism is used to provide an explanation about the preferences of the user. The self-attention-based RS demonstrates prominence performance for the sequential recommendation [16]. Several studies claim that reducing the influence of relevant features does not critically influence the output distributions, which omits the interpretation of the model's decision-making process. These studies, as mentioned earlier, explainability relies on the weight preference demonstration provided to the relevant features after completing the training procedure. On the other hand, the existing methods are trained on a single-objective function in general mean squared error (MSE), resulting in a poor interpretation of the training process.

1.2   Motivation and Contribution

To analyse the behavioural characteristics of the model, the interpretation of its decisions is crucial for enhancing the recommendation performance. The model's decisions are significantly influenced by the predefined features for personalised recommendations, such as genres, release years, and tags, which demonstrate the user-item interaction. Furthermore, existing recommendation models often rely on a single objective function, which lacks explainability and is susceptible to the initialisation of weights during the training process. This approach does not yield fair decisions across different user groups. Identifying the most relevant features is a key objective to establish an explainable training procedure that utilises the predictive performance for the RS. Considering these issues, this study endeavours to develop a feature importance aware deep neural network model for explainable recommender systems, aiming to predict users' ratings. The proposed method considers the combinations of loss functions to determine the optimal solutions for the deep neural network (DNN) model, and incorporates explainability during the training procedure of the model. It is accomplished by introducing a penalty term with the combination of loss functions, which aims to maximize the marginal contribution of positive impacted features, simultaneously minimizing the influence of negative impact on the prediction, and acts as a regularizer. This strategy penalizes the negative Shapley values and allows the model an explainable training procedure, resulting in a more transparent decision approach. This transition can steer the learning procedure towards trustworthy and interpretable decisions. It overcomes the existing limitation by providing an explainable training procedure and determining the optimal solution for generating the user's preference lists. A comprehensive experimental study is performed over the publicly accessible FilmTrust [3] and MovieLens-100K [4] datasets for rating prediction and presents the advantages of the proposed method.

The remainder of this study is organized as follows. Section 2 presents a detailed methodology followed by the experimental framework provided in Sect. 3. Section 4 outlines the concluding remarks.

2   Methodology

This section briefly introduces the feature importance aware deep neural network model for explainable recommender systems, followed by a detailed optimization process. It is composed of three steps: (1) feature extraction and processing, (2) an embedded feature-wise linear modulation DNN model, and (3) an explainable stochastic gradient training process. The proposed method is designed for rating prediction tasks, and a diagrammatic illustration is shown in Fig. 1.

2.1 Feature Extraction and Processing

Consider that the system consists of M_u users, M_{it} items, and M_r total ratings. $m_r^{(i,j,t)}$ represents the rating of ith user for jth item at time t. The pre-defined features associated with the users, items, and movies are extracted in the first step. For users, the user ID and the demographic features, for instance, age, gender, and occupation, are considered. Simultaneously, the item ID and associated features of the item are extracted, for instance, title, release date, and genres, etc. The extracted features include numerical, categorical, and text types of data, which are segregated. The user ID and item ID can consist of a numerical and categorical data combination. Therefore, the user ID and item ID are converted to the contiguous integers. Additionally, the system may comprise text-type features, for instance, movie title and IMDb URL. For the text processing, a bag-of-words style vector strategy is used to map into a vector representation format [11]. All the relevant features are extracted using the above-mentioned processing steps, and a collection of data samples is acquired, say $\{X^{(k)}, Y^{(k)}\}_{k=1}^{M_r}$, where $X^{(k)} \in \mathbb{R}^{M_r \times D}$ is a D-dimensional feature vector. The acquired data samples are divided into train and test samples. Assume that the $\{X^{(k)}, Y^{(k)}\}_{k=1}^{m_1}$ is the notion of training dataset consisting m_1 number of training samples and $\{X^{(k)}, Y^{(k)}\}_{k=1}^{m_2}$ is the notion of the testing samples, where $M_r = m_1 + m_2$. $X^{(k)} = \{\tilde{x}_1^{(k)}, \tilde{x}_2^{(k)}, ..., \tilde{x}_{m_1}^{(k)}\}$ represents the kth training sample. These training samples are used as input for the proposed model. To compute the predictive rating, an embedded feature-wise linear modulation DNN module is designed, which follows the strategy of collaborative filtering to apprehend the user and item interaction.

2.2 Embedded Feature-Wise Linear Modulation (FiLM) DNN Model

This section introduces the proposed module for computing the predictive rating derived from known ratings. The training dataset $\{X^{(k)}, Y^{(k)}\}_{k=1}^{m_1}$ includes user ID and item ID, which defines the semantic relationship among users and items. Consequently, learnable embedding layers are integrated into the neural network to represent user and item IDs in a vector representation, as illustrated in Fig. 1. Each user and item Ids are converted to a fixed-size embedding vector, say $\tilde{e}_{u_i} \in \mathbb{R}^{1 \times d_1}$, $\forall i \in \{1, 2, ..., M_u\}$ and $\tilde{e}_{it_j} \in \mathbb{R}^{1 \times d_2}$, $\forall j \in \{1, 2, ..., M_{it}\}$. The collection of embedding vectors of user and item IDs is represented as $\tilde{E}_u \in \mathbb{R}^{M_u \times d_1}$ and $\tilde{E}_{it} \in \mathbb{R}^{M_{it} \times d_2}$, respectively. This approach follows a similar structure to natural language processing, where each word in text is mapped to a fixed-size vector [11]. The embedded features are concatenated and defined in the following way,

$$\tilde{E}_{uit} = \tilde{E}_u \oplus \tilde{E}_{it} \tag{1}$$

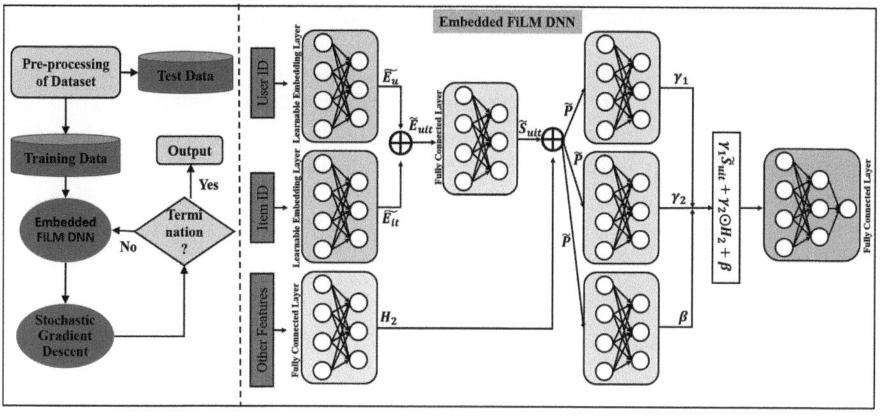

Fig. 1. Diagrammatic illustration of the proposed model.

where each vector of $\tilde{E}_{uit}$ is a $d_1 + d_2$-dimensional vector and $\oplus$ represents the concatenation operation. Simultaneously, the other relevant features $S \in \mathbb{R}^{m_1 \times D-2}$ are fed into the NN model to capture the hidden complex patterns from the side information of the training data and acquire H_2 outcome. Here, $D - 2$ represents the relevant feature dimension other than user and item ID. In order to determine the dependency relying on user and item interaction based on features, the acquired outcome $\tilde{E}_{uit}$ and H_2 needs to be concatenated to determine a linear relationship among the feature information. Before the concatenation operation, $\tilde{E}_{uit}$ is fed to another embedded layer to match the dimension to the H_2 as shown in Fig. 1, and obtain $\tilde{S}_{uit}$. The embedded features $\tilde{S}_{uit}$ and H_2 are concatenated, defined as

$$\tilde{P} = \tilde{S}_{uit} \oplus H_2 \tag{2}$$

Incorporating the outcomes $\tilde{S}_{uit}$ of the item and user, followed by side feature information H_2, leads to bias towards the dominating information. Therefore, an extended form of the FiLM [12] method is integrated to provide the conditional control over the relevant features. It facilitates enhancing the features of $\tilde{P}$ that are more important to the user and item. This method learns scalings (γ_1, γ_2) and shifting (β) parameters using $\tilde{P}$ and aggregate the information in the following way,

$$FiLM(\tilde{P}; \gamma_1, \gamma_2, \beta) = \gamma_1 \odot \tilde{S}_{uit} + \gamma_2 \odot H_2 + \beta \tag{3}$$

where, $\gamma_1, \gamma_2, \beta = f_{NN}(\tilde{P})$ are learning parameters. $\odot$ is the notion of element-wise dot product and f_{NN} is the notion of the NN model. The combined features $FiLM(\tilde{P}; \gamma_1, \gamma_2, \beta)$ are transferred to the fully connected NN model to compute the predictive rating, $\hat{Y}^{(k)}$ for kth sample as shown in Fig. 1. The ReLU activation function is employed during the training process to learn complex feature characteristics.

2.3 Explainable Stochastic Gradient Training Process

This section introduces an explainable stochastic gradient approach to determine the optimal solution by penalizing the objective space using an explainable regularizer term. As the number of features increases, capturing the beneficial information that can influence the recommendations becomes crucial. But the existing study [2,8] optimizes the model using the single objective function only, for instance, MSE or root mean squared error (RMSE), or mean absolute error (MAE), defined as

$$MSE = \frac{1}{m_1}\sum_{k=1}^{m_1}(Y^{(k)} - \hat{Y}^{(k)})^2, \quad RMSE = \sqrt{MSE} \tag{4}$$

$$MAE = \frac{1}{m_1}\sum_{k=1}^{m_1}|Y^{(k)} - \hat{Y}^{(k)}| \tag{5}$$

It is notable from Eqs. 4 and 5, the model considers all features irrespective of the nature of the direction of contribution in prediction. Inclusion of the marginal contribution of features facilitates an explainable training procedure. It determines the best set of weights in an optimal region and enhances the predictive performance by increasing the influence of potential features on the model. The SHAP mechanism is utilized to compute the features' importance [13], defined as

$$\psi_l^{(k)} = \sum_{B \subseteq X^{(k)}\setminus x_l^{(k)}} \frac{(D-2-|B|)!|B|!}{(D-2)!}\{O_{p^{(g)}}(B \cup x_l^{(k)}) - O_{p^{(g)}}(B)\}. \tag{6}$$

where $\psi_l^{(k)}$ represents the marginal contribution of lth feature ($l \in \{1, 2, ..., D-2\}$). $O_{p^{(g)}}$ represents the evaluation measure, which is assumed here as the model's prediction. Response of $\psi_l^{(k)}$ represents how much the lth feature is contributing to the model's prediction for the kth sample. The positive sign indicates that the particular feature contributes more to reducing the error between the actual and the predicted. However, a negative sign leads to a lower prediction corresponding to the actual or detrimental influence on the prediction. Therefore, imposing the penalty term over the model's loss can reduce the effect of negative impact of feature importance. With this view, the proposed embedded FiLM DNN model is optimized using the $\mathcal{RSO}$ loss function defined as,

$$\mathcal{RSO} = \alpha_1 RMSE + \alpha_2 MAE + \alpha_3 \left(\frac{\lambda}{m_1 C}\sum_{k=1}^{m_1}\sum_{l=1}^{D-2}\max(0, -\psi_l^{(k)})\right) \tag{7}$$

where $\alpha_1, \alpha_2, \alpha_3 \in [0,1]$ are the weighting parameter to adjust the weights of the $\mathcal{RSO}$ loss function. $\lambda(> 0)$ is a hyperparameter for the penalty term. C represents the number of negative Shapley values. In Eq. 7, the last term functions as a penalty over the objective functions to control the behavior of the negative

Shapley value. Whenever any feature provides a negative marginal contribution towards the prediction, the objective functions are penalized as presented in Eq. 7. Here, the stochastic gradient descent optimization algorithm is employed, which randomly selects the mini-batches across each epoch and minimizes the $\mathcal{RSO}$ loss function to acquire an explainable training procedure. This strategy facilitates the acquisition of an understandable training procedure with an optimal set of parameters for RS. The empirical study of the proposed model is given in the following section.

3   Experimental Framework

This section outlines the experimental validation of the proposed method performed over the publicly accessible datasets with a detailed comparative analysis.

3.1   Datasets

For the experimental validation of the proposed method, two publicly accessible FilmTrust [3] and MovieLens-100K [4] datasets are selected (refer to Table 1). FilmTrust is a collection of datasets for a movie recommender comprising 35,497 ratings acquired from a well-known movie sharing and ratings website. The range of the ratings is on a scale of 1 to 5. MovieLens-100K is a popular benchmark dataset that is highly sparse and used for research purposes. This dataset comprises tuples of the form '<user, item, rating, timestamp>' as shown in Fig. 2, which establish ratings assigned by users for movies at a specific timestamp on a scale of 1 to 5. These ratings are acquired from the MovieLens website, which facilitates personalized recommendations using the user's ratings. The MovieLens-100K dataset consists of 100,836 ratings of 610 users for 9742 items. This dataset includes side feature information of items such as movie genres, title, user, and tags.

X MovieLens 100K Ratings Viewer with Genres				— □ X
User ID	**Movie Title**	**Genres**	**Rating**	**Timestamp**
498	Casino (1995)	Drama	★★★	881957625
642	Pocahontas (1995)	Animation, Children, Musical, Rom	★★★★★	885606609
58	2001: A Space Odyssey (1968)	Drama, Mystery, Sci-Fi, Thriller	★★★★	884305150
495	Cat People (1982)	Horror	★★★	886635995
618	Philadelphia (1993)	Drama	★★★	891306571
725	Air Force One (1997)	Action, Thriller	★★★★	876106729
794	James and the Giant Peach (1996)	Animation, Children, Musical	★★★★	891036222
678	Chungking Express (1994)	Drama, Mystery, Romance	★	879544915
43	Top Gun (1986)	Action, Romance	★★★★	883955467
752	Fallen (1998)	Action, Mystery, Thriller	★★★★	891208357

Fig. 2. Visualization of MovieLens-100K data information.

Table 1. Description of the datasets utilized to perform the experimental study using the proposed method.

Dataset	# Users	# Items	# Ratings	Scale	# Hidden Neurons in Layers	# Epochs
FilmTrust	1508	2071	35,497	1–5	[20,25,10]	200
MovieLens-100K	610	9742	100,836	1–5	[30,35,15]	200

3.2 Baselines and Implementation Details

The comparative analysis is performed with the following methods to validate the superiority of the proposed method over the considered dataset presented in Table 1.

– *Singular Value Decomposition (SVD):* It is a classical matrix factorization technique used in RS to improve users' ratings [14].
– *DP-DAE:* This model is designed for differentially private RS by proposing a dual semi-autoencoder recommender technique [2].
– *RBM-Q-Learning:* This work [18] presents RBM-Q-learning1 and RBM-Q-learning2 algorithms to improve representation and selection process.
– *LAMAR:* This model combines the advantageous properties of LLMs and conventional RS to improve collaborative and semantic information [10].

From the FilmTrust dataset, the userID, itemID, aggregated trust of the trustor, aggregated trust of the trustee information, and interaction count of the userID associated with the itemID are taken as feature information. Subsequently, userID, itemID, title, genres, and user demographic information are considered from the MovieLens-100K dataset. For computing the Shapley value, userID and itemID are not taken into account. The comprehensive experimental study is performed on a Linux desktop Intel Xeon Gold 6326 processor, NVIDIA RTX A5000 GPUs, with a programming environment in Jupyter Notebook. The PyTorch library is used to train the proposed model. The model runs for 200 epochs, and a detailed description of hyperparameters is presented in Tables 1 and 2. The model is optimized using the Adam optimizer with the presented loss function with a learning rate of 0.005. The embedding dimension is chosen to be 25, with the batch size for the dataset being 128. RMSE and MAE evaluation metrics are utilized for the empirical study.

To acquire the best combination of parameters $\lambda, \alpha_1, \alpha_2$, and α_3, the training of the proposed model is conducted by taking different parameter settings presented in Table 2. The parameter settings corresponding to the lowest RMSE and MAE values are taken into consideration. For the FilmTrust dataset $(\alpha_1, \alpha_2, \alpha_3, \lambda) = (0.25, 0.45, 0.30, 5.0)$ is chosen as shown in Table 2. Simultaneously, $(\alpha_1, \alpha_2, \alpha_3, \lambda) = (1/3, 1/3, 1/3, 1.0)$ is for MovieLens-100K dataset.

3.3 Experimental Results

The experimental results are reported in Tables 3 and 4 for the FilmTrust and MovieLens-100K datasets, respectively. We have performed analysis by segre-

Table 2. Tuning mechanism of hyperparameters for FilmTrust and MovieLens-100K datasets when volume of training data = 80%.

FilmTrust						MovieLens-100K					
α_1	α_2	α_3	λ	RMSE ↓	MAE ↓	α_1	α_2	α_3	λ	RMSE ↓	MAE ↓
1/3	1/3	1/3	0.5	0.8020	0.6022	1/3	1/3	1/3	0.5	0.9352	0.7320
1/3	1/3	1/3	1.0	0.8046	0.6057	1/3	1/3	1/3	1.0	**0.9328**	**0.7253**
0.35	0.25	0.40	2.0	0.8014	0.6019	0.35	0.25	0.40	2.0	0.9345	0.7271
0.25	0.45	0.30	5.0	**0.7973**	**0.6005**	0.25	0.45	0.30	5.0	0.9532	0.7312

gating the training data into different sizes as presented in Tables 3 and 4. The visualization $\mathcal{RSO}$ training loss, RMSE, MAE, and penalty induced from negative Shapley value across the epochs are shown in Figs. 3 and 4 for FilmTrust and MovieLens-100K datasets, respectively. Notably, at the initial stage, the penalty imposed over the objective functions is higher, whereas after 50 epochs, it decreases rapidly for the FilmTrust dataset. The impact of the adverse feature is lower on the prediction, leading to an improvement in the performance. For the MovieLens-100K dataset, the initial penalty is higher and decreases. Although there is a comprehensive reduction in penalty, oscillations indicate a trade-off between the loss and maintaining fairness, as MovieLens-100K is a complex dataset. Figure 5 represents the top-10 recommended movies to user 877 with predicted rating through the proposed method and actual rating. The deviation between the actual and predicted rating is lower, as shown in Fig. 5.

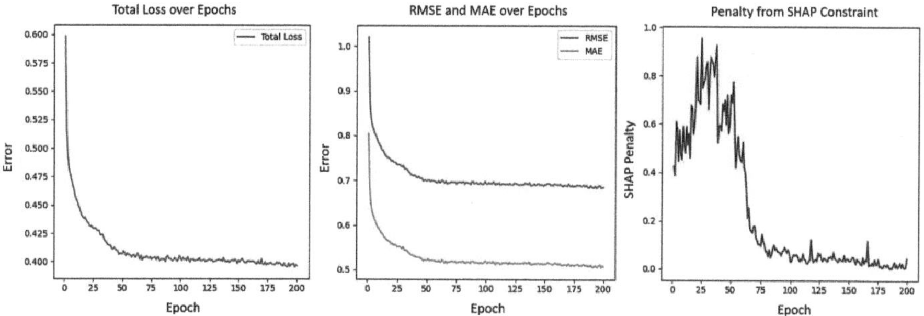

Fig. 3. Visualization of RMSE and MAE metrics of FilmTrust datasets, followed by penalty while training the model using an explainable approach.

A quantitative analysis is presented in Tables 3 and 4, followed by a comparison with existing methods as mentioned in Sect. 3.2. The proposed method attains overall adequate performance compared to the current methods over the FilmTrust and MovieLens-100K datasets with varying training data sizes, as shown in Tables 3 and 4. This implies that the proposed method can utilise

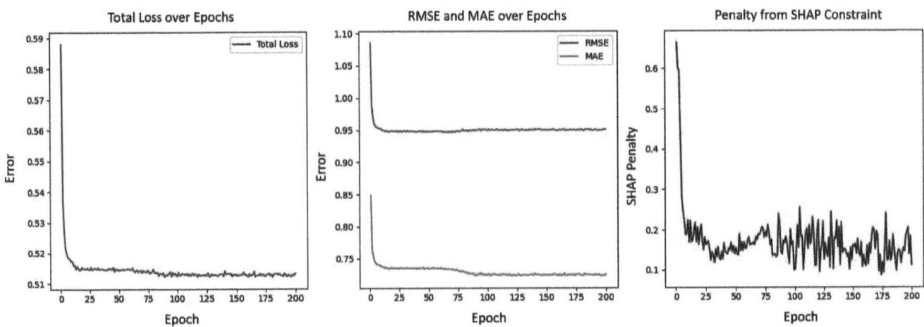

Fig. 4. Visualization of RMSE and MAE metrics of MovieLens-100K datasets, followed by penalty while training the model using an explainable approach.

User ID	Movie Title	Actual Rating	Predicted Rating
877	Cinema Paradiso (1988)	5.0	4.26
877	Farewell My Concubine (1993)	4.0	4.22
877	Pulp Fiction (1994)	5.0	4.14
877	Madness of King George, The (1994)	4.0	4.05
877	Four Weddings and a Funeral (1994)	5.0	4.03
877	Muriel's Wedding (1994)	4.0	3.98
877	Game, The (1997)	4.0	3.97
877	Last of the Mohicans, The (1992)	4.0	3.90
877	Farinelli: il castrato (1994)	4.0	3.86
877	Gattaca (1997)	4.0	3.85

Fig. 5. Visualization of top-10 recommendations, predicted, and actual rating of movies for the user from the validation set of MovieLens-100K dataset.

Table 3. Evaluation metrics for the proposed method compared with the existing methods over the FilmTrust test dataset.

Evaluation Metrics	Volume of training data	SVD	DP-DAE [2]	Ours
RMSE ↓	70%	0.8145	0.8203	**0.8049**
	80%	0.8085	0.8152	**0.7973**
	90%	**0.7913**	0.8130	<u>0.7942</u>
MAE ↓	70%	0.6293	0.6313	**0.6081**
	80%	0.6212	0.6247	**0.6005**
	90%	0.6022	0.6219	**0.5913**

predictive capabilities by incorporating explainability into the training process to find the optimal solution.

Table 4. Evaluation metrics for the proposed method compared with the existing methods over the MovieLens-100K test dataset.

Evaluation Metrics	Volume of training data	SVD	RBM-Q-learning2 [18]	LAMAR [10]	Ours
RMSE ↓	70%	0.9416	–	–	**0.9407**
	80%	0.9334	1.190	1.0778	**0.9328**
	90%	0.9378	–	–	**0.9335**
MAE ↓	70%	0.7431	–	–	**0.7288**
	80%	0.7349	0.9930	0.8399	**0.7253**
	90%	0.7391	–	–	**0.7283**

4   Conclusion

In this study, we designed a feature importance-aware deep neural network model for explainable recommender systems, an interpretable RS for rating prediction. Our primary motivation is to capture the feature importance of the item and user-item interaction that influences the recommendation performance. We therefore introduce an explainable training procedure to improve the embedded FiLM DNN model performance for computing the optimal solution by introducing a Shapley value-based penalty term over the objective functions, which controls the negative impact on the model's prediction. This strategy allows the model to capture the user-item interaction using side feature information and enhances the positive effects of features towards evaluating the predictive probability. Experimental results demonstrate the transparency and superiority of the proposed method over the publicly accessible FilmTrust and MovieLens-100K datasets and validate in handling of larger real-life datasets.

References

1. Abdollahi, B., Nasraoui, O.: Explainable restricted boltzmann machines for collaborative filtering. arXiv preprint arXiv:1606.07129 (2016)
2. Deng, Y., Zhou, W., Haq, A.U., Ahmad, S., Tabassum, A.: Differentially private recommender framework with dual semi-autoencoder. Expert Syst. Appli. **260**, 125447 (2025)
3. Guo, G., Zhang, J., Yorke-Smith, N.: A novel Bayesian similarity measure for recommender systems. In: Proceedings of the 23rd International Joint Conference on Artificial Intelligence (IJCAI), pp. 2619–2625 (2013)
4. Harper, F.M., Konstan, J.A.: The movielens datasets: history and context. Acm Trans. Interact. Intell. Syst. (TIIS) **5**(4), 1–19 (2015)
5. Hidasi, B., Karatzoglou, A.: Recurrent neural networks with top-k gains for session-based recommendations. In: Proceedings of the 27th ACM International Conference on Information and Knowledge Management, pp. 843–852 (2018)
6. Jendal, T.E., Le, TH., Lauw, H.W., Lissandrini, M., Dolog, P., Hose, K.: Hypergraphs with attention on reviews for explainable recommendation. In: Goharian, N., et al. (ed.) Advances in Information Retrieval. ECIR 2024. LNCS, vol. 14608. Springer, Cham (2024). https://doi.org/10.1007/978-3-031-56027-9_14

7. Khan, Z., Khan, Z., Iltaf, N.: ConvSeq-MF: convo-sequential matrix factorization for recommender system. Neurocomputing **618**, 128932 (2025)
8. Kiran, R., Kumar, P., Bhasker, B.: DNNRec: a novel deep learning based hybrid recommender system. Expert Syst. Appli. **144**, 113054 (2020)
9. Koren, Y.: Collaborative filtering with temporal dynamics. In: Proceedings of the 15th ACM SIGKDD International Conference on Knowledge Discovery and Data Mining, pp. 447–456 (2009)
10. Luo, S., Wang, J., Zhou, A., Ma, L., Song, L.: Large language models augmented rating prediction in recommender system. In: ICASSP 2024-2024 IEEE International Conference on Acoustics, Speech and Signal Processing (ICASSP), pp. 7960–7964. IEEE (2024)
11. Mikolov, T., Chen, K., Corrado, G., Dean, J.: Efficient estimation of word representations in vector space. arXiv preprint arXiv:1301.3781 (2013)
12. Perez, E., Strub, F., De Vries, H., Dumoulin, V., Courville, A.: Film: Visual reasoning with a general conditioning layer. In: Proceedings of the AAAI Conference on Artificial Intelligence, vol. 32 (2018)
13. Shapley, L.S.: A value for n-person games. In: Kuhn, H.W. (eds.) Classics in Game Theory, pp. 69–79. Princeton University Press (1997). Originally published in *Contributions to the Theory of Games II* (1953), pp. 307–317
14. Sharifi, Z., Rezghi, M., Nasiri, M.: New algorithm for recommender systems based on singular value decomposition method. In: ICCKE 2013, pp. 86–91. IEEE (2013)
15. Shulman, E., Wolf, L.: Meta decision trees for explainable recommendation systems. In: Proceedings of the AAAI/ACM Conference on AI, Ethics, and Society, pp. 365–371 (2020)
16. Singh, N.K., et al.: Self-attention mechanism-based federated learning model for cross context recommendation system. IEEE Trans. Consumer Electr. **70**(1), 2687–2695 (2023)
17. Smith, B., Linden, G.: Two decades of recommender systems at amazon. com. IEEE Internet Comput. **21**(3), 12–18 (2017)
18. Yelği, A.: Q-learning improved RBM for rate prediction in recommendation systems. Soft Comput., 1–13 (2025)

Special Session on Artificial Intelligence Applied to Image Analysis

Severity Classification of Lumbar Spine Degeneration Using 3D Voxel Grids and Vision Transformers

Le Dinh Huynh[1,2], Truong Cong Doan[2], and Phan Duy Hung[1(✉)] (iD)

[1] FPT University, Hanoi, Vietnam
{huynhld3,hungpd2}@fe.edu.vn, 23078026@vnu.edu.vn
[2] International School, Vietnam National University, Hanoi, Vietnam
tcdoan@vnu.edu.vn

Abstract. Lumbar degenerative spine disease is a prevalent age-related condition, particularly affecting older adults. It is primarily manifested through persistent lower back pain and progressive muscular stiffness, often leading to severe functional impairment and restricted mobility. A comprehensive pipeline was designed, beginning with the transformation of MRI-based DICOM images from the RSNA 2024 Lumbar Spine Degenerative Classification dataset into 3D voxel representations to enhance anatomical structure preservation. A Vision Transformer architecture was employed to model spatial dependencies and capture global contextual features more effectively than conventional convolutional approaches. The model was trained to classify degeneration severity across multiple categories relevant to clinical assessment. Experimental evaluation demonstrated strong predictive capability, with the model attaining a training accuracy of 98.7% (F1 score: 0.981) and maintaining robust generalization on the test set with an accuracy of 96.7% (F1 score: 0.970), indicating strong potential for real-world clinical application in supporting diagnostic decisions and improving workflow efficiency.

Keywords: Machine Learning · Degenerative Disease Diagnosis · 3D Voxel Representation · Vision Transformer

1 Introduction

The global burden of lumbar degenerative spine disease (DSD) and low back pain (LBP) continues to escalate. As of 2021, approximately 628.8 million individuals suffer from LBP worldwide, with 266.9 million new incident cases annually [1, 2]. Epidemiological reports indicate especially high prevalence in Southeast Asia (about 69 million cases) [2], and projections estimate a rise to 843 million prevalent cases by 2050, up from 619 million in 2020 [3, 4]. In Vietnam, a community-based radiographic study involving lumbar spine degeneration affects up to 89% of individuals aged 60–69. The incidence rate among people aged 25–45 is also significant, reaching 30%. On average, spine degeneration in general accounts for approximately 35% of the Vietnamese population [5].

© The Author(s), under exclusive license to Springer Nature Switzerland AG 2026
L. Martínez et al. (Eds.): IDEAL 2025, LNCS 16239, pp. 123–134, 2026.
https://doi.org/10.1007/978-3-032-10489-2_11

Spinal degeneration commonly presents chronic back pain, radiating leg pain, nausea, and vomiting. These symptoms are often accompanied by functional impairment, significantly restricting mobility and daily activities. As the disease progresses, neurological deficits such as muscle weakness and poor coordination may develop, further worsening the condition and diminishing quality of life. Patients often endure persistent pain, rigorous rehabilitation, and emotional distress [6]. Diagnosis relies on clinical evaluation, conventional radiography, and advanced imaging techniques such as MRI and CT. However, accurate classification and localization of degenerative lesions remain challenging due to overlapping clinical presentations and variability in imaging findings. Figure 1 illustrates MRI scans of the lumbar spine: the left image shows a normal sagittal T1-weighted view, the right displays a degenerative spine, and the center provides an axial T2-weighted image for additional perspective [7].

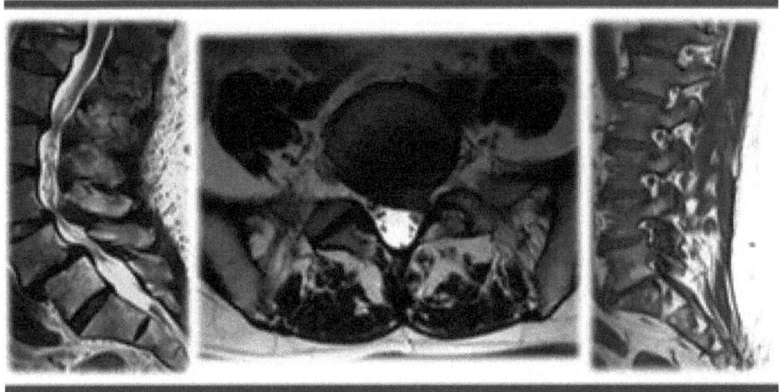

Fig. 1. MRI Scan of Lumbar Spine Degenerative Disorders [7]

In response to the diagnostic challenges associated with lumbar spine degeneration, recent studies have explored the application of advanced computational approaches, including computer-aided diagnosis (CAD), deep learning, and artificial intelligence. Convolutional neural networks (CNNs) have demonstrated high efficacy in medical imaging tasks such as classification, segmentation, and abnormality detection [8, 9]. More recently, Vision Transformers (ViT) have emerged as a powerful alternative due to their ability to model long-range dependencies via self-attention mechanisms [10]. This is particularly beneficial for capturing subtle radiographic features often indicative of early or complex spinal degenerative conditions.

Nonetheless, MRI-based classification of lumbar spine disorders (LSDs) remains challenging due to image variability, feature similarity among disease stages, and class imbalance. ViTs, with their flexible attention-based structure, offer potential for improving classification robustness and clinical interpretability. The objective of this study is to classify the severity of lumbar spine degeneration using 3D voxel grids and Vision Transformers, aiming to address the limitations of 2D CNN-based methods and improve clinical interpretability. DICOM images are transformed into 3D voxel grids to facilitate effective spatial learning.

The remainder of this paper is organized as follows: Sect. 2 reviews related work in spine degeneration imaging and deep learning applications; Sect. 3 outlines the proposed ViT-based classification method; Sect. 4 presents experimental results and evaluation; and Sect. 5 concludes with discussions and future directions.

2 Related Works

Early automated approaches to spinal analysis used Support Vector Machines (SVMs) and other pattern-recognition techniques [11, 12], which showed the feasibility of computer-assisted diagnosis but were constrained by handcrafted features. Classical methods such as k-Nearest Neighbors and Random Forests achieved only moderate results [13, 14]. To improve accuracy, Beulah et al. [15] introduced hybrid feature extraction for lumbar MR images, while Chiu et al. [16] demonstrated the use of machine learning for prognosis after nucleoplasty. Deep learning, particularly convolutional neural networks (CNNs), marked a breakthrough by enabling hierarchical feature extraction and delivering strong performance in disc grading and stenosis detection [17, 18]. However, CNNs require large annotated datasets, high computational resources, and remain limited in modeling global anatomical context [19]. Vision Transformers (ViTs) address these challenges through self-attention and long-range dependency modeling, but their application to lumbar spine degeneration remains relatively scarce, with only a few recent studies reporting initial results [31].

In medical image classification, deep learning, particularly CNNs has emerged as a state-of-the-art approach. These models excel at processing visual data and often outperform traditional methods in terms of precision and accuracy. A key advantage of CNNs lies in their ability to automatically extract high-level features from raw image inputs, making them well-suited for spinal disease diagnosis. Applications include intervertebral disc grading and the detection of specific pathologies such as foraminal stenosis and spinal canal narrowing [17, 18]. However, CNNs also present certain limitations. They rely heavily on large, annotated datasets, which are scarce in medical imaging due to the high cost and effort of expert labeling. Additionally, CNNs may struggle to capture global context, as they lack mechanisms for long-range dependency modeling. Furthermore, training these models often demands substantial computational resources, including GPUs or TPUs [19].

Viral H. Borisagar et al. [7] proposed a CNN-based approach for diagnosing various spinal abnormalities, demonstrating high diagnostic accuracy in cases of degenerative pathology. Similarly, Abuhayi et al. [20] introduced the Involutional Neural Network based on VGG architecture (INVGG), which integrates involution and convolution layers with shared weights to classify LSD. This hybrid model enhances diagnostic performance by leveraging the strengths of both CNNs and the VGG framework, offering potential for future clinical applications. While CNNs and other traditional deep learning models have been widely explored, the application of ViTs to LSD classification remains relatively underexplored.

In the context of lumbar spine degeneration classification, the attention mechanism inherent in Vision Transformers offers a significant advantage. Unlike conventional 2D CNNs that rely on local receptive fields, ViTs can model long-range dependencies and global contextual relationships within 3D voxelized data [21]. This is particularly valuable in medical imaging, where subtle anatomical variations, such as changes in disc height or vertebral alignment, are critical for accurate diagnosis. When applied to 3D voxel grids, ViTs transform each voxel into query, key, and value vectors to compute attention scores. These scores determine the relevance of each voxel, allowing the model to selectively focus on important regions while suppressing irrelevant features. This selective attention enhances the model's discriminative ability to detect complex and subtle pathological patterns characteristic of LSD.

Self-attention mechanisms have been increasingly integrated into medical imaging applications [22]. For instance, Vakanski et al. [23] incorporated attention modules into a U-Net architecture to improve breast tumor segmentation in ultrasound images by enabling the network to focus on semantically important spatial regions. Similarly, Sangjoon Park et al. [24] applied ViT architectures within a federated learning framework for COVID-19 diagnosis using chest CT scans. Their work highlighted the viability of ViTs for secure, collaborative learning across institutions without compromising patient data. These developments suggest that ViTs, with their capacity for contextual learning and feature prioritization, are well-suited for LSD classification. By enabling the model to attend to diagnostically relevant features in a 3D space, such as disc bulging, narrowing, or vertebral misalignment, ViTs hold promise for improving the accuracy and interpretability of automated diagnostic systems in spinal radiography [21].

3 Methodology

To ensure robust performance evaluation, we adopted k-fold cross-validation, a widely accepted method in machine learning. The dataset was divided into k subsets (folds), where in each iteration one fold was used for validation and the remaining $k - 1$ folds for training. This process was repeated k times, allowing each subset to serve once as the validation set. Final performance metrics were computed by averaging the results across all folds, thereby reducing variance and improving reliability. The dataset used in this study was extracted from DICOM-format medical images. DICOM files contain not only grayscale image data but also metadata, such as patient demographics, spatial resolution, slice thickness, and orientation, which are unsuitable for direct input into Vision Transformer models. To prepare the data for ViT processing, a two-step preprocessing pipeline was applied: (1) conversion of DICOM images into 3D point clouds, and (2) construction of voxel grids from these point clouds [25].

3.1 Convert DICOM to 3D Point Cloud

DICOM slices were transformed into a set of 3D spatial coordinates, collectively forming a point cloud representation. This conversion involved extracting pixel values from each 2D slice, where only pixels with intensities greater than zero were retained for processing. These pixels were then projected into 3D space using key metadata attributes embedded in the DICOM files, including PixelSpacing, SliceThickness, and ImagePositionPatient. These parameters ensured accurate spatial scaling and alignment of the reconstructed anatomical structures. The resulting visualization is illustrated in Fig. 2, which showcases the 3D point cloud using 30%, 60%, and 90% retained sample points extracted from DICOM [26].

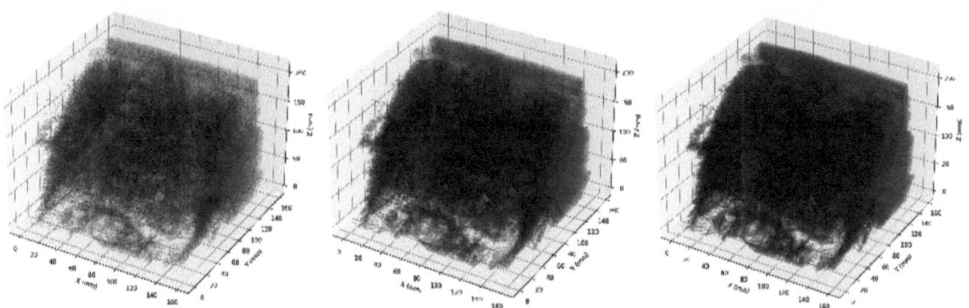

Fig. 2. Example of 3D point cloud derived from DICOM images, shown with different proportions of retained points (30%, 60%, 90%) to illustrate the effect of sampling density.

To enhance visualization and distinguish between different MRI modalities, a color-coding scheme was applied based on the MRI sequence type: T1-weighted points were colored red (1, 0, 0), T2-weighted points green (0, 1, 0), and T2/STIR-weighted points blue (0, 0, 1). This visual encoding facilitated differentiation of imaging techniques within a unified 3D model. The resulting point cloud provided a continuous and anatomically faithful 3D reconstruction of the imaged region, enabling improved spatial interpretation and analysis. This method not only preserved critical volumetric information but also offered a robust basis for downstream tasks such as pathology identification or preoperative planning. The integration of computational methods with clinical imaging workflows underscores the potential for enhanced diagnostic accuracy and treatment planning.

3.2 Create Voxel Grids from 3D Point Cloud

For ViT processing, DICOM-derived 3D point clouds are transformed into voxel grids. This process begins with normalizing the point cloud using its bounding box, scaling all coordinates into a fixed-size voxel space, typically of dimensions (128, 128, 128) [27]. This ensures consistent spatial representation regardless of the original anatomy's size or orientation. Voxel values are then determined by aggregating the RGB color information of the points within each voxel, where color encodes the MRI modality: red for T1-weighted, green for T2-weighted, and blue for STIR images. The final voxel grid is stored as a 4D NumPy array with shape (3, 128, 128, 128), with the first dimension corresponding to the R, G, B channels.

To ensure numerical stability and improve learning efficiency, the voxel intensities are normalized to the [0, 1] range [28]. For computational efficiency, especially in large-scale experiments, the preprocessed voxel grids are compressed and cached in npy.gz format, reducing runtime during model training. This transformation enables ViT to directly process volumetric representations of anatomical structures, bridging raw imaging data with deep learning-based spatial analysis. Figure 3 provides a visual representation of the voxel grid converted from 3D point cloud data, highlighting the spatial encoding of anatomical features.

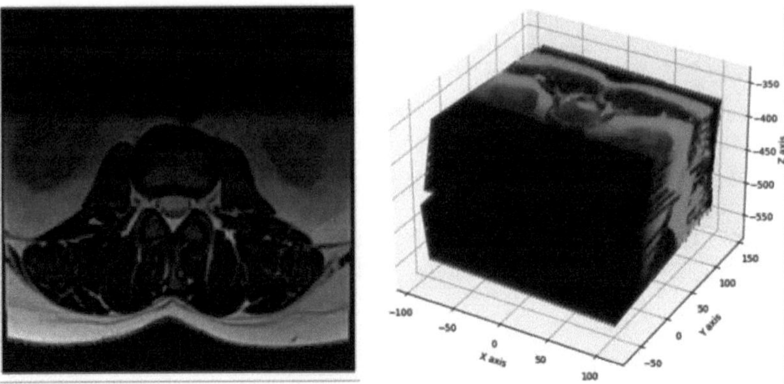

Fig. 3. Visualization of Data Conversion from DICOM to Voxel Grid. Left: Original DICOM slice; right: Resulting 3D voxel representation.

3.3 Modeling

A Vision Transformer - based architecture is adapted to process 3D voxel grids derived from medical imaging data. While the original ViT is designed for 2D image classification, several modifications are applied to support 3D volumetric data and address the multi-class classification of lumbar spine degeneration. The main components adjusted include:

- Input and Patch Embedding: The input is a 3D RGB volume $X_{input} \in R^{B \times C \times D \times H \times W}$ where B is the batch size and D, H, W are the spatial dimensions. A 3D convolutional layer with kernel size and stride P is employed to divide the volume into non-overlapping patches of size $P \times P \times P$. Each patch is embedded into a feature vector of dimension E, resulting in a sequence of embeddings suitable for transformer input.
- Positional Encoding and Class Token: To incorporate spatial information, learnable positional encodings are added to the patch embeddings. A learnable class token is also prepended to the sequence, enabling the model to aggregate global features during attention operations.
- Transformer Encoder: The sequence of embeddings, along with the class token and positional encodings, is processed through multiple standard transformer blocks. Each block consists of multi-head self-attention (MHSA) and a feed-forward network (FFN), both surrounded by residual connections and layer normalization. The depth of the encoder, number of attention heads, and hidden dimensions are configured according to the characteristics of the 3D input and available computational resources.
- Classification Head: The output corresponding to the class token is passed through a layer normalization step, followed by a linear layer to produce class probabilities. The final output includes three classes representing the severity levels: Normal/Mild, Moderate, and Severe.

4 Experiments and Results

4.1 Dataset

The experiments were conducted using the publicly available RSNA Lumbar Spine Degenerative Classification dataset [28], which contains approximately 2,000 magnetic resonance (MR) studies. All image files are stored in the DICOM format, which is widely used in medical imaging due to its ability to preserve full volumetric information and rich metadata. This includes acquisition parameters, modality details, and patient-related information - features that support both diagnostic interpretation and automated processing. The dataset includes expert annotations provided by spine radiologists, indicating the vertebral level and location of any observed lumbar spine stenosis. MR imaging of the spine is typically acquired in two main planes: the axial plane (cross-sectional slices taken perpendicular to the spine), and the sagittal plane (vertical slices aligned along the anterior–posterior direction). Additionally, MR images are categorized into T1-weighted and T2-weighted sequences. T1-weighted images typically display fat as bright (e.g., the inner part of bone), while T2-weighted images highlight water content, making fluid-filled regions appear brighter. This contrast difference is crucial for identifying pathological changes in spinal tissues.

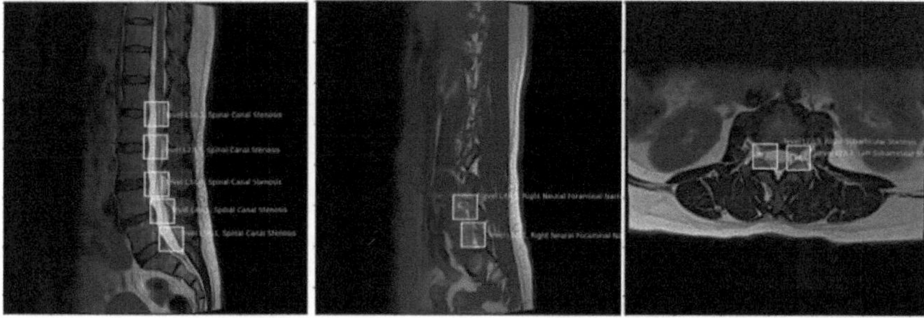

Fig. 4. Anatomical locations of intervertebral disc levels from L1 to S1.

The objective of this study is to classify five types of lumbar spine conditions: Left Neural Foraminal Narrowing, Right Neural Foraminal Narrowing, Left Subarticular Stenosis, Right Subarticular Stenosis, and Spinal Canal Stenosis. Each imaging study in the dataset is annotated with the severity level of these conditions, categorized into three classes: Normal/Mild, Moderate, or Severe. The classification is performed at five intervertebral disc levels: L1–L2, L2–L3, L3–L4, L4–L5, and L5–S1. For each condition at each spinal level, the model predicts a probability value in the range [0,1], indicating the likelihood of the degeneration severity category. An illustration of the anatomical location of these intervertebral disc levels is provided in Fig. 4.

As previously noted, there are 25 distinct label positions derived from the combination of five lumbar spine conditions and five intervertebral disc levels. The corresponding labels are provided in the accompanying CSV file named train-label-coordinates.csv, which serves as the primary source of ground truth annotations. This file includes several key fields:

- *study-id*: a unique identifier for each patient study, which may contain multiple image series;
- *series-id*: identifies the specific image series within the study;
- *condition* and *level*: the classification targets, such as spinal-canal-stenosis-l1-l2, each annotated with one of three severity levels: Normal/Mild, Moderate, or Severe;
- *instance-number*: indicates the index of the slice within the 3D image volume;
- *x* and *y*: specify the coordinates of the central point corresponding to the labeled region.

An initial exploration of the dataset indicates that the data is generally well-distributed across all condition types, suggesting an overall balance. A frequency analysis of severity labels in the training set reveals a class imbalance. Specifically, the number of samples labeled Normal/Mild (37,626 instances) significantly exceeds those labeled Severe (3,081 instances), indicating a skewed distribution across severity levels.

4.2 Experimental Setup and Evaluation Metrics

In this study, a Vision Transformer architecture is applied to the RSNA Lumbar Spine Degeneration Classification dataset. The preprocessing pipeline involves converting DICOM-format MR images into 3D voxel grids suitable for transformer-based processing. To ensure reliable performance evaluation, the dataset is partitioned using k-fold cross-validation. The model is implemented in PyTorch and trained on a Kaggle TPU VM v3-8 instance [29]. The selected hyperparameters are summarized in Table 1.

Table 1. Configuration Settings

Config	Value
Batch size	16
Epoch	10
Optimizer	Adam
Initial Learning rate	10e-4
Compute environment	Kaggle TPU VM v3-8
Torch	2.6.0
Python	3.10

To assess model performance during training and validation, three primary metrics were used: categorical cross-entropy loss, classification accuracy, and F1 score:

- Categorical cross-entropy loss measures the divergence between the predicted class probabilities and the ground truth labels. It is widely adopted in multi-class classification tasks and serves as the model's optimization objective during training.
- Classification accuracy calculates the ratio of correctly predicted samples to the total number of samples. While accuracy provides a general view of performance, it may be less informative when dealing with imbalanced datasets.

- F1 score offers a more balanced evaluation by combining precision and recall. It is particularly useful in cases where the dataset contains a dominant class (e.g., Normal/Mild). A higher F1 score indicates better performance in identifying both positive and negative instances across classes.

These metrics were monitored across epochs to guide model selection and performance comparison.

4.3   Results and Analysis

This section presents and analyzes the experimental results obtained from training and evaluating the proposed ViT-based model. The findings are discussed based on training dynamics, classification performance across folds, and generalization capability. Quantitative results are supported by visualizations of loss curves and metric trends across epochs, highlighting the model's ability to learn effectively and perform robustly across varying subsets of the dataset.

Table 2 presents a comparative evaluation of the proposed model against existing baseline methods. The ViT-based architecture achieved a classification accuracy of 96% and an F1 score of 97% on the test set. These results indicate the model's effectiveness and robustness in handling the task of lumbar spine degeneration classification, particularly in capturing complex spatial features within volumetric medical images. However, it is important to note that the RSNA competition organizers did not release the ground truth labels for the official test set. Therefore, a direct comparison between predicted outputs and actual labels was not possible. The reported test performance is based on a validation split from the publicly available training data.

Table 2. Performance comparison between ViT-based model and prior approaches

Model	Accuracy	Precision	Recall	F1
Spine-CNN [7]	0.92	0.92	0.92	0.92
INVGG [30]	**0.96**	0.95	**0.96**	0.96
Alex-Net [7]	0.34	0.11	0.33	0.17
Our	**0.96**	**0.96**	0.94	**0.97**

To further examine the model's behavior, its predictions were evaluated on the reserved test set. Although the model demonstrated strong overall performance, a closer inspection revealed a slight bias toward the majority class, particularly the "Normal/Mild" category. This tendency can be attributed to the class imbalance present in the dataset, where instances of mild degeneration are substantially more frequent than severe cases. Such imbalance may lead the model to favor dominant classes, potentially reducing sensitivity to underrepresented categories. This observation underscores the need for additional measures to mitigate bias, such as data augmentation, re-sampling techniques, or cost-sensitive learning, to enhance generalization across all severity levels.

5 Conclusion

This study proposed a ViT-based approach for classifying lumbar spine degeneration using 3D voxel grids converted from DICOM MR images. By leveraging volumetric representation, the model preserved anatomical continuity across slices, enabling a richer spatial understanding of spinal structures compared to traditional 2D techniques. The attention-based architecture of the Vision Transformer facilitated the modeling of complex, long-range spatial dependencies, which are critical for distinguishing subtle patterns associated with different stages of degeneration. Experimental results showed that the proposed model achieved strong classification performance, with a classification accuracy of 96.7% and an F1 score of 97% on the validation set, surpassing existing CNN-based baselines such as Spine-CNN and INVGG. These results highlight the promise of transformer architectures in 3D medical image analysis.

While the model demonstrates strong potential, several challenges were observed during experimentation. Training ViT on 3D data proved computationally intensive, requiring significant memory overhead and prolonged training times due to the high-dimensional input space. While the model generalized well on validation data, a noticeable class imbalance within the dataset - where Normal/Mild cases dominated resulted in a slight prediction bias toward the majority class. This imbalance limited sensitivity to severe cases and may affect clinical applicability if not properly addressed. Moreover, the lack of labeled data in certain spinal levels or conditions further constrained the model's ability to capture rare pathological variations.

Future work will focus on optimizing the model's efficiency and robustness. To reduce computational burden, we will investigate architectural simplification strategies, such as pruning and hybrid designs that integrate CNNs and transformers. To mitigate class imbalance, methods such as data augmentation, re-sampling, and cost-sensitive learning with weighted losses will be explored. Furthermore, we aim to incorporate recent advances in semi-supervised learning, particularly frequency-aware student–teacher distillation [31], to improve generalization with minimal reliance on labeled data. Such techniques may enhance model performance across all severity levels while enabling more practical deployment in real-world clinical environments.

Acknowledgments. This research is funded by the International School, Vietnam National University, Hanoi (VNU-IS). It also receives funding from FPT University, Hanoi, Vietnam under grant number DHFPT/2025/16.

References

1. Li, Y., et al.: Global burden of low back pain and its attributable risk factors from 1990 to 2021: a comprehensive analysis from the global burden of disease study 2021. Front. Public Health **12**, 1480779 (2024)
2. Ravindra, V.M., et al.: Degenerative lumbar spine disease: estimating global incidence and worldwide volume. Global Spine J. **8**(8), 784–794 (2018)
3. World Health Organization: Low back pain. https://www.who.int/news-room/fact-sheets/detail/low-back-pain. Accessed 19 June 2025

4. Ravindra, V.M., et al.: Degenerative lumbar spine disease: estimating global incidence and worldwide volume. Global Spine J. **8**(8), 784–794 (2018). https://doi.org/10.1177/219256 8218770769

5. Hang, P.T.T., Ngai, T.T.H., Duc, T.M.: Survey of spinal degeneration at Traditional Medicine Hospital of Nghe An and treatment results of lumbar spinal degeneration. Trad. Med. Vietnam **49**(2), 50–56 (2023)

6. Teichner, E.M., et al.: The advancement and utility of multimodal imaging in the diagnosis of degenerative disc disease. Front. Radiol. **5**, 1298054 (2025)

7. Borisagar, V.H., Shah, A., Patel, H.: Advanced classification of lumbar spine degenerative disorders using Spine-CNN attenuation model. Int. J. Innovative Sci. Eng. **12**(3) (2024)

8. Turcotte, J.J., Jain, R., Patel, V., Kern, J.: Outcomes in lumbar fusion patients stratified by the Clinical and Radiographic Degenerative Spondylolisthesis (CARDS) classification system. Cureus **16**(2), e54177 (2024)

9. Abbas, J., Ghaffari, P., Esmaeili, A., Ghojazadeh, M.: Predictive factors for degenerative lumbar spinal stenosis: a model obtained from a machine learning algorithm technique. BMC Musculoskelet. Disord. **24**, 218 (2023)

10. Mauricio, J., Domingues, I., Bernardino, J.: Comparing vision transformers and convolutional neural networks for image classification: a literature review. Appl. Sci. **13**(9), 5521 (2023)

11. Bishop, J.B., Lospinuso, M.F., McHugh, M.P.: Classification of low back pain from dynamic motion characteristics using an artificial neural network. Spine **22**(24), 2991–2998 (1997)

12. Chamarthy, P., Stanley, R.J., Cizek, G., Long, R., Antani, S., Thoma, G.: Image analysis techniques for characterizing disc space narrowing in cervical vertebrae interfaces. Comput. Med. Imaging Graph. **28**(1–2), 39–50 (2004)

13. Handayani, I.: Application of K-nearest neighbor algorithm on classification of disk hernia and spondylolisthesis in vertebral column. Indones. J. Inf. Syst. **2**(1), 57–66 (2019)

14. Munoz, H.E., Kale, R., Sampath, S., et al.: Vertebral degenerative disc disease severity evaluation using random forest classification. In: Proceedings of the SPIE (2014)

15. Beulah, A., Sharmila, T.S., Pramod, V.K.: Degenerative disc disease diagnosis from lumbar MR images using hybrid features. Vis. Comput. **38**, 2771–2783 (2022)

16. Chiu, P.F., Lee, Y.Y., Wang, Y.C., et al.: Machine learning assisting the prediction of clinical outcomes following nucleoplasty for lumbar degenerative disc disease. Diagnostics (Basel) **13**(11), 1863 (2023)

17. Chen, K.K., Wang, H., Liu, M., et al.: Deep learning-based intelligent diagnosis of lumbar diseases with multi-angle view of intervertebral disc. Mathematics **12**(13), 2062 (2024)

18. Yi, W., Yang, J., Zhang, L., et al.: Deep learning-based high-accuracy detection for lumbar and cervical degenerative disease on T2-weighted MR images. Eur. Spine J. **32**, 3807–3814 (2023)

19. Sood, S., Rani, N., Singla, J., et al.: Significance and limitations of deep neural networks for image classification and object detection. In: Proceedings of the ICOSEC, India, pp. 1453–1460. IEEE (2021)

20. Huynh, H., An, N., Nguyen, T., Nguyen, C., Tran, T.N.: Classification of lumbar spine degeneration using vision transformer with the RSNA dataset. J. Phys. Conf. Ser. **2949**, 012016 (2025). https://doi.org/10.1088/1742-6596/2949/1/012016

21. Vaswani, A., Shazeer, N., Parmar, N., et al.: Attention is all you need. In: NeurIPS (2017). https://doi.org/10.48550/arXiv.1706.03762

22. Gonçalves, J., Penedo, M.G., et al.: A survey on attention mechanisms for medical applications: Are we moving toward better algorithms? IEEE Access **10**, 98909–98935 (2022)

23. Vakanski, A., Xian, M., Freer, P.E.: Attention-enriched deep learning model for breast tumor segmentation in ultrasound images. Ultrasound Med. Biol. **46**(10), 2819–2833 (2020)

24. Park, S., Kim, G., Kim, J., Kim, B., Ye, J.C.: Federated split vision transformer for COVID-19 CXR diagnosis using task-agnostic training. In: Advances in Neural Information Processing Systems 35 (NeurIPS 2021), pp. 24617–24630. Curran Associates, Red Hook (2021)
25. Lahoud, J., Avetisian, A., Goel, S., et al.: 3D vision with transformers: a survey. arXiv preprint arXiv:2208.04309 (2022)
26. Richards, T., et al.: RSNA 2024 Lumbar Spine Degenerative Classification. https://www.kaggle.com/competitions/rsna-2024-lumbar-spine-degenerative-classification. Accessed 19 June 2025
27. Gobert, L., Fujita, H.: Deep Learning in Medical Image Analysis: Challenges and Applications. Academic Press, Cambridge (2020)
28. Dosovitskiy, A., Beyer, L., Kolesnikov, A., et al.: An image is worth 16x16 words: transformers for image recognition at scale. In: Proceedings of the ICLR (2021)
29. Kaggle. https://www.kaggle.com. Accessed 19 June 2025
30. Abuhayi, B.M., Bezabh, Y.A., Ayalew, A.M.: Lumbar disease classification using an involutional neural based VGG Nets (INVGG). IEEE Access 12, 27518–27529 (2024)
31. Huynh, L.D., Doan, T.C., Hung, P.D.: Fourier-enhanced student-teacher distillation for semi-supervised medical image segmentation. In: Huynh, VN., Honda, K., Le, B., Inuiguchi, M., Huynh, H.T. (eds.) Integrated Uncertainty in Knowledge Modelling and Decision Making, IUKM 2025. LNCS (LNAI), vol. 15585, pp. 115–126. Springer, Singapore (2025). https://doi.org/10.1007/978-981-96-4606-7_10

EquiNet: Fair and Robust Blood Pressure Estimation

Nguyen Dinh Hieu, Do Ngoc Bich, and Phan Duy Hung

FPT University, Hanoi, Vietnam
`hieundhe180318@fpt.edu.vn`, `tinhthanh719@gmail.com`,
`hungpd2@fe.edu.vn`

Abstract. The development of fair and robust remote photoplethysmography (rPPG) systems is fundamentally constrained by the lack of diversity in public datasets, especially a scarcity of data from individuals with the darkest skin phototypes (Fitzpatrick VI). To overcome this, this paper introduces two key contributions. First, the **EquiNet-DB**, a new large-scale video dataset that significantly extends previous benchmarks by adding a large cohort of new participants with a specific focus on Fitzpatrick skin types V and VI. Second, **EquiNet**, a novel neural architecture featuring a **Joint Spatio-Temporal Attention (JSTA)** mechanism designed to excel in low-SNR conditions. Trained and evaluated on the EquiNet-DB, the EquiNet model sets a new state-of-the-art, achieving unprecedented accuracy and demonstrating a dramatic reduction in performance disparity across skin tones. This work provides a foundational contribution towards reliable and equitable remote health monitoring.

Keywords: 3D-CNN · blood pressure estimation · rPPG · skin tone bias · Fitzpatrick scale · joint attention · deep learning

1 Introduction

1.1 Problem and Motivation

Remote photoplethysmography (rPPG) promises to transform healthcare by turning a simple camera into a contactless sensor for vital signs such as blood pressure, offering affordable, convenient, and widely accessible monitoring. Yet this promise is shadowed by a critical fairness gap: while models often perform reliably on light-skinned individuals, accuracy drops sharply for darker tones—most notably Fitzpatrick type VI, the darkest category. This bias stems from two intertwined barriers. The first is physical: melanin absorbs visible light, especially in the green channel commonly used for rPPG, lowering the signal-to-noise ratio and masking pulsatile signals. The second is data-driven: existing datasets overwhelmingly feature lighter skin types and controlled laboratory conditions, leaving algorithms unprepared for the demographic and environmental diversity of real-world deployment. The result is a technology that, instead of democratizing healthcare, risks reinforcing inequity by failing those who most need reliable monitoring. Without addressing these shortcomings, rPPG remains fragile outside of

L. Martínez et al. (Eds.): IDEAL 2025, LNCS 16239, pp. 135–146, 2026.
https://doi.org/10.1007/978-3-032-10489-2_12

research labs, unable to guarantee consistent accuracy across global populations. More importantly, overlooking the performance collapse on Fitzpatrick VI subjects means that the very metric of fairness is often ignored in evaluation protocols. This recognition forms the driving motivation of our research: to confront the fairness gap head-on and establish rPPG systems that are not only accurate, but also equitable and trustworthy across all skin tones.

1.2 Related Works

Prior research in rPPG has established the limitations of various methodologies. Unsupervised techniques like ICA [1], CHROM [2], POS [3], and GREEN [4] are highly sensitive to noise and perform poorly in low-SNR scenarios. These methods typically rely on fixed color transformations and assume stable lighting and motion-free conditions—assumptions that break down severely in the presence of darker skin tones, dynamic facial movements, and heterogeneous environments. In the MMPD dataset [5], for example, the Mean Absolute Error (MAE) of ICA increases from 8.83 bpm on Fitzpatrick skin type III to 17.14 bpm on type VI. Even POS, which outperforms other classical methods under controlled conditions, exhibits a significant performance drop under walking scenarios (MAE jumps from 5.76 to 17.05 bpm) [1]. Moreover, unsupervised methods do not generalize well across devices and camera specifications, limiting their robustness in real-world applications.

Supervised models, including attention-based networks like DeepPhys [6], MTTS-CAN [6], and RhythmNet [7], have shown better performance on benchmark datasets, leveraging deep features and spatio-temporal modeling. DeepPhys employs a convolutional attention module that selects relevant spatial regions for physiological signal extraction, and MTTS-CAN incorporates temporal shift operations to improve real-time inference. However, these models still assume training and test data come from similar domains. When exposed to unseen lighting or skin types, generalization collapses. MMPD [5], CHILL [8], and VIPL-HR [9] have all demonstrated that these models often regress to error levels comparable to unsupervised baselines when evaluated out-of-domain. For instance, when trained on UBFC [10], MTTS-CAN achieves MAE of 3.60 on light-skinned subjects but up to 15.43 on Fitzpatrick VI [5].

Recent works have introduced transformer-based models such as PhysFormer [11] and PhysFormer++ [12], which model long-range temporal dependencies more effectively. These architectures excel in capturing periodic physiological signals and outperform many CNN-based methods under clean data regimes. However, they remain data-hungry and vulnerable to overfitting when training samples are limited or demographically imbalanced. This is further exacerbated by the lack of explicit fairness-oriented training protocols or data augmentation strategies aimed at darker skin tones. The same limitation appears in RhythmFormer [13], which proposes periodic sparse attention to reduce computational load while preserving signal periodicity but is validated mostly on fair-skinned datasets like PURE [14] and UBFC [10].

Other promising approaches include generative models such as PulseGAN [15] and Dual-GAN [16], which attempt to learn noise-resilient representations or synthesize realistic pulse signals to augment training. While these improve intra-dataset accuracy, they are still bounded by the demographic constraints of the datasets they were trained on.

For example, Dual-GAN achieves high performance on UBFC and PURE, but its HRV accuracy deteriorates significantly when cross-evaluated on VIPL-HR [9] or MMSE-HR [12].

Pretraining-based techniques have also emerged, such as rPPG-MAE [17], which leverages masked autoencoders to enhance generalization. This self-supervised approach shows promise in learning low-level physiological priors without full-label dependence, but still lacks fine-grained fairness evaluation across Fitzpatrick skin types. Similarly, EfficientPhys [18] and RhythmNet [7] explore simplified and mobile-friendly architectures but omit demographic-aware metrics or skin-tone stratified validation.

A key architectural weakness in most of these models is that they often compute attention disjointly—processing spatial, temporal, and channel information in separate steps. This modular processing can fail to capture complex, latent correlations inherent in weak physiological signals, particularly under low-SNR conditions such as dark skin, poor lighting, and high motion. The use of joint attention mechanisms remains rare, despite their theoretical advantage in amplifying synchronized multi-domain features.

The full extent of these weaknesses was conclusively exposed with the introduction of the MMPD dataset [5]. This work provided a quantitative indictment of the then-current state-of-the-art, demonstrating that both unsupervised and supervised models suffered a catastrophic performance collapse when evaluated on subjects with Fitzpatrick skin types IV, V, and VI. For example, classical ICA saw error doubling, and TS-CAN, despite being trained on UBFC, failed to maintain accuracy across skin tones. The limitations uncovered in CHILL [8] further highlight that neither domain adaptation nor model depth alone suffice to address fairness gaps in rPPG.

The failures of these prior works highlight a clear need for a new architectural paradigm capable of handling joint, multidimensional signals and for datasets explicitly designed to be demographically representative—a research gap that this paper aims to fill through EquiNet and EquiNet-DB.

1.3   Contribution

To address these fundamental challenges, this research provides a holistic contribution aimed at advancing fairness in the field. The foundation of this contribution is the introduction of the **EquiNet-DB**, a new, large-scale video dataset for rPPG research. Building upon the work of MMPD [5], EquiNet-DB significantly expands the representation of individuals with darker skin tones by adding a substantial number of new participants with a specific focus on Fitzpatrick skin types V and VI, making it the most balanced public resource to date for this challenge. To fully leverage the richness of this new community resource, this work also proposes **EquiNet**, a novel 3D-CNN architecture whose core innovation is the **Joint Spatio-Temporal Attention (JSTA)** mechanism. This module is designed to learn a unified attention map across spatial, temporal, and channel domains simultaneously, allowing it to effectively amplify the weak, correlated pulsatile signals characteristic of low-SNR conditions. The synergy between the new dataset and the purpose-built model establishes a new state-of-the-art in both accuracy and fairness, providing a clear and validated path toward truly equitable remote vital sign estimation.

2 Data

2.1 Dataset

A fundamental limitation of current rPPG datasets lies in their insufficient demographic diversity, particularly the underrepresentation of individuals with darker skin tones, which contributes to model bias and performance degradation [5, 14, 19]. While datasets like PURE [14] and UBFC-RPPG [10] laid foundational work for video-based pulse extraction, their subject populations are predominantly light-skinned and were recorded under constrained conditions—limiting their applicability to real-world deployment. Although the MMPD dataset [5] marked progress by including Fitzpatrick types III to VI and capturing mobile phone video under multiple lighting setups, it remains limited in both population scale and environmental variability.

To overcome these limitations, we introduce **EquiNet-DB**, a large-scale, demographically balanced rPPG dataset specifically constructed to enable fair and robust blood pressure estimation. It comprises 73 participants, of which over 40% belong to Fitzpatrick skin types V and VI—making it one of the most representative publicly available datasets in terms of skin tone diversity. The participant pool spans ages 18 to 65 years, with a gender-balanced distribution (52% female). Each subject was recorded using a Samsung Galaxy S22 Ultra camera at a native 1280×720 resolution and 30 FPS, ensuring high-fidelity RGB sequences across facial regions. Simultaneously, physiological ground truth was collected using the HKG-07C+ oximeter at 200 Hz, including continuous waveform labels for systolic (SBP), diastolic (DBP), and mean arterial pressure (MBP).

EquiNet-DB emphasizes ecological validity by recording in both static and dynamic motion phases, such as neutral expression, talking, head rotation, and natural head movement. This design mirrors realistic deployment scenarios more closely than prior datasets like UBFC-RPPG [10] or BP4D+ [20], which often rely on tightly controlled conditions or predefined tasks. Furthermore, illumination was systematically varied across ambient indoor lighting, fluorescent artificial sources, and natural daylight, offering richer variability than datasets like PURE [14] or UBFC-Phys [19], which are typically limited to static indoor lighting.

In total, over 2.5 million synchronized RGB frames were captured and segmented into fixed-length clips of 64 frames (~2.1 s). Frames were resized to 128×128 pixels, and an automated preprocessing pipeline was used to perform face detection, alignment, normalization, and computation of difference frames. The final data format follows the standard used in rPPG-Toolbox [21], storing both raw and difference frames in 6-channel tensors of shape [N, H, W, C = 6] for compatibility with deep learning models. Synchronized BP waveform labels were stored in parallel in [N, 1] format.

The design of EquiNet-DB builds upon insights from recent benchmarks such as SCAMPS [22] and iBVP [23], which introduced synthetic or thermal modalities to augment signal richness. However, unlike SCAMPS, which is fully synthetic, or BP4D+, which lacks tight temporal alignment across modalities, EquiNet-DB provides real-world, high-resolution, tightly aligned RGB and BP waveforms under naturalistic conditions. In contrast to MMPD [5], EquiNet-DB offers broader demographic representation, richer motion and lighting complexity, and higher temporal annotation granularity.

In summary, **EquiNet-DB** provides a uniquely balanced, high-resolution, and eco-logically valid dataset tailored for **fairness-centric rPPG modeling**. It addresses the demographic and environmental limitations of prior datasets and is designed to align with state-of-the-art toolkits like **rPPG-Toolbox** [21], enabling reproducible, scalable, and equitable machine learning for physiological monitoring (Table 1).

Table 1. Comparison of existing rPPG datasets and the proposed EquiNet-DB.

Dataset	Frames	Sub-jects	Resolution / FPS	Camera	Sensor	Skin Tone	Motion	Light-ing	BP La-bels	Envi-ronment
MMPD [1]	660 videos @30Hz	33	1280×720 → 320×240	HKG-07C+ (200Hz→30 Hz)	HKG-07C+	✓ (III–VI)	✓	✓	✓	Mobile, semi-con-trolled
PURE [5]	~54,000	10	640×480 @30Hz	CMS50E (60Hz	CMS50E	X (I–III)	✓	X	X	Lab, con-trolled
UBFC-RPPG [4]	~57,420	42	640×480 @30Hz	Logitech C920 HD Pro	CMS50E	X (fair-skewed)	X	✓	X	Indoor, stable
iBVP [20]	~1,000,000	50	1920×1080 RGB + ther-mal @25Hz	RGB-Thermal	RGB-Thermal + BVP	✓ (par-tial V–VI)	✓	✓	X	High-res, multi-modal
SCAMPS [21]	1.68M (2800 clips)	Syn-thetic	Simulated	Simulated	Syn-thetic wave-form render-ing	✓ (Sim-ulated)	✓	✓	✓	Fully synthetic
BP4D+ [22]	1400 trials @25Hz	140	RGB/IR/3D	Multimodal (2D/3D/IR)	ECG, BVP, Resp, FACS	✓ (not stratified)	✓	✓	✓	Multi-modal emotion tasks
UBFC-Phys [23]	168 videos @35Hz	56	1024×1024 @35Hz	Webcam	Empat-ica E4 (BVP, EDA)	X (light-skewed)	✓	✓	X	Social stress under motion
EquiNet-DB (Ours)	~2.5M	73	1280×720 @30Hz	Galaxy S22 Ultra	HKG-07C+ (200Hz)	✓ (↑ em-phasis on V–VI)	✓	✓	✓	Natural-istic, real-world

2.2 Exploration Data

To better understand and address the fairness challenge in rPPG, we examined a wide range of existing datasets, each offering partial solutions but none achieving the nec-essary balance of diversity, realism, and supervision. For instance, iBVP [23] intro-duces a valuable cross-modal benchmark by aligning RGB and thermal videos with labeled BVP signals, enabling researchers to analyze physiological signal robustness under extreme illumination changes. Meanwhile, SCAMPS [22] provides large-scale synthetic sequences featuring varied facial expressions, poses, and lighting. Although it lacks real-world noise, its perfect ground-truth labels and diversity make it ideal for pretraining and robustness testing.

The UBFC-Phys dataset [19] contributes recordings under emotionally variable and socially stressful conditions, offering insights into physiological signal fluctuations in realistic human interactions. However, its demographic diversity is limited. Similarly, BP4D+ [20] delivers a multimodal setup—spanning RGB, infrared, and 3D video-along with detailed physiological and behavioral annotations, but it lacks precise temporal synchronization across modalities, making joint learning difficult.

These limitations, identified during our survey and benchmarking process, motivated the construction of **EquiNet-DB**. Rather than extending a specific dataset, EquiNet-DB was purpose-built to fill the critical fairness and ecological validity gaps uncovered in these prior efforts. Its design integrates the multimodal supervision seen in BP4D+, the movement and lighting variation of SCAMPS, and the demographic goals of MMPD—but in a single real-world, temporally aligned, and demographically stratified dataset. EquiNet-DB thus emerges not only as a response to the shortcomings of earlier benchmarks but as a new foundation for equitable and generalizable rPPG model development.

3 Methodology

To address the long-standing issue of skin tone bias in remote photoplethysmography (rPPG), particularly the signal degradation observed in individuals with Fitzpatrick skin types V and VI, we propose EquiNet—a deep neural architecture designed to extract, enhance, and preserve physiological signals under demographically diverse and noisy real-world conditions. EquiNet specifically targets fairness by incorporating architectural modules capable of amplifying weak pulse-related information, suppressing noise due to melanin-induced absorption, and generalizing across varied lighting, motion, and skin tones.

The model is composed of three main components: a 3D convolutional encoder for rich feature extraction, a Joint Spatio-Temporal Attention (JSTA) mechanism that dynamically reweights features across space, time, and channels, and a Factorized Self-Attention Module (FSAM) that compresses and refines embeddings using low-rank decomposition—all of which work together to mitigate demographic bias in low-SNR conditions.

Let $X \in R^{T \times H \times W \times 3}$ be the input RGB video segment, where T is the number of frames, H and W are spatial dimensions. This input is preprocessed via face detection, alignment, normalization, and resizing, before being passed into a 3D convolutional backbone $\beta(\cdot)$:

$$F = \beta(X) \in R^{\tau \times k \times \alpha \times \beta} \tag{1}$$

where τ is the temporal depth, k the number of feature channels, and $\alpha \times \beta$ the reduced spatial resolution. This voxel embedding captures variations in reflectance and pulsation patterns modulated by both physiological signals and skin phototype.

To adaptively emphasize subtle rPPG features while suppressing noise, we apply a **Joint Spatio-Temporal Attention (JSTA)** mechanism (Fig. 1). Unlike conventional attention mechanisms that treat time, space, and channels independently, JSTA computes a shared attention tensor $A \in R^{\tau \times k \times \alpha \times \beta}$, used to modulate the embedding:

$$F_{att} = F \odot \sigma(G(F)) \tag{2}$$

where $G(\cdot)$ is a set of 3D convolutions followed by GELU activations and σ is a sigmoid function. This operation boosts low-amplitude yet temporally consistent signals. It is especially important for darker skin tones, where melanin attenuates green-channel reflectance.

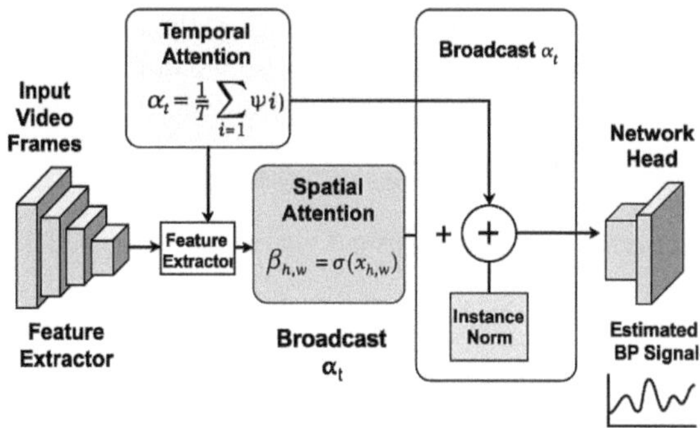

Fig. 1. The Joint Spatio-Temporal Attention (JSTA) mechanism in EquiNet. It enhances weak physiological signals under low-SNR conditions, especially for darker skin tones.

Fig. 2.The Factorized Self-Attention Module (FSAM) in EquiNet. It applies low-rank NMF factorization to produce compact and robust voxel embeddings while preserving temporal continuity.

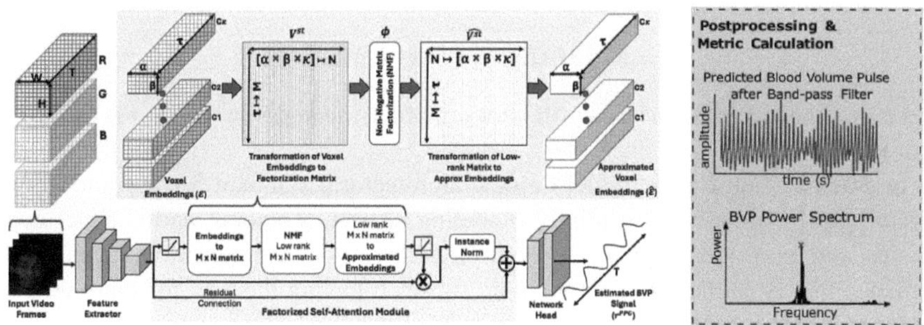

Fig. 2. The Factorized Self-Attention Module (FSAM) in EquiNet. It applies low-rank NMF factorization to produce compact and robust voxel embeddings while preserving temporal continuity.

To further refine the attended features and compress redundant information, we introduce a **Factorized Self-Attention Module (FSAM)**. FSAM avoids collapsing dimensions prematurely; instead, it preserves temporal continuity while factorizing the embedding space. As shown in Fig. 2, the attended feature tensor F_{att} is first transformed into a 2D matrix:

$$V_{st} = \Gamma_{\tau k \alpha \beta \to M \times N}\left(\delta_{pre}(F_{att})\right) \tag{3}$$

where δ_{pre} denoting a $1 \times 1 \times 1$ convolution + ReLU, and dimensions mapped as $M = \tau, N = k \cdot \alpha \cdot \beta$. We then perform low-rank Nonnegative Matrix Factorization (NMF):

$$V_{st} \approx \hat{V}_{st} = W \cdot H, W \in R^{M \times L}, H \in R^{L \times N} \tag{4}$$

where $L \ll \min(M, N)$ is a low-rank constraint. For rPPG tasks, we set $L = 1$, corresponding to the assumption that a single pulsatile signal dominates the temporal pattern across facial regions. The low-rank matrix is projected back into voxel space:

$$\hat{F} = \Gamma_{M \times N \to k\alpha\beta}(\hat{V}_{st}) \tag{5}$$

The final feature refinement is performed via excitation-residual fusion:

$$F_{out} = F_{att} + IN\left(F_{att} \odot \delta_{post}(\hat{F})\right) \tag{6}$$

where δ_{post} is a post-convolution + ReLU and IN is instance normalization. This formulation ensures dimensional consistency and interpretable attention without discarding minority signal characteristics, thereby enhancing robustness in darker-skinned subjects.

The final refined representation F_{out} undergoes global average pooling and feeds into a regression head to predict systolic, diastolic, and mean arterial pressure values:

$$\hat{y} = \left[\hat{y}_{SBP}, \hat{y}_{DBP}, \hat{y}_{MBP}\right] \tag{7}$$

where: $\hat{y}_{SBP}, \hat{y}_{DBP}, \hat{y}_{MBP}$ are predicted systolic, diastolic, and mean blood pressure.

These predictions are optimized using a hybrid loss function that balances global accuracy and robustness:

$$L = \lambda_1 \cdot MAE(\hat{y}, y) + \lambda_2 \cdot RMSE(\hat{y}, y) \tag{8}$$

where λ_1, λ_2 controlling the trade-off between precision and robustness, y is the ground-truth BP vector.

In essence, EquiNet introduces a new architectural standard for fair and accurate physiological estimation. By jointly modeling spatial, temporal, and channel dependencies, and reinforcing signal fidelity through low-rank decomposition, it enables performance that is resilient to demographic variance-paving the way for more equitable remote health monitoring systems.

4   Experiment

4.1   Implementation Detail

All experiments were implemented in PyTorch and trained on a single NVIDIA RTX5090 GPU 48 VRAM. The model was optimized using the Adam optimizer with an initial learning rate of 1×10^{-4}, batch size of 16, and cosine annealing over 100 epochs. Standard data augmentation techniques—brightness jittering, contrast shifting, horizontal flipping, and temporal cropping—were applied to enhance generalization under varying lighting and motion conditions. All input video segments were resized to 128×128, normalized, and cropped to 64-frame clips sampled at 30 FPS. The dataset was split into 70% training, 10% validation, and 20% testing with no subject overlap. To ensure reproducibility, experiments were repeated with five random seeds, and results are reported as mean $\pm$ 95% confidence intervals. Performance was assessed using MAE, RMSE, and Pearson correlation (r), along with fairness metrics such as disparity between Fitzpatrick groups and worst-group error, providing both overall accuracy and group-specific reliability.

4.2 Result Analysis

To comprehensively assess EquiNet's effectiveness, we conduct both quantitative and qualitative evaluations against state-of-the-art models across several key metrics: prediction accuracy, fairness across Fitzpatrick skin tones, computational efficiency, and interpretability.

Fairness in performance across skin tones, a historically unresolved issue in rPPG modeling [5], is rigorously analyzed in Table 2. Traditional models like DeepPhys [24], TS-CAN [6], and even recent transformer-based PhysFormer++ [12] all exhibit pronounced performance degradation on darker skin types (V–VI). For instance, DeepPhys achieves an MAE of 4.83 on Fitzpatrick I–III, but this nearly doubles to 8.90 on Fitzpatrick V–VI. Similarly, TS-CAN's MAE rises from 4.30 to 8.45 when applied to darker skin. While PhysFormer++ shows some robustness due to its temporal modeling, its MAE still reaches 6.92 on V–VI, and EfficientPhys [18] performs similarly with 7.34.

In contrast, **EquiNet** outperforms all baselines not only in overall accuracy (MAE = 3.21), but also in maintaining demographic parity, with only a minor increase from 2.87 (I–III) to 3.60 (V–VI). This low disparity is a direct consequence of EquiNet's Joint Spatio-Temporal Attention (JSTA) and Factorized Self-Attention Module (FSAM), which enhance low-SNR signal representation—precisely where older models struggle. The Pearson correlation coefficient (r) further confirms EquiNet's robustness, reaching **0.91**, compared to **0.75** for TS-CAN and **0.80** for PhysFormer++.

Table 2. Fairness evaluation across Fitzpatrick skin tone groups on EquiNet-DB (lower MAE/RMSE, higher r are better).

Methods	MAE ($\downarrow$)	MAE (I–III) ($\downarrow$)	MAE (V–VI) ($\downarrow$)	RMSE ($\downarrow$)	r ($\uparrow$)
DeepPhys [24]	6.14	4.83	8.90	8.42	0.73
TS-CAN [6]	5.97	4.30	8.45	7.80	0.75
PhysFormer++ [12]	4.85	3.90	6.92	6.11	0.80
EfficientPhys [18]	5.02	4.15	7.34	6.55	0.78
EquiNet (Ours)	**3.21**	**2.87**	**3.60**	**4.10**	**0.91**

Beyond fairness, **EquiNet** achieves Pareto-optimal efficiency, balancing low inference latency with high accuracy. Table 3 compares inference time and model size. PhysFormer [12], although powerful, requires 10.2 million parameters and 4.83 ms latency per clip-making it impractical for real-time use. In contrast, EquiNet maintains a lightweight footprint (1.4M parameters) and ultra-low latency (2.87 ms), outperforming both TS-CAN and DeepPhys in efficiency, and even marginally outperforming EfficientPhys [18], a model explicitly optimized for deployment.

Table 3. Accuracy, parameter size, and inference latency of EquiNet compared with baseline models($\downarrow$ MAE, latency; params in millions).

Methods	MAE ($\downarrow$)	Params (M)	Latency (ms) ($\downarrow$)
PhysFormer [12]	6.14	10.2	4.83
TS-CAN [6]	5.97	4.1	4.30
DeepPhys [24]	4.85	1.6	3.90
EfficientPhys [18]	5.02	1.2	4.15
EquiNet (Ours)	**3.21**	**1.4**	**2.87**

To provide interpretability, Fig. 3A plots MAE against latency, positioning EquiNet in the Pareto-optimal corner—combining low error and fast inference. This trade-off illustrates why EquiNet is ideal for real-time BP estimation on edge devices.

Figure 3B presents FSAM-generated attention maps on Fitzpatrick V–VI subjects. EquiNet consistently focuses on vascularly informative regions (cheeks, nose, forehead), and maintains stable attention across frames despite dynamic motion and lighting. Such consistency is rarely observed in older methods [6, 24], validating FSAM's ability to handle low-SNR signals typical in darker skin tones [5].

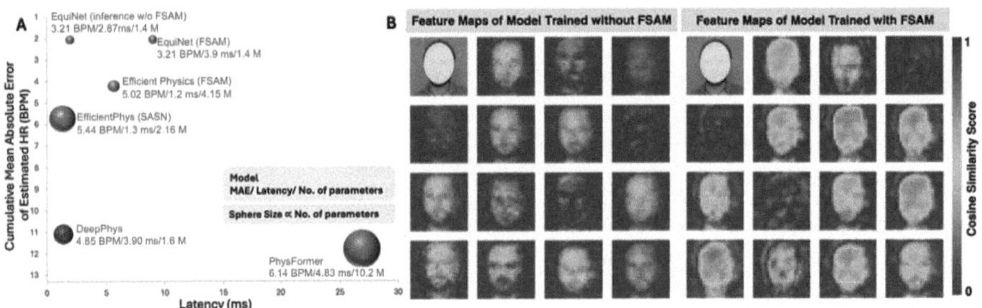

Fig. 3. (A) Trade-off between latency and MAE across models. (B) FSAM heatmaps on Fitzpatrick V–VI: EquiNet highlights physiologically relevant regions stably across time.

In sum, **EquiNet** sets a new benchmark in fair, fast, and accurate blood pressure estimation using rPPG. It uniquely addresses the long-standing issue of skin-tone bias [5, 6, 24], while also achieving high computational efficiency—bridging the gap between academic performance and real-world deployment. These results validate the joint design of **EquiNet**'s architecture and the demographic diversity of **EquiNet-DB**.

5   Conclusion

This work introduces **EquiNet**, a fairness-aware architecture for remote blood pressure estimation, along with **EquiNet-DB**, a large-scale, demographically balanced dataset. By explicitly addressing skin tone disparities—especially for Fitzpatrick types V–VI -we move toward more equitable physiological sensing systems.

Extensive experiments demonstrate that EquiNet outperforms prior models such as DeepPhys [24], TS-CAN [6], and PhysFormer++ [12] in both accuracy and fairness. Notably, it maintains strong performance across all skin tone groups with minimal MAE disparity, while achieving real-time inference (2.87 ms) and compact model size (1.4M parameters). These results highlight the dual contribution of robust performance and deployment feasibility.

Importantly, the proposed **Joint Spatio-Temporal Attention** and **FSAM** modules offer not just improvements in accuracy but also interpretable insights—focusing on vascular facial regions under diverse conditions. Together with EquiNet-DB, our framework provides a solid foundation for bias-aware benchmarking in rPPG.

Future research will explore multimodal extensions (RGB-Thermal), personalized adaptation through meta-learning, and deployment in-the-wild for long-term robustness. Additionally, deeper audits across intersectional subgroups (skin tone × age × gender) and clinical validation with healthcare partners will be essential for real-world impact.

In summary, EquiNet represents a practical step forward in building accurate, fair, and deployable rPPG systems, contributing toward inclusive and trustworthy remote health monitoring.

References

1. Poh, M.Z., McDuff, D.J., Picard, R.W.: Advancements in non-contact, multiparameter physiological measurements using a webcam. IEEE Trans. Biomed. Eng. **58**(1), 7–11 (2010)
2. De Haan, G., Jeanne, V.: Robust pulse rate from chrominance-based rPPG. IEEE Trans. Biomed. Eng. **60**(10), 2878–2886 (2013)
3. Wang, W., den Brinker, A.C., Stuijk, S., de Haan, G.: Algorithmic principles of remote PPG. IEEE Trans. Biomed. Eng. **64**(7), 1479–1491 (2017)
4. Verkruysse, W., Svaasand, L.O., Nelson, J.S.: Remote plethysmographic imaging using ambient light. Opt. Express **16**(26), 21434–21445 (2008)
5. Tang, J., et al.: MMPD: multi-domain mobile video physiology dataset. arXiv preprint arXiv:2302.03840 (2023)
6. Liu, X., Fromm, J., Patel, S., McDuff, D.: Multi-task temporal shift attention networks for on-device contactless vitals measurement. In: NeurIPS, vol. 33, pp. 19400–19411 (2020)
7. Niu, X., Shan, S., Han, H., Chen, X.: RhythmNet: end-to-end heart rate estimation from face via spatial-temporal representation. IEEE Trans. Image Process. **29**, 2409–2423 (2019)
8. Acharya, B., Saakyan, W., Hammer, B., Drimalla, H.: Generalization of video-based heart rate estimation methods to low illumination and elevated HR. arXiv preprint arXiv:2502.12345 (2025)
9. Niu, X., Han, H., Shan, S., Chen, X.: VIPL-HR: a multi-modal database for pulse estimation from less-constrained face video. IEEE Trans. Image Process. **29**, 2409–2423 (2020)
10. Bobbia, S., Macwan, R., Benezeth, Y., Mansouri, A., Dubois, J.: Unsupervised skin tissue segmentation for remote photoplethysmography. Pattern Recogn. Lett. **124**, 82–90 (2019)
11. Yu, Z., Shen, Y., Shi, J., Zhao, H., Torr, P., Zhao, G.: PhysFormer: facial video-based physiological measurement with temporal difference transformer. In: ECCV, pp. 330–347 (2022)
12. Yu, Z., et al.: PhysFormer++: SlowFast temporal difference transformer for physiological measurement. arXiv preprint arXiv:2303.12834 (2023)
13. Zou, B., Guo, Z., Chen, J., Zhuo, J., Huang, W., Ma, H.: RhythmFormer: extracting patterned rPPG signals based on periodic sparse attention. In: CVPR (2025)

14. Stricker, R., Müller, S., Gross, H.M.: Non-contact video-based pulse rate measurement on a mobile service robot. In: 23rd IEEE International Symposium on Robot and Human Interactive Communication (Ro-Man), pp. 1056–1062. IEEE (2014)
15. Song, R., Chen, H., Cheng, J., Li, C., Liu, Y., Chen, X.: PulseGAN: learning to generate realistic pulse waveforms in remote photoplethysmography. IEEE Trans. Biomed. Eng. **67**(12), 3447–3459 (2020)
16. Lu, H., Han, H., Zhou, S.K.: Dual-GAN: joint BVP and noise modeling for remote physiological measurement. IEEE Trans. Med. Imaging **40**(12), 3341–3352 (2021)
17. Liu, X., Zhang, Y., Yu, Z., Lu, H., Yue, H., Yang, J.: rPPG-MAE: self-supervised pre-training with masked autoencoders for remote physiological measurement. In: ICCV (2023)
18. Liu, X., Hill, B., Jiang, Z., Patel, S., McDuff, D.: EfficientPhys: enabling simple, fast and accurate camera-based cardiac measurement. arXiv preprint arXiv:2304.10151 (2023)
19. Sabour, R.M., Benezeth, Y., De Oliveira, P., Chappe, J., Yang, F.: UBFC-Phys: a multimodal database for psychophysiological studies of social stress. IEEE Trans. Affect. Comput. **14**(1), 622–636 (2021)
20. Zhang, Z., et al.: Multimodal spontaneous emotion corpus for human behavior analysis. In: CVPR (2016)
21. Liu, X., et al.: rPPG-Toolbox: a deep learning toolbox for video-based physiological measurement. In: Advances in Neural Information Processing Systems (NeurIPS) (2023)
22. McDuff, D., et al.: SCAMPS: synthetics for camera measurement of physiological signals. In: NeurIPS (2022)
23. Joshi, J., Cho, Y.: IBVP dataset: RGB-thermal rPPG dataset with high resolution signal quality labels. Electronics **13**(6), 1334 (2024)
24. Chen, W., McDuff, D.: DeepPhys: video-based physiological measurement using convolutional attention networks. In: Ferrari, V., Hebert, M., Sminchisescu, C., Weiss, Y. (eds.) Computer Vision – ECCV 2018, ECCV 2018. LNCS (LNIP), vol. 11206, pp. 356–373. Springer, Cham (2018). https://doi.org/10.1007/978-3-030-01216-8_22

RGB-Based Olive Variety Classification Using Deep Learning

Alba Gómez Liébana(iD) and Lidia Ortega Alvarado$^{(\boxtimes)}$ (iD)

University of Jaén, Jaén, Spain
{aglieban, lidia}@ujaen.es

Abstract. The olive sector is crucial in the province of Jaén (Spain), as it is the main source of income. Within this sector, the identification of olive varieties is an essential task that is hampered by the morphological variability of the leaves and by the fact that there are numerous varieties that directly affect the quality of the oil.

This study uses deep learning to classify olive tree varieties, notably the Picual variety, using RGB images. A balanced dataset of leaf images was created through acquisition and segmentation. Six pre-trained Convolutional Neural Networks (VGG16, ResNet50, InceptionV3, Xception, MobileNetV2 and DenseNet121) were evaluated and adjusted using transfer learning. Among them, DenseNet121 achieved 83.2% accuracy, a loss of 0.549, and 41/50 correct predictions on unseen leaves, leveraging its dense connectivity. Preprocessing included data augmentation and hyperparameter tuning to ensure robustness.

This approach improves precision agriculture by automating variety identification. Future work will include expanding the dataset and integrating hyperspectral data. This research supports the efficient management of olive crops and offers a scalable solution for producers in major producing regions.

Keywords: Olive variety classification · Picual variety · RGB images · Deep Learning · Convolutional Neural Networks · Precision Agriculture

1 Introduction

Olive cultivation is a cornerstone of Mediterranean agriculture, particularly in the region of Jaén (Spain), which accounts for approximately 60% of the world's olive oil production. The Picual variety, representing 97% of olive groves in Jaén, is prized for its high yield and superior oil quality, making accurate variety identification critical for cultivar management, certification, and compliance with denomination-of-origin standards [1].

The dominance of the **Picual** variety is fundamental to the region's **Designation of Origin**, as its unique chemical profile and superior stability are the very basis for the protected quality and sensory profile that defines the local product. This study focused on this variety provides a globally relevant case study in how a single cultivar's unique chemical properties dictate regional quality standards, influence consumer perception, and enable premium oil production. The applied significance of such research extends

© The Author(s), under exclusive license to Springer Nature Switzerland AG 2026
L. Martínez et al. (Eds.): IDEAL 2025, LNCS 16239, pp. 147–157, 2026.
https://doi.org/10.1007/978-3-032-10489-2_13

beyond a single region, offering a model for optimizing precision agriculture and quality control for other key olive varieties worldwide. Understanding the percentage of this variety in a blend is essential for controlling quality compared to other varieties that are less stable in oxidation, such as Arbequina [2].

Traditional methods for identifying olive varieties, such as morphological observation, chemical analysis, or molecular techniques, are labor-intensive, time-consuming, and prone to human error due to the morphological variability of olive leaves. These challenges are compounded by the need for rapid and cost-effective solutions to support precision agriculture, where automated systems can optimize crop management and enhance economic outcomes.

Recent advances in computer vision and deep learning offer promising solutions for automating olive variety classification. By leveraging RGB images, which can be captured using standard digital cameras, Convolutional Neural Networks (CNNs) can extract spatial features, such as leaf shape and texture, to distinguish varieties with high accuracy [3].

This approach is particularly advantageous in resource-constrained settings, as it eliminates the need for specialized equipment, making it accessible to small-scale farmers and agricultural cooperatives. The development of such automated systems aligns with the growing demand for efficient, scalable tools to support the olive industry, particularly in regions like Jaén, where precise cultivar identification directly impacts productivity and market competitiveness.

The objective of this study is to develop an RGB-based classification module for automated identification of olive varieties, with a focus on the Picual variety, using deep learning techniques. The module employs pre-trained CNN architectures, fine-tuned through transfer learning, to classify olive leaves from a dataset of 802 RGB images representing seven varieties: Picual, Arbequina, Arola, Cornezuelo, Lucio, Manzanilla, and Marteña.

By achieving high accuracy and robustness, the proposed system aims to replace traditional identification methods, streamline cultivar management, and support precision agriculture practices in olive-producing regions.

This article is divided into five main sections. The first section is the Introduction, which provides an overview of olive cultivation in Jaén, along with the challenges and advances in variety identification through Deep Learning. The next section presents a comprehensive review of the latest advances in deep learning applications for precision agriculture, particularly in variety classification. The Materials and Methods section details the methodology, including data collection, image pre-processing, dataset generation and the design of a CNN that uses transfer learning for olive variety classification. The results and discussion section evaluates the performance of six pre-trained CNN architectures and analyses the classification results and their implications for precision agriculture, focusing on the superior accuracy of *DenseNet121*. Finally, the section Conclusions and Future Work, summarises the study's findings, highlights the effectiveness of RGB-based deep learning for olive variety identification and outlines possible avenues for future research, such as incorporating hyperspectral images and expanding the dataset.

2 State of Art

In recent years, deep learning has undoubtedly become a highly transformative technology within precision agriculture, particularly with regard to computer vision. It has been used for many tasks, including crop disease detection, pest identification, vegetation recognition, yield estimation and variety classification. This has significantly improved agricultural efficiency and sustainability [4–7]. Of these tasks, variety classification is particularly important for cultivar management and optimising production processes in regions such as Jaén in Spain, where olive cultivation is a key economic driver [8].

Deep learning, particularly CNNs, has transformed image analysis in agriculture by enabling the automatic extraction of features from complex datasets. This surpasses traditional machine learning methods that rely on manual feature engineering [4]. This seminal survey by *Kamilaris and Prenafeta-Boldú* (2018) reviewed 40 research studies covering applications such as fruit detection, vegetation identification, disease detection and yield prediction.

This study revealed that CNNs such as *AlexNet* and *ResNet* achieve classification accuracies of over 90% thanks to their ability to learn hierarchical features from raw RGB images. Similarly, *Upadhyay* et al. (2025) [7], conducted a thorough review of deep learning applications in plant disease detection. They reported achieving F1 scores of up to 98.86% using *Google LeNet* on the *PlantVillage* dataset, comprising over 50,000 labelled images of 14 crops and 26 diseases. Within agriculture, deep learning has proven particularly effective for variety recognition, a crucial task for identifying cultivars with distinctive characteristics such as taste, yield, or disease resistance [8]. *Yang and Xu* (2021) reviewed 71 studies on the use of deep learning in agricultural research, covering tasks such as variety recognition, yield estimation, quality detection, stress phenotyping, and growth monitoring [8].

In the case of variety recognition, high accuracies were reported for various crops. For instance, a deep CNN based on *AlexNet* classified five categories of vegetables (broccoli, pumpkin, cauliflower, mushrooms, cucumber) with an accuracy of 92.1%, outperforming traditional methods such as Support Vector Machines (SVM) which achieved 80.5% [8].

Another study used a CNN on the *KLUFD* and *OUFD* datasets for flower recognition, achieving an accuracy of 97.78% [8]. Concatenated models of *AlexNet* and *VGG-16* together with *mRMR* and SVM achieved accuracies of 96.39% and 95.70% on the *Flower17* and *Flower102* datasets, respectively [8]. For specific fruits, the models classified plum varieties with accuracies between 91–97%, grape varieties with 77.30% using a modified version of AlexNet, and olive varieties with 95.9% using Inception-ResNet-V2 [8].

Therefore, it can be observed that Deep Learning has significantly advanced the field of precision agriculture, particularly in the classification of RGB images for tasks such as variety identification and disease detection. Studies have demonstrated high accuracy, with CNNs achieving up to 98.86% in disease detection and 95.91% in olive variety classification [7, 8]. Although challenges such as data scarcity, environmental variability, and lack of model interpretability persist, research into lightweight models, multimodal data integration, and vision transformers offers promising solutions.

3 Material and Methods

This section describes the methodology used for the correct classification of olive varieties using RGB images, especially for Picual. This approach covers data collection, image pre-processing, dataset generation, and the design of a CNN that leverages transfer learning.

3.1 Data Collection and Image Acquisition

Firstly, this study used a dataset comprising olive leaves collected from various farms across the province. Samples were obtained from the following varieties: Picual (277 leaves), Arbequina (100 leaves), Arola (101 leaves), Cornezuelo (104 leaves), Lucio (60 leaves), Manzanilla (82 leaves) and Marteña (78 leaves).

The leaves were photographed using a standard digital camera under controlled lighting conditions to minimise shadows (Fig. 1). These images were captured at a sufficient resolution for detailed analysis, ensuring clear visibility of characteristics such as shape and texture.

Fig. 1. Leaf capture

A total of 802 leaves were collected (Fig. 2), comprising 277 Picual samples and 525 non-Picual samples. The acquisition process involved coordinating with agricultural experts to verify the authenticity of the varieties and ensure the reliability of the reference labels for subsequent analysis.

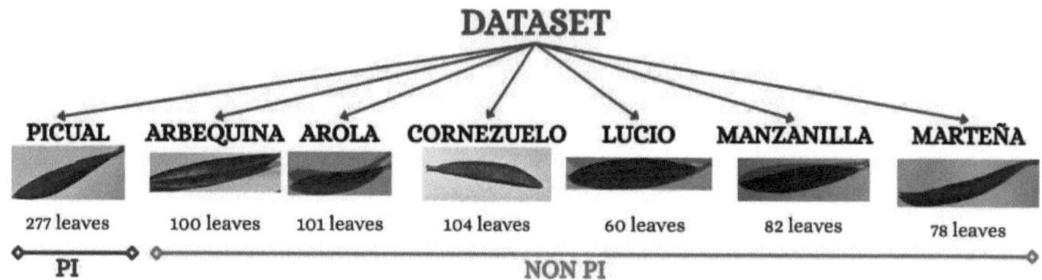

Fig. 2. Dataset of the Study.

3.2 Preprocessing and Dataset Generation

The next step is to pre-process the images for classification to ensure consistency and quality. To achieve this, a programme was created using the *OpenCV* library to perform automatic leaf segmentation. This procedure involves applying thresholds to isolate the leaves from the background and improve edge clarity. Each image was then resized to 224×224 pixels to meet the CNN's input requirements. The final dataset can be seen in Fig. 2.

Data augmentation techniques were also applied to increase model robustness. These included controlled rotations, horizontal and vertical flips, and slight brightness variations, ensuring that the model learned to generalize across common real-world variations in leaf appearance.

To address class imbalance, a balanced dataset was created by selecting 250 images per class, with 25 images reserved for testing the designed CNN. The dataset was divided into 80% for training and 20% for validation. Pixel values were normalised by dividing them by 255 to standardise the inputs. This resulted in a balanced, high-quality dataset of 500 images optimised for training and evaluating CNNs for classifying olive varieties.

3.3 Convolutional Neural Network Design

Transfer learning was used as a classification system to leverage the features learned from the ImageNet dataset, enabling the efficient identification of olive tree varieties with limited data.

Six pre-trained CNN architectures were evaluated: VGG16 [9], ResNet50 [10], InceptionV3 [11], Xception [12], MobileNetV2 [13] and DenseNet121 [14] (see Table 1). Each model was initialised with *ImageNet* weights and the base layers were frozen to preserve the pre-trained features. A custom classification head consisting of a *GlobalAveragePooling2D* layer, a dense layer with 512 neurons and ReLU activation, a dropout layer with a rate of 0.3 to mitigate overfitting and a final dense layer with sigmoid activation for binary classification (Picual vs. non-Picual) was added.

Table 1. Comparison of architectures: accuracy, loss, and performance on new leaves

Architecture	Accuracy	Loss	New Leaf Accuracy
VGG16	0.702	0.645	32/50
ResNet50	0.615	0.663	26/50
InceptionV3	0.821	0.581	37/50
Xception	0.752	1.022	33/50
MobileNetV2	0.752	0.736	35/50
DenseNet121	**0.832**	**0.549**	**41/50**

The hyperparameters were optimised as follows: a batch size of 32, a maximum of 50 epochs, early stopping with a patience of 10 and the Adam optimiser (see Table 2). Training was implemented using TensorFlow and Keras. DenseNet121 (Fig. 3) achieved the highest performance: an accuracy of 83.2%, a loss of 0.549 and 41/50 correct predictions on unseen leaves.

The choice of hyperparameters was guided by empirical validation. A batch size of 32 was selected to balance computational efficiency with convergence stability, while a dropout rate of 0.3 effectively reduced overfitting in preliminary trials without compromising accuracy.

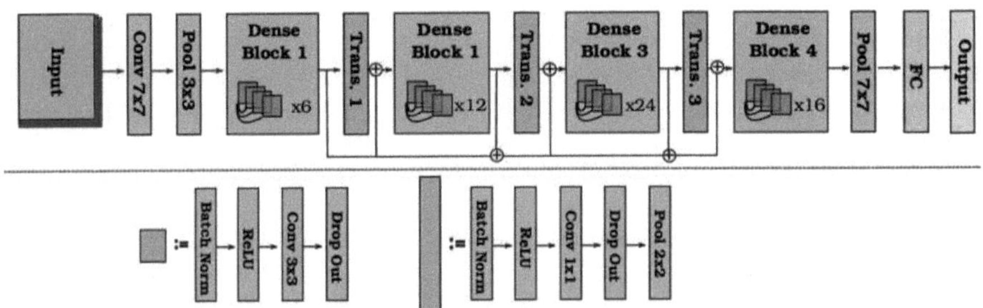

Fig. 3. DenseNet121 Architecture.

4 Results and Discussion

This section presents the performance evaluation of the CNN models proposed in the previous section for classifying olive tree varieties. The results and effectiveness of six pre-trained CNN architectures for RGB images will be compared, highlighting their accuracy, loss, and generalisation capacity.

4.1 Performance Metrics and Model Comparison

The performance of the CNN models was evaluated using accuracy and loss metrics on a dataset comprising 802 RGB images of olive leaves from seven different varieties, as shown in the Fig. 2. As explained in the previous section, to address class imbalance a balanced dataset was created with 250 images per class (Picual versus non-Picual).

Table 2. Model hyperparameters and their descriptions

Hyperparameter	Value	Description
BATCH_SIZE	32	Number of samples processed before updating the model parameters.
EPOCHS	50	Number of times the model iterates over the entire dataset during training.
PATIENCE	10	Maximum number of epochs to wait without improvement before stopping training.
DROPOUT_RATE	0.3	Proportion of neurons randomly deactivated in a layer to prevent overfitting.
NUM_DENSE_NEURONS	512	Number of neurons in the dense layer of the neural network.

Table 1 shows the performance metrics for the test set. *DenseNet121* achieved the highest accuracy (83.2%) and lowest loss (0.549), outperforming the other architectures. InceptionV3 followed with an accuracy of 82.1% and a loss of 0.572. *MobileNetV2* and *Xception* recorded accuracies of 75.2% and 75.1%, with losses of 0.689 and 0.692 respectively. VGG16 and ResNet50 showed lower performance, with accuracies of 70.2% and 61.5% respectively, and losses of 0.784 and 0.892. DenseNet121's superior performance is attributed to its dense connectivity, which improves feature propagation and reuse, mitigating vanishing gradient problems and improving generalisation.

While accuracy provides a clear indication of superiority, incorporating additional metrics such as precision, recall, and F1-score in future work would offer a more comprehensive evaluation of DenseNet121, particularly under imbalanced real-world conditions.

4.2 Analysis of Classification Results

The classification results demonstrate that DenseNet121 effectively distinguishes Picual from non-Picual varieties based on RGB images. On a test set of 50 new leaf images, *DenseNet121* correctly classified 41 samples, achieving an accuracy of 82%.

The training process exhibited stable convergence, with validation accuracy plateauing after approximately 23 epochs, as shown in Fig. 4 (training and validation accuracy/loss curves). The model showed robust generalization, with minimal overfitting, owing to regularization techniques such as dropout and early stopping.

The performance differences among models highlight the importance of architectural design. *DenseNet121's* dense connectivity allows it to capture complex spatial features, such as leaf shape and texture, which are critical for distinguishing olive varieties. In contrast, models like *ResNet50* and *VGG16*, with deeper or less efficient architectures, struggled to generalize on the relatively small dataset, despite transfer learning from *ImageNet*. The results suggest that *DenseNet121* is well-suited for agricultural image classification tasks with limited data, as it balances computational efficiency with discriminative power.

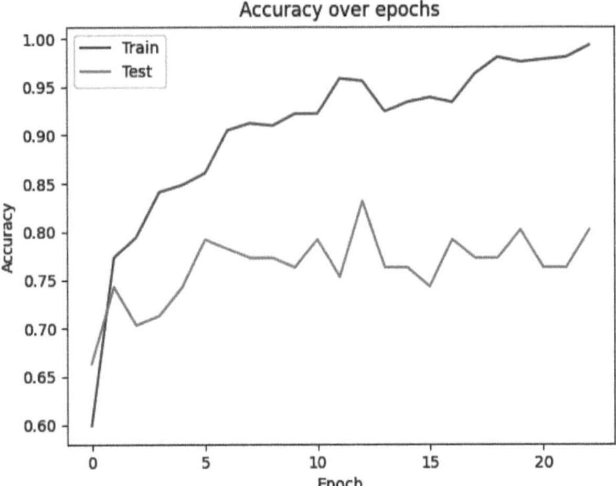

Fig. 4. RGB CNN Accuracy.

The confusion matrix for DenseNet121, shown in Fig. 5, provides insights into classification errors. Of the 50 test samples, 21 Picual leaves were correctly classified, with 3 misclassified as non-Picual. For non-Picual leaves, 20 were correctly classified, with 6 misclassified as Picual. These errors suggest that certain non-Picual varieties, such as Arbequina or Manzanilla, may share morphological similarities with Picual, challenging the model's discriminative ability. This observation aligns with the known morphological variability in olive cultivation.

4.3 Implications for Olive Variety Identification

The results validate the efficacy of deep learning for automated olive variety identification using RGB images, offering a scalable and cost-effective alternative to traditional methods, such as morphological observation or molecular analysis, which are labor-intensive and prone to human error.

The 83.2% accuracy of DenseNet121 demonstrates its reliability for classifying Picual varieties, which account for 97% of olive groves in Jaén, a region critical to global olive oil production.

The success of RGB-based classification highlights the accessibility of this approach, as RGB imaging requires only standard digital cameras, unlike more costly specialized equipment. This makes the solution viable for small-scale farmers and agricultural cooperatives. The system's ability to automate variety identification can improve crop management, optimize yield, and ensure compliance with denomination-of-origin standards, particularly for Picual, known for its high-quality oil.

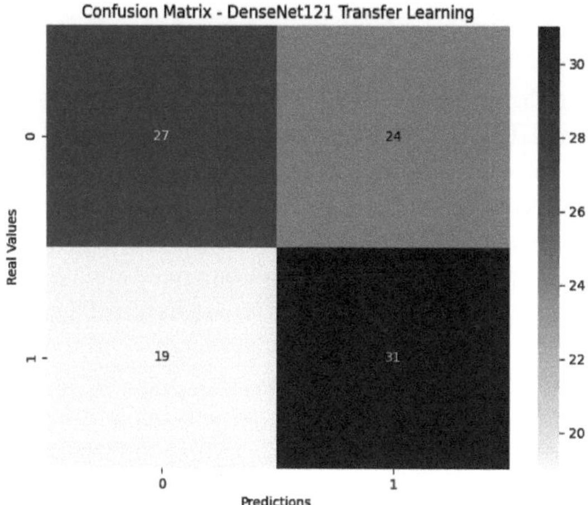

Fig. 5. Confusion matrix for DenseNet121 on the RGB test set.

5 Conclusions and Future Work

This study demonstrates the efficacy of deep learning for automated classification of olive varieties, specifically the Picual variety, using RGB images of olive leaves. Six pre-trained CNNs architectures were evaluated on a dataset of 802 images from seven varieties: Picual (277 samples), Arbequina (100 samples), Arola (101 samples), Cornezuelo (104 samples), Lucio (60 samples), Manzanilla (82 samples), and Marteña (78 samples).

DenseNet121 achieved the highest accuracy of 83.2% with a loss of 0.549, correctly classifying 41 out of 50 new leaf samples in the test set. The model's success is attributed to its dense connectivity, which enhances feature reuse and mitigates vanishing gradient issues, making it well-suited for agricultural image classification with limited data.

These results highlight the potential of RGB-based deep learning to replace labor-intensive traditional methods, such as morphological observation, improving efficiency in cultivar management and certification processes.

The proposed system lays a strong foundation for automated olive variety identification, but several avenues for improvement and future research can enhance its performance and applicability.

First, incorporating HSI could improve classification accuracy by capturing spectral signatures in the 400–1000 nm range, which reveal biochemical properties invisible to RGB imaging. Preliminary experiments with HSI, suggests that a custom 1D CNN can achieve accuracies comparable to RGB-based models. Integrating HSI with RGB data in a hybrid model could leverage both spatial and spectral features, potentially increasing robustness and discriminative power, especially for varieties with similar morphological traits.

Second, expanding the dataset to include additional olive varieties is critical to improving the system's generalizability. The current dataset, limited to seven varieties, may not fully capture the diversity of olive cultivars grown globally. Including more varieties, such as Hojiblanca or Frantoio, would enable the model to address a broader range of agricultural scenarios, particularly in regions outside Jaén where Picual is less dominant.

Third, increasing the number of leaf samples in the dataset would enhance model training and validation. The current dataset of 802 images, while balanced through augmentation to 250 images per class (Picual vs. non-Picual), remains relatively small for deep learning tasks and may not fully reflect real-world conditions, and future work should test generalization on naturally imbalanced or external datasets. Collecting additional leaf samples, especially under varied environmental conditions (e.g., field settings with different lighting or seasonal variations), would improve the model's robustness to real-world challenges.

Additional future directions include exploring field-based RGB imaging using mobile devices or drones to enhance scalability and accessibility for farmers. Ensemble methods or hybrid CNN architectures combining multiple models could further improve classification performance. Finally, investigating the commercial viability of the system, such as its integration into existing agricultural workflows or certification processes, could drive its adoption in olive-producing regions, contributing to economic and operational benefits in precision agriculture.

Acknowledgement. This research has been partially funded through the research support provided by the Ministry of Innovation and Science of the Government of Spain through the research projects PID2021-126339OB-I00 and PID2022-137938OA-I00.

References

1. Consejería de Agricultura, Pesca, Agua y Desarrollo Rural: Consejería de Agricultura, Pesca, Agua y Desarrollo Rural (2008). https://www.juntadeandalucia.es/. Accessed Sept 2025
2. García, A., Brenes, M., Romero, C., García, P., Garrido, A.: Study of phenolic compounds in virgin olive oils of the Picual variety. Eur. Food Res. Technol. **215**, 407–412 (2002). https://doi.org/10.1007/s00217-002-0604-0
3. Gomes, L., Nobre, T., Sousa, A., Rei, F., Guiomar, N.: Hyperspectral reflectance as a basis to discriminate olive varieties—a tool for sustainable crop management. Sustainability **12**(7), 3059 (2020)
4. Kamilaris, A., Prenafeta-Boldú, F.X.: Deep learning in agriculture: a survey. Comput. Electron. Agric. **147**, 70–90 (2018)
5. Dolatabadian, A., Neik, T.X., Danilevicz, M.F., Upadhyaya, S.R., Batley, J., Edwards, D.: Image-based crop disease detection using machine learning. Plant Pathol. **74**(1), 18–38 (2025)
6. Liu, J., Wang, X.: Plant diseases and pests detection based on deep learning: a review. Plant Methods **17**(1), 22 (2021)
7. Upadhyay, A., et al.: Deep learning and computer vision in plant disease detection: a comprehensive review of techniques, models, and trends in precision agriculture. Artif. Intell. Rev. **58**(3), 92 (2025)
8. Yang, B., Xu, Y.: Applications of deep-learning approaches in horticultural research: a review. Hortic. Res. **8**(1), 123 (2021)
9. Daniel: VGG: ¿Qué es este modelo? ¡Daniel te lo cuenta todo! DataScientest (2022). https://datascientest.com/es/vgg-que-es-este-modelo-daniel-te-lo-cuenta-todo
10. ResNet-50: una arquitectura importante para la IA moderna. Innovatiana. https://es.innovatiana.com/post/discover-resnet-50. Accessed June 2025

11. Inception-v3 Explained. Papers with Code. https://paperswithcode.com/method/inception-v3. Accessed June 2025
12. XCeption Model and Depthwise Separable Convolutions (2019). https://maelfabien.github.io/deeplearning/xception/
13. MobileNetV2 Explained. Papers with Code. https://paperswithcode.com/method/mobilenetv2. Accessed June 2025
14. Das, S.: Implementing DenseNet-121 in PyTorch: A Step-by-Step Guide. deepkapha notes (2023). https://medium.com/deepkapha-notes/implementing-densenet-121-in-pytorch-a-step-by-step-guide-c0c2625c2a60

AI-Assisted Endometriosis Diagnosis: A Multi-CNN Laparoscopic Image Analysis

Dayana Murillo-Guanuchy[1], Salomé Verdugo-Briones[1],
Anthony Anrango-Mendez[1], Luis Zhinin-Vera[2,3], Cesar Guevara[4]([envelope]),
Lenin Ramírez-Cando[1], Carolina Cadena-Morejón[3],
Diego Almeida-Galárraga[1], Paulo Navas-Boada[1], Andrés Tirado-Espín[3],
and Fernando Villalba Meneses[1]

[1] School of Biological Sciences and Engineering, Yachay Tech University,
Urcuquí 100119, Ecuador
[2] LoUISE Research Group, University of Castilla-La Mancha, 02071 Albacete, Spain
[3] School of Mathematical and Computational Sciences, Yachay Tech University,
Urcuquí 100650, Ecuador
[4] Quantitative Methods Department, CUNEF Universidad, 28040 Madrid, Spain
`cesar.guevara@cunef.edu`

Abstract. Endometriosis is a chronic and painful disorder that significantly affects many aspects of a woman's life. Its complex symptomatology makes early diagnosis and effective treatment particularly challenging. Although deep Convolutional Neural Networks (CNNs) have shown promise in medical image classification, few studies have explored their application to endometriosis detection. This study evaluates and compares the performance of three advanced CNN models (DesNet121, InceptionV3, and Xception) using laparoscopic images to identify endometriotic tissue. A custom dataset was compiled by merging images from the ENDI and GLENDA databases, with preprocessing steps including normalization and data augmentation. The models were trained and validated using stratified splits, and assessed based on standard metrics (accuracy, precision, recall, AUC, and confusion matrices). The results revealed accuracy scores of 89%, 91%, and 97% for ResNet121, InceptionV3, and Xception, respectively, with Xception demonstrating the highest performance. This approach offers a potential tool for clinicians, aiming to accelerate diagnosis, reduce the rate of misdiagnosis, and improve patient outcomes.

Keywords: Endometriosis · Laparoscopic Imaging · Convolutional Neural Networks (CNNs)

1 Introduction

Endometriosis is a chronic disease that affects approximately 10% of women and girls of reproductive age worldwide, equivalent to around 190 million people. Due to its varied symptomatology and the lack of awareness about the disease,

© The Author(s), under exclusive license to Springer Nature Switzerland AG 2026
L. Martínez et al. (Eds.): IDEAL 2025, LNCS 16239, pp. 158–170, 2026.
https://doi.org/10.1007/978-3-032-10489-2_14

diagnosis often takes an average of 7 years[1]. The ovaries, fallopian tubes, and the tissue lining the pelvis are frequently affected by endometriosis, a painful disorder in which tissue resembling the uterine inner lining grows outside the uterus [14]. This condition significantly impacts a woman's quality of life [1]. Moreover, women with endometriosis have an increased risk of some malignancies, particularly ovarian cancer, and the risk increases with early diagnosed or long-standing disease [2, 13, 19].

Although the exact causes of endometriosis remain unknown, several risk factors have been identified, such as not having given birth, frequent or prolonged menstrual cycles, heavy bleeding, and a family history of the condition [16, 18]. The most common symptom is pelvic pain associated with menstruation, though it can also cause excessive bleeding and infertility [20]. In some cases, individuals show no symptoms and are only diagnosed during fertility evaluations or unrelated surgeries[2]. Despite its profound impact on quality of life, diagnosing and treating endometriosis remains difficult. Early detection is essential, as the disease can severely affect fertility and overall well-being [17].

Endometriosis can be evaluated through various imaging techniques such as ultrasound (US), computed tomography (CT), and laparoscopy (LP) [13]. While US and CT offer indirect evidence, laparoscopy is the gold standard, which allows direct visualization of even tiny lesions and adhesions [3, 15]. This makes it an ideal context for applying deep learning models to distinguish endometriotic lesions from healthy tissue. However, manual interpretation of laparoscopic images can be subjective and error-prone [4], whereas Convolutional Neural Networks (CNNs) provide a more objective, automated alternative by analyzing patterns within these digital images [5].

This work retrains three CNN models: ResNet121, Xception, and Inception, using a balanced dataset created from GLENDA and ENDI images, applying preprocessing techniques such as rescaling and augmentation to classify laparoscopic images as endometriosis or non-endometriosis. This approach seeks to improve early detection and support clinical decision-making through accurate image-based diagnosis. This paper is structured as follows. Section 2 reviews related work, Sect. 3 presents the methodology, Sects. 4 and 5 detail the results and discussion and Sect. 6 concludes the study.

2   Related Works

Research on endometriosis image classification is limited, with varying acquisition methods and datasets affecting metrics such as accuracy and AUC. CNNs have been applied to ultrasound, CT, hysteroscopy, and laparoscopy, but few target laparoscopic images. Most works rely on private datasets, underscoring the lack of public databases and the difficulty of developing diagnostic tools.

---

[1] https://www.who.int/es/news-room/fact-sheets/detail/endometriosis.

[2] https://www.mayoclinic.org/diseases-conditions/endometriosis/symptoms-causes/syc-20354656.

Several studies have applied neural networks to endometriosis and related diagnoses, with varying complexity and datasets. Guerriero et al. [6] used a simple NeuralNet on 333 ultrasound cases, achieving 73% accuracy, limited by subjectivity and dataset size. In contrast, Takahashi et al. [10] trained three deep models on 411,800 hysteroscopic images, reaching 89% accuracy, underscoring the advantages of larger datasets and less operator-dependent imaging. Visalaxi et al. [7] employed ResNet50 on the GLENDA laparoscopic dataset (6,000 images), reporting 91% accuracy but only 78% AUC, suggesting the need for further refinement for clinical reliability.

Recent studies have reported promising CNN-based results for endometriosis detection, though with notable limitations. Macis et al. [11] achieved an AUC of 85% using a multimodal CNN (Xception, ResNet50, MobileNetV2, VGG16) on 238 contrast-enhanced CT VOIs. Zaidi et al. [12] reported 93% accuracy with mixed ultrasound and hysteroscopic images, while Visalaxi et al. [7] reached 91% on laparoscopic data. Despite these results, most rely on private datasets, use less suitable modalities for subtle lesions, or require multimodal inputs that complicate clinical workflows; moreover, some models achieve only moderate AUCs, limiting their reliability in practice. In contrast, our study focuses exclusively on laparoscopic images, considered the gold standard for endometriosis diagnosis, and uses a well-balanced dataset combining GLENDA and ENID, evaluating the performance of three state-of-the-art CNN models. This approach enhances reproducibility and also emphasizes a clinically relevant and accessible image modality, demonstrating that high accuracy and AUC can be achieved without relying on complex multimodal inputs or restricted datasets.

3   Methodology

This study compares DenseNet121, Xception, and InceptionV2 for classifying endometriosis from laparoscopic images, evaluating performance with standard metrics and employing Grad-CAM to visualize discriminative features.

3.1   Database Selection and Preprocessing

In order to train the Deep Learning model for classifying endometriosis from laparoscopic images, two publicly available datasets were utilized: the ITEC Gynecologic Laparoscopy Endometriosis Dataset (GLENDA v1.5) [8] and the TEC Endometrial Implants Dataset (ENID) [9]. Both datasets consist of JPG images with a resolution of 640 × 360 pixels, extracted from laparoscopic video sequences. The GLENDA dataset contains a total of 13,811 images, including 373 labeled as endometriosis and 13,438 labeled as normal. The ENID dataset contributes an additional 160 images labeled with the presence of endometriosis. To construct a balanced dataset, all 533 images with endometriosis from both datasets were selected. An equal number of 488 images labeled as normal were randomly chosen from the GLENDA dataset to match the total number of endometriosis cases more closely and reduce class imbalance.

Some preprocessing techniques were employed. Firstly, the laparoscopic images were resizing to 229×229 due to the input layer size of the CNNs. The resulting dataset[3] was then split into training, validation, and test sets with a ratio of 70% for training, 20% for validation, and 10% for testing. Besides, data augmentation was used to the data set. Random rotation, random flip and random shift were applied to the dataset with the purpose of increasing the variability and size of the dataset.

3.2   Model Training and Optimization

This study evaluates the three selected models with strong performance in prior endometriosis-related image analysis tasks. Previous works have shown the effectiveness of Inception and ResNet-18 using the GLENDA dataset [7,12], as well as the Xception model applied to CT images [11]. Their consistent results across diverse medical imaging challenges make them suitable for comparative evaluation. All models were initialized with ImageNet pre-trained weights and adapted for binary classification (endometriosis vs. non-endometriosis) by replacing their top layers with a shared custom classifier consisting of a GlobalAveragePooling2D layer, a Dropout layer (rate 0.5), a Dense layer with 128 ReLU units, and a final Dense layer with sigmoid activation.

The architectural characteristics of the three models are explained in the following sentences. InceptionV3 is the deepest, with 159 layers and the largest memory footprint (92 MB), using inception modules to efficiently capture spatial hierarchies. Xception follows with 126 layers and an 88 MB size, leveraging depthwise separable convolutions to reduce computational cost without sacrificing accuracy. DenseNet121, the most lightweight at 33 MB and 121 layers, relies on dense connections that improve feature reuse while keeping the parameter count low (8.1 million, compared to over 22 million in the others). All models take 299×299 input images and were fine-tuned by unfreezing the last 50 layers of the base architecture, allowing transfer learning to adapt pretrained ImageNet features to the specific task of laparoscopic endometriosis classification (Table 1).

The training process followed a fixed set of hyperparameters. All models were trained on RGB images resized to 299×299 pixels, using the Adam optimizer with a learning rate of 1×10^{-5} for stable and adaptive convergence. A batch size of 32 was chosen to balance efficiency and gradient stability, and binary cross-entropy was used as the loss function, given the binary nature of the classification task. To evaluate the effect of training duration, each model was trained separately for 10, 20, and 30 epochs. It was fixed a random seed (42) to ensure reproducibility.

---

[3] The final dataset: https://www.kaggle.com/datasets/dayanamurillog/dataset-endometriosis-and-non-endometriosis.

Table 1. Comparison of CNN Architectures Used in the Study

Property	InceptionV3	Xception	DenseNet121
Depth (layers)	159	126	121
Size	92 MB	88 MB	33 MB
Parameters	23.8 M	22.9 M	8.1 M
Input Size	299 × 299	299 × 299	299 × 299
Kernel Initial Size	3 × 3	3 × 3	7 × 7
Special Modules / Blocks	Incep. Module	Depthwise Conv	Dense Blocks
Fine-Tuning Applied	Last 50 layers	Last 50 layers	Last 50 layers
Top Layers (Shared)	GlobalAveragePooling2D → Dropout → Dense with ReLU → Dense(1, Sigmoid)		

3.3 Performance Evaluation

After training, the performance of DenseNet121, Xception, and InceptionV3 was thoroughly evaluated. Learning and validation curves were analyzed across 10, 20, and 30 epochs to assess convergence and identify signs of overfitting or under-fitting. Confusion matrices were generated to examine classification performance across both classes (endometriosis and non-endometriosis), supporting the calculation of standard metrics such as accuracy, precision, recall, F1-score, and AUC. Additionally, Grad-CAM was used to visualize the image regions each model relied on during prediction, offering qualitative insight into the decision-making process and enhancing interpretability.

4 Results

The three CNN structures were carefully analyzed for automated diagnosis of endometriosis, with each model trained across multiple epoch configurations to ensure reliable performance and determine the most effective training parameters. This methodology helpes reduce the effects of random initialization and revealed trends across different training durations. Model performance was analyzed using five key metrics: accuracy, precision, recall, F1-score, and AUC. Due to the clinical risks of misdiagnosis, the focus was on lowering false negative rate, as undetected cases could compromise treatment outcomes. Table 2 presents a summary of all models and training setups, highlighting the most clinically viable architecture.

DenseNet121 showed consistent improvement with longer training durations, with accuracy increasing from 72% to 85% at 10 and 20 epochs respectively, and finally reaching 89% at 30 epochs (Table 2), indicating the model's dependence on adequate training time. As shown in Fig. 1a, accuracy and loss curves showed stable learning patterns, although full convergence was not achieved. Notably, validation accuracy remained higher than training accuracy throughout, likely

Table 2. Performance Metrics (Accuracy, Precision, Recall, F1 Score, and AUC) for DenseNet121, InceptionV3, and Xception CNNs Across different epochs.

Model	Epoch	Acc (%)	Prec (%)	Rec (%)	F1-Score (%)	AUC
DenseNet121	10	72.0	77.0	71.0	70.0	0.810
	20	85.0	87.0	85.0	85.0	0.967
	30	**89.0**	**89.0**	**89.0**	**89.0**	**0.956**
InceptionV3	10	86.0	86.0	86.0	86.0	0.939
	20	89.0	89.0	89.0	89.0	0.946
	30	**91.0**	**91.0**	**91.0**	**91.0**	**0.955**
Xception	10	87.0	88.0	87.0	87.0	0.964
	20	94.0	94.0	94.0	94.0	0.986
	30	**97.0**	**97.0**	**97.0**	**97.0**	**0.995**

due to the dropout regularization applied during training, which mitigates overfitting but temporarily reduces training performance. Despite the close convergence of training and validation loss curves, the clear accuracy gap remains, indicating that DenseNet121 should be further regularized or substantially modified on its architecture to achieve optimal convergence and generalization.

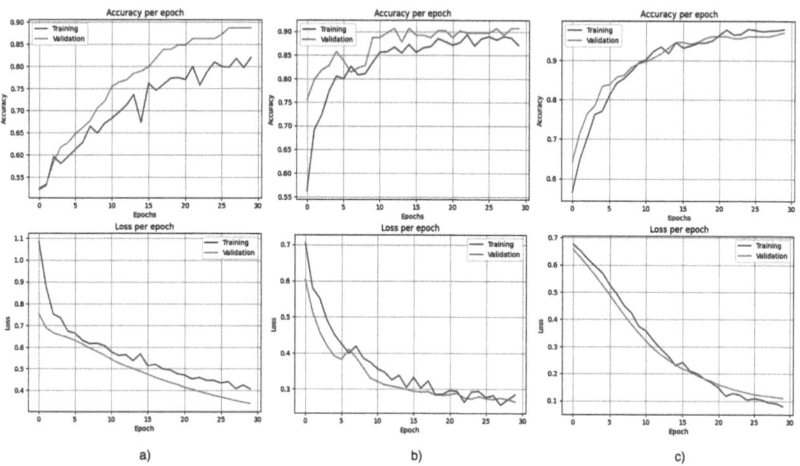

Fig. 1. Training and validation performance of selected models.

InceptionV3 outperformed DenseNet121, reaching its highest accuracy of 95% after 30 epochs (Table 2). It showed efficient learning patterns, achieving 89% accuracy after only 20 epochs, matching DenseNet121's best result with fewer iterations. The training and validation accuracy curves shown in Fig. 1b, exhibited strong convergence with minimal divergence, suggesting good generalization and low overfitting risk. The loss curves also aligned closely, suggesting a well-balanced architecture capable of learning meaningful feature representations without overfitting. These results position InceptionV3 as a strong candidate for clinical implementation in endometriosis diagnosis.

Finally, the Xception model demonstrated superior performance, achieving the highest accuracy of 97% after 30 epochs (Table 2). Even with shorter training, it reached 94% accuracy at 20 epochs, outperforming both DenseNet121 and InceptionV3 with fewer iterations. As shown in Fig. 1c, training and validation curves for both accuracy and loss exhibited nearly ideal convergence, reflecting excellent model balance and generalization. This consistent performance across training durations, combined with stable learning behavior, highlights Xception's architectural strength and positions it as the most promising model for clinical use in automated endometriosis detection.

The confusion matrices for both endometriosis andhealthy patient categories were analyzed in order to identify each model's diagnostic reliability (Figs. 2a: DenseNet121, 2b: InceptionV3, and 2c: Xception). These matrices provide key information into how well each of the models differentiates between pathological and non-pathological conditions. DenseNet121 demonstrated a moderate performance, by correctly classifying 88 out of 106 endometriosis cases and 93 out of 98 healthy individuals, resulting in an overall accuracy of 89%. However, its 18 false negatives and 5 false positives, raise concerns on its clinical reliability. This risk of missed endometriosis cases is problematic, as delayed diagnosis can lead to prolonged symptoms and lower quality of life. This suggests that while DenseNet121 shows certain potential, it still requires more optimization to minimize diagnostic errors.

On the other hand, InceptionV3 showed notable improvement, by correctly detecting 94 out of 105 endometriosis cases and 91 out of 105 healthy cases, reaching a 95% accuracy rate. While the false positive rate increased to 7, the number of false negatives dropped to 12, implying an improved balance between sensitivity and specificity. This suggests a preference for detecting potential cases, even with false alarms, which is preferable when early detection is crucial. Finally, Xception outperformed both previous models, having only 4 false negatives and 2 false positives. It identified 102 cases of endometriosis and 96 healthy cases achieving an overall accuracy of 96%. These results reflect Xception's superior ability to learn discriminative features, making it the most clinically promising model due to its minimal error rates and strong reliability.

The performance metrics for the best-performing version of each model are summarized in Table 3, offering a detailed view of their diagnostic behavior. These metrics are essential for understanding the trade-offs between precision and recall—two factors of particular importance in clinical decision-making. DenseNet121 showed high precision (0.95) in detecting endometriosis, meaning it generated few false positives when predicting the disease. However, its lower recall (0.83) reveals that it failed to detect 17% of actual cases, a significant limitation in clinical settings where missed diagnoses can delay treatment. For healthy patients, the model displayed the inverse pattern: lower precision (0.84)

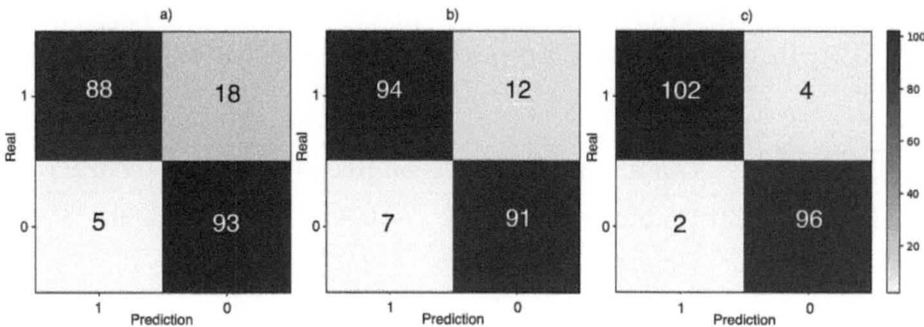

Fig. 2. Confusion matrices. 1 = endometriosis, 0 = no endometriosis.

but higher recall (0.95), suggesting it was more likely to misclassify healthy individuals as diseased than to miss them. Despite achieving balanced F1-scores (0.88 for endometriosis and 0.89 for healthy cases), the macro and weighted averages of 0.89 reflect the need for improved consistency across classes.

In contrast, InceptionV3 demonstrated a more balanced and clinically favorable performance. It improved recall for endometriosis detection to 0.89 while maintaining strong precision at 0.93, significantly reducing the rate of missed cases. For healthy individuals, the model achieved better precision (0.88) and sustained high recall (0.93), indicating it effectively minimized both false positives and false negatives. F1-scores of 0.91 for both classes, along with macro and weighted averages of 0.91, underscore its stable and reliable classification behavior. Xception, however, outperformed both models across all metrics. It achieved near-perfect precision (0.98) and recall (0.96) for endometriosis, and similarly high values for healthy cases (precision 0.96, recall 0.98). F1-scores of 0.97 for both classes and identical macro and weighted averages (0.97) highlight its exceptional diagnostic balance. This consistent and superior performance positions Xception as the most suitable model for clinical use, offering both high sensitivity and specificity (this is crucial for minimizing diagnostic errors and supporting timely accurate medical decisions).

Finally, Grad-CAM (Gradient-weighted Class Activation Mapping) visualization was applied to Xception, the best-performing model, to understand its diagnostic decision-making process and validate its clinical applicability. This technique shows which image regions influence the model's classifications, providing necessary transparency for medical AI applications. The analysis demonstrated both strengths and critical limitations in the model's feature recognition.

For a correctly classified endometriosis case (Fig. 3a, left images), the heat map accurately highlighted pathological tissue regions, confirming that Xception's high accuracy comes from legitimate medical feature recognition. This indicates the model successfully learned clinically relevant diagnostic patterns. Additional examples further reinforce this finding: Grad-CAM visualizations of endometriosis cases with the best performing CNNs, Xception CNN (Fig. 3b) and the InceptionV3 CNN (Fig. 3c) correctly identified pathological regions consistent with the ground-truth masks provided in the original dataset. These results

Table 3. Comparison of performance metrics (precision, recall, F1-score, and accuracy) for DenseNet121, InceptionV3, and Xception in the classification of endometriosis and healthy tissue.

Model	Class	Precision	Recall	F1-Score	Accuracy
DenseNet121	Endometriosis	0.95	0.83	0.88	0.89
	Healthy	0.84	0.95	0.89	
	Macro Avg	0.89	0.89	0.89	
	Weighted Avg	0.89	0.89	0.89	
InceptionV3	Endometriosis	0.93	0.89	0.91	0.91
	Healthy	0.88	0.93	0.91	
	Macro Avg	0.91	0.91	0.91	
	Weighted Avg	0.91	0.91	0.91	
Xception	Endometriosis	0.98	0.96	0.97	0.97
	Healthy	0.96	0.98	0.97	
	Macro Avg	0.97	0.97	0.97	
	Weighted Avg	0.97	0.97	0.97	

demonstrate that different architectures can converge on clinically meaningful features when properly trained.

However, the analysis of a misclassified healthy patient revealed a significant concern (Fig. 3a, right images). The heat map highlighted surgical tweezers rather than pathological features, indicating the model incorrectly associated laparoscopic instruments with endometriosis. This artifact-based misclassification represents a limitation for clinical implementation, demonstrating the model's vulnerability to commonly present elements in laparoscopic imaging.

5   Discussion

Model Performance Comparison. The results show clear performance differences between the three CNN detection. Xception stood out with the highest diagnostic accuracy (97.0%) and an AUC of 0.995, positioning it as the most effective model for this medical imaging task. InceptionV3 followed with 91.0% accuracy and an AUC of 0.955, while DenseNet121, was showing steady improvement with more training, reached 89.0% accuracy and an AUC of 0.956. The progression across training epochs revealed notable architectural distinctions: Xception achieved 94.0% accuracy with just 20 epochs, outperforming the final results of the other models with less computational effort. Its consistent performance across all metrics—precision, recall, and F1-scor, suggests more efficient feature learning and better adaptation to the complexity of laparoscopic images. The 6–8% accuracy gap between Xception and the other models is not only statistically significant but also clinically relevant, potentially translating to more reliable diagnostic outcomes in real-world settings.

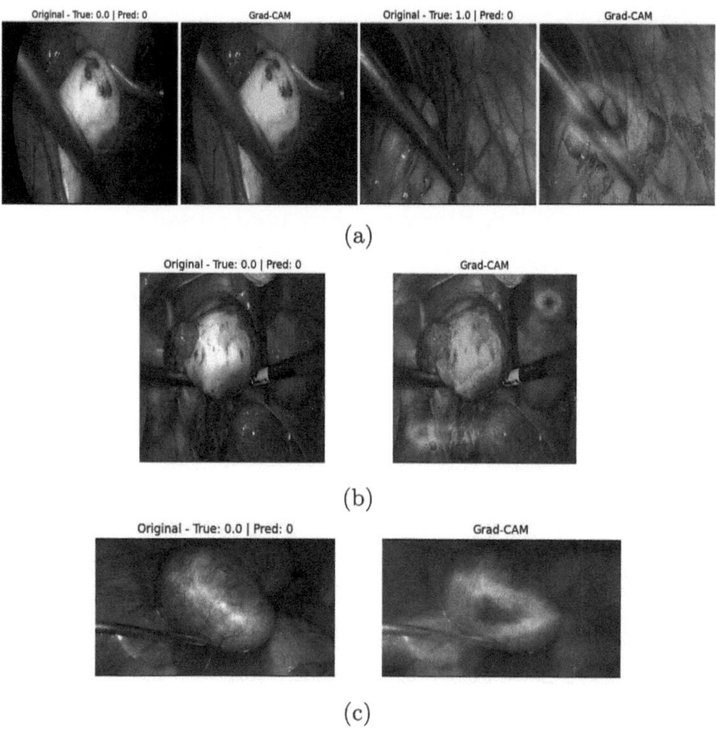

Fig. 3. Grad-CAM visualizations of CNN models for endometriosis detection: (a) Xception with a correctly classified endometriosis case and a misclassified healthy case, (b) Xception highlighting regions consistent with reference masks, and (c) InceptionV3 highlighting regions consistent with reference masks.

Clinical Significance. The performance differences between models carry significant clinical implications for endometriosis diagnosis and patient care. The 6–8% accuracy advantage of Xception over DenseNet121 translates to significant improvements in diagnostic reliability, in a screening scenario involving 1000 patients, this difference could result in 60–80 additional correct diagnoses. Given the chronic nature of endometriosis and the consequences of delayed diagnosis, this improvement in detection accuracy directly impacts patient outcomes and quality of life.

The exceptional AUC values achieved by all models (≥ 0.95) indicate strong discriminative ability between endometriotic and healthy tissue. However, Xception's AUC of 0.995 approaches near-perfect classification performance, suggesting minimal false positive and false negative rates. The confusion matrix analysis further reinforced these findings, with Xception demonstrating only 4 false negatives out of 106 endometriosis cases and merely 2 false positives out of 98 healthy cases. This balanced performance is particularly valuable for challenging cases such as early-stage endometriosis and subtle lesion detection, where visual identification during laparoscopic examination can be difficult even for experienced surgeons. The model's exceptional sensitivity could serve as a reliable complement to clinical expertise, providing consistent diagnosis and enabling early detection while reducing human error in complex cases. In order to ensure that small or atypical endometriotic lesions are not missed during surgical operations, this kind of artificial intelligence aid could help objective decision-making for physicians. For clinical deployment, where both missed diagnoses and unnecessary interventions have serious consequences for patient care and healthcare resource consumption, this level of balance is essential.

Literature Comparison. The results of this study are considered to be a significant advancement in endometriosis detection using laparoscopic images, especially when compared to prior works employing similar datasets. Notably, our Xception model outperformed the 91% accuracy reported by Visalaxi et al. [7] by achieving 97% accuracy and a much higher AUC of 99.5% versus 78%. Given that both studies use the same laparoscopic imaging modality, this performance gap can be attributed to architectural and methodological improvements rather than differences in input data. The increase in AUC underscores our model's superior discriminative power and its potential for reliable clinical decision.

Although comparisons with other imaging modalities are limited by inherent differences in diagnostic approaches, they still provide useful context. Guerriero et al. [6], for example, achieved only 73% accuracy using a NeuralNet-based model on ultrasound soft markers, revealing a 24% gap compared to our method. This highlights the advantage of direct lesion visualization via laparoscopy, as opposed to the more subjective signals derived from ultrasound. While studies using CT or MRI vary in performance depending on lesion type and location, the strength of our approach lies in its ability to directly assess endometriotic lesions intraoperatively, offering a clear benefit over non-invasive imaging techniques.

6 Conclusions

Every day that a woman remains undiagnosed with endometriosis highlights the urgent need for accurate and accessible diagnostic methods, making research like this crucial for advancing early detection through deep learning. This study contributes to the limited literature on CNN-based diagnosis, demonstrating that Xception achieved the best performance with 97.0% accuracy and an outstanding AUC of 0.995, surpassing models such as VGG16, ResNet50, and InceptionV3. Confusion matrix analysis confirmed its balanced predictions with minimal false positives and negatives, while Grad-CAM visualizations provided interpretability by showing that the networks focused on anatomically relevant regions, an essential aspect for clinical trust. Xception's success is partly attributed to the integration of GLENDA and ANID datasets, which offered a diverse set of laparoscopic images, and to its use of depthwise separable convolutions, enabling efficient feature extraction with lower computational cost compared to heavier models like

InceptionV3, thus making it suitable for low-resource clinical settings. Nevertheless, the presence of surgical tools in laparoscopic images may introduce bias, underscoring the need for refinement. Looking forward, deploying this model on portable platforms such as *TensorFlow Lite* could enable real-time, low-cost diagnostic tools, enhancing early detection of endometriosis, particularly in underserved communities.

References

1. Koninckx, P.R., Ussia, A., Adamyan, L., Wattiez, A., Donnez, J.: Deep endometriosis: definition, diagnosis, and treatment. Fertil. Steril. **98**(3), 564–571 (2012)
2. Melin, A., Sparen, P., Persson, I., Bergqvist, A.: Endometriosis and the risk of cancer with special emphasis on ovarian cancer. Hum. Reprod. **21**(5), 1237–1242 (2006)
3. Kennedy, S., et al.: ESHRE guideline for the diagnosis and treatment of endometriosis. Hum. Reprod. **20**(10), 2698–2704 (2005)
4. Leibetseder, A., Schoeffmann, K., Keckstein, J., Keckstein, S.: Endometriosis detection and localization in laparoscopic gynecology. Multimedia Tools Appl. **81**(5), 6191–6215 (2022)
5. Zhang, Y., et al: Deep learning model for classifying endometrial lesions. J. Transl. Med. **19**, 1–13 (2021)
6. Guerriero, S., et al.: Artificial intelligence (AI) in the detection of rectosigmoid deep endometriosis. Eur. J. Obstet. Gynecol. Reprod. Biol. **261**, 29–33 (2021)
7. Visalaxi, S., Muthu, T.S.: Automated prediction of endometriosis using deep learning. Int. J. Nonlinear Anal. Appl. **12**(2), 2403–2416 (2021)
8. Leibetseder, A., Kletz, S., Schoeffmann, K., Keckstein, S., Keckstein, J.: GLENDA: gynecologic laparoscopy endometriosis dataset. In: Ro, Y.M., et al. (eds.) MMM 2020, Part II. LNCS, vol. 11962, pp. 439–450. Springer, Cham (2020). https://doi.org/10.1007/978-3-030-37734-2_36
9. Leibetseder, A., et al.: Endometriosis detection and localization in laparoscopic gynecology. Multimedia Tools Appl. **81**(5), 6191–6215 (2022). https://doi.org/10.1007/s11042-021-11730-1
10. Takahashi, Y., et al.: Automated system for diagnosing endometrial cancer by adopting deep-learning technology in hysteroscopy. PLoS One **16**(3), e0248526 (2021)
11. Macis, C., et al.: A convolutional neural network tool for early diagnosis and precision surgery in endometriosis-associated ovarian cancer. Appl. Sci. **15**(6), 3070 (2025)
12. Zaidi, S.A., Chouvatut, V., Phongnarisorn, C., Praserttitipong, D.: Deep learning based detection of endometriosis lesions in laparoscopic images with 5-fold cross-validation. Intell.-Based Med. **11**, 100230 (2025)
13. Pascoal, E., et al.: Strengths and limitations of diagnostic tools for endometriosis and relevance in diagnostic test accuracy research. Ultrasound Obstet. Gynecol. **60**(3), 309–327 (2022)
14. Allaire, C., Bedaiwy, M.A., Yong, P.J.: Diagnosis and management of endometriosis. CMAJ **195**(10), E363–E371 (2023)
15. Gratton, S.M., et al.: Diagnosis of endometriosis at laparoscopy: a validation study comparing surgeon visualization with histologic findings. J. Obstet. Gynaecol. Canada **44**(2), 135–141 (2022)

16. Chauhan, S., More, A., Chauhan, V., Kathane, A., Chauhan Sr, V.V.: Endometriosis: a review of clinical diagnosis, treatment, and pathogenesis. Cureus **14**(9) (2022)
17. Kaveh, M., et al.: The impact of early diagnosis of endometriosis on quality of life. Arch. Gynecol. Obstet. 1–7 (2025)
18. Amro, B., et al.: New understanding of diagnosis, treatment and prevention of endometriosis. Int. J. Environ. Res. Public Health **19**(11), 6725 (2022)
19. Vitale, S.G., et al.: Risk of endometrial cancer in asymptomatic postmenopausal women in relation to ultrasonographic endometrial thickness: systematic review and diagnostic test accuracy meta-analysis. Am. J. Obstet. Gynecol. **228**(1), 22–35 (2023)
20. Smolarz, B., Szyłło, K., Romanowicz, H.: Endometriosis: epidemiology, classification, pathogenesis, treatment and genetics (review of literature). Int. J. Mol. Sci. **22**(19), 10554 (2021)

Defect Segmentation in OCT Scans of Ceramic Parts for Non-destructive Inspection Using Deep Learning

Andrés Laveda-Martínez[1]([⊠]), Natalia P. García-de-la-Puente[1],
Fernando García-Torres[1], Niels Møller Israelsen[2,3], Ole Bang[2,3],
Dominik Brouczek[4], Niels Benson[5], Adrián Colomer[1], and Valery Naranjo[1]

[1] Instituto Universitario de Investigación en Tecnología Centrada en el Ser Humano,
Universitat Politècnica de València, Valencia, Spain
andreslaveda@gmail.com, napegar@upv.es
[2] Department of Electrical and Photonics Engineering, Technical University of
Denmark, Kongens Lyngby, Denmark
[3] NORBLIS ApS, Virum, Denmark
[4] Lithoz GmbH, Vienna, Austria
[5] airCode UG, Duisburg, Germany

Abstract. Non-destructive testing (NDT) is essential in ceramic manufacturing to ensure the quality of components without compromising their integrity. In this context, Optical Coherence Tomography (OCT) enables high-resolution internal imaging, revealing defects such as pores, delaminations, or inclusions. This paper presents an automatic defect detection system based on Deep Learning (DL), trained on OCT images with manually segmented annotations. A neural network based on the U-Net architecture is developed, evaluating multiple experimental configurations to enhance its performance. Post-processing techniques enable both quantitative and qualitative evaluation of the predictions. The system shows an accurate behavior of 0.979 Dice Score, outperforming comparable studies. The inference time of 18.98 s per volume supports its viability for detecting inclusions, enabling more efficient, reliable, and automated quality control.

Keywords: Non-Destructive Inspection · Ceramics · Defects · Segmentation · Optical Coherence Tomography

1 Introduction

Quality control in industrial manufacturing processes is essential to ensure the proper condition of the final product. Currently, most of these processes are carried out manually through visual inspection of the manufactured parts. This approach presents significant limitations related to operator subjectivity, visual fatigue, and the inability to detect certain subsurface defects that are not visible to the naked eye. Therefore, there is a growing need in the industry to develop

© The Author(s), under exclusive license to Springer Nature Switzerland AG 2026
L. Martínez et al. (Eds.): IDEAL 2025, LNCS 16239, pp. 171–182, 2026.
https://doi.org/10.1007/978-3-032-10489-2_15

quality control techniques with real-time monitoring systems that enable early and automatic defect detection during the manufacturing process [33].

To improve this process, Non-Destructive Testing (NDT) methods have gained particular importance in recent years. These techniques are highly valuable for assessing product integrity without causing damage and allow the detection of defects or anomalies in the early stages of production, thus preventing their propagation to later phases. There are numerous NDT methods, among which some stand out as the most commonly used. Firstly, industrial radiography is notable, as it employs short-wave X-rays, gamma rays, or neutrons to penetrate materials [1]. Another widely used technique is ultrasonic inspection, which is based on the transmission of mechanical waves to identify internal irregularities [12]. The liquid penetrant method is also one of the most common methods for detecting surface defects in dense materials [2]. Finally, thermography is used to analyse abnormal thermal distributions in materials [22].

Besides the techniques already described, there are other methods, originally developed for use in fields such as medicine, that are now gaining traction in industrial NDT [6,26,29]. One notable example is Optical Coherence Tomography (OCT), a technique that uses a low-coherence near-infrared light beam ($\sim$800 nm or $\sim$1300 nm) to obtain high-resolution cross-sectional images of both the surface and internal layers of a material [13]. Its ability to acquire three-dimensional information and its high resolution have made OCT widely used in medicine [30]. However, due to its potential, it has also begun to be applied in defect detection processes in industrial manufacturing [5,28]. In this work, we focus on its application to ceramic components manufactured using Lithography-based Ceramic Manufacturing (LCM), a process that enables the fabrication of complex, high-resolution ceramic parts. These components are particularly sensitive to internal defects, making them ideal candidates for the application of high-resolution, non-destructive imaging and automated defect detection [18,34].

Deep Learning (DL) is an advanced technique within the field of Artificial Intelligence (AI), based on deep neural networks, that enables automatic learning of complex patterns from data. Specifically, supervised learning utilises labelled datasets to train models that can predict on new samples. In this context, the combination of DL tools with NDT techniques enables the automatic analysis of images and the detection of defects without damaging the part, aiming to minimise material waste. DL has revolutionised industrial inspection by facilitating automatic detection of defects in images. Specifically, the U-Net architecture has been successfully used for defect detection tasks in industrial settings [31,32].

The U-Net architecture was initially developed to segment medical images [27] and has proven to be very useful in identifying structures or anomalies, such as tumours or cells, in various types of clinical images [10]. Its design allows for the combination of fine details with contextual information, making it highly effective for pixel-level segmentation, as demonstrated in the referenced studies. When used in conjunction with imaging techniques such as OCT, this network has successfully segmented internal structures in great detail in medical fields such as ophthalmology [7,17,23,25] or cancer diagnosis [20]. The results obtained in these studies, with Dice Similarity Coefficient values close to 98% or 99%, confirm that this combination can automatically detect regions of interest with high precision.

In the industrial sector, U-net architectures have been applied to detect defects in materials using X-ray or thermographic images [14,16]. In the context of ceramic parts, U-Net architectures have been used to detect cracks through visual inspection imaging and to segment the 3D microstructure in electron microscopy volumes [11,15]. Although the combination of U-Net and OCT has not yet been widely applied in industrial settings, some recent efforts have explored the use of machine learning techniques to analyse OCT data in LCM processes. In particular, Heise et al. [8] employed a mid-infrared OCT system alongside a pre-trained ResNet architecture to classify individual B-scans of ceramic components produced by LCM. Their network was able to distinguish between different types of defects, such as voids, inclusions, and contamination. This work highlights the potential of combining OCT imaging and DL for defect detection in additive manufacturing. However, this study was limited to 2D classification and did not address volumetric segmentation.

This study explores the application of a U-Net architecture for defect segmentation in OCT scans of ceramic components manufactured by LCM. Our contribution lies in adapting and validating this approach within an industrial context, demonstrating its potential to improve defect detection accuracy, reduce inspection time, and support the automation of quality control processes in ceramic manufacturing.

2   Methodology

The proposed approach for automatic segmentation of subsurface defects in ceramic parts using images obtained by OCT is described in this section. The method is based on the training of convolutional neural networks of type U-Net to identify defective regions in OCT volumes.

Problem Formulation: The problem was formulated as a supervised segmentation task, where each input image was paired with a binary mask representing the locations of internal defects. These image-mask pairs were used to train a neural network. Various model configurations were evaluated to identify the one that most accurately segmented defects in previously unseen images.

2.1   Generation of Binary Masks

Before training the models, it was necessary to generate binary masks through pixel-wise segmentation to train the network. As a first step, the OCT volumes were loaded into MATLAB's VolumeViewer tool. Manual segmentation was chosen due to the low visibility and lack of homogeneity of the defects present in this type of images, which made their reliable automatic identification difficult.

Several types of defects were considered, but finally, only inclusions were segmented, discarding the rest because they did not present sufficiently clear or consistent visual characteristics to be used for model training. By segmenting a single type of defect, a binary segmentation (defect/no defect) was performed.

In Fig. 1, we can observe the inclusion defect surrounded by a bounding box and its corresponding segmentation. The inclusion typically appears as a small bright region in the OCT image.

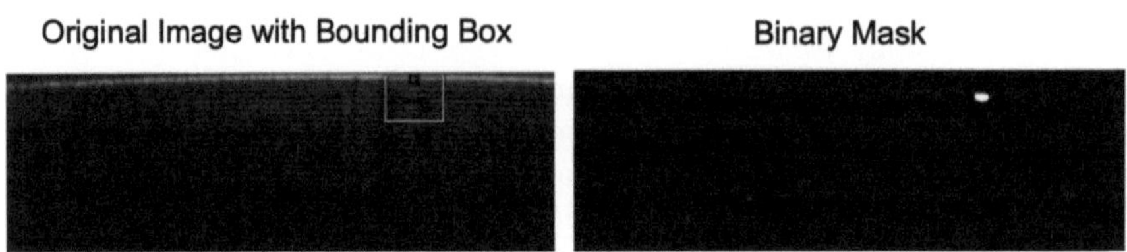

Fig. 1. Detail of the original OCT image showing the inclusion defect with bounding box and binary mask.

2.2 Preprocessing

Once the binary masks were created, and before training the network, it was necessary to preprocess the data by applying transformations to the original images and their masks. Some images initially have a size of 1024 × 700 pixels, others 2048 × 700, but it was observed that the lower part did not contain relevant information for training in any of the images in the entire dataset.

U-Net networks require input dimensions divisible by 32 to avoid decoding errors. To meet this requirement without losing information, horizontal padding was applied using the Albumentations library, adjusting the width from 700 to 704 pixels, along with two additional transformations.

Furthermore, a Crop transformation of 352 × 704 was applied to satisfy the divisible-by-32 condition and to crop the image, removing irrelevant parts outside the indicated region. At this stage, the irrelevant lower part was removed. This preprocessing not only allowed for standardising the input sizes, but also prevented common structural errors.

Therefore, once the corresponding masks were created and organised, they were preprocessed along with the images using the indicated transformations and input together with the images into the U-Net neural network for training.

2.3 U-Net Architecture

For the task of segmenting the defects in the OCT images of ceramic parts, the U-Net architecture was used [24]. This network is specially designed for segmentation, combining a compression phase that extracts relevant features with an expansion phase that reconstructs the segmentation pixel by pixel. The U-Net architecture is named after the "U" shape of its structure. It is made up of two main paths: a contraction phase (encoder) and an expansion phase (decoder).

Contraction Phase: In this first stage, the network extracts features from the input image by applying consecutive convolutional layers followed by max pooling. This process reduces spatial resolution while capturing the global context of the image.

Expansion Phase: The network reconstructs the original resolution of the feature map using upsampling operations and convolutions. At each step, the features of the encoder are combined, allowing the network to recover fine spatial details and improve edge accuracy in the segmentations. Thanks to this combination of encoding and decoding, U-Net can make very accurate predictions even with a small number of training images.

The implementation was carried out using SMP (Segmentation Models PyTorch). This tool allows loading architectures such as U-Net with different encoders (backbones). In this work, ResNet34 was used as the backbone, loaded with pretrained weights on ImageNet. This means that instead of starting training from scratch with random weights, prior knowledge is leveraged from a network that has already learned to identify basic visual patterns, which is very useful when working with a small dataset. In this case, a U-Net with an encoder depth parameter set to 5 was used. This indicates that the encoder consists of five main stages or blocks that extract features from the image at different resolutions. Each stage progressively reduces the spatial resolution of the image while increasing the number of channels to capture more complex information. This allows the model to understand patterns that range from very fine details to global structures.

2.4   Network Training and Model Evaluation

At this point, we created the binary masks through segmentation, preprocessed the images and masks, and built the neural network. The network was then trained and used to generate predictions.

In training mode, the images used belong to the training and validation sets. First, preprocessing is applied, which in this case consists of a crop. Then, the images and their ground truth (GT) masks are fed into the network. The GT refers to the manually segmented masks that serve as the reference during training. The network generates a prediction that is compared with the GT using a loss function. This loss is used to perform backpropagation, allowing the network's weights to be updated. Once all images have been processed over multiple epochs, the model with the trained weights is saved.

In evaluation mode, the images used belong to the test set. Just as in training, preprocessing is applied to adjust their size. The images are fed into the network, where the previously trained weights are used. The network outputs an image with per-pixel probability values. A threshold of 0.5 is applied to convert these probabilities into final binary predictions [4]. Once a model has been trained, it is necessary to objectively evaluate its performance. For this purpose, different quantitative metrics were used to evaluate the predictions, which allowed us to measure the performance and compare it with other models.

3   Experimental Settings

3.1   Dataset and Evaluation Metrics

The MIR-OCT Scans dataset, available in the Zenodo repository (https:// zenodo.org/records/15165755), contains 30 volumetric OCT scans of ceramic parts in PNG format, including annotations in the form of bounding boxes (BBs), which are coordinates that delimit the defect region and assist in manual segmentation of the volumes.

The volumes consist of between 700 and 1499 slices, with image resolutions of either 1024 × 700 or 2048 × 700 pixels, depending on the scan. They include three types of internal defects: pores, delaminations, and inclusions.

To train and evaluate the model, the 30 volumes were divided into three subsets: 21 for training, 6 for validation, and 3 for test. This split allowed the model to learn from the training data and be evaluated on previously unseen images. Each volume corresponds to a single ceramic part. Therefore, the test set includes volumes from parts different from those used in training and validation, ensuring a proper evaluation of the model's generalization capability. The division is presented in Table 1, which shows the number of volumes used in each subset, as well as the corresponding number of B-scans per set.

Table 1. Dataset split statistics: number of volumes and B-Scans in each set.

Set	No. of Volumes	No. of B-Scans
Train	21	16033
Validation	6	4449
Test	3	2069

Metrics such as Precision, Recall, and Dice Similarity Coefficient (DSC) were used [21]. The combined use of Precision, Recall, and DSC made it possible to evaluate not only the number of defects detected, but also the accuracy of the segmentations and their overlap with the ground truth.

3.2   Implementation Details

Fixed hyperparameters were defined for training, such as a batch size of 16, a learning rate of 0.001, and the Adam optimizer. A model checkpointing strategy was implemented based on improvements in the validation loss, combined with early stopping with a patience of 15 epochs. This approach stops training when no improvement is observed, helping to prevent overfitting and reduce computational cost. After training, the final model and its weights were saved.

To improve performance, five loss configurations were evaluated: BCE, DLS, their equal combination, and two weighted variants (BCE 0.7 + DLS 0.3 and BCE 0.3 + DLS 0.7), as summarized in Table 2.

Table 2. Summary of training configurations and loss functions used.

N.º Train	Loss Function
1	BCE
2	Dice Loss (DLS)
3	BCE + DLS
4	BCE (0.7) + DLS (0.3)
5	BCE (0.3) + DLS (0.7)

Regarding the hardware, we used an NVIDIA DGX system, equipped with 8× NVIDIA A100 GPUs (40 GB each), an AMD EPYC 7742 processor (64 cores, 2.25 GHz, 256 threads), and 1 TB of DDR4 RAM (16 × 64 GB, 3200 MHz). The software used included PyTorch for model training, Albumentations for data augmentation, OpenCV and NumPy for image preprocessing, and Matplotlib for results visualization.

4   Results

The validation results for each training configuration are presented in Table 3, showing the performance on the validation set corresponding to each trained model. It can be observed that the first model, trained using the BCE loss function, achieves the best performance metrics during training. To verify the generalization capability of the trained models, they were evaluated on the test set, and the corresponding performance metrics are reported in Table 4. The reported test metrics correspond to the average values across these volumes.

Table 3. Segmentation performance metrics for each training configuration in the validation set.

N.º Train	DSC	Precision	Recall
1	**0.944**	**0.967**	0.923
2	0.938	0.938	**0.937**
3	0.926	0.920	0.932
4	0.938	0.940	0.936
5	0.914	0.898	0.930

Table 4. Segmentation performance metrics for each training configuration in the test set and inference time.

N.º Train	DSC	Precision	Recall	Inference (s)
1	0.974	0.978	0.973	**18.526**
2	0.959	0.963	0.958	18.543
3	0.978	0.982	0.977	18.546
4	**0.979**	**0.983**	**0.978**	18.983
5	0.974	0.979	0.973	18.593

Training configuration 4 was selected as the final model due to its superior performance in the test set, achieving the highest values of DSC (0.979), Precision (0.983), and Recall (0.978).

To demonstrate the effectiveness of our approach, we compared it with a similar study on microvessel segmentation in IVOCT images [19], which reported a DSC of 0.73±0.10. In contrast, our model achieved a DSC of 0.979, demonstrating higher segmentation accuracy and robustness in defect detection.

Although the other study primarily focuses on defect detection rather than segmentation, we also mention it because it uses OCT imaging to detect inclusions in ceramic components [9], reporting an F1 Score of 0.76. In contrast, our work addresses segmentation-level analysis based on OCT, which is crucial for more detailed industrial inspection. This further highlights the importance of OCT-based segmentation analysis for industrial applications, especially given the superior results achieved by our method.

Our model processes volumes of $1024 \times 700 \times 700$ in 18.98 s, with an average time of 0.027 s per slice. In comparison, a similar study on OCT-based cancer organ segmentation [3] uses volumes of $1200 \times 800 \times 698$, reporting an inference time of 30 s per volume and 0.043 s per slice. This indicates that our method is more efficient both in processing the entire volume and on a per-image basis.

Figure 2 showcases representative segmentation results in comparison with manual annotations. The model demonstrates high accuracy in detecting and localizing defects, with outputs that closely match expert labels. These results provide strong evidence of the effectiveness and reliability of the proposed approach. Nonetheless, as shown in Fig. 3, occasional failure cases occur, where certain defects remain undetected. Such instances, however, are relatively infrequent and highlight opportunities for further refinement.

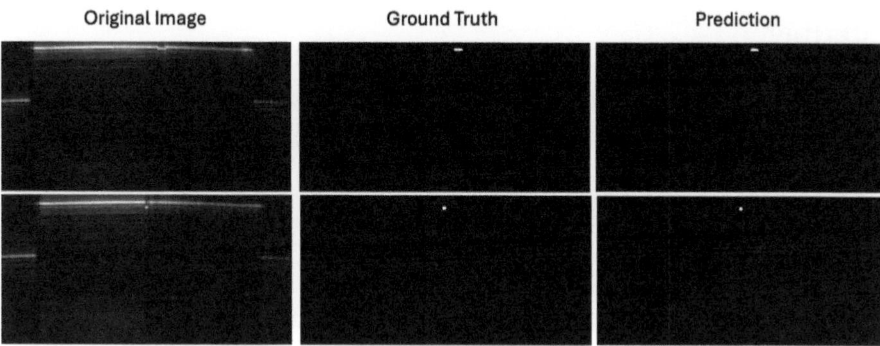

Fig. 2. Representative segmentation results (Prediction) compared with manual annotations (Ground Truth). The model consistently achieves accurate defect localization, showing strong concordance with expert labels.

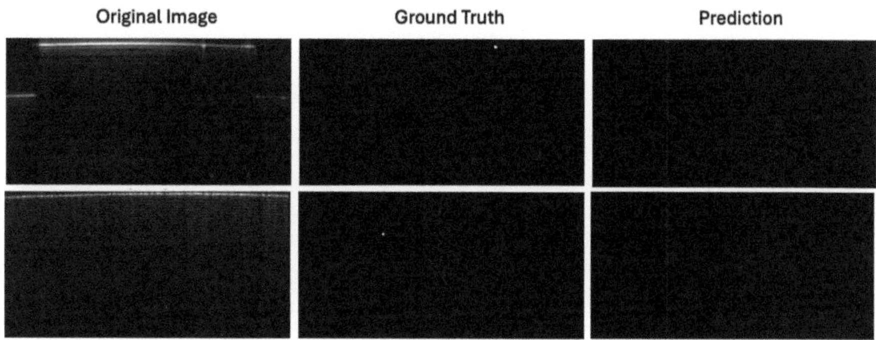

Fig. 3. Examples of a segmentation outcome (no Prediction) where the model did not capture the annotated defect, illustrating a failure case.

5   Conclusion

This work demonstrates the successful application of a U-Net architecture for automatic defect segmentation in OCT scans of ceramic parts. The selected model achieved high accuracy, with a DSC of 0.979, outperforming comparable studies. The efficient inference time of 18.98 s per volume supports its viability for industrial quality control. The combination of precise segmentation and fast processing enables reliable automatic defect detection, with the potential to reduce inspection costs and improve manufacturing standards.

Despite the good results of the trained model, this study has some important limitations. First, the dataset is relatively small, with only 30 volumes, which may limit the model's ability to generalize to diverse situations. Second, the analysis focused exclusively on inclusion defects, as a preliminary step toward evaluating the feasibility of automated defect detection, while extending the approach to additional defect types remains an opportunity for future work.

Funding

This project has received funding from Horizon Europe, the European Union's Framework Programme for Research and Innovation, under Grant Agreement No. 101057404 (ZDZW), and Grant Agreement No. 101058054 (TURBO). This project has received funding from Villum Fonden (2021 Villum Investigator project No. 00037822: Table-Top Synchrotrons). Generalitat Valenciana partially funded this work through the project CIPROM/2022/20. The work of NPG was supported by the grant PID2022-140189OB-C21 funded by MICIU/AEI/10.13039/501100011033 ERDF/UE and FSE+.

References

1. Andersson, C., Ingman, J., Varescon, E., Kiviniemi, M.: Detection of cracks in multilayer ceramic capacitors by x-ray imaging **64**, 352–356. https://doi.org/10.1016/j.microrel.2016.07.110, https://www.sciencedirect.com/science/article/pii/S0026271416302542
2. Berroa, R.P.B.: Inspección visual y líquidos penetrantes en uniones soldadas. http://repositorio.unsa.edu.pe/handle/UNSA/2601 (2015), accedido: 26 de mayo de 2025
3. Branciforti, F., et al.: Segmentation and multi-timepoint tracking of 3d cancer organoids from optical coherence tomography images using deep neural networks. Diagnostics **14**(12) (2024). https://doi.org/10.3390/diagnostics14121217, https://www.mdpi.com/2075-4418/14/12/1217
4. Brownlee, J.: Threshold-moving for imbalanced classification (2021), https://www.machinelearningmastery.com/threshold-moving-for-imbalanced-classification/, Accessed 10 July 2025
5. Czajkowski, J., Prykäri, T., Alarousu, E., Palosaari, J., Myllylä, R.: Optical coherence tomography as a method of quality inspection for printed electronics products **17**(3), 257–262. https://doi.org/10.1007/s10043-010-0045-0, https://doi.org/10.1007/s10043-010-0045-0
6. Fu, M.Y., et al.: The progress of optical coherence tomography in industry applications **5**, 0053. https://doi.org/10.34133/adi.0053, https://spj.science.org/doi/full/10.34133/adi.0053, publisher: American Association for the Advancement of Science
7. García-Torres, F., et al.: Using diffusion models for data augmentation on limited rodent OCT datasets. In: Intelligent Data Engineering and Automated Learning – IDEAL 2024: 25th International Conference, Valencia, Spain, 20–22 November 2024, Proceedings, Part I, pp. 313–324. Springer-Verlag, https://doi.org/10.1007/978-3-031-77731-8_29
8. Heise, B., et al.: Mid-infrared optical coherence tomography and machine learning for inspection of 3d-printed ceramics at the micron scale **11**. https://doi.org/10.3389/fmats.2024.1441812, publisher: Frontiers
9. Heise, B., et al.: Mid-infrared optical coherence tomography and machine learning for inspection of 3d-printed ceramics at the micron scale. Front. Mater. **11** (2024). https://doi.org/10.3389/fmats.2024.1441812, https://www.frontiersin.org/journals/materials/articles/10.3389/fmats.2024.1441812

10. Hill, C., et al.: Three-dimension epithelial segmentation in optical coherence tomography of the oral cavity using deep learning. https://www.mdpi.com/2072-6694/16/11/2144, accedido: 25 de mayo de 2025
11. Hirabayashi, Y., Iga, H., Ogawa, H., Tokuta, S., Shimada, Y., Yamamoto, A.: Deep learning for three-dimensional segmentation of electron microscopy images of complex ceramic materials **10**(1), 46. https://doi.org/10.1038/s41524-024-01226-5, https://www.nature.com/articles/s41524-024-01226-5, publisher: Nature Publishing Group
12. Ibáñez, D.E.: Control de calidad de puntos de soldadura mediante inspección por ultrasonidos. https://academica-e.unavarra.es/handle/2454/2107 (2010), accedido: 26 de mayo de 2025
13. Israelsen, N.M., et al.: Real-time high-resolution mid-infrared optical coherence tomography. Light: Sci. Appl. **8**(1), 11 (2019)
14. Jha, S.B., Babiceanu, R.F.: Deep cnn-based visual defect detection: survey of current literature. Comput. Ind. **148**, 103911 (2023). https://doi.org/10.1016/j.compind.2023.103911
15. Junior, G.S., Ferreira, J., Millán-Arias, C., Daniel, R., Junior, A.C., Fernandes, B.J.T.: Ceramic cracks segmentation with deep learning **11**(13), 6017. https://doi.org/10.3390/app11136017, https://www.mdpi.com/2076-3417/11/13/6017, number: 13 Publisher: Multidisciplinary Digital Publishing Institute
16. Konovalenko, I., Maruschak, P., Brezinová, J., Prentkovskis, O., Brezina, J.: Research of u-net-based cnn architectures for metal surface defect detection. Machines **10**(5), 327 (2022). https://doi.org/10.3390/machines10050327
17. Kugelman, J., et al.: A comparison of deep learning u-net architectures for posterior segment oct retinal layer segmentation. Sci. Rep. **12**(1), 14888 (2022). https://doi.org/10.1038/s41598-022-18646-2
18. Lapre, C., et al.: Rapid non-destructive inspection of sub-surface defects in 3d printed alumina through 30 layers with 7 m depth resolution **18**, 100611. https://doi.org/10.1016/j.oceram.2024.100611, https://www.sciencedirect.com/science/article/pii/S2666539524000750
19. Lee, J., et al.: Automated segmentation of microvessels in intravascular oct images using deep learning. Bioengineering (Basel) **9**(11), 648 (2022). https://doi.org/10.3390/bioengineering9110648, https://www.ncbi.nlm.nih.gov/pmc/articles/PMC9687448/, pMCID: PMC9687448
20. Liu, X., Ouellette, S., Jamgochian, M., Liu, Y., Rao, B.: One-class machine learning classification of skin tissue based on manually scanned optical coherence tomography imaging. Sci. Rep. **13**(1), 1–9 (2023). https://doi.org/10.1038/s41598-023-28155-5
21. Mahmoudi, R., Benameur, N., Mabrouk, R., Mohammed, M., Garcia-Zapirain, B., Bedoui, M.: A deep learning-based diagnosis system for covid-19 detection and pneumonia screening using ct imaging. Appl. Sci. **12**(10), 4825 (2022). https://doi.org/10.3390/app12104825, https://doi.org/10.3390/app12104825
22. Maldague, X.P.: Nondestructive evaluation of materials by infrared thermography. Springer (2012), iSBN: 9781447119951, Accedido: 26 de mayo de 2025
23. Morales, S., et al.: Retinal layer segmentation in rodent OCT images: local intensity profiles & fully convolutional neural networks **198**, 105788. https://doi.org/10.1016/j.cmpb.2020.105788, https://www.sciencedirect.com/science/article/pii/S0169260720316217
24. Mota, A.G.: TFG_ALEJANDRO_GARCIA_MOTA.pdf. https://oa.upm.es/75220/3/TFG_ALEJANDRO_GARCIA_MOTA.pdf (2025), Accessed 26 May 2025

25. Oh, D., et al.: Gcn-assisted attention-guided unet for automated retinal oct segmentation. Expert Syst. Appl. **249**, 123620 (2024). https://doi.org/10.1016/j.eswa.2024.123620
26. Petersen, C.R., et al.: Non-destructive subsurface inspection of marine and protective coatings using near- and mid-infrared optical coherence tomography **11**(8), 877. https://doi.org/10.3390/coatings11080877, https://www.mdpi.com/2079-6412/11/8/877, number: 8 Publisher: Multidisciplinary Digital Publishing Institute
27. Ronneberger, O., Fischer, P., Brox, T.: U-net: convolutional networks for biomedical image segmentation. In: Navab, N., Hornegger, J., Wells, W.M., Frangi, A.F. (eds.) Medical Image Computing and Computer-Assisted Intervention – MICCAI 2015, pp. 234–241. Springer. https://doi.org/10.1007/978-3-319-24574-4_28
28. Su, R., Kirillin, M., Chang, E.W., Sergeeva, E., Yun, S.H., Mattsson, L.: Perspectives of mid-infrared optical coherence tomography for inspection and micrometrology of industrial ceramics **22**(13), 15804–15819. https://doi.org/10.1364/OE.22.015804, https://opg.optica.org/oe/abstract.cfm?uri=oe-22-13-15804, publisher: Optica Publishing Group
29. Su, R., Kirillin, M., Ekberg, P., Roos, A., Sergeeva, E., Mattsson, L.: Optical coherence tomography for quality assessment of embedded microchannels in alumina ceramic **20**(4), 4603–4618. https://doi.org/10.1364/OE.20.004603, https://opg.optica.org/oe/abstract.cfm?uri=oe-20-4-4603, publisher: Optica Publishing Group
30. Swanson, E.A., Fujimoto, J.G.: The ecosystem that powered the translation of oct from fundamental research to clinical and commercial impact [invited]. Biomed. Opt. Express **8**(3), 1638–1664 (2017). https://doi.org/10.1364/BOE.8.001638
31. Usamentiaga, R., Lema, D.G., Pedrayes, O.D., Garcia, D.F.: Automated surface defect detection in metals: a comparative review of object detection and semantic segmentation using deep learning **58**(3), 4203–4213. https://doi.org/10.1109/TIA.2022.3151560, https://ieeexplore.ieee.org/abstract/document/9713940
32. Wang, W., Li, Q., Zhang, D., Fu, J.: Image segmentation of adhesive ores based on msba-unet and convex-hull defect detection. Eng. Appl. Artif. Intell. **123**, 106185 (2023). https://doi.org/10.1016/j.engappai.2023.106185
33. Yun, J.P., Shin, W.C., Koo, G., Kim, M.S., Lee, C., Lee, S.J.: Automated defect inspection system for metal surfaces based on deep learning and data augmentation. J. Manuf. Syst. **55**, 317–324 (2020). https://doi.org/10.1016/j.jmsy.2020.03.009, https://www.sciencedirect.com/science/article/pii/S027861252030042X
34. Zorin, I., et al.: Mid-infrared optical coherence tomography as a method for inspection and quality assurance in ceramics additive manufacturing **12**, 100311. https://doi.org/10.1016/j.oceram.2022.100311, https://www.sciencedirect.com/science/article/pii/S2666539522000943

Hybrid Machine VisionDigital Twin Approach for Quality Control

Sergio Illana Rico, Silvia Satorres Martínez(✉), Elisabet Estévez Estévez, Alejandro Sánchez García, and Diego M. Martínez Gila

Robotics, Automation and Computer Vision Group,Electronic and Automation Engineering Department, University of Jaén, Jaén, Spain
satorres@ujaen.es

Abstract. This paper presents a hybrid framework that combines a Machine Vision System (MVS) and a Digital Twin (DT), to enable proactive quality control in industrial environments. Grounded in the ISO 23247 standard, the proposed architecture provides support to transform traditional reactive inspection systems into intelligent, collaborative platforms capable of early defect detection and technical factor analysis. For that, the DT integrates a deep learning-based defect identification using the YOLO architecture. Then, a multi-agent platform, developed with SPADE, orchestrates the synchronization between the physical system (MVS) and its DT, supporting feedback mechanisms and enabling data-driven interventions.

The methodology is validated through an industrial use case involving the automated inspection of headlamp lenses, a process characterized by multiple defect sources across several production stages. Results demonstrate that the integration of DT and AI technologies significantly improves defect traceability, minimizes rework, and enhances process adaptability, aligning the system with Industry 5.0 principles.

Keywords: Machine Vision · Digital Twins · Industry 5.0 · Proactive Quality Control · Zero Defect Manufacturing

1 Introduction

The current industrial landscape is shaped by the simultaneous presence of two transformative paradigms: Industry 4.0 and Industry 5.0. While Industry 4.0 focuses primarily on automation, data analysis, and cyber-physical systems, Industry 5.0 introduces a human-centric view that emphasizes the synergistic collaboration between human operators and advanced digital technologies [15]. Rather than replacing human input, this new industrial vision seeks to reintroduce human value at the core of production systems paving the way for more sustainable and high-quality production.

Expanding the industrial paradigm has a direct impact on manufacturing processes, driving a redefinition of their structure, goals, and modes of operation.

© The Author(s), under exclusive license to Springer Nature Switzerland AG 2026
L. Martínez et al. (Eds.): IDEAL 2025, LNCS 16239, pp. 183–194, 2026.
https://doi.org/10.1007/978-3-032-10489-2_16

Among these, quality control stands out as one of the areas most significantly influenced by this conceptual shift. Under Industry 4.0, quality assurance primarily relied on automated inspection systems and statistical analysis embedded within cyber-physical architectures. In contrast, Industry 5.0 embraces a more proactive approach, where human expertise complements data-driven insights to ensure product excellence and enhance process adaptability [3].

One of the most widely used technologies for quality control in the manufacturing sector is Machine Vision (MV), a branch of computer vision specifically designed for industrial applications that enables automated inspection. MV systems are also undergoing a notable transformation in the transition from Industry 4.0 to Industry 5.0. Under the Industry 4.0 paradigm, MV systems function primarily as autonomous tools for high-speed defect detection, leveraging pattern recognition and deep learning algorithms to replace traditional manual inspection. In contrast, their role in Industry 5.0 is being redefined to encompass not only automation, but also meaningful collaboration with human operators [13]. As a result, MV technologies evolve from isolated detection units, known as reactive, into intelligent, interactive components of human-centric quality assurance processes, designed as proactive.

The transition from reactive to proactive quality control processes entails a structural transformation and one paradigm that enables this change is Zero Defect Manufacturing (ZDM). As defined by Psarommatis et al. [9], ZDM is a holistic approach aimed at ensuring both process and product quality by minimizing defects through corrective, preventive, and predictive strategies, primarily supported by data-driven technologies. Its objective is to guarantee that no defective products leave the production system, thereby enhancing manufacturing sustainability. Given that ZDM not only seeks to eliminate defects but also to optimize resource utilization and reduce waste, it constitutes an ideal framework for quality assurance within the human-centric and sustainability-oriented vision of Industry 5.0.

To operationalize ZDM, the use of Digital Twins (DTs) has been identified as one of the enabling technologies. Unlike their role in Industry 4.0, where DTs primarily supported autonomous process control and predictive maintenance, their function in Industry 5.0 is extended to facilitate effective human-machine collaboration. By providing a continuously updated data from the physical asset, DTs allow human operators to interact with, interpret, and influence production processes based on contextual data and expert knowledge.

The inherently interdisciplinary nature of DTs, along with the wide array of technologies they integrate, poses significant challenges for manufacturers. There is a clear need for standardized definitions of terms, concepts, and reference models. In 2022, ISO 23247 series of standards [4] established a generic framework for DT in manufacturing. Building upon this generic framework, a highly effective approach for its implementation involved the adoption of Multi-Agent Systems (MAS) [14]. MAS architectures naturally align with the decentralized, intelligent, and flexible requirements of modern manufacturing environments, offering distinct advantages over traditional paradigms.

This paper explores a framework based on ISO23247 and MAS for giving support to a DT of a Machine Vision System. The DT transforms the MVS from a reactive into a proactive one, enabling identification of the production stage where the defect originated, allowing an early intervention to prevent defects before they occur. We detail the methodology for DT implementation and present the framework for an industrial use case: the automated inspection of headlamps.

2   Tools for Proactive Quality Control

Proactive quality control focuses on preventing defects and process deviations before they occur, rather than detecting and correcting them afterwards. Unlike traditional reactive quality control methods, which typically rely on post-production inspection to identify non-conformities, this strategy emphasizes early detection through continuous monitoring and feedback mechanisms [1].

One industry known for its rigorous quality control is the automotive sector. Given the critical safety requirements and intense competitive pressures, quality assurance is embedded at every stage of the production process. In Saadallah et al. [11] a convolutional neural network (CNN) was trained using a computer vision sensor data from electronic circuit boards. The system generated heatmaps that highlighted defective regions in PCB images, enabling early prediction of faults during the electronic assembly process. In another study, Cavaliere et al., [16] deep learning was applied to inspect the surfaces of die-cast components in hybrid vehicles. The model was trained using images labeled with visual defects and demonstrated the ability to generalize to new parts with similar characteristics. Finally, Shahin et al. [12] introduced a machine-learning-based inspection strategy with edge computing on a PCB production line, showing that machine learning models could predict final product quality in real time.

Research has also extended to specialized domains such as quality control of the car chassis painting process or the assembling, where quality issues can be costly. For example, one study [6] focused on proactive quality control in automotive coating processes by using hierarchical decision-making algorithms to reduce paint defects. By analysing environment sensor data (humidity, particulate count) and process parameters, the system could predict runs or fish-eyes in paint before they occurred, allowing adjustments in ventilation or paint mix. Another work by Leberruyer et al. [5] presented an industrial application of AI for ZDM in automotive, achieving near-zero defects in a certain assembly process through real-time monitoring and automated corrective feedback. They noted challenges such as data preparation and model training time, but overall demonstrated that AI can be an effective enabler for ZDM in manufacturing.

In summary, modern automotive manufacturing systems are evolving into intelligent ecosystems that detect anomalies and implement corrective actions in a closed-loop manner. Although most of the solutions are ad-hoc, our proposal goes one step further. It is independent of the purpose of the DT, providing support to the definition of DT and its physical counterpart.

3   Methodology

The proposed framework to provide a proactive quality control system is fully aligned with the ISO 23247 standard for Digital Twins (DT) in manufacturing. So, this section begins identifying the fundamental principles and core components defined by the ISO 23247 framework. Building on this foundation, it is introduced a proposed architecture leveraging SPADE (Smart Python Agent Development Environment), a robust open-source platform for developing a Multi-Agent System (MAS) compliant with ISO 23247 requirements.

3.1   The Conceptual Framework

Our approach adheres to the 2020 definition provided by the Digital Twin Consortium, which defines a DT as "a virtual representation of real-world entities and processes, synchronized at a specified frequency and fidelity" [4]. It is important to emphasize that the DT is not merely a digital model or replica of the Machine Vision System (MVS); rather, it serves as an extension that enhances the inspection system by enabling additional value-added functionalities.

The first step is the design of a conceptual framework capable of:

- Ensuring real-time synchronization between physical assets (e.g. MVS) and their digital counterparts (e.g. DT).
- Facilitating interoperability among heterogeneous systems and devices.
- Strengthening data-driven decision-making processes to support the deployment of advanced value-added functionalities.

The physical asset, represented in our case by the MVS, is defined in the ISO 23247 standard as an Observable Manufacturing Element (OME). It is necessary to ensure the synchronization of both: the OME and the DT. For that, the framework comprises the following layers and entities (Fig. 1):

- Domain layers:
  - Observable Manufacturing Domain: It contains the physical OMEs.
  - Digital Twin Domain: It houses the digital entities representing the OMEs.
  - Device Communication Domain: Acting as an interface, this domain contains entities for sensors and actuators/controllers, facilitating bidirectional links between the physical and digital realms.
  - User Domain: This upper layer facilitates interaction between users, including software systems like Product Life-cycle Management(PLM), Manufacturing Execution System (MES), Enterprise Resource Planning (ERP), and Manufacturing Operations Management (MOM), as well as human-machine interfaces, and both the OMEs and their digital twins.
- Functional entities:
  - Device Communication Entity: This entity contains two sub-entities:
    * Data Collection Sub-entity: Collects data from OMEs utilizing sensors.

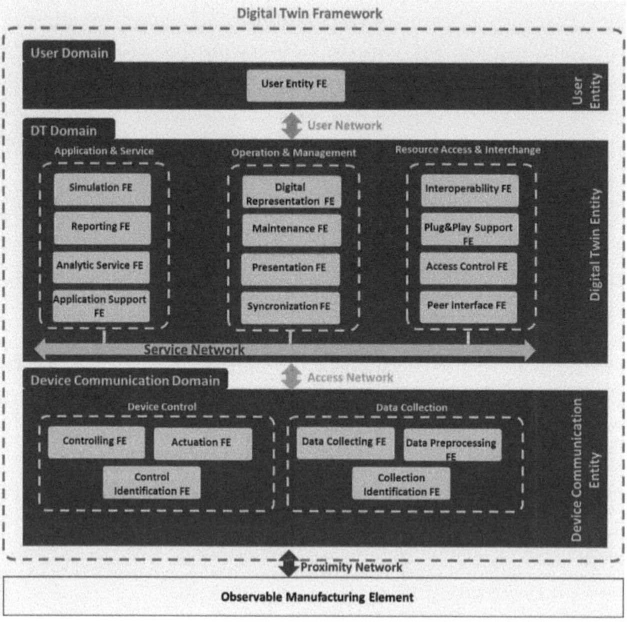

Fig. 1. General Scenario of ISO 23247 proposed framework (adapted from [4]).

      * Device Control Sub-entity: Controls and acts upon the OMEs.
- Digital Twin Entity: This entity digitally represents the OMEs and is further composed of three sub-entities:
  - * Operation and Management Sub-entity: Responsible for maintaining information regarding OMEs, digital modelling, presentation, and synchronization.
  - * Application and Service Sub-entity: Provides functionalities such as simulation, data analysis, and production status reporting.
  - * Resource Access and Exchange Sub-entity: Enables the exchange of information between the digital twin entity and the user entity, with support for interoperability.
- User Entity: This entity accommodates application software systems and human-machine interfaces.

3.2 Proposed Architecture

To bring the conceptual framework to life, it is proposed an architecture based on Multi-Agent-Systems (MAS). In MAS agents can independently monitor, analyse, and react to dynamic process conditions, moving away from monolithic control paradigms. This inherent flexibility and reconfigurability allows for easy addition, removal, or modification of agents, enabling manufacturing lines to quickly adapt to changes in production or equipment. MAS promotes enhanced interoperability through adherence to standardized communication protocols like FIPA, facilitating seamless information exchange between diverse systems and vendors, which is crucial for meeting ISO 23247 data exchange requirements. The decentralized design ensures robustness and fault tolerance, allowing the system to operate seamlessly despite individual agent failures, and thus reducing downtime.

Given these benefits, SPADE (Smart Python Agent Development Environment) was selected as the robust open-source framework for developing and deploying this multi-agent system. The proposed architecture implements ISO 23247 s Functional Entities (FE) through specialized agents:

- DataCollecting Agent: Maps to Data Collecting FE (partially Actuation FE), responsible for real-time OME data acquisition via OPC-UA and storage in In-fluxDB (this DB contains data of OPC-UA information model).
- DeviceControl Agent: Maps to Device Control / Actuation FE, for executing control actions on the OME via OPC-UA based on messages from the ManagementAgent.
- DTRepresentation Agent: Maps to Digital Representation FE, maintaining and updating a coherent digital state of the OME.
- MachineStateUpdater Agent: Maps to Synchronization FE, comparing real and digital states to detect inconsistencies.
- Viewer Agent: Maps to Presentation FE, visualizing device states in the user interface via UDP.
- Management Agent: Maps to Maintenance FE and Interoperability Support FE, orchestrating message routing, validating operations, and loading XML models.

In the proposed Platform, the OPC-UA Information Model plays a fundamental role. Thanks to this model it is known which data and with which permissions will be available from the OME. The structure of InfluxDB is also fixed by this information model. In addition to the system's own agents described above, Application Agent is also essential. This agent is specifically tasked with implementing the DT purpose. Current version of platform, this Application Agent is a skeleton serving to seamlessly integrate specific applications with the broader agent ecosystem of the system.

4   Use Case: the Automated Inspection of Headlamp Lenses

The production of headlamp lenses involves a multi-stage process that includes injection molding, coating, and drying (Fig. 2). Each stage can generate different types of defects, affecting the quality of the final product.

During injection molding, typical defects include black dots (from degraded material), burrs (due to mold mismatch or excessive pressure), sink marks (from uneven cooling), sprue cutting marks, and mold damage transfer (such as nicks or

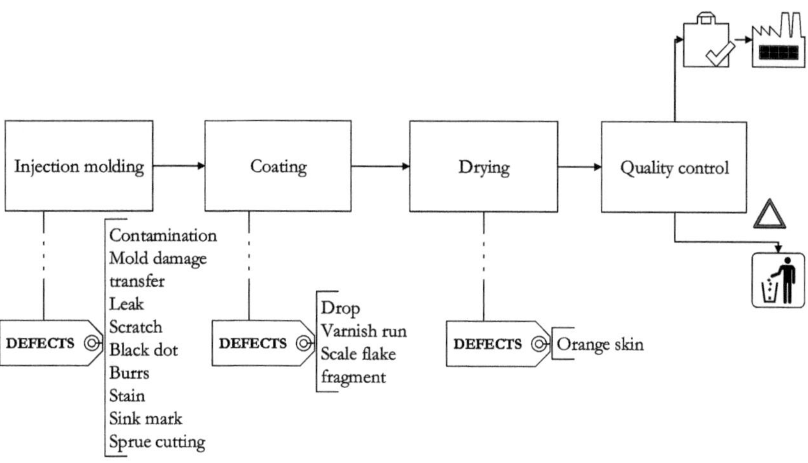

Fig. 2. Overview of the headlamp lens manufacturing process, outlining the key production stages and the common defects that may arise at each step.

particle impressions) are replicated on the part. Contamination from embedded particles can also result in raised surface defects.

In the coating phase, which applies a protective varnish layer, defects like varnish run, drops and scale flake fragment may occur due to improper application, surface tension issues, or over-spraying. Finally, during the drying/curing phase, issues such as orange skin can arise from poor solvent evaporation control or airborne contamination.

In the quality control phase, the inspection of the lens is carried out. Defective parts go to scrap while good parts go to the assembly line. The Machine Vision System (MVS) is the core of the quality control stage and enables the automated inspection of the lens.

Fig. 3 shows an example of these defects. Once produced, they cannot be eliminated; therefore, it is essential to prevent them or at the very least minimize their occurrence, whenever possible. The following subsections present the current reactive MVS and detail the necessary upgrades to transform it into a proactive quality control system.

4.1   Reactive Quality Control MVS

To enable automated inspection of headlamp lenses, it is essential to have an industrial system that supports real-time inspection capabilities. The authors of this work developed in [8] the basis for the development of industrial systems capable of inspecting this type of parts. The lighting system and the computer vision algorithms, core of the MVS, were also deeply detailed in [7].

The reactive quality control MVS detects common defects in headlamp lenses above 0.5 mm in diameter, since smaller flaws are typically not visible to the human eye. A capability analysis of the system was carried out according to the

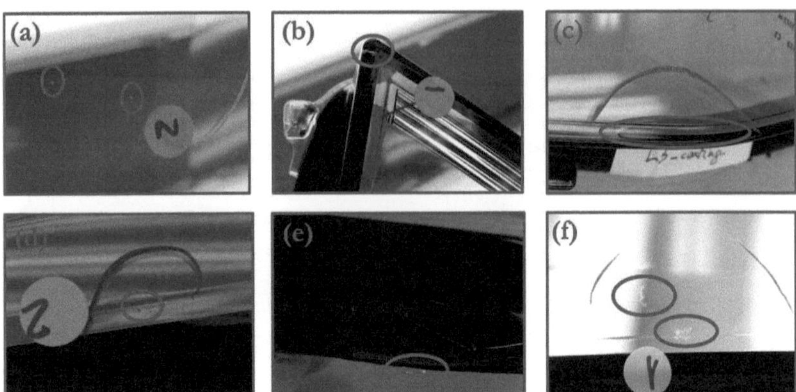

Fig. 3. Samples of defective headlamp lenses: (a) Mold damage transfer; (b) Leak; (c) Varnish run; (d) Black dot; (e) Burr; (f) Stains.

Automotive Industry Action Group [2] demonstrating that the MVS precision, repeatability and reproducibility, was acceptable. The cycle time met industry requirements, performing a part inspection in less than 30 s.

Fig. 4a shows the inside of the MVS while an inspection of a headlamp lens, affected by a varnish run, is being carried out. The mobile stripes of the lighting system enhance the defect (Fig. 4b) and the image preprocessing, designed as aspect image (Fig. 4c), provides an image where the defect can be located.

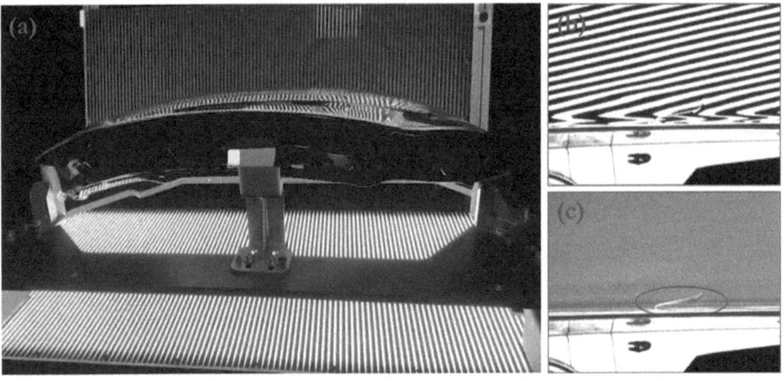

Fig. 4. Reactive quality control MVS: (a) Inside the MVS during the inspection; (b) Image acquisition; (c) Aspect image.

Achieving full inspection of a headlamp lens at the required spatial resolution necessitates the use of multiple cameras and a synchronization between the lighting device and cameras. A detailed explanation of the image acquisition and processing is available in [7]. In summary, the MVS operational cycle initiates with the automated placement of the lens. Image acquisitions are synchronized with the lighting system displacement and once the image sequence is completed, the processing starts. The first step is the preprocessing, and as result the *aspect image* is available. In the *aspect image*, defects appear well contrasted on an homogeneous background. Global thresholding algorithms, specifically adapted to images that present unimodal histograms, has a good performance for image segmentation. The image processing ends with the classification of defective and non-defective of the extracted regions. If one camera detects a defective region, the whole part is labelled as KO. The operational cycle then finishes with the inspection result presentation and the automated uploading of the lens.

In addition, in each cycle, the MVS is responsible for saving all the aspect and segmented images to an SFTP server. Concurrently, it stores the metadata of these images (name, size, resolution, date, tags, camera ID, analysis result, and storage path) in a MongoDB database. Precisely these data constitute the information model of the OPC-UA server.

4.2 Proactive Quality Control MVS

Achieving proactive quality control with the MVS requires a thorough analysis of technical factors. This process involves two key steps: first, identifying the types of defects, and second, mapping them to the corresponding stages of the manufacturing process. Both are the application support of the DT.

So, for the first step, defect identification, the Label Studio application has been utilized. Label Studio provides an intuitive graphical interface where users can draw bounding boxes, assign defect categories, and add metadata to each annotation, ensuring consistency and traceability across the dataset. This process not only streamlines the creation of high-quality training data for machine learning models but also allows for collaborative review and validation by domain experts. Figure 5 shows a screenshot of the Label Studio interface during the annotation of headlamp lens defects.

Once the ground-truth is generated, advanced deep learning models have to be applied to automatically identify the defect type. In this work, the YOLO (You Only Look Once) architecture [10] has been used for this purpose. YOLO is a single-stage object detection framework that divides the input image into a grid and predicts bounding boxes and class probabilities for each cell in a single forward pass, enabling high-speed inference suitable for industrial environments.

Main YOLO components are: the backbone, the neck and the head and in this application the YOLOv8 model has been used. The backbone extracts multi-level visual features from the input aspect image of a car headlamp lens using a sequence of convolutional layers, reducing spatial resolution while capturing increasingly abstract patterns relevant to defect detection. The neck then merges features across different scales, through structures like PANet or FPN, to ensure that both small and large defects can be accurately identified. Finally, the head processes the aggregated feature maps to predict bounding boxes, objectness scores, and class probabilities. These predictions are overlaid onto the original

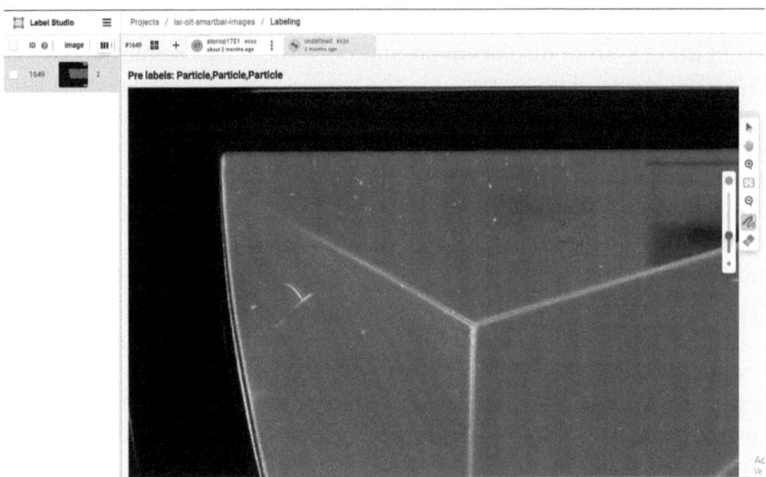

Fig. 5. Screenshot of the Label Studio interface during the annotation of headlamp lens defects.

image, producing an annotated output where each visible defect on the headlamp lens is highlighted in real time.

For the second step, mapping to the manufacturing stage, a decision tree has been implemented. Based on the information presented in Fig. 3, once the defect type is identified, it becomes straightforward to infer the manufacturing stage in which it was originated.

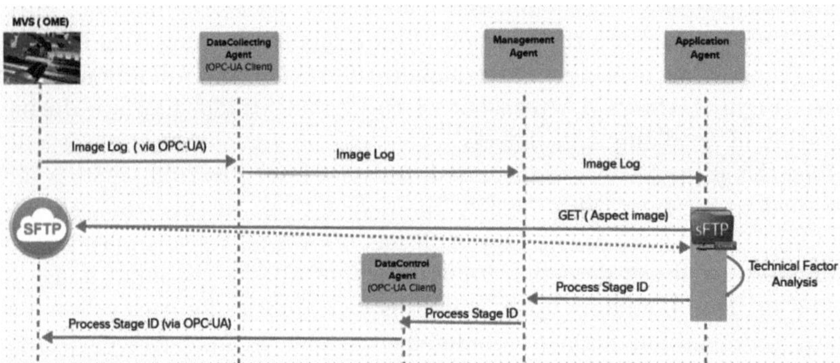

Fig. 6. Sequence diagram illustrating data exchange between the MVS and the DT.

Finally, data exchange between the MVS and the DT is illustrated through a sequence diagram in Fig. 6. It is comprised of four agents: *DataCollecting, Management, Application* and *DataControl*. Every time the MVS gives a KO result, it is stored in the MongoDB database, the OPC-UA server transmits this information to the OPC-UA client integrated within the *Data Collecting* Agent. Subsequently, this agent relays the information to the system's *Management* Agent, which then forwards it to the *Application* Agent. To perform the comprehensive technical factor analysis, the *Application* Agent requires the image segmentation. This image is retrieved via its integrated SFTP client, which requests it from the corresponding server. The outcome of this analysis identifies the process phase responsible for the KO. The *Application* Agent then sends this result to the *Management* Agent, which, in turn, provides it to the *DataControl* Agent. The *DataControl* Agent, equipped with an OPC-UA client, communicates this information back to the MVS. This sequence effectively closes the communication loop between the MVS and the DT, ensuring the operator is informed of which manufacturing stage originated the defect before the MVS initiates a new cycle.

5   Conclusion

This work presents a hybrid framework that leverages Digital Twin (DT) technology, Multi-Agent Systems (MAS), and Machine Vision Systems (MVS) to enable the transition from reactive to proactive quality control in the manufacturing of headlamp lenses. The proposed system, fully aligned with the ISO 23247 standard, integrates real-time monitoring, defect classification, and feedback control mechanisms to close the loop between detection and intervention.

The deployment of an MAS architecture based on SPADE allows for modular and interoperable agent-based communication, ensuring robust synchronization between the physical asset (Observable Manufacturing Element) and its digital counterpart. Through automated image acquisition, annotation via Label Studio, and real-time defect detection using advanced deep learning models such as YOLO, the system not only identifies defects but traces them back to the specific production stage in which they were generated, known as technical factor analysis.

Through the technical factor analysis, the framework enhances manufacturing sustainability and supports the human-centric goals of Industry 5.0. It provides a scalable and adaptable solution for industrial inspection systems, capable of improving product quality, reducing waste, and empowering human operators through transparent and data-driven decision-making.

This study confirms the viability and advantages of integrating DT and AI into existing MVS infrastructures to create intelligent, proactive quality control systems aligned with the demands of modern, resilient manufacturing ecosystems. The technical factor provided by the DT plays a key role in determining the root cause of failures. Future work will continue to build upon this capability.

Acknowledgments. This research was partially funded by the Spanish Ministry of Science and Innovation under the project of the national plan PID2023-150832OB-I00.

References

1. Azamfirei, V., Psarommatis, F., Lagrosen, Y.: Human factors in the design of advanced quality inspection systems in the era of zero-defect manufacturing. In:

Lecture Notes in Mechanical Engineering, pp. 797–804. Springer Science and Business Media Deutschland GmbH (2024). https://doi.org/10.1007/978-3-031-38165-2_92

2. Corporation, G.M.: Measurement systems analysis : reference manual. [Daimler-Chrysler]: [Ford Motor :] [General Motors] (2010)

3. Frick, J., Grudowski, P.: Quality 5.0: a paradigm shift towards proactive quality control in industry 5.0. Inter. J. Bus. Adm. **14**, 51 (2023). https://doi.org/10.5430/ijba.v14n2p51

4. Geneva, S.I.O.f.S.: ISO 23247-1:2021 - automation systems and integration digital twin framework for manufacturing part 1: overview and general principles. https://www.iso.org/standard/75066.html

5. Leberruyer, N., Bruch, J., Ahlskog, M., Afshar, S.: Toward zero defect manufacturing with the support of artificial intelligenceinsights from an industrial application. Comput. Ind. **147** (5 2023).https://doi.org/10.1016/j.compind.2023.103877

6. Lou, H.H., Huang, Y.L.: Hierarchical decision making for proactive quality control: System development for defect reduction in automotive coating operations. Eng. Appl. Artif. Intell. **16**, 237–250 (2003). https://doi.org/10.1016/S0952-1976(03)00060-5

7. Martínez, S.S., Ortega, J.G., García, J.G., García, A.S.: A machine vision system for defect characterization on transparent parts with non-plane surfaces. Mach. Vision Appl. **23**, 1–13 (2012). https://doi.org/10.1007/s00138-010-0281-0

8. Martínez, S.S., Ortega, J.G., García, J.G., García, A.S., Estévez, E.E.: An industrial vision system for surface quality inspection of transparent parts. Int. J. Adv. Manufact. Technol. **68**, 1123–1136 (2013). https://doi.org/10.1007/s00170-013-4904-2

9. Psarommatis, F., May, G., Dreyfus, P.A., Kiritsis, D.: Zero defect manufacturing: state-of-the-art review, shortcomings and future directions in research. Inter. J. Prod. Res. **58**, 1–17 (2020). https://doi.org/10.1080/00207543.2019.1605228

10. Redmon, J., Divvala, S., Girshick, R., Farhadi, A.: You only look once: unified, real-time object detection (2016). http://arxiv.org/abs/1506.02640

11. Saadallah, A., Abdulaaty, O., Büscher, J., Panusch, T., Morik, K., Deuse, J.: Early quality prediction using deep learning on time series sensor data. In: Procedia CIRP. vol. 107, pp. 611–616. Elsevier B.V. (2022). https://doi.org/10.1016/j.procir.2022.05.034

12. Shahin, M., Chen, F.F., Bouzary, H., Krishnaiyer, K.: Integration of lean practices and industry 4.0 technologies: smart manufacturing for next-generation enterprises. Inter. J. Adv. Manufact. Technol. 2927–2936 (2020). https://doi.org/10.1007/s00170-020-05124-0

13. Tzampazaki, M., Zografos, C., Vrochidou, E., Papakostas, G.A.: Machine vision-moving from industry 4.0 to industry 5.0. Appl. Sci. (Switzerland) **14** (2024). https://doi.org/10.3390/app14041471

14. Wooldridge, M.: An introduction to multiagent systems - 2nd edition, vol. 41 (2009)

15. Xu, X., Lu, Y., Vogel-Heuser, B., Wang, L.: Industry 4.0 and industry 5.0inception, conception and perception. J. Manufact. Syst. **61**, 530–535 (2021). https://doi.org/10.1016/j.jmsy.2021.10.006

16. Cavaliere, G., Borgianni, Y., Savio, E.: Performances of an in-line deep learning-based inspection system for surface defects of die-cast components for hybrid vehicles. Procedia CIRP **126**, 999–1004 (2024). https://doi.org/10.1016/j.procir.2024.08.379

Assessing Olive Tree Detection Performance in UAV Images

P. Latorre-Hortelano[1], F. D. Pérez-Cano[2(✉)], D. Jurado-Rodriguez[1],
and G. Parra-Cabrera[1]

[1] Computer Graphics and Geomatics Group,Computer Science Department,
University of Jaén, Campus Las Lagunillas S/N, 23071 Jaén, Spain
[2] Computer Graphics and Geomatics Group,Department of Computer Languages
and Systems, University of Granada, 18071 Granada, Spain
`fdperez@ujaen.es`

Abstract. The automatic identification of crops in images plays a crucial role in the digitization of agriculture, especially related to monitoring and preserving the agroecosystem. This study presents an analysis and comparison of traditional and modern methods for identifying olive trees in RGB images. Among these approaches, we examine classical computer vision algorithms using OpenCV as well as advanced deep learning models, such as U-Net, which leverage convolutional neural networks. Identifying crops in olive groves using only RGB images relies exclusively on color information, which can lead to ambiguities when distinguishing vegetation from surrounding elements in the ecosystem. Factors such as varying lighting conditions, seasonal changes in foliage and different types of terrain further complicate the identification process. This study aims to evaluate the strengths and limitations of each approach by testing RGB images under different brightness, resolution and distance conditions and to examine its potential impact on optimizing agricultural processes. The correct selection of these segmentation techniques could represent a significant advance in agricultural automation, allowing more efficient and reliable management of olive groves through real-time integration of artificial vision tools.

Keywords: Olive identification · Deep learning · Segmentation · UAV · Image processing

1 Introduction

The digitization of agriculture is crucial for enhancing efficiency and sustainability in modern farming. In Spain, one of the main crops is the olive grove. In 2019, the Undersecretariat of Agriculture, Fisheries and Food published an analysis in which it estimated the area of olive groves occupies to be 2,733,620ha, which represents 16.1% of the total cultivated area [3]. The relevance of studying olive groves in Spain can be understood through several key themes: environmental

L. Martínez et al. (Eds.): IDEAL 2025, LNCS 16239, pp. 195–206, 2026.
https://doi.org/10.1007/978-3-032-10489-2_17

sustainability, economic importance, technological innovation and cultural significance [4, 15].

Segmentation of olive groves from RGB images faces multiple challenges. Factors such as variability in illumination, seasonal changes in vegetation cover and terrain heterogeneity can make accurate tree identification difficult. Accurate segmentation of olive trees is essential for monitoring crop health, optimizing resource allocation, and improving precision agriculture. Traditional computer vision methods have been widely used. However, their effectiveness is often limited by environmental factors. In contrast, deep learning-based approaches have the potential to learn more robust features from large volumes of data, improving segmentation accuracy under varying conditions.

This study aims to compare the effectiveness of three different approaches for olive grove segmentation in RGB images: a classical OpenCV-based method, a optimized U-Net semantic segmentation model and the YOLO extension for segmentation. The performance of each method will be evaluated in terms of accuracy, computational efficiency and robustness to variations in different conditions and image resolution. The findings aim to enhance real-time computer vision applications in olive grove management, improving decision-making and increasing agricultural automation efficiency.

The structure of the article is as follows: firstly, a review of previous literature has been carried out to analyze the latest scientific advances in vegetation identification techniques for improved precision agriculture. The methodologies used to identify the olive trees on RGB images will be described in the following section. We show the results obtained and discuss about them in the following section. Finally, the conclusions and future work are presented.

2   Previous Work

Precision agriculture can be defined as the application of technology to crops with the aim of improving production and optimizing resources. The authors establish five levels of evolution over time, which progress according to the degree of automation of the processes, from traditional methodology to the use of artificial intelligence [1, 8]. The incorporation of advanced technologies in agriculture has made it possible to optimize the use of resources, improve productivity and quality and minimize environmental impact. The use of sensors, geographic information systems or disease detection systems allow to more efficient crop management by providing real-time data on the state of soil, vegetation and climatic conditions [11].

The mapping of olive groves is an important step in remote sensing applications for precision agriculture. Correct identification of individual olive trees to predict disease and physiological needs is essential. Traditional aerial platforms such as planes and satellites are not suitable for these applications due to their low spatial and temporal resolutions [21]. In this context, the proliferation of Unmanned Aerial Vehicles (UAVs) have emerged as a powerful alternative for high-resolution remote sensing in precision agriculture [10, 25].

Crop and tree species segmentation, in RGB images, has been studied in recent years using both computer vision and deep learning-based approaches [2,22]. Traditional methods have been widely adopted due to their low computational cost, whereas neural networks have emerged as powerful alternatives to traditional computer vision techniques [20]. As for traditional methodologies for automatic vegetation segmentation, they do not require data for training and can be used quickly. There are different approaches in the literature:

- Color threshold-based. This method segments vegetation using thresholds in color values to distinguish vegetated areas from the background [16,24].
- Vegetation indexes. Spectral bands are combined to highlight the presence of vegetation [7].
- Clustering-based. Pixels with similar characteristics, such as color or intensity, are grouped together to identify regions of vegetation [14,18].

The advent of deep learning and convolutional neural networks (CNNs) has revolutionized agricultural image segmentation. Neural networks have enabled substantial improvement in the classification and segmentation of different types of trees and crops, opening up new opportunities for precision agriculture.

One of the most widely used models in crop segmentation is U-Net, an architecture based on CNNs that has demonstrated outstanding performance in semantic segmentation tasks [5]. Liu et al. [13] identified crop species from remotely sensed images obtained with UAVs, improving accuracy through an optimized versions of the U-Net architecture. Silva et al. [19] also use these images with the similar architecture reaching the conclusion that U-Net especially useful for working with high-resolution images in precision agriculture. These features are essential for obtaining detailed and accurate segmentation, which facilitates real-time crop monitoring and analysis. MCAC-UNet for segmentation and coverage measurement of maize canopy images in order to optimize the fertilization process [9]. Wang et al. [23] also presents MDE-UNet, a multi-tasking deformable U-Net for accurate segmentation of cropland boundaries in precision agriculture. Despite its many advantages, U-Net presents some challenges, such as its high computational consumption and the need for a large volume of labeled data to adequately train the model.

In addition to U-Net, others models have gained popularity in crop identification and monitoring. In particular, the You Only Look Once (YOLO) series of models which has been used for real-time crop detection and segmentation [17]. YOLO is designed to detect specific objects, such as individual olive trees in large-scale agricultural imagery. Its speed and accuracy make it ideal for real-time applications, such as crop inspection with UAVs. Other authors also studied the performance of YOLO versus U-Net variation models in areas with weeds and beans, reaching the conclusion that although YOLO obtained better results in terms of identification and accuracy, it also gave rise to quite a few false positives [19]. La et al. [12] compare YOLO with other classical models in order to obtain a precise segmentation of the trunk and branches of fruit trees. They conclude that YOLO produces the best average precision, especially in the trunk area. Despite its advantages, it also has some limitations, such as its difficulty in accurately identifying specific areas.

3   Material and Methods

This section describes the methodologies developed for olive tree identification in RGB images using different segmentation techniques. Each approach will be evaluated in terms of segmentation accuracy, robustness to environmental variations and computational efficiency to determine its effectiveness in olive grove analysis. Figure 1 visually summarizes the three approaches analyzed for olive tree identification.

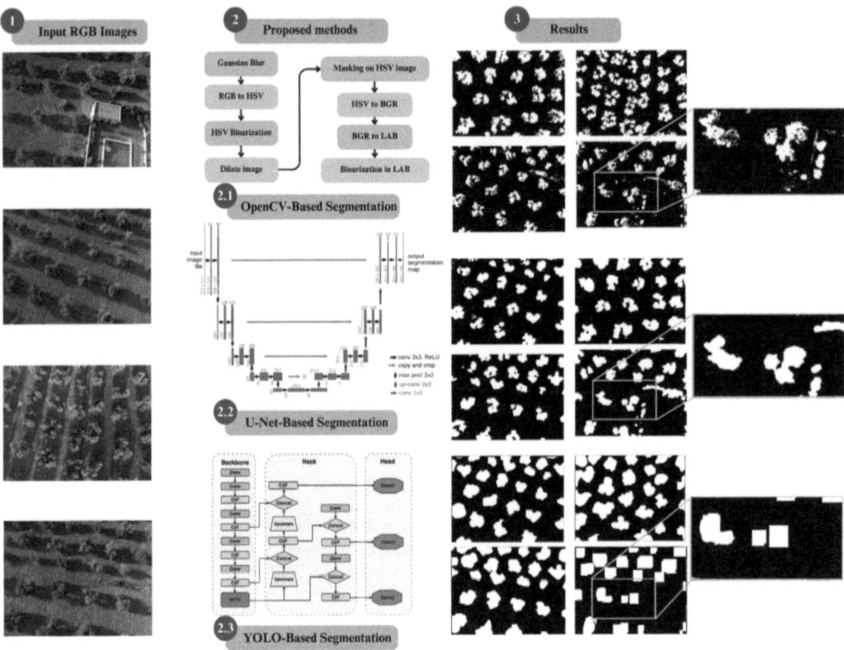

Fig. 1. Comparison of OpenCV, an optimized U-Net version and YOLOv11 approaches for detecting and analyzing olive trees in aerial images.

3.1   Dataset

The dataset used to train architectures consists of 2650 RGB images captured by UAVs flights at different heights, using different sensors, on multiple olive grove plots in the province of Jaén (Spain). The areas included zones fully covered with olive groves, partially covered, and regions without olive groves or intermingled with other environmental elements.

For the composition of the dataset, 2 flights were made at low altitude (50 m) and 4 flights at a higher altitude (100 m). In addition, we have used the DJI ZH20T sensor which gives us a resolution of 4056×3040, as well as the DJI M3M and DJI FC6520 sensors, which have a resolution of 5280×3956. For the training of both neural networks, it was necessary to crop the images to a square shape and downsample them to a resolution of 640×640, which is the resolution that the YOLO architecture is recommended for and which allows for greater training and prediction efficiency.

Additionally, data augmentation techniques such as rotation, flipping and brightness adjustments were applied to increment the dataset to 4000 images while avoiding excessive similarity among images to prevent overfitting.

To mark the olive trees in the images and thus label the dataset, 3D and multispectral information was processed to detect olive groves, and this information was extrapolated to the image sets used to generate the models.

3.2 Traditional Approach

Traditional image processing techniques rely on manually designed features to segment and identify objects within an image. In this study, we have implemented a combined version of different traditional methodologies, through the OpenCV library [6], to evaluate their effectiveness in the segmentation of olive trees. First, a Gaussian filter is applied to the UAV image. This initial step helps reduce image noise, which in turn improves the effectiveness of subsequent detection procedures. The image is then transformed into the HSV (Hue, Saturation, Value) color space. Within this space, vegetation generally presents as various green hues with comparable intensity and brightness, simplifying its differentiation from other features. Next, a thresholding operation is executed to produce a binary image. This image then undergoes morphological operations, including dilation and erosion, which help to emphasize vegetation contours. Following this, the binary image is superimposed onto the original HSV image to identify and eliminate the shadows cast by the vegetation. This action enhances the precision of vegetation detection by lessening the influence of shadows, leading to a more distinct and accurate identification of vegetated zones. Finally, an additional color space conversion is performed, changing the image to the LAB color space to further distinguish vegetation from the surrounding terrain. In the LAB color space, each pixel is characterized by three components: luminance (L), which represents the brightness of the vegetation, and the chromatic components (A and B), which capture differences in green tones. Through the analysis of these components, the method effectively differentiates vegetation from other land features, yielding a binary image where vegetation is clearly separated from non-vegetated regions as shown in Figure 2.

3.3 U-Net

A deep learning-based approach has been implemented for olive tree identification using an improved version of U-Net. This model was selected due to its

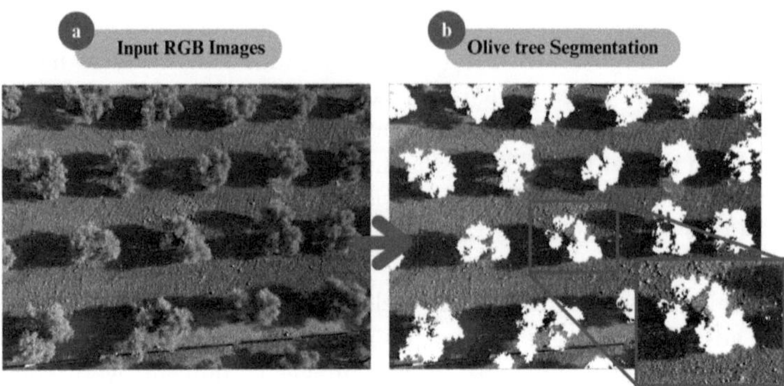

Fig. 2. Olive tree segmentation process using traditional techniques: (a) Input RGB aerial image, (b) Segmented output highlighting the detected tree crowns in white.

effectiveness in image segmentation tasks, particularly when precise localization of objects is required. The U-Net architecture has been specifically designed to address challenges related to varying illumination, complex backgrounds and occlusions that traditional image processing techniques struggle with.

The input consists of RGB images on olive grove obtained from UAVs, while the output is a binary mask indicating the presence or absence of olive trees at each pixel. Each isolated region within this mask allows the identification of each olive tree individually. To prepare the data for training, images were pre-processed by resizing and normalizing pixel values. The training was conducted using PyTorch, leveraging its flexibility and GPU acceleration capabilities. The dataset was divided into training, validation and test subsets following an 80-10-10 split to ensure a balanced evaluation. The training process involved optimizing the network using the Adam optimizer with a weight decay regularization to prevent overfitting.

To enhance segmentation performance, the model was trained using a combined loss function comprising the Dice loss and Binary Cross-Entropy (BCE) loss. The Dice loss, defined as:

$$\mathcal{L}_{\text{Dice}} = 1 - \frac{2 \sum_i y_i \hat{y}_i}{\sum_i y_i + \sum_i \hat{y}_i}, \tag{1}$$

was utilized to maximize the overlap between predicted and actual tree regions. Additionally, BCE loss was employed to refine pixel-wise classification:

$$\mathcal{L}_{BCE} = -\frac{1}{N} \sum_{i=1}^{N} \left[y_i \log(\hat{y}_i) + (1 - y_i) \log(1 - \hat{y}_i) \right]. \tag{2}$$

These loss functions together ensured precise boundary segmentation and reduced false positives and negatives.

3.4 YOLO

YOLO offers a versatile family of models covering a wide range of computer vision tasks—including image classification, object detection, multi-object tracking and even instance segmentation. In this subsection, we focus specifically on the YOLO segmentation extensions that generate pixel-level masks around trees.

Instance segmentation techniques extend object detection by generating precise pixel-level masks for each olive tree in the image. In this study, we employed the YOLOv11 segmentation extension to leverage its real-time performance while providing detailed outlines of individual trees. Unlike pure bounding-box detectors, the segmentation variant of YOLO predicts both the location and shape of each tree, outputting a mask that delineates its exact canopy region alongside confidence scores (see Figure 3).

The YOLO segmentation extension divides the input image $I \in \mathbb{R}^{h \times w \times c}$, where h, w and c denote the image height, width and number of channels, respectively, into a regular grid. YOLO predicts both instance masks and class probabilities for each cell. This approach preserves the real-time efficiency by sharing convolutional features across the entire grid while providing the detailed, while also delivering the per pixel details that classical bounding box detection cannot provide. As a result, multiple trees can be segmented accurately in a single pass, combining the speed of YOLO grid based inference with the spatial precision of instance segmentation.

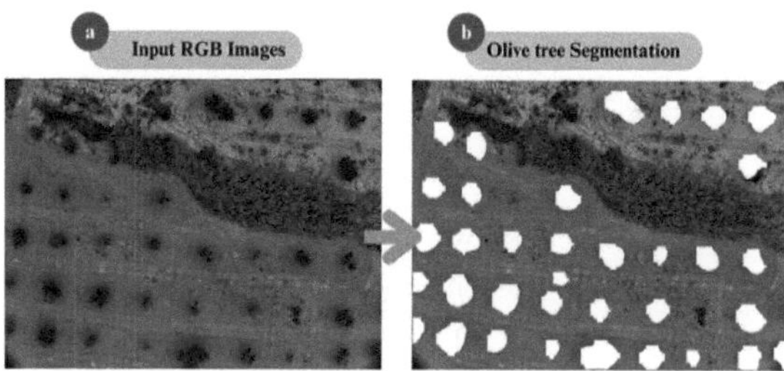

Fig. 3. Olive tree detection using YOLO: (a) Input RGB aerial image, (b) Output as a segmentation masks.

The loss function used for training consists of three primary components:

$$\mathcal{L}_{\text{total}} = \lambda_{coord}\mathcal{L}_{coord} + \mathcal{L}_{conf} + \mathcal{L}_{class}, \tag{3}$$

where $\mathcal{L}_{coord}$ is the localization loss, ensuring accurate bounding box predictions by penalizing deviations in predicted center coordinates and dimensions. $\mathcal{L}_{conf}$ is the confidence loss, penalizing incorrect tree presence predictions, ensuring that the model learns to distinguish olive trees from other vegetation or background noise. $\mathcal{L}_{class}$ is the classification loss, optimizing category assignments for detected objects when different tree varieties are considered.

4   Results and Discussion

This section reports the performance of the evaluated segmentation and detection methods for olive tree identification, using metrics such as precision, recall, Intersection over Union (IoU), mean Average Precision (mAP) and processing time. It also addresses key challenges encountered during testing, including illumination variability, image resolution and the models' ability to generalize across diverse environmental conditions. All experiments were conducted on a system equipped with 32 GB of RAM and an NVIDIA RTX 3060 GPU.

4.1   Training Procedures and Model Evaluation

The dataset was divided into training (80%), validation (10%) and test (10%) subsets to ensure robust performance and minimize overfitting under varying environmental conditions.

The traditional approach combined adaptive thresholding, contour detection and clustering. Optimal thresholds were determined via histogram analysis and morphological operations improved tree separation. Different color spaces (RGB, HSV, LAB) were tested to enhance discrimination.

The U-Net model, based on an encoder-decoder architecture with skip connections, was trained with a combination of Binary Cross-Entropy and Dice loss. Regularization included batch normalization and 30% dropout. Training used the Adam optimizer (10^{-4} learning rate.)

Table 1. Training settings used for each approach

Parameter	Traditional	U-Net	YOLOv11
Optimizer	–	Adam	SGD + Momentum
Learning Rate	–	Fixed (10^{-4})	Cosine Annealing
Batch Size	–	32	16
Regularization	Tuning	Dropout (30%), BN	Weight Decay, CV
Augmentation	Color space	Rot., Bright., Scale	Flip., Crop., Resize

For YOLOv11, the training process focused on detecting and localizing olive trees via bounding boxes. Training was performed using stochastic gradient descent with momentum and a cosine annealing learning rate schedule. A pre-training phase on a large-scale dataset initialized the feature extractor, followed by fine-tuning on domain-specific UAV imagery. Five-fold cross-validation was employed to improve robustness. Model performance was evaluated using mean Average Precision (mAP) and Intersection over Union (IoU) metrics.

Table 1 summarizes the key hyperparameters and training techniques used for each approach, ensuring that models were optimized for both accuracy and efficiency under real-world agricultural conditions.

4.2 Evaluation Metrics

To evaluate model performance, standard segmentation metrics—precision, recall, IoU, and mAP—were used. These provide a comprehensive assessment of accuracy, efficiency, and adaptability across varying conditions, allowing objective comparison in olive tree identification tasks. Results are shown in Table 2.

Additionally, the number of olive trees detected by each method was compared, with results summarized in Table 3, including average success rates and false positives. Ground truth values were obtained via visual inspection of RGB images. For OpenCV and U-Net, each white object was counted as a tree, while YOLO provided tree counts via bounding boxes.

Table 2. Performance comparison of segmentation and detection methods

Method	Precision	Recall	F1-Score	IoU/mAP	Time Per Image (ms)
Traditional Segmentation	0.781	0.821	0.8	0.673	1025
U-Net Segmentation	0.857	0.87	0.863	0.839	775
YOLO Segmentation	0.916	0.887	0.901	0.943	90

Table 3. Performance comparison from quantity of trees segmented

Method	True positive	False Positives
Traditional Methods	58.67%	45%
U-Net	84.8%	28.75%
YOLO	90.75%	30.15%

As shown in Table 3, the YOLO approach yields the best overall performance, achieving a high true positive rate as well as high precision and IoU values. In contrast, this architecture also tends to over-segment tree crowns, often generating multiple bounding boxes per tree due to the multi-trunk structure of olive trees, which leads to an inflated false positive rate and a reduced tree count. The OpenCV-based traditional method performs worst in this regard, as it cannot effectively differentiate between tree canopies and background vegetation, resulting in noisy masks and poor contour closure.

Despite its higher computational demands and the need for extensive labeled data, U-Net remains an accurate and robust solution but lacks a more advanced evolution of the basic architecture. Its trained model offers faster inference than the traditional method and only slightly lags behind YOLO in processing speed. Both U-Net and YOLO are suitable for real-time monitoring systems.

Overall, the comparison highlights the trade-offs among the three approaches. OpenCV is lightweight and fast but lacks robustness under varying environmental conditions. YOLO offers the fastest inference, making it ideal for real-time applications, though its accuracy decreases when labeling multi-trunk olive trees, often producing incorrect detections. U-Net achieves the lowest false positive rate being consistent and, despite requiring substantial training resources and annotated data, delivers efficient inference once trained. Ultimately, the optimal method depends on the application's requirements—whether prioritizing speed, precision, or adaptability.

5   Conclusions and Future Work

This study compared different approaches for olive tree identification using RGB images: a traditional OpenCV-based segmentation algorithm, an optimized U-Net model and YOLO. The results highlight trade-offs of accuracy, computational cost and adaptability to diverse environmental conditions.

The traditional approach, based on OpenCV, is useful for accurate segmentation, computationally efficient and easy to implement and suitable for preliminary vegetation analysis in low-resource settings. However, it lacks robustness against lighting, shadow and seasonal variations and its real-time performance is inferior to deep learning models.

U-Net provides the fewest false positives, making it ideal for precision agriculture. While slightly slower than YOLO, it remains viable for real-time drone-based monitoring. Its main limitations are the need for large annotated datasets and high computational demands, which may hinder adoption by small-scale farmers. This model suffers when the flight height is very high and the leaf density is very low, confusing olive trees with the ground or shadows.

YOLO is the fastest method processing olive grove images and achieves a high true positive rate, but it suffers from false negatives when detecting olive trees with multi-trunk structures and irregular canopies.

Future work could explore incorporating multispectral or hyperspectral imagery that could enhance model performance by leveraging spectral vegetation features. Domain adaptation and self-supervised learning may improve generalization, while lightweight models optimized for UAV-based edge computing would enable real-time, per-tree analysis to support resource optimization and sustainable yield management.

References

1. Abbasi, R., Martinez, P., Ahmad, R.: The digitization of agricultural industry – a systematic literature review on agriculture 4.0. Smart Agric. Technol. **2**, 100042 (2022). https://doi.org/10.1016/j.atech.2022.100042, http://dx.doi.org/10.1016/j.atech.2022.100042
2. Abozeid, A., Alanazi, R., Elhadad, A., Taloba, A.I., Abd El-Aziz, R.M.: A large-scale dataset and deep learning model for detecting and counting olive trees in satellite imagery. Comput. Intell. Neurosci. **2022**, 1–8 (2022). https://doi.org/10.1155/2022/1549842, http://dx.doi.org/10.1155/2022/1549842

3. Ministerio de Agricultura, P.y.A.: Análisis de las plantaciones de olivar en españa. Tech. rep., Encuesta sobre Superficies y Rendimientos de Cultivos (ESYRCE) (2019). https://www.mapa.gob.es/es/estadistica/temas/estadisticas-agrarias/olivar2019_tcm30-122331.pdf, Último acceso: 25 de febrero de 2025

4. Anastasiou, E., Balafoutis, A.T., Fountas, S.: Trends in remote sensing technologies in olive cultivation. Smart Agric. Technol. **3**, 100103 (2023). https://doi.org/10.1016/j.atech.2022.100103, http://dx.doi.org/10.1016/j.atech.2022.100103

5. Bhatti, M.A., et al.: Utilizing convolutional neural networks (CNN) and u-net architecture for precise crop and weed segmentation in agricultural imagery: A deep learning approach. Big Data Res. **36**, 100465 (2024). https://doi.org/10.1016/j.bdr.2024.100465

6. Bradski, G.: The OpenCV Library. Dr. Dobb's Journal of Software Tools (2000)

7. Casella, A., et al.: Segmentación de imágenes spot a partir de íÂŋndices de vegetación para la cuantificación de cultivo de cebolla bajo riego en el valle inferior del ríÂŋo colorado. In: Sione, W. (ed.) SELPER 2016: XVII Simposio Internacional en Percepción Remota y Sistemas de Información Geográfica. p. 387. EdUnLu, Luján, Buenos Aires, Argentina (2017). https://www.researchgate.net/publication/326655195_SEGMENTACION_DE_IMAGENES_SPOT_A_PARTIR_DE_INDICES_DE_VEGETACION_PARA_LA_CUANTIFICACION_DE_CULTIVO_DE_CEBOLLA_BAJO_RIEGO_EN_EL_VALLE_INFERIOR_DEL_RIO_COLORADO, libro digital, PDF

8. Fountas, S., Espejo-GarcíÂŋa, B., Kasimati, A., Gemtou, M., Panoutsopoulos, H., Anastasiou, E.: Agriculture 5.0: cutting-edge technologies, trends, and challenges. IT Professional **26**(1), 40–47 (2024). https://doi.org/10.1109/mitp.2024.3358972

9. Gong, H., Xiao, L., Wang, X.: Segmentation and coverage measurement of maize canopy images for variable-rate fertilization using the MCAC-UNET model. Agronomy **14**(7), 1565 (2024). https://doi.org/10.3390/agronomy14071565

10. Jurado, J.M., López, A., Pádua, L., Sousa, J.J.: Remote sensing image fusion on 3d scenarios: a review of applications for agriculture and forestry. Int. J. Appl. Earth Obs. Geoinformation **112**, 102856 (2022). https://doi.org/10.1016/j.jag.2022.102856

11. Kamilaris, A., Prenafeta-Boldú, F.X.: Deep learning in agriculture: a survey. Comput. Electron. Agric. **147**, 70–90 (2018). https://doi.org/10.1016/j.compag.2018.02.016

12. La, Y.J., Seo, D., Kang, J., Kim, M., Yoo, T.W., Oh, I.S.: Deep learning-based segmentation of intertwined fruit trees for agricultural tasks. Agriculture **13**(11), 2097 (2023). https://doi.org/10.3390/agriculture13112097

13. Liu, Z., Su, B., Lv, F.: Intelligent identification method of crop species using improved u-net network in UAV remote sensing image. Sci. Program. **2022**, 1–9 (2022). https://doi.org/10.1155/2022/9717843

14. Marino, S., Alvino, A.: Vegetation indices data clustering for dynamic monitoring and classification of wheat yield crop traits. Remote Sensing **13**(4), 541 (2021). https://doi.org/10.3390/rs13040541

15. Noguera, M., Millan, B., Moro, R., Andújar, J.M.: Monitorización del contenido de fósforo del olivo mediante robot terrestre. Jornadas de Automática (45) (2024). https://doi.org/10.17979/ja-cea.2024.45.10918

16. Otsu, N.: A threshold selection method from gray-level histograms. IEEE Trans. Syst. Man Cybern. **9**(1), 62–66 (1979). https://doi.org/10.1109/TSMC.1979.4310076

17. Redmon, J., Divvala, S., Girshick, R., Farhadi, A.: You only look once: unified, real-time object detection. In: 2016 IEEE Conference on Computer Vision and Pattern Recognition (CVPR), pp. 779–788. IEEE (2016). https://doi.org/10.1109/cvpr.2016.91

18. Seidl, J., KačmaříÂŋk, M., Klimánek, M.: A tree segmentation algorithm for airborne light detection and ranging data based on graph theory and clustering. Forests **15**(7), 1111 (2024). https://doi.org/10.3390/f15071111

19. Silva, J.A.O.S., et al.: Deep learning for weed detection and segmentation in agricultural crops using images captured by an unmanned aerial vehicle. Remote Sensing **16**(23), 4394 (2024). https://doi.org/10.3390/rs16234394

20. Sultana, F., Sufian, A., Dutta, P.: Evolution of image segmentation using deep convolutional neural network: A survey. Knowledge-Based Systems **201–202**, 106062 (2020). https://doi.org/10.1016/j.knosys.2020.106062

21. Torres-Sánchez, J., Peña, J., de Castro, A., López-Granados, F.: Multi-temporal mapping of the vegetation fraction in early-season wheat fields using images from UAV. Comput. Electron. Agric. **103**, 104–113 (2014). https://doi.org/10.1016/j.compag.2014.02.009

22. Vasilakos, C., Verykios, V.S.: Burned olive trees identification with a deep learning approach in unmanned aerial vehicle images. Remote Sensing **16**(23), 4531 (2024). https://doi.org/10.3390/rs16234531

23. Wang, Y., Gu, L., Jiang, T., Gao, F.: Mde-unet: A multitask deformable unet combined enhancement network for farmland boundary segmentation. IEEE Geosci. Remote Sens. Lett. **20**, 1–5 (2023). https://doi.org/10.1109/LGRS.2023.3252048

24. Xu, S., et al.: A robust index to extract paddy fields in cloudy regions from SAR time series. Remote Sens. Environ. **285**, 113374 (2023). https://doi.org/10.1016/j.rse.2022.113374, https://www.sciencedirect.com/science/article/pii/S0034425722004801

25. Yao, H., Qin, R., Chen, X.: Unmanned aerial vehicle for remote sensing applications–a review. Remote Sensing **11**(12), 1443 (2019). http://dx.doi.org/10.3390/rs11121443

Dynamic Feature Filtering for Robust Monocular Visual SLAM

Marton Gonczy$^{(\boxtimes)}$ (ORCID), Kieran Wood (ORCID), and Hujun Yin (ORCID)

The University of Manchester, Manchester M13 9PL, UK
`marton.gonczy@manchester.ac.uk`

Abstract. Visual SLAM systems often suffer from degraded performance in dynamic environments due to the inclusion of feature points on moving objects, which violate the static world assumption. In this paper, we propose a dynamic object filtering framework designed to enhance the performance of monocular ORB-SLAM3 by excluding features associated with dynamic regions. Our method integrates deep learning-based object detection, multi-object tracking, and dense optical flow analysis to generate per-frame binary masks that identify and suppress dynamic content. A motion similarity metric, which combines directional- and magnitude-based flow differences with spatial weighting, is used to assess object motion relative to the background. Temporal consistency is enforced through a probabilistic state filter to smooth the classification over time. We evaluated our approach on the KITTI Odometry dataset, and it shows that our solution significantly reduced pose estimation errors in highly dynamic sequences, while maintaining performance in static scenes. The results demonstrate the effectiveness of the proposed filtering strategy as a lightweight and general-purpose enhancement to existing SLAM systems when operating in real-world environments.

Keywords: Visual SLAM · Monocular · Dynamic object filtering · Optical flow · Motion segmentation

1 Introduction

Simultaneous Localisation and Mapping (SLAM) has become a foundational component in autonomous robotics, augmented reality, and navigation systems. Traditional SLAM pipelines assume a static world, leveraging geometric consistency to track the camera and reconstruct a map of the environment. However, this assumption is often invalid in real-world environments where dynamic objects, such as people or vehicles, introduce inconsistencies in the feature correspondences and degrade the performance of visual SLAM systems. In response, an increasing number of studies has focused on enhancing SLAM robustness in dynamic environments by detecting and filtering out features associated with objects moving independently. These approaches vary widely in terms of their

L. Martínez et al. (Eds.): IDEAL 2025, LNCS 16239, pp. 207–218, 2026.
https://doi.org/10.1007/978-3-032-10489-2_18

sensor input (monocular, stereo, RGB-D), filtering strategies (semantic segmentation, object detection, motion consistency), and computational design (real-time and offline systems). Early solutions predominantly relied on geometric constraints, such as epipolar consistency and depth disparity, to detect moving features. Later, the integration of semantic information enabled systems to selectively ignore or down-weight features associated with known dynamic classes. The majority of existing solutions utilise sensor inputs that not only provide frames but also additional information such as depth.

2   Related Work

A foundational approach to detecting dynamic elements in SLAM relies on consistency checks between observed motion and expected camera motion. Alcantarilla et al. [4] fuse stereo disparity with optical flow to compute per-pixel 3D motion, comparing it with the egomotion estimate from visual odometry. Discrepancies, measured via Mahalanobis distance, return residual-motion likelihoods that flag dynamic pixels for removal during SLAM optimisation. Zhang et al. [22] improve on this approach by using deep learning-based optical flow. They use PWC-Net [18] to compute dense optical flow between RGB-D frames and classify pixels as dynamic when observed flow diverges from predicted egomotion, considering both photometric and depth errors.

Geometric consistency checks are another common approach. Cheng et al. [10] use the epipolar line distance to remove residual dynamic features not filtered by segmentation, applying monocular or depth-based checks depending on sensor input. Similarly, Yu et al. [20] combine SegNet-based semantic segmentation with motion checks using epipolar geometry and optical flow residuals. Regions with high concentrations of inconsistent features are labelled dynamic and discarded from the map. Ma et al. [15] mask out likely dynamic objects (e.g., people, vehicles) using SegNet on one stereo frame while extracting features from the other, filtering points that violate epipolar constraints. In RGB-D settings, depth-aware backprojection and reprojection further improve consistency validation. Methods like DDL-SLAM [3] integrate temporal and spatial consistency by predicting dynamic likelihoods per pixel using prior frames and refining this through multi-view depth checks and reprojection based outlier rejection.

Another major category of dynamic-aware SLAM approaches leverages semantic segmentation or object detection to identify and exclude features on potentially moving objects. These methods typically rely on pretrained deep networks to generate pixel-wise masks or bounding boxes for dynamic classes. Mask-SLAM [12] employs DeepLabv2 [9] to segment semantic classes like vehicles and sky. By removing features associated with these classes, the system reduces RANSAC failures. RDS-SLAM [14] extends ORB-SLAM3 [8] by integrating semantic information from Mask R-CNN and SegNet in a parallel processing thread. Each map point receives a Bayesian probability of being static, dynamic, or unknown. During pose estimation, priority is given to static points, but fallback mechanisms ensure continuity in uncertain cases.

Other approaches incorporate real-time object detection into the pipeline. Ai et al. [2] use YOLOv4 to detect moving object candidates via bounding boxes. Feature points within these regions receive a dynamic probability score that is updated over time. Su et al. [17] run a detection thread to isolate dynamic classes. A homography is estimated using points outside these boxes, which is then used to warp the images. Features inside the boxes are compared against the predicted motion to detect inconsistencies. DynaSLAM [7] builds on ORB-SLAM2 [16] by integrating dynamic object segmentation using Mask R-CNN. In RGB-D mode, it further refines dynamic masks using multi-view depth consistency and inpaints occluded backgrounds to preserve map cleanliness. DynaSLAM II [6] extends this idea with joint camera-object trajectory optimisation using stereo/RGB-D input. Dynamic object tracks are gradually improved over time, while a semantic map records both static and dynamic elements to enhance scene understanding.

Many recent methods combine semantic segmentation or object detection with motion consistency checks, such as optical flow or epipolar geometry. VDO-SLAM [21] fuses instance segmentation from Mask R-CNN with scene-flow estimation. Object masks are generated first, and 3D points within these regions are classified as dynamic if their scene-flow magnitudes exceed a predefined threshold. Detect-SLAM [23] takes a keyframe-based approach where object detections are performed only periodically to preserve real-time performance. Detected bounding boxes are used to assign moving probabilities to nearby feature points. These probabilities are propagated across frames and spatially to neighbouring features, enhancing robustness to occlusion and viewpoint changes. Ballester et al. [5] estimate object motion by fitting individual 6-DoF poses to segmented objects based on RGB-D input. After segmenting candidate objects using Mask R-CNN, the system minimises photometric reprojection error to compute each object's motion. When object motion diverges significantly from camera motion, those regions are labelled dynamic. Xiao et al. [19] detect dynamic candidates using SSD [13] and further refine detection by assuming temporal continuity of motion. Each region is given a class-dependent dynamicity score, and tracking selectively discards features with motion exceeding that of neighbouring static areas. SG-SLAM [11] follows a similar principle, but prioritises computational efficiency. Instead of full segmentation, it uses a modified SSD detector and assigns each bounding box a class-weight reflecting its likelihood of being dynamic (e.g., high for people, low for buildings). Epipolar geometry checks are applied to features within each box, and if the weighted count of inconsistent features exceeds a threshold, all features in that region are discarded.

Most existing dynamic SLAM methods rely on dense depth maps, RGB-D input, or stereo vision to detect and suppress dynamic objects. While effective, these approaches often require expensive sensors or high computational overhead. Others use dense segmentation networks or scene flow estimation, which can be unsuitable for real-time use in monocular setups. In contrast, our method is designed to operate using only monocular RGB input, without reliance on depth sensors, stereo data, or full-scene segmentation.

3 System Overview

This section details the approach used to detect and filter dynamic objects from a sequence of frames. The proposed method integrates object detection, appearance-based tracking, optical flow estimation, motion similarity analysis and probabilistic temporal smoothing to robustly segment dynamic content from static background. Most existing methods utilise extra information such as depth to identify dynamic objects or features in the frames, however, our approach aims to achieve this with only monocular frames and deep learning. Figure 1 shows the flowchart of the proposed system.

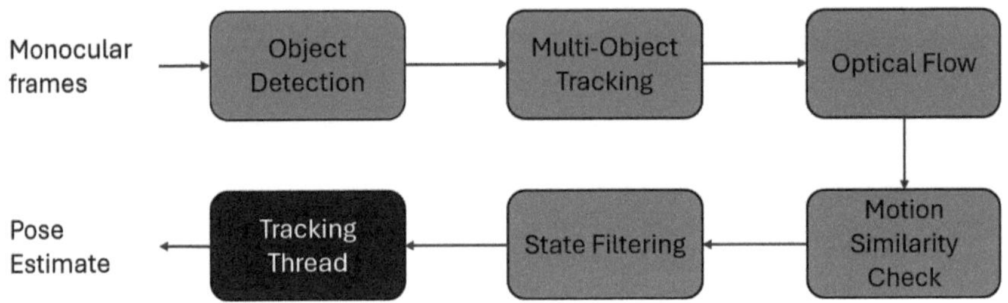

Fig. 1. System flowchart. Light boxes represent our system, while the dark box indicates the tracking thread input to ORB-SLAM3.

3.1 Object Detection

We employ object detection to identify candidate dynamic regions in each frame. Object detection provides both semantic labels and bounding boxes for individual objects, enabling spatially localised filtering. This semantic prior narrows the search space for identifying dynamic elements in visual SLAM. We use YOLOv11, a real-time object detector known for high accuracy and computational efficiency. Given an input frame at time t, the detector outputs a set of n bounding boxes:

$$\mathcal{D}_t = \{d_1, d_2, \ldots, d_n\}, \quad d_i = (x_1, y_1, x_2, y_2, c_i, s_i)$$

where (x_1, y_1, x_2, y_2) are bounding box coordinates, c_i is the class label, and s_i is the confidence score. A threshold $\theta_c = 0.5$ is applied to discard low-confidence detections. To reduce unnecessary computation, we define a fixed set of objects, such as **person**, **dog**, and other typically mobile categories, as *always dynamic*. Bounding boxes belonging to these classes are filtered out immediately, without further motion analysis. All remaining detections are passed to the tracking and motion evaluation modules for downstream processing.

3.2   Multi-object Tracking

To maintain consistent object identities across frames, we incorporate BoT-SORT [1], a robust tracking-by-detection algorithm. Given per-frame detections from YOLOv11, BoT-SORT links bounding boxes over time by combining motion prediction with appearance-based association. Each detection above a confidence threshold is used to initialise or update a tracker. Kalman filtering is applied to model object motion, with explicit handling of bounding box position and scale. To compensate for camera-induced motion, BoT-SORT includes a global motion compensation module that aligns detections across frames, improving track stability. Data association is performed via a combination of IoU and appearance feature distances, followed by Hungarian matching. Tracks are confirmed after consistent matches and retained for a short time if temporarily unmatched. The resulting trajectories are then passed to the motion analysis module for dynamic/static classification.

3.3   Dense Optical Flow

To assess motion dynamics, we compute dense optical flow between adjacent grayscale frames using the Farneback algorithm, selected for its efficiency and accuracy in monocular video. This method estimates a per-pixel flow vector field $\mathbf{F}_t$, where each vector represents the 2D displacement of a pixel between consecutive grayscale images I_{t-1} and I_t. For each detected object bounding box, we compute the dominant object flow $\mathbf{v}_{\text{obj}}$ as the mean of all flow vectors within the box:

$$\mathbf{v}_{\text{obj}} = \frac{1}{N} \sum_{(x,y) \in \text{bbox}} \mathbf{F}_t(x, y) \tag{1}$$

To model the background motion, we define a surrounding neighbourhood by extending the bounding box dimensions by 50% in width and height. The dominant background flow $\mathbf{v}_{\text{neigh}}$ is computed as the mean of flow vectors in this region. While our method operates solely on 2D optical flow and does not incorporate explicit depth information, we partially mitigate this limitation by comparing each object's motion to that of its immediate surroundings. Under the assumption that a dynamic object and its neighbouring background lie at similar depths, especially in structured environments like driving scenes. This comparison highlights motion inconsistencies while reducing false positives caused by projection effects. Additionally, because the surrounding region is drawn from the same part of the image as the object, this local comparison also helps mitigate parallax effects that arise from depth variation across the image plane.

3.4   Motion Similarity Check

Once the dominant optical flow vectors are computed for each object and its surrounding region, we evaluate whether the object is moving independently of the background using a motion similarity score. The core idea is that objects whose motion differs significantly in magnitude or direction from their surroundings are likely to be dynamic. Let $\mathbf{v}_{\mathrm{obj}} \in \mathbf{R}^2$ denote the dominant flow vector inside an object's bounding box, and $\mathbf{v}_{\mathrm{neigh}} \in \mathbf{R}^2$ the dominant flow in the surrounding neighbourhood. We compute a combined similarity score based on two metrics: a normalised Euclidean distance and a Cosine distance.

Normalised Euclidean Distance. The Euclidean distance quantifies the magnitude of difference between two flow vectors:

$$D_E = \frac{\|\mathbf{v}_{\mathrm{obj}} - \mathbf{v}_{\mathrm{neigh}}\|}{D_{\mathrm{max}}} \tag{2}$$

where D_{max} is the maximum Euclidean distance between the two flow vectors in the region of interest. This normalisation serves two purposes. First, it restricts the value of D_E to the range $[0, 1]$, preventing large errors and enabling consistent thresholding when combined with the Cosine distance. Second, it adapts the metric to scene-specific motion intensity, for instance, differentiating motion in fast- and slow-moving camera scenarios, making the score less sensitive to global motion magnitude.

Cosine Distance. The Cosine distance captures the angular difference between the object and background motion direction:

$$D_C = \frac{1}{2}\left(1 - \frac{\mathbf{v}_{\mathrm{obj}} \cdot \mathbf{v}_{\mathrm{neigh}}}{\|\mathbf{v}_{\mathrm{obj}}\| \, \|\mathbf{v}_{\mathrm{neigh}}\|}\right) \tag{3}$$

This formulation maps D_C to the range $[0, 1]$, where 0 indicates perfect directional alignment and 1 indicates complete opposition. Compared to magnitude-based metrics, cosine distance is less sensitive to depth-induced projection effects and is more robust in monocular settings.

Position-Based Weighting. To account for perspective distortion (e.g., smaller flow vectors near the image centre due to forward camera motion), we apply a spatial weighting factor $\alpha \in [0, 1]$ based on the object's position in the image:

$$\alpha = 1 - \frac{\|\mathbf{c}_{\mathrm{obj}} - \mathbf{c}_{\mathrm{img}}\|}{d_{\mathrm{max}}} \tag{4}$$

Here, $\mathbf{c}_{\mathrm{obj}}$ is the centre of the object bounding box, $\mathbf{c}_{\mathrm{img}}$ is the image centre, and d_{max} is the distance from the image centre to the corner. This heuristic prioritises directional similarity (cosine distance) in the image centre and magnitude-based difference (Euclidean distance) near the edges. It assumes a typical forward-facing camera setup, such as in driving datasets, where such perspective effects are most prominent.

Combined Similarity Score. The final similarity score $S \in [0, 1]$ is computed as a weighted combination of the two metrics:

$$S = \alpha \cdot D_C + (1 - \alpha) \cdot D_E \tag{5}$$

Higher values of S indicate greater motion dissimilarity between the object and its background. An object is classified as dynamic if $S > \theta_{th}$, where we empirically set the threshold $\theta_{th} = 0.1$.

3.5   Probabilistic State Filtering

To ensure temporal consistency in motion classification and suppress frame-to-frame label flicker, we introduce a *probabilistic filtering* mechanism. Instead of relying solely on per-frame binary decisions, each tracked object maintains a *moving probability* that reflects its likelihood of being dynamic over time. Let $p_{t-1} \in [0, 1]$ denote the dynamic probability of an object at the previous frame, and let $s_t \in \{0, 1\}$ represent the current binary classification from the motion similarity module, where $s_t = 1$ indicates dynamic and $s_t = 0$ indicates static. The update is governed by two parameters: a learning rate $\alpha \in [0, 1]$ that controls responsiveness to new evidence, and a bias term $\beta \in [0, 1]$ that modulates the strength of updates based on current confidence. The probability is recursively updated as:

$$p_t = s_t \cdot [(1 - \beta)p_{t-1} + \alpha] + (1 - s_t) \cdot [(1 + \beta)p_{t-1} - \alpha] \tag{6}$$

This equation implements a *soft temporal filter*. When an object is classified as dynamic ($s_t = 1$), its dynamic probability increases gradually; if static ($s_t = 0$), the score decays, but not immediately. This allows the system to smooth out occasional misclassifications while still adapting to persistent motion patterns. To keep the value in the valid range, the updated probability p_t is clamped to $[0, 1]$. An object is considered dynamic if $p_t > 0.5$, at which point all associated features are removed from SLAM processing. We empirically chose $\alpha = 0.2$ and $\beta = 0.3$, based on qualitative testing across multiple scenes. Although no ablation study is included, these values offered a good trade-off between rapid adaptation and resistance to noisy decisions. Newly detected or re-identified objects (i.e., those assigned a new ID after tracking loss) are always initialised with a neutral probability $p_0 = 0.5$, reflecting maximum uncertainty and allowing the filter to quickly adjust based on motion consistency.

4   Results

We evaluated our method on monocular visual SLAM to demonstrate its robustness in dynamic environments, even without additional sensors that aid drift correction. The KITTI Odometry dataset was used, and performance was measured using absolute trajectory error (ATE). We focused on 11 sequences for which ground truth poses were available. For the experiments, an Intel i9 and an

Table 1. Comparison of RMSE and standard deviation between baseline and proposed method on KITTI Odometry dataset.

Seq.	RMSE (ORB-SLAM3)	std (ORB-SLAM3)	RMSE (Ours)	std (Ours)
00	**6.99**	**3.24**	9.37	4.92
01	513.41	244.62	**100.37**	**58.79**
02	26.63	17.50	**24.39**	**16.59**
03	**1.05**	**0.65**	1.22	0.72
04	0.90	0.54	**0.76**	**0.40**
05	**6.02**	**2.19**	10.29	4.27
06	16.52	8.03	**11.01**	**5.20**
07	**2.38**	**1.23**	3.54	1.15
08	63.16	39.46	**58.40**	**36.14**
09	10.74	8.02	**7.49**	**2.84**
10	8.98	5.23	**8.86**	**4.75**

Nvidia RTX 4080 were used (with YOLO11x). The algorithm takes around 41 ms per frame, depending on the type of sequence and is proportional to the number of objects being tracked in the individual frames. Trajectories were aligned using Sim(3) Umeyama alignment via the **evo** toolbox[1], which also computes the Root Mean Square Error (RMSE) and standard deviation (std) of the aligned poses. These metrics reflect absolute trajectory accuracy and temporal stability, respectively.

Table 1 presents a quantitative comparison between the baseline ORB-SLAM3 [8] and our proposed dynamic filtering method. In low-dynamic sequences such as 03, 04, and 07, both methods achieved comparable results, with ORB-SLAM3 performing slightly better. In such scenes, the static world assumption holds, so additional filtering is unnecessary. In contrast, our method excels in highly dynamic sequences, such as 01, 08, and 09, where moving vehicles and lack of loop closures cause accumulated errors for ORB-SLAM3. Notably, on sequence 01 (recorded on a motorway with persistent high-speed motions), our method reduced RMSE from 513.41 to 100.37 and improved stability by more than 4 times. These improvements stem from explicitly removing unstable features from dynamic regions that otherwise would corrupt pose estimation.

Some sequences, such as 00 and 05, showed better performance with ORB-SLAM3. These contain multiple successful loop closures, enabling the backend optimisation to compensate for earlier drift. In such cases, the benefit of dynamic filtering is less pronounced and, in some frames, may marginally degrade feature availability.

Our method outperformed ORB-SLAM3 in several sequences even with monocular input and without any depth or stereo support. Figure 2 illustrates

---

[1] https://github.com/MichaelGrupp/evo.

trajectory alignment in two representative sequences. In sequence 2a, recorded on a highway with sustained motion and without loop closures, ORB-SLAM3 suffered significant drift. Our method better aligned with the ground truth, showing resilience to dynamic features. In sequence 2b, where multiple loop closures are present, ORB-SLAM3 performed comparably; however, dynamic elements still introduced localised inconsistencies, which our method mitigated by filtering out unstable regions. These results demonstrate that our method improves robustness and accuracy in monocular SLAM pipelines, especially under real-world dynamic conditions. Figure 3 presents a qualitative example of dynamic feature suppression on a frame from the KITTI dataset. The top image (3a) shows correctly identified dynamic objects in the frame where the boxes are provided from the filtering algorithm and feature points inside these are removed from tracking. The bottom image (3b) shows a failure case where parked cars are misclassified as dynamic due to the lighting conditions present in the image. Importantly, this filtering occurs prior to pose estimation, ensuring that only reliable features are passed to the RANSAC module. By increasing the inlier ratio and removing geometric outliers, this pre-filtering step improves the accuracy and stability of motion estimation, particularly in highly dynamic scenes.

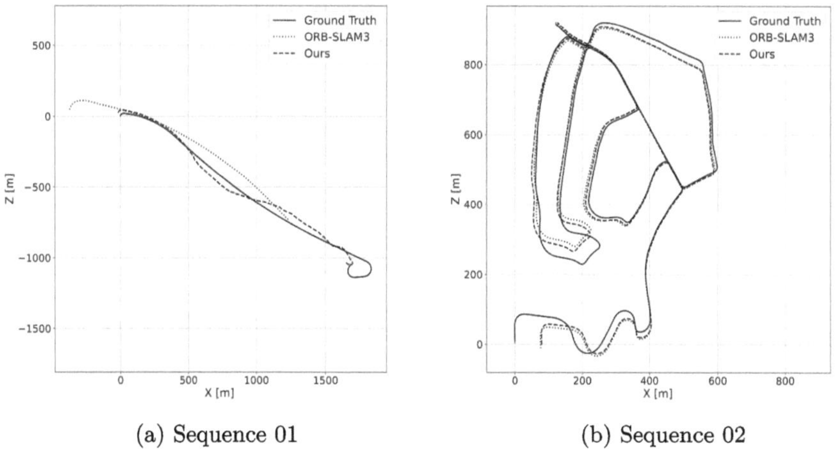

(a) Sequence 01   (b) Sequence 02

Fig. 2. Trajectory comparison between the baseline and our method on two KITTI sequences. The solid line represents the ground truth, the dotted line corresponds to ORB-SLAM3, and the dashed line denotes our method.

(a) Correctly filtered objects

(b) Incorrectly filtered objects

Fig. 3. Results from different sequences where the bounding boxes are created by the algorithm. Top: correctly classifying dynamic objects. Bottom: misclassifying parked cars as dynamic.

5   Conclusions and Future Work

In this paper, we presented a dynamic object filtering pipeline for feature-based monocular visual SLAM operating in dynamic environments. Our method combines deep object detection, appearance-based multi-object tracking, dense optical flow analysis, motion similarity evaluation, and probabilistic temporal filtering to segment and suppress unstable features associated with dynamic regions. These binary masks are integrated into the SLAM frontend, allowing the system to exclude dynamic features prior to pose estimation. This early filtering improves the inlier ratio in RANSAC, leading to more robust and accurate motion estimation. However, the proposed approach also has several limitations. First, the accuracy of optical flow can degrade significantly in the presence of motion blur, leading to unreliable motion cues may cause misclassification of dynamic regions. Second, in cluttered scenes with a high density of overlapping or occluding objects, detection, tracking, and classification performance may deteriorate, reducing the effectiveness of dynamic feature suppression. Finally, the use of manually tuned thresholds for motion classification introduces sensitivity to dataset-specific characteristics, limiting the generalisability of the system across diverse environments without additional adaptation or learning. Quantitative evaluations on the KITTI dataset show that the proposed approach significantly reduces absolute trajectory error (RMSE) and standard deviation in dynamic scenes while maintaining competitive performance in predominantly static environments. These results demonstrate that dynamic object suppression enhances tracking reliability without degrading performance in low-motion scenarios. Qualitative results further show that our method successfully removes features from moving objects while preserving background structure, contributing to smoother and more stable SLAM trajectories. While our current method relies on modular components with manually defined heuristics, future work could explore an end-to-end learning-based approach. Recent advances in deep optical flow and motion segmentation provide an opportunity to train neural networks that directly predict dynamic regions from frame pairs. Such models could learn to generalise motion patterns across diverse environments and eliminate the need for explicit geometric checks.

References

1. Aharon, N., Orfaig, R., Bobrovsky, B.Z.: BoT-SORT: robust associations multi-pedestrian tracking, July 2022, arXiv:2206.14651 [cs]
2. Ai, Y.B., et al.: Visual SLAM in dynamic environments based on object detection. Def. Technol. **17**(5), 1712–1721 (2021)
3. Ai, Y., Rui, T., Lu, M., Fu, L., Liu, S., Wang, S.: DDL-SLAM: a robust RGB-D SLAM in dynamic environments combined with deep learning. IEEE Access **8**, 162335–162342 (2020)
4. Alcantarilla, P.F., Yebes, J.J., Almazan, J., Bergasa, L.M.: On combining visual SLAM and dense scene flow to increase the robustness of localization and mapping in dynamic environments. In: 2012 IEEE International Conference on Robotics and Automation, pp. 1290–1297. IEEE, St Paul, MN, USA, May 2012
5. Ballester, I., Fontan, A., Civera, J., Strobl, K.H., Triebel, R.: DOT: dynamic object tracking for visual SLAM. In: 2021 IEEE International Conference on Robotics and Automation (ICRA), pp. 11705–11711. IEEE, Xi'an, China, May 2021
6. Bescos, B., Campos, C., Tardos, J.D., Neira, J.: DynaSLAM II: tightly-coupled multi-object tracking and SLAM. IEEE Robot. Autom. Lett. **6**(3), 5191–5198 (2021)
7. Bescos, B., Fácil, J.M., Civera, J., Neira, J.: DynaSLAM: tracking, mapping and inpainting in dynamic scenes. IEEE Robot. Autom. Lett. **3**(4), 4076–4083 (2018), arXiv:1806.05620 [cs]
8. Campos, C., Elvira, R., Rodríguez, J.J.G., M. Montiel, J.M., D. Tardós, J.: ORB-SLAM3: an accurate open-source library for visual, visual–inertial, and multimap SLAM. IEEE Trans. Robot. **37**(6), 1874–1890 (2021), conference Name: IEEE Transactions on Robotics
9. Chen, L.C., Papandreou, G., Kokkinos, I., Murphy, K., Yuille, A.L.: DeepLab: semantic image segmentation with deep convolutional nets, Atrous convolution, and fully connected CRFs. IEEE Trans. Pattern Anal. Mach. Intell. **40**(4), 834–848 (2018)
10. Cheng, J., Wang, Z., Zhou, H., Li, L., Yao, J.: DM-SLAM: a feature-based SLAM system for rigid dynamic scenes. ISPRS Int. J. Geo Inf. **9**(4), 202 (2020)
11. Cheng, S., Sun, C., Zhang, S., Zhang, D.: SG-SLAM: a real-time RGB-D visual SLAM toward dynamic scenes with semantic and geometric information. IEEE Trans. Instrum. Meas. **72**, 1–12 (2023)

12. Kaneko, M., Iwami, K., Ogawa, T., Yamasaki, T., Aizawa, K.: Mask-SLAM: robust feature-based monocular SLAM by masking using semantic segmentation. In: 2018 IEEE/CVF Conference on Computer Vision and Pattern Recognition Workshops (CVPRW), pp. 371–3718. IEEE, Salt Lake City, UT, USA, June 2018

13. Liu, W., et al.: SSD: single shot multibox detector. In: Leibe, B., Matas, J., Sebe, N., Welling, M. (eds.) ECCV 2016. LNCS, vol. 9905, pp. 21–37. Springer, Cham (2016). https://doi.org/10.1007/978-3-319-46448-0_2

14. Liu, Y., Miura, J.: RDS-SLAM: real-time dynamic SLAM using semantic segmentation methods. IEEE Access **9**, 23772–23785 (2021)

15. Ma, T., Zhang, J., Gao, S., Chen, C.: Dyna-VO: a semantic visual odometry in dynamic environment. In: 2021 China Automation Congress (CAC), pp. 477–482. IEEE, Beijing, China, October 2021

16. Mur-Artal, R., Tardós, J.D.: ORB-SLAM2: an open-source SLAM system for monocular, stereo, and RGB-D cameras. IEEE Trans. Robot. **33**(5), 1255–1262 (2017), conference Name: IEEE Transactions on Robotics

17. Su, P., Luo, S., Huang, X.: Real-time dynamic SLAM algorithm based on deep learning. IEEE Access **10**, 87754–87766 (2022)

18. Sun, D., Yang, X., Liu, M.Y., Kautz, J.: PWC-net: CNNs for optical flow using pyramid, warping, and cost volume. In: 2018 IEEE/CVF Conference on Computer Vision and Pattern Recognition, pp. 8934–8943. IEEE, Salt Lake City, UT, USA, June 2018

19. Xiao, L., Wang, J., Qiu, X., Rong, Z., Zou, X.: Dynamic-SLAM: semantic monocular visual localization and mapping based on deep learning in dynamic environment. Robot. Auton. Syst. **117**, 1–16 (2019)

20. Yu, C., et al.: DS-SLAM: a semantic visual SLAM towards Dynamic Environments. In: 2018 IEEE/RSJ International Conference on Intelligent Robots and Systems (IROS), pp. 1168–1174. IEEE, Madrid, October 2018

21. Zhang, J., Henein, M., Mahony, R., Ila, V.: VDO-SLAM: a visual dynamic object-aware SLAM system, December 2021, arXiv:2005.11052 [cs]

22. Zhang, T., Zhang, H., Li, Y., Nakamura, Y., Zhang, L.: FlowFusion: dynamic dense RGB-D SLAM based on optical flow. In: 2020 IEEE International Conference on Robotics and Automation (ICRA), pp. 7322–7328. IEEE, Paris, France, May 2020

23. Zhong, F., Wang, S., Zhang, Z., Chen, C., Wang, Y.: Detect-SLAM: making object detection and SLAM mutually beneficial. In: 2018 IEEE Winter Conference on Applications of Computer Vision (WACV), pp. 1001–1010. IEEE, Lake Tahoe, NV, March 2018

Special Session on Machine Learning Under Concept Drift in Highly Dynamic Environments: Advances and Challenges

Memory Combination-Based Approaches
for Multi-label Classification
on Non-Stationary Data Streams

Xihui Wang[1,2], Hugo Peuzet[1,2(✉)], Pascale Kuntz[1], Frank Meyer[2],
and Vincent Lemaire[2]

[1] Laboratory of Digital Sciences of Nantes, Nantes, France
{xihui.wang,hugo.peuzet,pascale.kuntz}@ls2n.fr
[2] Orange Research, Lannion, France
{xihui.wang,hugo.peuzet,franck.meyer,vincent.lemaire}@orange.com

Abstract. Due to their ever-increasing number of applications, multi-label classification algorithms are facing a major challenge: learning from evolving data streams with distribution changes over time, under limited computational and memory resources. In this paper, we first revisit existing works through a meta-model of low-complexity multi-scale algorithms combining short-term memory for fast drift adaptation and long-term memory to integrate distribution changes over time. Then, we develop a very strong baseline from this family, called A2ML (Adaptive Memories for Multi-Label stream classification), specifically designed for non-stationary streams. Its long-term memory is managed via adaptive label clustering and biased reservoir sampling, ensuring linear-time model updates. A2ML is compared to 7 state-of-the-art algorithms on 15 stationary and 4 non-stationary streams with over 100,000 examples, generated to test various data changes and concept drifts. Results show A2ML performs well in both settings and has lower computation times than competitors.

Keywords: Data Stream · Multi-Label Classification · Concept Drift

1 Introduction

Usual implementations of multi-label classification models assume full access to training data while a growing number of applications generate massive, fast, continuous, and virtually infinite data streams. Thus, considering that computational and storage resources are liable to be limited by energy cost reduction policies [1] or the application environment (e.g. IoT) [2], traditional models no longer meet current requirements. State-of-the-art reflects an increasing activity in Multi-Label Stream Classification (MLSC) [3] motivated by these new objectives and highlights a new challenge: efficiently handling complex concept drifts over time, which are more intricate than in mono-label streams. In the mono-label case, a concept drift is associated to a temporal evolution of the

L. Martínez et al. (Eds.): IDEAL 2025, LNCS 16239, pp. 221–233, 2026.
https://doi.org/10.1007/978-3-032-10489-2_19

joint probability distribution $\mathbb{P}_t(\mathbf{x}, \mathbf{y})$ where $\mathbf{x}$ (resp. $\mathbf{y}$) is an attribute (resp. a label) vector. From the joint probability decomposition $\mathbb{P}_t(\mathbf{x}, \mathbf{y}) = \mathbb{P}_t(\mathbf{x}|\mathbf{y})\mathbb{P}_t(\mathbf{y})$, a concept drift can be generated by a significant variation of each of the component, and the learning model needs to adapt between two time steps t_1 and t_2, $t_1 < t_2$, to three possible situations: (i) $\mathbb{P}_{t_1}(\mathbf{y}) \neq \mathbb{P}_{t_2}(\mathbf{y})$ but $\mathbb{P}_{t_1}(\mathbf{x}|\mathbf{y}) = \mathbb{P}_{t_2}(\mathbf{x}|\mathbf{y})$, (ii) $\mathbb{P}_{t_1}(\mathbf{x}|\mathbf{y}) \neq \mathbb{P}_{t_2}(\mathbf{x}|\mathbf{y})$ but $\mathbb{P}_{t_1}(\mathbf{y}) = \mathbb{P}_{t_2}(\mathbf{y})$, (iii) $\mathbb{P}_{t_1}(\mathbf{x}|\mathbf{y}) \neq \mathbb{P}_{t_2}(\mathbf{x}|\mathbf{y})$ and $\mathbb{P}_{t_1}(\mathbf{y}) \neq \mathbb{P}_{t_2}(\mathbf{y})$.

Generalization to the multi-label data in non-stationary streams is recent and most research still focuses on stationary streams, which significantly differ from real-world non-stationary scenarios. Few evaluation protocols for non-stationary streams exist [4], but they often struggle with large attribute and label spaces or do not account for different drift types. This limits understanding of how drift affects adaptation mechanisms, which is crucial for new real-world applications like cybersecurity, such as web attack detection [5].

This article makes two main contributions to multi-label classification of non-stationary streams. First, we introduce a meta-model of low-complexity multi-scale algorithms based on a combination of memories. Second, we propose A2ML (Adaptive Memories for Multi-Label stream classification), a very strong baseline from this family, specifically designed for non-stationary streams. With its lower memory complexity guaranteed by a theoretical bound, A2ML manages distribution changes over time using a new mechanism based on dynamic clustering and Biased Reservoir Sampling [6].

We compared A2ML with another memory-based algorithm (ODM [7]) and six state-of-the-art top strategies. Prior experiments on stationary streams confirmed memory-based methods competitiveness, and led us to retain a subset of the best four algorithms. For the evaluation on non-stationary streams we developed a new evaluation protocol adressing two types of data distribution changes: "abrupt drift", where one concept replaces another instantly, and "gradual drift", where two concepts coexist during a transition period.

Tests on four non-stationary streams with over $100,000$ examples revealed that memory-based approaches balance plasticity and stability effectively and run faster than competitors. Based on experimental outcomes, A2ML can be considered a very strong baseline that underscores the value of well-designed memory combinations for MLSC in applicative contexts.

The remainder of this article is organized as follows: Sect. 2 introduces the state-of-the-art approaches developed for MLSC; the meta-model and the new A2ML algorithm are respectively described in Sect. 3 and 4; Sect. 5 presents the results of the experimental comparisons for stationary and non stationary data streams with our new protocol.

2   Related Work

2.1   Background

Let $\mathcal{X} = \mathcal{R}^d$ be the attribute space and $\mathcal{Y} = \{0,1\}^l$ the label space. We consider a potentially infinite data stream $D = \{(\mathbf{x}_1, \mathbf{y}_1), ..., (\mathbf{x}_t, \mathbf{y}_t), ...\}$, where each example $(\mathbf{x}_t, \mathbf{y}_t) \in \mathcal{X} \times \mathcal{Y}$ consists of a vector of d attributes $\mathbf{x}_t = [x_t^1, ..., x_t^d]$ and a vector of l labels $\mathbf{y}_t = [y_t^1, ..., y_t^l]$ where $y_t^i = 1$ (resp. 0) if the label λ_i, $i = 1$ to l, is present (resp. absent). For each example, at least one label is present.

In supervised MLSC, the goal is to predict $\mathbf{y}_t$ from $\mathbf{x}_t$. The common evaluation is the Interleaved-Test-Then-Train, which involves three stages at each time step t: (i) making a prediction from $\mathbf{x}_t$ using the previous model $\hat{\mathbf{y}}_t = h_{t-1}(\mathbf{x}_t)$, (ii) re-evaluating the model using the real label vector $\mathbf{y}_t$, and (iii) computing a new model h_t from the previous model h_{t-1} and $(\mathbf{x}_t, \mathbf{y}_t)$. Assuming $(\mathbf{x}_t, \mathbf{y}_t)$ is generated by an unknown joint probability distribution $\mathbb{P}_t(\mathbf{x}, \mathbf{y})$, the stream D is stationary when $\mathbb{P}_t(\mathbf{x}, \mathbf{y})$ remains stable, and non-stationary otherwise [8]. Concept drift occurs when $\mathbb{P}_t(\mathbf{x}, \mathbf{y})$ between two points in time t_1 and t_2, $t_1 < t_2$, is different. The model is thus required not only to learn incrementally when $\mathbb{P}_t(\mathbf{x}, \mathbf{y})$ is stable, but also to adapt itself quickly when $\mathbb{P}_t(\mathbf{x}, \mathbf{y})$ changes. In MLSC, the label space can be very large, 2^l, and label distributions are more complex, with correlations and varying co-occurrence frequencies. Concept drift, caused by changes in label prior probabilities, is a key factor in performance decline, as noted by [8].

2.2   Methods

Recent approaches to MLSC fall into three categories: (i) transformation approaches converting MLSC into multiple single-label problems; (ii) adaptation approaches modifying single-label algorithms for MLSC; and (iii) ensemble approaches combining several simpler models to boost performance.

- Transformation approaches vary in how they handle label relationships. The Binary Relevance (BR) method decomposes MLSC into binary problems but ignores label correlations. Several implementations of this strategie, like AMW [9], have been proposed. Other methods, like PS [10] and MINAS-PS [11], attempt to leverage label dependencies, with PS extending the multi-class classification label combination algorithm [12] to include multi-labels, and MINAS-PS associating each distinct label vector with clusters. However, PS can be computationally costly for datasets with many labels.
- Adaptation approaches are often based on Hoeffding trees, such by using a majority vote classifier incorporating the PS strategy (HTps, [13]), or by integrating multi-target regression perceptrons (ISOUPT, [14]) into the tree's leaves. However, HTps struggles with real-time label vector frequency, and ISOUPT doesn't handle concept drift well. Other strategies include neural networks (OSML-ELM [15]) and rule-based systems (ML-AMRules [16]), but they also face drift issues. Some algorithms, like MLSAMPkNN [17] and MLSAkNN [18], incorporate memory with adaptive windowing or auto-tuning of k-NN parameters to better handle drift.
- Ensemble approaches often use the classical ADWIN Bagging, replacing the worst-performing model after drift detection, as in EaHTps [13] and AESAKNNS [19]. The "weighted majority" strategy which weighs the classifier predictions at each time (e.g. GOOWE-ML [20]), show good performances

at the cost of high complexity. Recently, ODM [7] combined short and long-term memory for better efficiency, but its drift adaptation weakens over time. These limitations motivated the development of our new algorithm, A2ML.

3 Memory-Based Algorithms for Multi-label Data Streams

In MLSC, "concept drift" encompasses various types: "abrupt drift", where a concept replaces the existing one instantly; "gradual drift", where concepts can overlap during the transition period; and "recurring drift", when previous concepts reappear. Addressing this complexity requires multi-scale learning strategies suited to resource constraints. We here suggest to revisit some existing works as a general framework of a low-complexity multi-scale classification framework combining short-term memory for quick drift adaptation and long-term memory to track distribution changes. Both are continuously updated, and the global model integrates their information, enabling efficient adaptation to various concept drifts with low time and memory costs. In the subsequent sections, we detail the structure of the two memories and their prediction and learning mechanisms. For simplicity, all components are described at any time t in the data stream.

3.1 Meta-Model

Let us denote by M_{STM} (resp. M_{LTM}) the short-term (resp. long-term) memory. M_{STM} is a sliding window of the m most recent examples to handle abrupt drifts, and M_{LTM} store "useful" information beyond these, capturing stable, gradual, and recurring concepts. M_{LTM} consists of s clusters $C_v = \{\mathbf{c}_v^x, \mathbf{c}_v^y, S_v\}$, $v = 1$ to s, each being described by a triplet made up of two prototypes $\mathbf{c}_v^x \in \mathcal{X}$ and $\mathbf{c}_v^y \in \mathcal{Y}$, and a reservoir S_v. These clusters summarize the distribution in attribute and label spaces. At each time t, the new learning example is assigned to its closest cluster based on similarity in label space. The cluster's prototypes are then updated, and its reservoir stores the example using sampling to maintain a limited memory of accumulated knowledge.

3.2 Prediction Process

The prediction $\hat{\mathbf{y}}_t$, associated with $\mathbf{x}_t$, is obtained by aggregating the predictions $\hat{\mathbf{y}}_{STM}$ and $\hat{\mathbf{y}}_{LTM}$ from M_{STM} and M_{LTM}.

$\hat{\mathbf{y}}_{STM} = [\hat{y}_{STM}^l, ..., \hat{y}_{STM}^l]$ is computed using a multi-label k-NN classifier: each label $\hat{y}_{STM}^i$, $i = 1$ to l, is set to 1 if the majority of the k closest examples using a given distance have label λ_i, otherwise 0.

$\hat{\mathbf{y}}_{LTM} = [\hat{y}_{LTM}^1, ..., \hat{y}_{LTM}^l]$ is computed by first identifying the cluster $C_p = \{\mathbf{c}_p^x, \mathbf{c}_p^y, S_p\}$ in M_{LTM} with the X-prototype closest to $\mathbf{x}_t$, then predicting labels using a multi-label k-NN classifier within that cluster's reservoir S_p, similar to $\hat{\mathbf{y}}_{STM}$.

And $\hat{\mathbf{y}}_t$ is a weighted aggregation of $\hat{\mathbf{y}}_{STM}$ and $\hat{\mathbf{y}}_{LTM}$: $\hat{\mathbf{y}}_t = ag(\hat{\mathbf{y}}_{STM}, w_{STM}, \hat{\mathbf{y}}_{LTM}, w_{LTM})$ where w_{STM} and w_{LTM} are the respective contributions of the memories to the prediction and ag is an aggregation function.

3.3   Learning Process

After the comparison of the prediction $\hat{\mathbf{y}}_t$ with the real label vector $\mathbf{y}_t$, both memories and their associated weights are updated with the information provided by the new example $(\mathbf{x}_t, \mathbf{y}_t)$.

For the Short-term Memory: The new example is inserted at the start of M_{STM}'s sliding window, shifting place of others by one. When full, the oldest example is removed to make space. A fixed window size m enables quick adaptation to recent data.

For the Long-term Memory: At each step, two clusters are updated: the best cluster $C_b = \{\mathbf{c}_b^x, \mathbf{c}_b^y, S_b\}$, with the closest Y-prototype to $\mathbf{y}_t$, and the predicted cluster $C_p = \{\mathbf{c}_p^x, \mathbf{c}_p^y, S_p\}$, with the closest X-prototype to $\mathbf{x}_t$. The new example is added to C_b's reservoir S_b via sampling, and its prototypes are adjusted to maintain cluster homogeneity: $(\mathbf{c}_b^x, \mathbf{c}_b^y) = (\mathbf{c}_b^x, \mathbf{c}_b^y) + ((\mathbf{x}_t, \mathbf{y}_t) - (\mathbf{c}_b^x, \mathbf{c}_b^y)) \times lr_b$.

If $C_b \neq C_p$, the predicted cluster C_p is moved away from the new example to increase heterogeneity between clusters and improve prediction: $(\mathbf{c}_p^x, \mathbf{c}_p^y) = (\mathbf{c}_p^x, \mathbf{c}_p^y) - ((\mathbf{x}_t, \mathbf{y}_t) - (\mathbf{c}_p^x, \mathbf{c}_p^y)) \times lr_p$ with a rate lr_p which is also decided by the sampling strategy.

For the Weights: Weights w_{STM} and w_{LTM} respectively associated with the memories M_{STM} and M_{LTM} are updated after each prediction:

$$w_{STM} = \frac{1}{m} \sum_{i=t-m+1}^{t} \frac{|\mathbf{y}_i \cap \hat{\mathbf{y}}_i|}{|\mathbf{y}_i \cup \hat{\mathbf{y}}_i|}$$ where $\hat{\mathbf{y}}_i$ is the prediction of M_{STM} and $\mathbf{y}_i$ is the true label vector at time i. w_{LTM} is updated similarly using the predictions $\hat{\mathbf{y}}_i$ of the M_{LTM}.

4   A New Strong Baseline for Non-stationary Data Streams: A2ML

The ODM algorithm [7] implements the described meta-model with a fixed number of long-term clusters and random reservoir sampling. While effective for stationary data and small streams in non-stationary settings, its performance declines over time. Additionally, setting the cluster number and initialization pose challenges. To address non-stationary streams, we developed A2ML, a new memory-based strategy following the meta-model. Its long-term memory uses a biased reservoir sampling for fast-changing data, and dynamic clustering to add new clusters, with a theoretical bound on their maximum number, eliminating the need for prior tuning. Details are provided in Algorithm 1.

Algorithm 1. LTM updating at each time t

Input: $M_{LTM} = \{\{\mathbf{c}_v^x, \mathbf{c}_v^y, S_v\}|v = 1, \ldots,, s\}$, the maximum number of examples stored in each reservoir r, new example $(\mathbf{x}_t, \mathbf{y}_t)$
Output: updated M_{LTM}

1: **for** $(\mathbf{c}_v^x, \mathbf{c}_v^y, S_v) \in M_{LTM}$ **do**
2:     # Calculate the cosine distance between the $\mathbf{y}_t$ (resp. $\mathbf{x}_t$) and the s Y-prototypes (resp. X-prototypes)
3:     $d_v^y = \mathcal{D}(\mathbf{c}_v^y, \mathbf{y}_t)$
4:     $d_v^x = \mathcal{D}(\mathbf{c}_v^x, \mathbf{x}_t)$
5: **end for**
6: $b = argmin(d_v^y, v \in 1, ..., s)$ # The index of the best cluster
7: $p = argmin(d_v^x, v \in 1, ..., s)$ # The index of the predicted cluster
8: **if** $d_b^y < 1$ **then**
9:     # Update the best and predicted cluster
10:     $(\mathbf{c}_b^x, \mathbf{c}_b^y) = (\mathbf{c}_b^x, \mathbf{c}_b^y) + \frac{1}{|S_b|} \times ((\mathbf{x}_t, \mathbf{y}_t) - (\mathbf{c}_b^x, \mathbf{c}_b^y))$
11:     **if** $b \neq p$ **then**
12:         $(\mathbf{c}_p^x, \mathbf{c}_p^y) = (\mathbf{c}_p^x, \mathbf{c}_p^y) - \frac{1}{|r|} \times ((\mathbf{x}_t, \mathbf{y}_t) - (\mathbf{c}_p^x, \mathbf{c}_p^y))$
13:     **end if**
14:     $S_b.update(\mathbf{x}_t, \mathbf{y}_t)$ # use the BRS strategie
15: **else**
16:     # Create a new cluster
17:     $(\mathbf{c}_{s+1}^x, \mathbf{c}_{s+1}^y) = (\mathbf{x}_t, \mathbf{y}_t)$
18:     $S_{s+1} = \{(\mathbf{x}_t, \mathbf{y}_t)\}$
19:     $M_{LTM}.add(\{\mathbf{c}_{s+1}^x, \mathbf{c}_{s+1}^y, S_{s+1}\})$
20: **end if**
21: **return** M_{LTM}

4.1 Cluster Creation

The first cluster is initialized with the stream's first example. At each step, a new example joins the closest cluster sharing the most labels; if none share labels, a new cluster is created. The closest cluster, called the best cluster in Algorithm 1, is identified via cosine distance between $\mathbf{y}_t$ and the s Y-prototypes. If none of the distances in the label space are less than 1, a new cluster is added. Cosine similarity is preferred for sparse multi-label data, though other metrics are possible [21][1].

4.2 Cluster Prototype Updating

While alternatives like Incremental Support Vector Machine [22] or Stochastic Gradient Descent [23] exist, we chose Learning Vector Quantization (LVQ) [24] for its robustness in non-stationary environments [25]. Among variants, LVQ 2.1 [26] offered the best balance of performance and complexity during preliminary tests. The detailed presentation of these experiments goes beyond the scope of this article. The adaptation of LVQ to our meta-model involves three steps:

---

[1] Exploring adaptive metrics based on data type is a promising future direction.

(i) reducing the distance between the best Y-prototype $\mathbf{c}_b^y$ and the new label vector $\mathbf{y}_t$, (ii) reducing the distance between the best X-prototype $\mathbf{c}_b^x$ and the new attribute vector $\mathbf{x}_t$, and (iii) increasing the distance between the X-prototype of the predicted cluster C_p and $\mathbf{x}_t$ if C_p is not the best. These steps maintain cluster homogeneity (i), allow a continual adaptation to new attributes (ii), and ensure sufficient separation between clusters (iii).

In A2ML, the best cluster C_b contains the Y-prototype closest to $\mathbf{y}_t$. Its reservoir S_b is updated with biased sampling, and prototypes $\mathbf{c}_b^x$ and $\mathbf{c}_b^y$ are "moved closer" to the new example. The learning rate lr_b depends on reservoir size: $lr_b = \frac{1}{|S_b|}$, starting at 1.0 and decreasing to $\frac{1}{r}$. The fixed $lr_p = \frac{1}{r}$ prevents cluster spacing from disrupting matching.

4.3 Reservoir Updating

Sampling strategies [27] ensure low complexity and depend on the stream properties, which are often unknown. Roughly speaking, for a stationary distribution a random sampling is sufficient whereas non stationary distributions require biased strategies [6]. In A2ML, a biased sampling updates the reservoir S_b: if reservoir size is below r, the new example is added; otherwise, it replaces a random example in S_b, preserving older data and recent stream information. The update rate starts at 1.0 and decreases to $\frac{1}{r}$ as the reservoir fills, matching the prototype learning rate.

4.4 Aggregation of the Memory Information for the Final Decision

While various aggregators exist [28], a simple weighted average of memory accuracies, as used in ODM, is both effective and efficient. A score f_i for each label λ_i, $i = 1$ to l, is computed as a weighted mean of $\hat{\mathbf{y}}_{STM}^i$ and $\hat{\mathbf{y}}_{LTM}^i$. Each label's final prediction $\hat{\mathbf{y}}_t^i$ is 1 if $f_i \leq 0.5$, else 0. Initially, with empty memories, $\hat{\mathbf{y}}_t$ is set to the null vector.

4.5 Complexity

For any given example, the prediction and learning processes have linear complexities: respectively equal to $\mathcal{O}(s + r)$ and $\mathcal{O}(s)$. Since r is fixed, A2ML's complexity depends on the number of clusters. Proposition 1 ensures that the insertion process requires a finite number l of clusters in the long-term memory.

Proposition 1. *At each time, the number s of clusters in the long-term memory M_{LTM} is less than or equal to the label cardinality l.*

A new cluster is only created when new labels are encountered, as existing Y-prototypes will have a cosine distance less than 1 with the new label vector.

Thus, the number of clusters s is bounded by the total labels, eliminating the need for cluster deletion[2]. Consequently, A2ML's complexity is $\mathcal{O}(l + r)$.

5   Experimental Results

This section compares our A2ML algorithm with ODM and other state-of-the-art methods. We focus on seven algorithms: three adaptation strategies (HTps, ISOUPT, MLSAkNN) and four ensemble methods (EaHTps, GOOWE, AESAKNNS, ODM). Preliminary tests in stationary settings identified the top four candidates for non-stationary scenarios, demonstrating the competitiveness of memory-based approaches. We then introduce a new experimental framework for multi-label classification in non-stationary data streams to evaluate their performance.

5.1   Preliminary Comparisons for Stationary Data Streams

We evaluated the algorithms on 15 stationary data streams using the Interleaved Test-Then-Train process which is well suited to data stream learning. Datasets from various applications in the well-known KDIS[3] repository were randomized to simulate streams with an independent and identical distribution (iid). For each, 20% of data was used for hyperparameter tuning [4], with the rest for performance evaluation. Performance was primarily measured by accuracy [29], with F-measure also summarized.

Experimental comparisons were performed by allowing public access to the algorithm codes, which are available either in the Github repository or on the Massive Online Analysis platform (MOA[4] 2021.07). Our source (code and experimental results) is also available in Github[5]. The computations were performed on an Intel Xeon processor E5-1630v3@3.70 GHZ with 128 GB of RAM on a Windows 10 system.

Hyperparameters: Hyperparameters were set based on original publications. For A2ML, additionnal experiments were carried out to provide a robust baseline for comparison. The aim of these experiments was to determine three parameters: the size m of the M_{STM}, the number k of nearest neighbors considered in the prediction phase, and the size r of the reservoir associated with the clusters in the M_{LTM}. The value of m was found to be identical ($m = 50$) to that of the

---

[2] In rare cases with many label combinations, setting the number of clusters higher than the known labels may be considered. This approach was not needed in our experiments.

[3] KDIS selected datasets: 20NG, Bibtex, Bookmarks, Cooking, Corel16k, Enron, Eukaryote, Human, IMDB, Mediamill, Reuters-K500, Scene, Slashdot, TMC2007, Yeast.

[4] MOA: https://moa.cms.waikato.ac.nz/

[5] A2ML source code: https://anonymous.4open.science/r/A2ML-8FEC.

MLSAMkNN algorithm where the short-term memory plays a similar role. The value of k was set to 3 according to the recommendations of k-NN multi-label methods [17]. The value r was determined by experiments carried out on the 15 stationary streams. The performance of A2ML was evaluated using the accuracy measure for 6 values of r: 25, 50, 100, 200, 400 and 800. The results (resp. 0.299, 0.308, 0.313, 0.314, 0.315, 0.315) indicate that the level of performance increases as r increases and then stabilizes from $r = 100$, which is the value selected in order to obtain a good compromise between result quality and computation time complexity.

Results: Table 1 shows the results, with the best in bold. Standard deviations are not reported as they are small (0.01–0.05). A Nemenyi test indicates A2ML and ODM have similar rankings, both outperforming others. A2ML performs significantly better than ISOUPT, GOOWE, and AESAKNNs, but differences with HTps, MLSAkNN, EaHTps, and ODM are not significant. A2ML ranks first in 8 of 15 datasets and second in 4 others. F-measure results are similar. A2ML also has lower average computation time on the 15 datasets: $26.7s$, compared to ODM ($32.8s$), MLSAkNN ($93.2s$), and others up to $3737.9s$.

Table 1. Average accuracy (%) of 8 algorithms on 15 datasets.

Datasets	HTps	ISOUPT	MLSAkNN	EaHTps	GOOWE	AESAKNNS	ODM	A2ML
20NG	32.3	15.0	11.0	27.1	12.7	0.1	60.9	**64.6**
Bibtex	12.4	2.0	8.3	11.1	2.5	6.3	22.3	**29.6**
Bookmarks	18.2	11.1	17.2	17.6	2.0	13.3	23.2	**26.6**
Cooking	4.9	0.0	0.8	0.3	1.3	0.5	**15.6**	15.1
Corel16k001	8.1	0.1	2.9	4.8	5.2	2.6	**11.2**	9.0
Enron	17.4	16.8	26.0	17.3	19.3	26.8	**33.2**	31.2
Eukaryote	**28.4**	2.5	15.0	**28.4**	15.9	14.2	25.5	23.1
Human	28.5	0.8	10.4	28.2	18.2	3.4	26.9	**30.9**
Imdb	**21.0**	0.4	8.4	**21.0**	16.1	1.3	15.4	13.0
Mediamill	32.6	33.1	36.6	33.7	18.0	36.8	**38.8**	30.0
Reuters-K500	21.7	0.2	28.6	19.6	3.8	25.8	37.3	**39.8**
Scene	38.7	0.9	52.5	7.9	24.2	9.2	52.3	**67.3**
SLASHDOT	13.2	0.3	17.3	12.2	10.2	0.7	33.6	**41.6**
TMC2007	41.5	33.1	42.8	43.0	29.9	35.2	**54.1**	44.1
Yeast	41.7	39.1	35.5	39.8	42.3	29.4	42.5	**44.6**
avg. value	24.0	10.3	20.9	20.8	14.8	13.7	32.9	**34.0**
avg. rank	3.60	7.33	4.80	4.47	5.60	6.00	**2.06**	2.13

5.2   Experimental Results for Non-stationary Data Streams

In non-stationary contexts, comparisons must consider the different types of concept drift from Sect. 3 for an accurate assessment of algorithm behaviors.

Non-stationary Scenarios: Comparisons were made on four non-stationary streams simulating real-life scenarios, with two types of drift (abrupt and gradual) and two distribution changes (prior and class-conditional shift). Each stream has two drifts: a new concept replaces the initial one, then it reappears. Starting from a real dataset D, we simulate non-stationarity by dividing D into two subsets with different distributions and recombining them to create various drift types (see Fig. 1).

- Prior shift. It is simulated by splitting the label vectors of n examples into two subsets $\{D_1, D_2\}$ using k-means. The examples are randomized to ensure each subset is iid, with different prior probabilities $\mathbb{P}_1(\mathbf{y})$ and $\mathbb{P}_2(\mathbf{y})$, while $\mathbb{P}_2(\mathbf{y})$, while $\mathbb{P}(\mathbf{x}|\mathbf{y})$ remains unchanged.
- Class-conditional shift. It is simulated by randomizing D, splitting it into two subsets D_1' and D_2' with the same distribution, then randomly swapping attribute values in D_1' using Fisher-Yates algorithm, preserving labels [30]. The streams have different $\mathbb{P}_1(\mathbf{x}|\mathbf{y})$ and $\mathbb{P}_2(\mathbf{x}|\mathbf{y})$, while $\mathbb{P}(\mathbf{y})$ remains constant. Recombining D_1 and D_2 (resp. D_1' and D_2') simulates non-stationary streams based on the drift type.
- Abrupt drift. Over the period $[0, n]$ and the period $[2n, 3n]$, all the data in the stream come from the same subset D_1 (resp. D_1') and are distributed identically and independently. Over the period $[n, 2n]$, all the data comes from another subset D_2 (resp. D_2') with an iid distribution.
- Gradual drift. We combine examples from D_1 with those from D_2 with the following sequence: generate (i) a stream with $\frac{3}{4}n$ examples from D_1 (first concept), (ii) a transition with $\frac{1}{2}n$ examples from D_1 and D_2, (iii) a stream with $\frac{3}{4}n$ examples from D_2 (second concept), (iv) a new transition with $\frac{1}{2}n$ from D_1 and D_2, and (v) a stream with $\frac{3}{4}n$ examples from D_1 (first concept reappears).

We used the Bookmarks dataset with $80,000$ examples for realistic stream simulation. Each bi-partition (D_1, D_2) and (D_1', D_2') generated two non-stationary streams: one with $120,000$ examples and two abrupt drifts, and another with $130,000$ examples and two gradual drifts.

Algorithm Comparisons: We compared ODM and A2ML with the other top four stationary algorithms (HTps, EaHTps) using the same previous hyperparameters. Figure 2 shows A2ML consistently outperforms the others. All algorithms' performance drops at the first drift, but A2ML recovers fastest and maintains the highest accuracy. Moreover, A2ML quickly handles recurring concepts, maintaining the highest accuracy. The average accuracies across four streams were 17.82% (HTps), 14.75% (EaHTps), 24.68% (ODM), and 29.53% (A2ML).

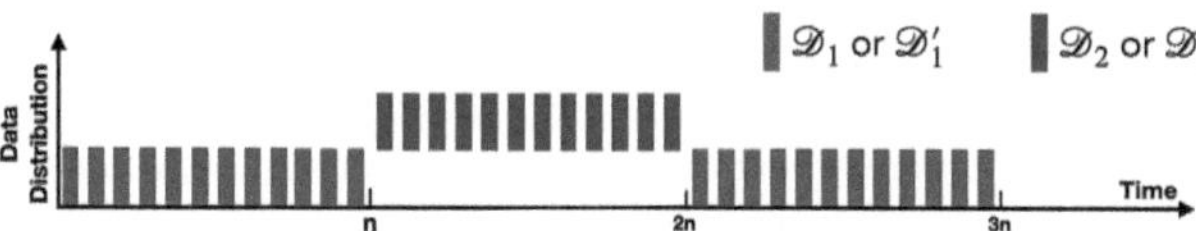

(a) Non-stationary case with two abrupt drifts: the first (resp. second) drift occurs at time n (resp. $2n$), and there is no transition period between the concepts.

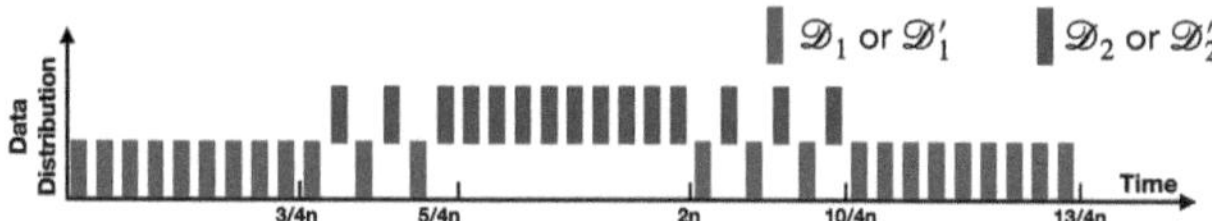

(b) Non-stationary case with two gradual drifts: the first (resp. second) drift occurs at time $\frac{3}{4}n$ (resp. $2n$) and the transition period between each concept is $1/2n$.

Fig. 1. Simulation of a non-stationary stream with different drifts.

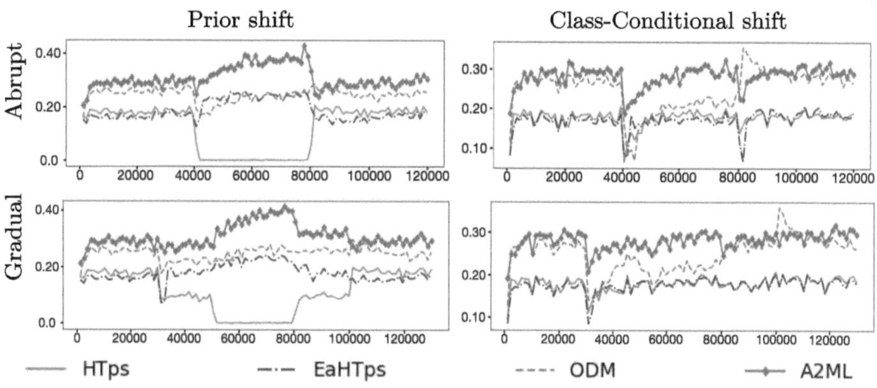

Fig. 2. Accuracy evolution in 4 non-stationary streams with two drifts: in the first drift a new concept replaced the initial one, and in the second drift the initial concept was recurring.

6   Conclusion

This article presents a framework for low-cost multi-label stream learning, featuring short-term memory for quick drift adaptation and long-term memory for past data. Based on this, we developed A2ML to adress non-stationary streams, which uses Biased Reservoir Sampling for efficient model updates and dynamic label partitioning to manage limited memory. Experiments against state-of-the-art methods confirm that A2ML serves as a very strong baseline. It demonstrates solid performance on both stationary and non-stationary streams. This is achieved using a novel protocol to incorporate different types of concept drift (prior shift, class-conditional, gradual and abrupt drifts) in a data stream made of real data.

Beyond its performance, A2ML and our new simulation protocol enhance understanding of MLSC algorithms on evolving data. The framework can be extended to continual learning scenarios [31], efficiently managing resources for infinite streams. In addition, from an application perspective, we plan to implement A2ML for analyzing URL streams in corporate cybersecurity. Its robustness and quick adaptation to abrupt and gradual drift make it a very promising candidate for this type of problem.

References

1. Strubell, E., Ganesh, A., McCallum, A.: Energy and policy considerations for modern deep learning research. In: Proceedings of the AAAI Conference on Artificial Intelligence, vol. 34, pp. 13693–13696 (2020)
2. Hua, H., Li, Y., Wang, T., Dong, N., Li, W., Cao, J.: Edge computing with artificial intelligence: a machine learning perspective. ACM Comput. Surv. **55**(9), 1–35 (2023)
3. Bakhshi, S., Can, F.: Balancing efficiency vs. effectiveness and providing missing label robustness in multi-label stream classification. Knowl.-Based Syst. **289**, 111489 (2024)
4. Losing, V., Hammer, B., Wersing, H.: Incremental on-line learning: a review and comparison of state of the art algorithms. Neurocomputing **275**, 1261–1274 (2018)
5. Riera, T.S., Higuera, J.-R.B., Higuera, J.B., Herraiz, J.-J.M., Montalvo, J.-A.S.: A new multi-label dataset for web attacks capec classification using machine learning techniques. Comput. Secur. **120**, 102788 (2022)
6. Aggarwal, C.C.: On biased reservoir sampling in the presence of stream evolution. In: Proceedings of the 32nd International Conference on Very large Data Bases, pp. 607–618 (2006)
7. Wang, X., Kuntz, P., Meyer, F., Lemaire, V.: Multi-label knn classifier with online dual memory on data stream. In: 2021 International Conference on Data Mining Workshops, pp. 405–413 (2021)
8. Gama, J., Žliobaitė, I., Bifet, A., Pechenizkiy, M., Bouchachia, A.: A survey on concept drift adaptation. ACM Comput. Surv. **46**(4), 1–37 (2014)
9. Spyromitros-Xioufis, E., Spiliopoulou, M., Tsoumakas, G., Vlahavas, I.: Dealing with concept drift and class imbalance in multi-label stream classification, Department of Computer Science, Aristotle University of Thessaloniki (2011)
10. Read, J., Pfahringer, B., Holmes, G.: Multi-label classification using ensembles of pruned sets. In: IEEE International Conference on Data Mining, pp. 995–1000 (2008)
11. Júnior, J.D.C., Faria, E.R., Silva, J.A., Gama, J., Cerri, R.: Pruned sets for multi-label stream classification without true labels. In: International Joint Conference on Neural Networks, pp. 1–8 (2019)
12. Boutell, M.R., Luo, J., Shen, X., Brown, C.M.: Learning multi-label scene classification. Pattern Recogn. **37**(9), 1757–1771 (2004)
13. Read, J., Bifet, A., Holmes, G., Pfahringer, B.: Scalable and efficient multi-label classification for evolving data streams. Mach. Learn. **88**, 243–272 (2012)
14. Osojnik, A., Panov, P., Džeroski, S.: Multi-label classification via multi-target regression on data streams. Mach. Learn. **106**, 745–770 (2017)

15. Venkatesan, R., Er, M.J., Wu, S., Pratama, M.: A novel online real-time classifier for multi-label data streams. In: International Joint Conference on Neural Networks, pp. 1833–1840 (2016)
16. Sousa, R., Gama, J.: Multi-label classification from high-speed data streams with adaptive model rules and random rules. Progress in Artif. Intell. **7**(3), 177–187 (2018). https://doi.org/10.1007/s13748-018-0142-z
17. Roseberry, M., Krawczyk, B., Cano, A.: Multi-label punitive knn with self-adjusting memory for drifting data streams. ACM Trans. Knowl. Disc. Data **13**(6), 1–31 (2019)
18. Roseberry, M., Krawczyk, B., Djenouri, Y., Cano, A.: Self-adjusting k nearest neighbors for continual learning from multi-label drifting data streams. Neurocomputing **442**, 10–25 (2021)
19. Alberghini, G., Junior, S.B., Cano, A.: Adaptive ensemble of self-adjusting nearest neighbor subspaces for multi-label drifting data streams. Neurocomputing **481**, 228–248 (2022)
20. Büyükçakir, A., Bonab, H., Can, F.: A novel online stacked ensemble for multi-label stream classification. In: Proceedings of the 27th ACM International Conference on Information and Knowledge Management, pp. 1063–1072 (2018)
21. France, S.L., Carroll, J.D., Xiong, H.: Distance metrics for high dimensional nearest neighborhood recovery: compression and normalization. Inf. Sci. **184**(1), 92–110 (2012)
22. Cauwenberghs, G., Poggio, T.: Incremental and decremental support vector machine learning. In: Advances in Neural Information Processing Systems, vol. 13 (2000)
23. Richtárik, P., Takáč, M.: Parallel coordinate descent methods for big data optimization. Math. Program. **156**, 433–484 (2016)
24. Nova, D., Estévez, P.A.: A review of learning vector quantization classifiers. Neural Comput. Appl. **25**, 511–524 (2014)
25. Straat, M., Abadi, F., Göpfert, C., Hammer, B., Biehl, M.: Statistical mechanics of on-line learning under concept drift. Entropy **20**(10), 775 (2018)
26. Kohonen, T.: Improved versions of learning vector quantization. In: 1990 International Joint Conference on Neural Networks, pp. 545–550 (1990)
27. El Sibai, R., Chabchoub, Y., Demerjian, J., Kazi-Aoul, Z., Barbar, K.: Sampling algorithms in data stream environments. In: 2016 International Conference on Digital Economy, pp. 29–36 (2016)
28. Seni, G., Elder, J.: Ensemble Methods in Data Mining: Improving Accuracy through Combining Predictions. Morgan & Claypool Publishers (2010)
29. Zhang, M.-L., Zhou, Z.-H.: A review on multi-label learning algorithms. IEEE Trans. Knowl. Data Eng. **26**(8), 1819–1837 (2013)
30. Ramamurthy S., Bhatnagar, R.: Tracking recurrent concept drift in streaming data using ensemble classifiers. In: Sixth International Conference on Machine Learning and Applications, pp. 404–409 (2007)
31. Gunasekara, N., Pfahringer, B., Gomes, H.M., Bifet, A.: Survey on online streaming continual learning. In: International Joint Conference on Artificial Intelligence (2023)

Special Session on Trustworthy and Explainable Artificial Intelligence: an Interdisciplinary Perspective

Feature Importance in Association Rule-Based Explanations for Time Series Forecasting

A. R. Troncoso-García[1]([✉]), M. Martínez-Ballesteros[2], F. Martínez-Álvarez[1], and A. Troncoso[1]

[1] Data Science and Big Data Lab, Pablo de Olavide University, 41013 Seville, Spain
{artrogar,fmaralv,atrolor}@upo.es
[2] Department of Computer Languages and Systems, University of Seville, 41012 Seville, Spain
mariamartinez@us.es

Abstract. Machine learning and deep learning are increasingly used in time series forecasting in critical fields like healthcare, finance, and agriculture, making the need for explainable artificial intelligence essential to ensure transparency and trust. In this work, a novel methodology is proposed to assess the influence of input features on the quality of explanations that describe the behavior of the models in the context of time series forecasting. To investigate this aspect, controlled perturbations are introduced to individual input features, and the resulting variations in the generated explanations are analyzed to assess their stability and reliability. The proposed methodology relies on the generation of association rules, which are inherently interpretable and capable of summarizing data patterns transparently. The results show that the proposed methodology can be applied to real-world time series data, particularly in the field of agriculture, where planning decisions frequently rely on forecasts across different time frames.

Keywords: Explainable artificial intelligence · feature importance · association rules · machine learning · deep learning · time series forecasting

1 Introduction

Nowadays, machine learning (ML) and deep learning (DL) techniques are widely applied across a broad range of domains [11,12]. These models have demonstrated remarkable performance in tasks such as classification, forecasting, and decision-making. However, one of the main disadvantages of complex ML and DL models is their inherent black-box nature, which makes it difficult to understand how predictions are generated. This lack of transparency raises concerns regarding trust and fairness, particularly in critical applications [5]. To address these challenges, the field of eXplainable Artificial Intelligence (XAI) has gained

L. Martínez et al. (Eds.): IDEAL 2025, LNCS 16239, pp. 237–248, 2026.
https://doi.org/10.1007/978-3-032-10489-2_20

significant attention in recent years. XAI aims to develop methods and tools that make the decision-making processes of ML and DL models more understandable and interpretable to human users. By providing insights into model behavior, XAI facilitates greater trust in AI systems, supports model validation, and enables the detection of biases or errors. Most explainability methods focus on identifying the feature or patterns that contribute to model predictions. However, the analysis of the sensitivity of explanations in the field of time series has received relatively little attention. [2]. Understanding how perturbations affect the stability of explanations is essential for evaluating their robustness and reliability.

One way to explore the relationships between input features and model predictions is through the use of association rules (ARs), which offer an interpretable and structured representation of these connections [24]. ARs are a popular data mining technique used to discover interesting relationships, patterns, or dependencies among variables in large datasets, typically expressed in the form if-then statements. Their inherent interpretability makes them particularly well-suited for explainability purposes [25]. In this work, AR mining techniques are applied to evaluate the robustness of explanations under controlled perturbations and to identify which features have a greater influence on the stability of the generated explanations [22].

The remainder of the paper is structured as follows: Sect. 2 reviews the literature related to this research. Then, Sect. 3 introduces the proposed methodology and its components. Section 4 presents and discusses the results and finally, Sect. 5 summarizes the main findings and outlines directions for future work.

2 Related Work

The growing demand for transparency in ML and DL models has driven extensive research in the field of XAI. A wide variety of techniques have been proposed to enhance model interpretability, ranging from model-agnostic approaches such as LIME [14] and SHAP [9] to ARs-based methods like RULEx [22] or gradient-based attribution methods like Integrated Gradients [18]. These techniques aim to clarify the influence of input features on individual predictions, thereby helping users understand the reasoning behind models that are otherwise difficult to interpret.

Several recent studies have shown that small perturbations to the input data can lead to substantial variations in the resulting explanations. For instance, the authors in [8] investigated the robustness of counterfactual explanations, which describe how small alterations in input data can impact model predictions, and the authors in [3] presented a unified framework for evaluating instance-based explanations, with a focus on counterfactuals. Furthermore, Schubert et al. [7] introduced Relative Feature Importance as a generalized framework for assessing feature relevance in ML models.

In parallel, ARs have emerged as a valuable tool for interpretable modeling and explanation. Traditionally used for uncovering hidden patterns in transactional datasets, ARs has been adapted to provide concise, human-readable insights into the behavior of classification models. Moreover, recent studies have explored the integration of ARs within the field of XAI [16]. For instance, the study in [17] proposed a method that combines ARs with SHAP values to improve both the predictive performance and interpretability of ML models. Similarly, the SHARQ framework [4] applies ARs to relational data to generate transparent and structured explanations.

Finally, several studies have explored the application of XAI techniques in agriculture to interpret the features influencing model predictions [13]. For instance, the authors in [15] investigated the impact of pests on crops and employed SHAP values to interpret the model's predictions, or the authors in [10] presented an XAI-based smart agriculture system designed to support precision farming by improving productivity and reducing environmental impact. Moreover, in the field of agriculture, reference evapotranspiration (ET_0) is a fundamental indicator that quantifies the rate at which water evaporates from the soil in crops. ET_0 provides a crucial benchmark for irrigation scheduling and water resource management. In this context, previous studies have utilized meteorological data to forecast ET_0. In fact, the same dataset employed in the present work has been studied in the literature. The work in [20] applied XAI techniques to interpret the contribution of individual features and, building on this, the same authors in [21] explored the use of DL models incorporating a Temporal Selection Layer to forecast ET_0, achieving improved accuracy and interpretability.

3   Methodology

The objective of this work is to assess the influence of input features on the quality of explanations that describe the behavior of ML and DL models, by analyzing the sensitivity of these explanations to changes in the input features. To investigate this aspect, controlled perturbations are introduced to individual input features, and the resulting variations in the generated explanations are analyzed to assess their stability and reliability. Then, the proposed methodology relies on the generation of ARs, which are inherently interpretable and capable of summarizing data patterns transparently. Figure 1 presents an overview of the workflow followed in the methodology.

The proposed methodology is conceived as a general framework in which diverse techniques can be applied at each step. A variety of DL or ML models can be used to generate predictions, different attribution-based XAI methods can be applied to obtain a ranked list of the most influential features, and multiple AR mining algorithms can then be applied to derive the rules. An example of how this methodology can be instantiated is provided in Sect. 4.

3.1   Learning Model and Explainability Technique

The process begins with the preprocessing of the input data to ensure its quality and suitability for analysis. Once the data is properly prepared, an ML or

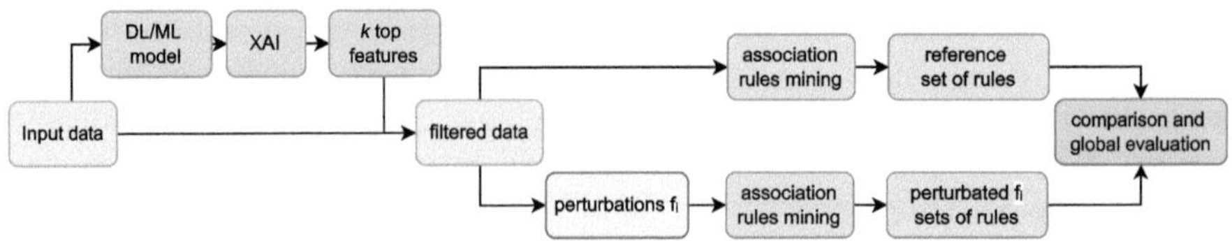

Fig. 1. Flowchart of the proposed methodology.

DL model is trained and applied to forecast the target variable, leveraging the underlying patterns within the input features.

After training the predictive model, a XAI technique is applied to estimate the importance of each input feature, serving as a basis for feature selection. The features are ranked in descending order according to their importance scores, reflecting their contribution to the model's predictions. To determine the most informative subset of features, the elbow method is employed, allowing for the selection of an optimal number of top features k by identifying the point at which additional features contribute marginal gains in importance. The original input dataset is then filtered to retain only these k selected features, resulting in a reduced and more interpretable representation for subsequent analysis [23].

3.2   Association Rules Mining

Once the relevant features are identified, ARs are extracted to capture the relationships between the selected k input features and the predicted target values. This process involves analyzing the interactions and dependencies between these features. Initially, ARs are extracted from the original (unperturbed) dataset to serve as a baseline or reference. This reference set of rules reflects the natural associations within the data without any external manipulation. The use of the original unperturbed dataset ensures that the extracted rules represent the genuine, underlying relationships between the input features and the target values.

Subsequently, controlled perturbations are introduced independently to each of the k input features, generating modified versions of the dataset. Since the input data are normalized to the range $[-1, 1]$, all perturbations are generated within this same interval. This ensures that the modified samples remain consistent with the original data distribution. These perturbations are applied in a systematic manner, where a constant value is either added or subtracted to each feature in every row of the dataset, effectively altering the feature values in a controlled way. Perturbations are added on each feature separately to evaluate the impact of each feature on the overall structure and behavior of the dataset. This process creates multiple perturbed versions of the dataset, allowing for a more comprehensive examination of how variations in the features influence the relationships between the input data and the predicted target values.

For each perturbed dataset, a set of ARs is extracted, following the same procedure used for the original unperturbed dataset, ensuring consistency in the comparison with the baseline rules. The goal is to assess how input changes affect the stability of the associations captured by the ARs.

3.3   Association Rules Evaluation

The comparison between the reference set of ARs and the perturbed set is conducted using three different metrics: the total number of rules that have been extracted and two global metrics, namely global support and global confidence. These global metrics have been introduced and discussed in previous research [23]. They evaluate the overall quality, robustness, and reliability of the generated rulesets. On the one hand, global support refers to the proportion of instances in the dataset for which the set of rules holds true. It provides a measure of how frequently the ruleset is observed across all data points, indicating relevance and general applicability. On the other hand, global confidence measures the likelihood that the consequent of the set of rules holds true given that the antecedent is satisfied. Essentially, it provides a measure of the predictive strength of the rule. Both global support and global confidence help assess the stability and robustness of the ARs in different dataset versions, especially after introducing perturbations. Finally, the number of rules reflects the scalability and simplicity of the explanation: fewer rules mean easier interpretation. These metrics determine whether the modified rules maintain their relevance and accuracy compared to the original ruleset, and whether any significant changes in the relationships between features and outcomes have occurred due to the perturbations.

4   Results

This section presents the results of applying the proposed methodology to real-world time series data, specifically in the field of agriculture. The experimental phase focuses on time series forecasting tasks; however, the methodology could be extended to other types of data, such as tabular data or images.

4.1   Input Data

The input data used in this study consists of multivariate time series collected from the agricultural domain in the field of ET_0 forecasting. The data were gathered from eight different monitoring stations located in the Alentejo region, in southern Portugal, as part of the Sistema Agrometeorológico para a Gestão da Rega no Alentejo (SAGRA) between 2012 and 2022. The Alentejo region features a Mediterranean climate, characterized by hot, dry summers. The average annual temperature is approximately 20 ºC, and the area is notable for its low levels of precipitation, making efficient water management essential. Each station recorded various environmental and meteorological features, which are included in the dataset. The features, along with their corresponding units, are detailed in Table 1.

Table 1. Features in the SAGRA dataset.

Feature	Full name	Unit
Tmed	Mean temperature	$^{\circ}$C
Tmax	Maximum temperature	$^{\circ}$C
Tmin	Minimum temperature	$^{\circ}$C
HRmed	Mean relative humidity	%
HRmax	Maximum relative humidity	%
HRmin	Minimum relative humidity	%
RSG	Global solar radiation	kj/m^2
DV	Wind direction	$^{\circ}$
VVmed	Mean wind speed	m/s
VVmax	Maximum wind speed	m/s
P	Precipitation	mm
Tmed Relva	Mean soil temperature	$^{\circ}$C
Tmax Relva	Maximum soil temperature	$^{\circ}$C
Tmin Relva	Minimum soil temperature	$^{\circ}$C
ET_0	Reference evapotranspiration	mm/day

As time series data, the feature values corresponding to three consecutive days are used as input. In previous studies, this input has been employed to predict the ET_0 values for the subsequent seven days. However, for simplicity, the present work focuses on a single prediction horizon, although the proposed approach can be readily extended to multiple horizons if required.

Furthermore, the input features have been structured across three different time lags to incorporate temporal information relevant for forecasting reference evapotranspiration (ET_0). Specifically, each of the 15 base variables listed in Table 1 is replicated for three preceding time steps, $t-3$, $t-2$, and $t-1$, where t denotes the current time step.

4.2 An LSTM Model and RULEX XAI Technique

A DL model, specifically an LSTM neural network, has been applied to predict future ET_0 values. The model's hyperparameters were optimized using the Corona Virus Optimization Algorithm (CVOA) method, as demonstrated in a previous study [20]. The optimized LSTM model achieved a mean accuracy of 0.765 across all stations, outperforming similar methods in the literature. However, this study prioritizes the extraction of robust explanations over the development of an accurate deep learning model. Thereafter, the RULEx XAI technique was applied to assess the relevance of the input features and identify the most informative subset for the prediction task. Based on prior research [23], among the 45 input features derived from the base variables across three time lags, 7 features have been selected as the most informative ones. The subsequent

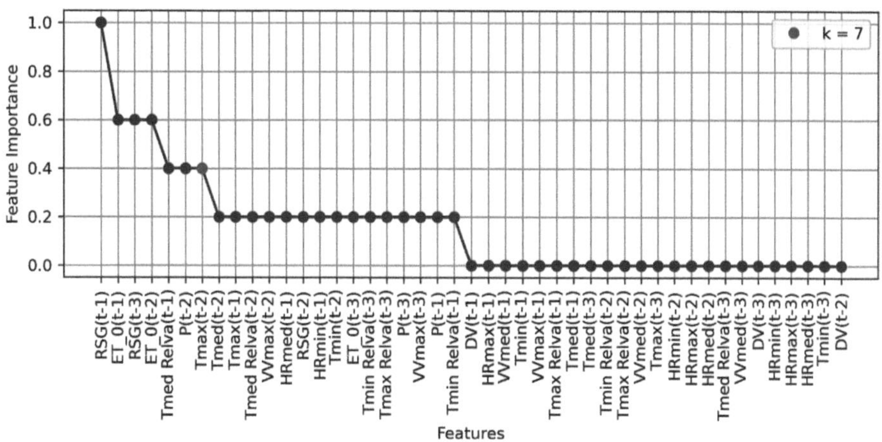

Fig. 2. Features ranked according to its importance level.

phases of the methodology are conducted using only these 7 selected features, which have been identified as the most relevant for the prediction task.

Figure 2 presents the features ranked according to their level of importance, as determined by RULEx [23]. The elbow method is used to identify the optimal number, denoted as k (red point) of the most relevant features selected for further analysis. The most pronounced inflection in Fig. 2 suggests selecting one characteristic. However, similar inflections can also be observed at four and seven characteristics. Building on prior research [20], where $k = 7$ was used for the same prediction horizon, this value was retained for consistency and comparability.

The selected variables include: solar radiation (RSG, in kJ/m^2) at $t - 3$ and $t - 1$; reference evapotranspiration (ET_0) itself at $t - 1$ and $t - 2$; mean soil temperature (Tmean Relva, in ºC) at $t - 1$; precipitation (P, in mm) at $t - 1$ (lag 3) and $t - 2$; and maximum temperature (Tmax, in ºC) at $t - 2$.

4.3   Association Rules Comparison

Subsequently, ARs are mined using the Apriori algorithm [1]. Since Apriori requires categorical data, the continuous input data was first discretized. The hyperparameters for the discretization process (specifically, the number of bins) and for the extraction of frequent itemsets (minimum support threshold) have been chosen based on prior experimentation [23], to obtain a reasonable number of rules, thus ensuring neither too few nor an excessive amount. The data have been discretized into five bins, and a minimum support threshold of 0.15 has been used. Figure 3 illustrates the differences between the rulesets generated from the perturbed data and the original unperturbed ruleset. The comparison has been conducted using three metrics: global confidence, global support, and the number of generated rules. Results have been normalized to allow proper

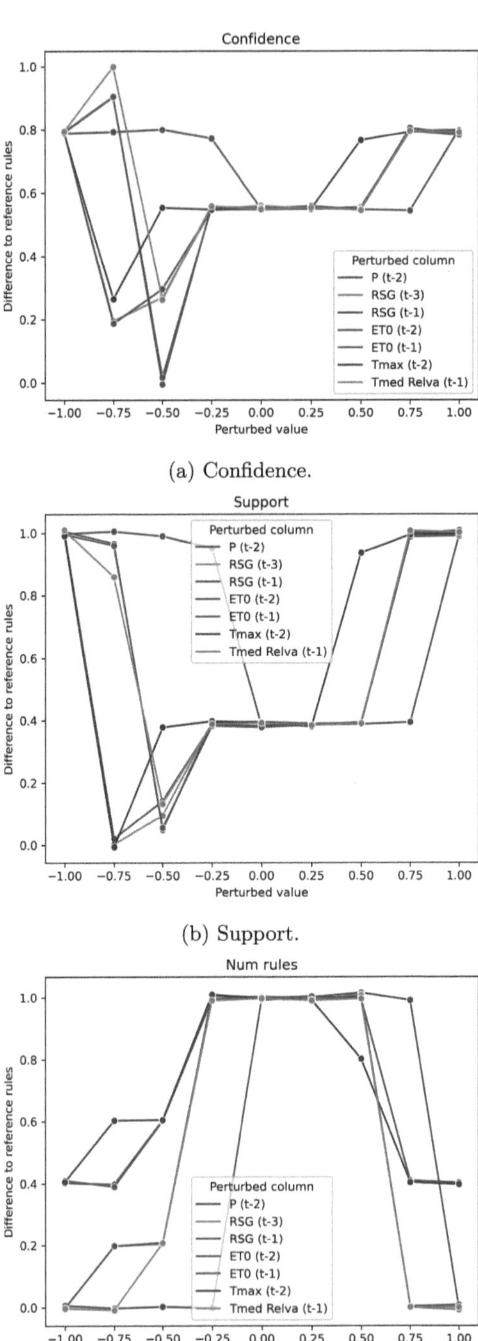

(a) Confidence.

(b) Support.

(c) Num rules.

Fig. 3. Differences between reference rules and perturbed rules.

comparison. The analysis of the three plots—illustrating the differences in confidence (Fig. 3a), support (Fig. 3b), and number of rules (Fig. 3c) between the reference and perturbed rule sets—reveals several key insights.

First, concerning the results for confidence (Fig. 3a), most features exhibit moderate changes under perturbations ranging from 0.25 to 0.75. When perturbations are below 0.5, all features show variations consistent with their confidence levels. The features demonstrating the greatest differences are $ET_0(t-1)$, $P(t-2)$, and $Tmed\ Relva(t-1)$. In contrast, variables like $RSG(t-1)$ and $RSG(t-3)$ demonstrate greater robustness, as perturbations to these features result in minimal changes. Subsequently, Fig. 3b illustrates the variation in global support values of the rule sets. Similarly, perturbations to $ET_0(t-1)$, $P(t-2)$, and $Tmed\ Relva(t-1)$ and also $ET0(t-2)$ result in the largest deviations, indicating that rules relying on these features are highly sensitive to input changes. Then, features like $RSG(t-1)$ show more stable behavior. Finally, Fig. 3c presents the normalized differences between the number of rules in the original rule set and those derived from perturbed data. The greatest deviations occur at perturbation values between -0.5 and 0.5. Almost all features demonstrate a comparable level of deviation.

Overall, the effects of perturbation are not always linear; in many cases, the largest deviations occur at intermediate values. While the differences in the number of rules remain relatively stable across variables, support and confidence metrics often display sharper variations, particularly for precipitation-related features. Furthermore, the features $ET_0(t-1)$, $P(t-2)$, and $Tmed\ Relva(t-1)$ stand out as the most sensitive, exhibiting the largest variations across all metrics, especially around perturbations near ±0.75. This is supported by existing research in the field, which indicates that precipitation and soil temperature are directly related to the amount of water evaporated from crops [6,19]. Moreover, past values of ET_0 have been consistently identified in the literature as the most essential features for forecasting future ET_0 [20].

5   Conclusions

Making complex models more transparent is crucial in time series forecasting, but XAI methods must also be rigorously evaluated for robustness and fidelity to the model's logic. In this work, a novel methodology has been proposed to assess the robustness of feature importance explanations generated by XAI techniques. This was achieved through the use of ARs, which serve as an interpretable, rule-based framework to analyze how the importance of features varies under controlled input perturbations.

The results highlight the differing stability and influence of individual input features, emphasizing the importance of understanding how each variable impacts the interpretability and consistency of rule-based explanations. Features such as $ET_0(t-1)$, $P(t-2)$, and $Tmed\ Relva(t-1)$ were identified as particularly sensitive, exhibiting the largest deviations across multiple evaluation metrics. These findings align with domain knowledge in agriculture and prior studies on

ET_0 forecasting, confirming the relevance of precipitation, soil temperature, and past evapotranspiration values.

Furthermore, this study opens new directions for extending the proposed methodology to other explainability approaches and to different time series tasks, such as classification. Moreover, future research could extend the current methodology by incorporating additional metrics or similarity measures to compare rulesets, enabling a more nuanced understanding of how ARs evolve under input perturbations. Additionally, the present experiments have been limited to a single prediction horizon (1-day ahead).

Expanding the analysis to multiple forecasting horizons would allow for a comprehensive evaluation of how the relevance and stability of explanatory features change as the prediction task becomes more temporally distant. This could be especially beneficial for practical agricultural uses, where planning decisions frequently rely on forecasts across different time frames. Additionally, the proposed method can serve as a strategy to evaluate the robustness of current explainability techniques in time series forecasting tasks.

Overall, this study provides a systematic approach to evaluate the robustness of current explainability techniques in time series forecasting, offering insights into the stability and reliability of rule-based explanations and paving the way for future research to enhance the interpretability of complex models.

Acknowledgments. This research has been supported by the grant PID2020-117954RB-C22, PID2020-117954RB-C21, PID2023-146037OB-C21 and PID2023-146037OB-C22 funded by MICIU/AEI/ 10.13039/501100011033. This work has also been supported by TED2021-131311B-C21 and TED2021-131311B-C22 funded by MICIU/AEI/10.13039/501100011033 and by the European Union NextGenerationEU/PRTR. The authors would like to also thank the Centro Operativo e de Tecnologia de Regadio (COTR) for giving access to data.

References

1. Agrawal, R., Srikant, R.: Fast algorithms for mining association rules. In: Proceedings of the International Conference on Very Large Data Bases, pp. 487–499 (1994)
2. Arrieta, A.B., et al.: Explainable artificial intelligence (XAI): concepts, taxonomies, opportunities and challenges toward responsible AI. Inf. Fusion **58**, 82–115 (2020)
3. Bayrak, B., Bach, K.: Evaluation of instance-based explanations: an in-depth analysis of counterfactual evaluation metrics, challenges, and the ceval toolkit. IEEE Access **12**, 137683 (2024)
4. Ben-Efraim, H., Davidson, S.B., Somech, A.: SHARQ: Explainability framework for association rules on relational data. Proc. ACM Manage. Data **3**(1), 1–25 (2025)
5. Cambria, E., Malandri, L., Mercorio, F., Mezzanzanica, M., Nobani, N.: A survey on XAI and natural language explanations. Inf. Process. Manage. **60**(1), 103111 (2023)
6. Ferreira, L.B., da Cunha, F.F.: Multi-step ahead forecasting of daily reference evapotranspiration using deep learning. Comput. Electron. Agric. **178**, 105728 (2020)

7. König, G., Molnar, C., Bischl, B., Grosse-Wentrup, M.: Relative feature importance. In: Proceedings of the International Conference on Pattern Recognition, pp. 9318–9325 (2021)
8. Leofante, F., Potyka, N.: Promoting counterfactual robustness through diversity. Proc. AAAI Conf. Artif. Intell. **38**(19), 21322–21330 (2024)
9. Lundberg, S.M., Lee, S.I.: A unified approach to interpreting model predictions. Proc. Int. Conf. Neural Inf. Process. Syst. **30**, 4765–4774 (2017)
10. Martin, R.J., et al.: XAI-powered smart agriculture framework for enhancing food productivity and sustainability. IEEE Access **12**, 168412–168427 (2024)
11. Melgar-García, L., Gutiérrez-Avilés, D., Rubio-Escudero, C., Troncoso, A.: A novel distributed forecasting method based on information fusion and incremental learning for streaming time series. Inf. Fusion **95**, 163–173 (2023)
12. Nepomuceno, J.A., Troncoso, A., Nepomuceno-Chamorro, I.A., Aguilar-Ruiz, J.S.: Integrating biological knowledge based on functional annotations for biclustering of gene expression data. Comput. Methods Programs Biomed. **119**(3), 163–180 (2015)
13. Fatimah, S., Razak, A., Yogarayan, S., Sayeed, M.S., Derafi, M.: Agriculture 5.0 and explainable AI for smart agriculture: a scoping review. Emerg. Sci. J. **8**(2), 744–760 (2024)
14. Ribeiro, M.T., Singh, S., Guestrin, C.: Why should i trust you?: Explaining the predictions of any classifier. In: Proceedings of the ACM SIGKDD International Conference on Knowledge Discovery and Data Mining, pp. 1135–1144 (2016)
15. Rodríguez-Díaz, F., Chacón-Maldonado, A.M., Troncoso-García, A.R., Asencio-Cortés, G.: Explainable olive grove and grapevine pest forecasting through machine learning-based classification and regression. Results Eng. **24**, 103058 (2024)
16. Roshanzamir, M., Alizadehsani, R., Moravvej, S.V., Joloudari, J.H., Alinejad-Rokny, H., Gorriz, J.M.: Enhancing interpretability in machine learning: a focus on genetic network programming, its variants, and applications. In: Proceedings of the International Work-Conference on the Interplay between Natural and Artificial Computation, pp. 98–107 (2024)
17. Sankar, S., Sathyalakshmi, S.: A study on the explainability of thyroid cancer prediction: shap values and association-rule based feature integration framework. Comput. Mater. Continua **79**(2), 3111–3138 (2024)
18. Sundararajan, M., Taly, A., Yan, Q.: Axiomatic attribution for deep networks. In: Proceedings of the International Conference on Machine Learning, pp. 3319–3328 (2017)
19. Trajkovic, S., Todorovic, B., Stankovic, M.: Forecasting of reference evapotranspiration by artificial neural networks. J. Irrig. Drain. Eng. **129**(6), 454–457 (2003)
20. Troncoso-García, A.R., Brito, I.S., Troncoso, A., Martínez-Álvarez, F.: Explainable hybrid deep learning and coronavirus optimization algorithm for improving evapotranspiration forecasting. Comput. Electron. Agric. **215**, 108387 (2023)
21. Troncoso-García, A.R., Jiménez-Navarro, M.J., Linares-Barrera, M.L., Brito, I.S., Martínez-Álvarez, F., Martínez-Ballesteros, M.: Time series forecasting in agriculture: explainable deep learning with lagged feature selection. Proc. Int. Conf. Soft Comput. Models Ind. Environ. Appl. **1**, 139–149 (2024)
22. Troncoso-García, A.R., Martínez-Ballesteros, M., Martínez-Álvarez, F., Troncoso, A.: A new approach based on association rules to add explainability to time series forecasting models. Inf. Fusion **94**, 169–180 (2023)
23. Troncoso-García, A.R., Martínez-Ballesteros, M., Martínez-Álvarez, F., Troncoso, A.: A new metric based on association rules to assess feature-attribution explain-

ability techniques for time series forecasting. IEEE Trans. Pattern Anal. Mach. Intell. **47**(5), 4140–4155 (2025)
24. Van der Waa, J., Nieuwburg, E., Cremers, A., Neerincx, M.: Evaluating XAI: a comparison of rule-based and example-based explanations. Artif. Intell. **291**, 103404 (2021)
25. Wang, Z., Koprinska, I., Martínez-Ballesteros, M., Troncoso, A., Jeffries, B.: Comparison of explainable machine learning methods for early prediction of student performance in programming courses. In: Proceedings of the International Conference on Artificial Intelligence in Education, pp. 161–168 (2025)

Synchronising Fuzzy Labels in Federated Evolutionary Systems: A Preliminary Study

María Asunción Padilla-Rascón[1,2](✉) , Ángel Miguel García-Vico[1,2] , and Cristóbal J. Carmona[1,2,3]

[1] Department of Computer Science, University of Jaén, Jaén 23071, Spain
mprascon@ujaen.es
[2] Andalusian Research Institute in Data Science and Computational Intelligence, University of Jaén, Jaén 23071, Spain
[3] Leicester School of Pharmacy, De Montfort University, Leicester LE1 7RH, UK

Abstract. The increasing demand for trustworthy and explainable artificial intelligence (AI) has driven the adoption of techniques like fuzzy logic and federated learning (FL), especially in domains requiring transparency and reliability. Evolutionary Fuzzy Systems (EFS) are particularly relevant in this context, as they combine the adaptive learning of evolutionary algorithms with the interpretability of fuzzy logic, making them well-suited for eXplainable AI (XAI) and FL applications. However, a key challenge in deploying EFS in federated settings is that clients independently evolve fuzzy sets based on local data, leading to semantic inconsistencies in linguistic labels. This heterogeneity undermines the global model's coherence and interpretability, affecting decision-making. To address this issue, we propose a synchronization mechanism that aligns fuzzy labels across clients. The approach involves centralizing fuzzy sets from all nodes, computing their average at a server, and redistributing a unified version to participants. This modification to standard EFS preserves data privacy while ensuring a consistent linguistic foundation, improving both robustness and explainability in federated EFS models.

Keywords: Fuzzy Labels · Explainability · Federated Learning

1 Introduction

In recent years, the development of accurate, explainable, robust and reliable artificial intelligence (AI) systems has become particularly relevant, especially in critical areas such as healthcare and finance. This interest is driven by both practical needs and regulations such as the AI Act [1], which require transparency and accountability in AI models. In this context, techniques such as fuzzy logic and linguistic labels [3] have proven to be effective in improving the human interpretability of AI decisions. Meanwhile, federated learning (FL) [4]

L. Martínez et al. (Eds.): IDEAL 2025, LNCS 16239, pp. 249–255, 2026.
https://doi.org/10.1007/978-3-032-10489-2_21

has emerged as a key solution for collaborative training that does not require direct data sharing while maintaining information privacy.

Within this context, Evolutionary Fuzzy Systems (EFS) [6] have positioned themselves as an effective tool for applications that require both precision and interpretability, such as the analysis of real-time data streams. However, their implementation in federated environments presents a significant challenge: evolving fuzzy sets independently in each client generates semantic inconsistencies between linguistic labels, which affects the coherence and explainability of the overall model. To address this problem, this paper proposes a semantic synchronisation mechanism that aligns fuzzy labels across participants. The strategy consists of centralising local fuzzy sets, averaging them on a central server and redistributing a unified version to all nodes. This approach not only preserves data privacy, but also establishes a common linguistic basis, thus improving the robustness and clarity of the model.

This paper is organised as follows: Sect. 2 reviews the state of the art. Section 3 outlines the proposed methodology for homogenising fuzzy sets. Section 4 describes the experimental design, and presents and analyses the results. Finally, Sect. 5 summarises the conclusions drawn from the study.

2   Preliminary

2.1   Trustworthy and Explainable Artificial Intelligence

The growing influence of artificial intelligence (AI) in sectors such as healthcare, industry, and finance has driven the development of more reliable and explainable systems. However, its rapid adoption introduces significant challenges, including model opacity, data privacy concerns, algorithmic bias, and environmental impact. In response, the concept of trustworthy AI has emerged, promoted by institutions such as the European Commission [5] [1], which outlines key principles including transparency, human oversight, fairness, and accountability, among others. Explainability is a central pillar of this approach, enabling users and auditors to understand how the model functions, improve validation processes, and support regulatory compliance. This need has prompted the development of specific proposals to assess trustworthiness in critical domains such as medicine, the environment, and finance, emphasizing the importance of interdisciplinary collaboration in building responsible AI systems.

2.2   Evolutionary Fuzzy Systems (EFS)

An Evolutionary Fuzzy System (EFS) is a type of fuzzy system that integrates evolutive algorithms (EAs) into its learning process [6]. These systems are typically structured through rules, making them rule-based fuzzy systems widely applied in fields such as finance, control, engineering, and medicine [11]. One of their main strengths lies in their ability to represent extracted knowledge in a comprehensible way and to effectively handle continuous variables through the use of fuzzy linguistic labels (LLs). This enhances interpretability compared to discretised approaches and avoids the information loss associated with discretisation [8].

The integration of evolutionary algorithms into EFSs provides an additional advantage by enabling efficient global searches across the solution space, thereby increasing the likelihood of identifying suitable or even optimal solutions within reasonable timeframes [7]. Furthermore, this approach allows the use of the same metrics employed to assess rule quality as guiding criteria during the search process, contributing to more effective learning aligned with the system's objectives.

2.3   Federated Learning (FL)

Federated learning is a distributed training paradigm that enables the construction of global models without the need to centralise data [4], thereby addressing issues related to privacy, network overload, and scalability inherent in traditional approaches. In FL, each device trains a local model using its own data, and these models are subsequently combined through an aggregation operator. This iterative cycle allows for the development of robust models without compromising individual privacy. Furthermore, by avoiding the continuous transfer of sensitive data, federated learning emerges as a promising solution for real-time applications and environments with connectivity constraints, while maintaining accuracy levels comparable to those of centralised learning.

3   Methodology

This work employs the EFS evolutionary algorithm TEFeS-SDR [6], based on emerging pattern mining [12], designed for federated learning scenarios over data streams. Although the algorithm has shown promising results in previous studies, a refinement is proposed to optimize its performance and enhance the explainability of the generated models. One of the main limitations identified is the semantic inconsistency among the linguistic labels generated locally by each client, as each defines its fuzzy sets based on the range of the first batch of data received. This can lead to divergences between local models.

To address this issue, a modification of the EFS algorithm is proposed, incorporating a synchronisation procedure for linguistic labels across clients. This strategy aims to standardise the fuzzy sets used by all nodes in the network, ensuring semantic consistency and enabling the construction of more comparable and coherent local models. The proposed improvement is outlined in Fig. 1 and summarised below:

1. **Local generation of fuzzy sets**: Upon receiving its first batch of data, each client generates its own fuzzy sets based on the observed distributions.
2. **Transmission to the server**: The client sends its fuzzy sets to the central server and waits for a response.
3. **Centralised averaging**: The server waits to receive information from all clients and computes an average fuzzy set for each linguistic label by averaging

the corresponding intervals. Formally, if n is the number of clients and $A_i = [a_i, b_i]$ represents the interval of the label at client i, the average fuzzy set is defined as:

$$A_{avg} = \left[\frac{1}{n} \sum_{i=1}^{n} a_i, \ \frac{1}{n} \sum_{i=1}^{n} b_i \right] \tag{1}$$

4. **Distribution of unified fuzzy sets**: The server sends the new unified fuzzy set to all clients.
5. **Local update**: Each client overwrites its original fuzzy set with the one received from the server and begins the *train-test* learning process.
6. **Adaptation to concept drift**: When a client detects concept drift, it generates a new fuzzy set along with its updated rule set and sends them to the server. The server then recalculates the global fuzzy set, taking into account the values from the other clients (which remain unchanged if no drift has occurred), and proceeds to merge the local models.
7. **Final synchronisation**: The server distributes the new fuzzy set to all clients and sends the merged model to the client that initiated the update.

4 Experimental Study and Results

The experimental study was conducted on a distributed architecture consisting of four Raspberry Pi 4 Model B and a central server running Ubuntu 23.04, simulating a realistic federated environment. For system evaluation, three datasets from the MOA repository [10] were used: SEA, Aggrawal, and Mixed—all binary-class streams exhibiting concept drift. Each stream contained one million instances, divided into blocks of 5000, with concept drift introduced randomly between blocks 10 and 15 of each segment, using different random seeds per client.

Regarding the algorithm parameters, three fuzzy linguistic labels were used, with triangular membership functions [11]. The evolutionary process employed a population of 50 individuals, a crossover probability of 0.7, a mutation probability of 0.05, and a maximum of 5000 evaluations per run. The objective function was defined as a weighted linear combination of three metrics: WRAccN, Support Difference, and Confidence. For concept drift detection, confidence and true positive rate (TPR) thresholds were set at 0.6 and 0.1, respectively.

The experimental results, summarised in Table 1, demonstrate significant improvements following the implementation of the linguistic label unification mechanism among clients. Three main aspects are analysed: interpretability, interest, and model reliability.

In terms of interpretability, there was a reduction of up to tenfold in the total number of rules generated, as well as a significant decrease in the average number of variables per rule, indicating a substantial increase in the model's interpretability, thereby facilitating user comprehension.

Regarding model interest, measured by the WRAccN metric, a 3% increase was observed in the aggregated model, indicating a higher degree of relevance in the rules generated.

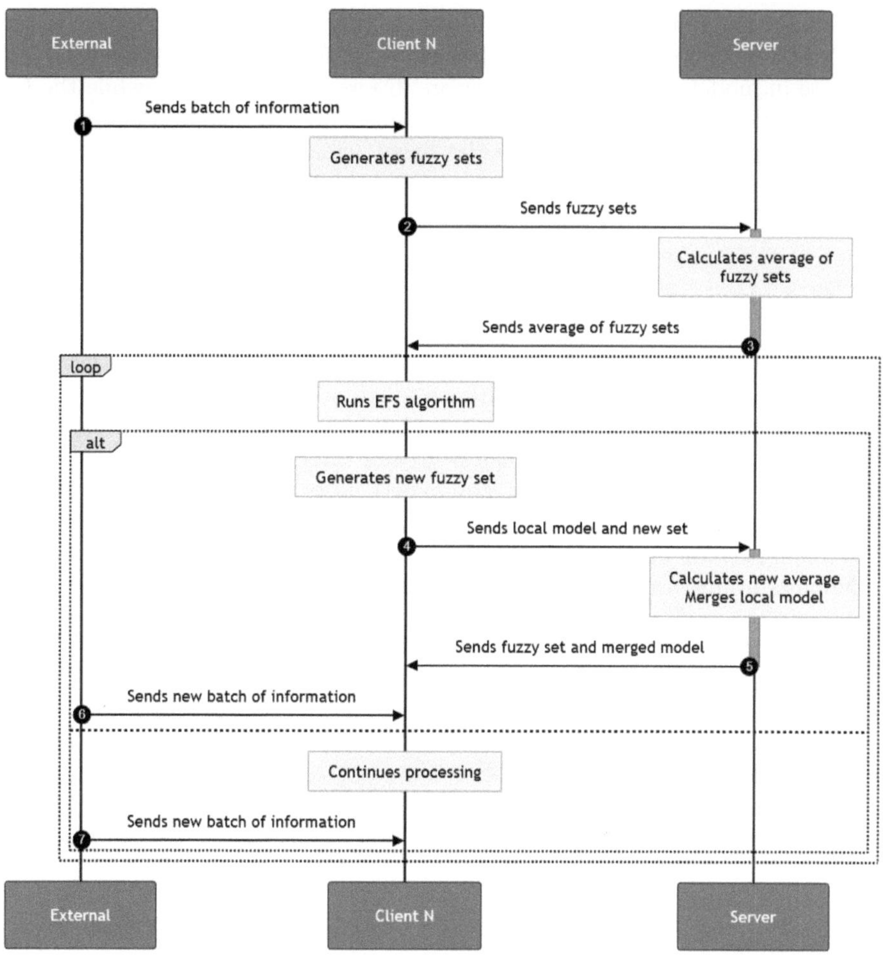

Fig. 1. Illustrates the process of exchanging and updating fuzzy sets between Client N and a central server. The client receives external data, generates fuzzy sets and sends them to the server, which calculates their average and returns it to the client. The client then runs an EFS algorithm. If concept drift is detected, the client generates and sends updated fuzzy sets and a local model to the server for merging. This cycle repeats with each new data batch.

Table 1. Average of the global model with and without aggregation of the labels.

	#Rules	#Vars	CONF	WRAccN	GR	FPR	TPR
W/O Aggregation	101.5	81.4995	0.6190	0.6195	0.5068	0.2682	**0.8542**
With Aggregation	**7.3333**	**3.5751**	**0.9036**	**0.6436**	**0.5078**	**0.0483**	0.3723

Concerning reliability, assessed through confidence (CONF), growth rate (GR), and false positive rate (FPR) metrics, results show an approximate 30% increase in model confidence and over an 80% reduction in FPR, while the GR remains practically unchanged. These findings confirm the effectiveness of the proposed approach to homogenise the fuzzy sets, demonstrating that greater semantic consistency among clients improves both the accuracy and explainability of the system.

5   Conclusion

In a context where explainable and trustworthy artificial intelligence is becoming increasingly important, this work makes a substantial contribution to both areas. An improved model has been developed that not only enhances the reliability of the obtained results but, also fosters greater interpretability of the generated knowledge, facilitating analysis by non-expert users.

This preliminary study highlights the importance of semantic synchronisation among clients in a federated learning system. The results demonstrate that differences in linguistic representations can significantly impact system quality, and that their unification enables the development of more coherent, robust, and interpretable models.

Acknowledgement. Financed by the Ministry of Science and Innovation with the project PID2023-149511OB-I00 and the FPU (Formación de Profesorado Universitario) grant programme with the code FPU23/02620.

Declaration of Competing Interest. The authors have no competing interests to declare that are relevant to the content of this article.

References

1. AI Act. https://digital-strategy.ec.europa.eu/en/policies/regulatory-framework-ai. Accessed 4 May 2025
2. van der Aalst, W.M., Bichler, M., Heinzl, A.: Responsible data science. Bus. Inf. Syst. Eng. **59**, 311–313 (2017)
3. Zadeh, L.A.: Fuzzy logic. Computer **21**(4), 83–93 (1988)
4. Brendan, M., et al.: Communication-efficient learning of deep networks from decentralized data. In: Artificial Intelligence and Statistics, pp. 1273–1282. PMLR (2017)
5. The Assessment List for Trustworthy Artificial Intelligence (ALTAI) for self assessment. https://digital-strategy.ec.europa.eu/en/library/assessment-list-trustworthy-artificial-intelligence-altai-self-assessment. Access 4 May 2025
6. Herrera, F.: Genetic fuzzy systems: taxomony, current research trends and prospects. Evol. Intell. **1**, 27–46 (2008)
7. Hullermeier, E.: Fuzzy sets in machine learning and data mining. Appl. Soft Comput. **11**(2), 1493–1505 (2011)
8. Hullermeier, E.: Fuzzy methods in machine learning and data mining: status and prospects. Fuzzy Sets Syst. **156**(3), 387–406 (2005)

9. Padilla-Rascón, M.A., García-Vico, A.M., Carmona, C.J.: Trustworthy and explainable federated system for extracting descriptive rules in a data streaming environment. Results Eng. **25**, 104137 (2025). https://doi.org/10.1016/j.rineng.2025.104137, ISSN 2590–1230

10. Bifet, A., et al.: MOA: massive online analysis, a framework for stream classification and clustering. In: Proceedings of the First Workshop on Applications of Pattern Analysis, of Proceedings of Machine Learning Research, vol. 11, pp. 44–50 (2010)

11. Zadeh, L.: The concept of a linguistic variable and its application to approximate reasoning–I. Inf. Sci. **8**, 199–249 (1975). https://doi.org/10.1016/0020-0255(75)90036-5

12. Dong, G.Z., Li, J.Y.: Efficient mining of emerging patterns: discovering trends and differences. In: Proceedings of the 5th ACM SIGKDD International Conference on Knowledge Discovery and Data Mining, San Diego, California, USA. ACM Press, pp. 43–52 (1999)

Toward Insightful Evaluation of Speech-To-Text Technology Using Explainable AI

Najla Althuniyan[1]([⊠]) [iD], Souad Larabi-Marie-Sainte[1] [iD], Lamia Berriche[1] [iD], Jalila Zouhair[1], Arwa A. Bawazir[1], and Lubaba R. Raed[2]

[1] Computer Science Department, College of Computer & Information Sciences, Prince Sultan University, Riyadh, Kingdom of Saudi Arabia
nthuniyan@psu.edu.sa
[2] Software Engineering Department, College of Computer & Information Sciences, Prince Sultan University, Riyadh, Kingdom of Saudi Arabia

Abstract. Speech-to-text technology, a subfield of Automatic Speech Recognition, has significant advancements across various domains. Automatic Speech Recognition tools continue to encounter limitations in transcribing Arabic speech, particularly Modern Standard Arabic. These limitations stem from the language's rich morphology, diverse dialects, and the lack of high-quality datasets. This study investigates the effectiveness of commercial speech-to-text tools in transcribing Arabic speech. A novel dataset was created, comprising recordings from native Arabic speakers of varying ages, genders, and speech durations. Six transcription tools were evaluated using five similarity metrics. The experiments showed that Clipto yielded the best similarity score using the Cosine metric. The tool yielding the worst score, namely Notta, was investigated using Explainable AI. It was demonstrated that the topic used in the transcription, the talk speed, the speaker age, and the recording length significantly affect the tool's transcription accuracy.

Keywords: Speech-to-text · Euclidean distance · Cosine · Jaccard · Manhattan · Jaro-Winkler · Explainable AI · Ridge Linear Regression · SHAP

1 Introduction

Speech-to-text (STT) refers to the process of converting spoken language into written text. STT is part of Automatic Speech Recognition (ASR), which encompasses the process of analyzing and converting spoken language into text or commands (Alharbi et al. 2021). STT technology has seen significant advancements in recent years across various fields, including education, business, accessibility, and healthcare (Liu et al. 2008). This constantly evolving technology is becoming increasingly important due to its ability to automate processes, increase productivity, and improve accessibility for people with disabilities. While STT systems have proven to have high accuracy in transcribing English, Arabic is still exhibiting low performance in automatic transcription. This concern arises due to Arabic's rich morphology, dialectal variations, and limited datasets (Ghoul et al. 2022). Modern Standard Arabic (MSA) is distinct from spoken dialects but still presents difficulties.

L. Martínez et al. (Eds.): IDEAL 2025, LNCS 16239, pp. 256–268, 2026.
https://doi.org/10.1007/978-3-032-10489-2_22

Several state-of-the-art transcription tools, such as Amazon's Alexa, Apple's Siri, Google Assistant, automated phone systems, transcription services like Otter.ai, and in-car voice assistants, claim to support Arabic speech recognition; however, their performance varies depending on the language. (Besdouri et al. 2024; Ghoul et al. 2022) focused on the state of the art of Arabic STT and the challenges exhibited by the Arabic language. Despite improvements in speech recognition technology, transcribing Arabic speech remains a challenging task. Recent advancements in Arabic ASR have explored various deep learning models and datasets. However, the proposed systems' accuracy is still below expectation, with word error rates greater than 12% (Nasef et al. 2024; Toyin et al. 2023). Additionally, ASR tools may struggle with variations in speaker age, gender, and speech duration.

This study aims to demonstrate the need for further exploration of SST using the Arabic language. Despite the existence of many tools, they are not very efficient in processing MSA. Notably, there is limited comparative research evaluating how well current commercial transcription tools handle MSA speech input. This study aims to analyze the performance of different state-of-the-art transcription tools by evaluating their transcription outputs using MSA. To achieve this purpose, a benchmark evaluation is performed, providing feedback about the capabilities and limitations of various ASR tools using a curated dataset of MSA speech samples. The dataset includes recordings from diverse speakers with different speech lengths. To evaluate the tools' performance, five comparison metrics are employed, namely the Euclidean, Manhattan, Cosine, Jaro-Winkler, and Jaccard distances. The findings can inform future improvements in Arabic STT technology. Consequently, the main contributions are:

- Provide a new MSA speech records dataset with various speakers and record lengths, along with the original texts.
- Evaluate the performance of the existing transcription tools in MSA using various metrics.
- Analyze the effect of different parameters, such as the speech length, on the tools' accuracy.
- Provide insight into future improvements in Arabic STT.

The article is organized as follows: Sect. 2 presents state-of-the-art studies in addition to the current transcription tools. Section 3 outlines the proposed methodology for the study. Section 4 explores experiments. Finally, Sect. 5 concludes this work.

2   Related Work

2.1   State of the Art

Arabic ASR research has shown considerable interest over the last decade, and various investigations have been conducted. The authors (Rahman et al. 2024) presented a comprehensive survey of AST. They began by examining recent studies and related systems. Then, they described the existing datasets and their preprocessing techniques, along with the feature extraction methods and Language Modeling. Later, they discussed the challenges and the strategies to overcome them. Also, the authors (Alqadasi et al. 2023) published a comprehensive review to recognize and examine the MAS datasets.

They identified 27 public datasets and 80 subjective datasets. The study aimed to guide researchers interested in this field. The authors concluded that the ASR corporation remains limited compared to other languages. Additionally, (Dhouib et al. 2022) have proposed a systematic review to emphasize the importance of research related to Arabic STT. The authors explored the recent studies published between 2011 and 2021. They discussed the types of Arabic language used, feature extraction techniques, ML techniques, and the four types of speech recognition (Isolated word recognition, Continuous speech recognition, Connected words, and Spontaneous speech). The authors concluded the study by highlighting the gaps and limitations, as well as some key perspectives. Moreover, the authors (Al-Anzi & AbuZeina 2022) investigated recent studies related to Arabic ASR. The authors focused on corpora that included both isolated words and continuous speech, as well as the feature extraction methods used, the classification techniques applied, and the language and acoustic models employed. The evaluation study used an off-the-shelf Mac Soundflower tool. The results achieved an accuracy of 54%. The authors concluded that the field warrants further investigation to enhance recognition performance. Furthermore, the authors (Hussein et al. 2021) presented a comprehensive study of Arabic ASR and its dialect. They discussed the entire process of the transformer, along with the combined Hidden Markov Model and Deep Neural Network, as well as Human Speech Recognition. The evaluation study was performed using three different datasets: MGB2 for the MSA and (MGB3, MGB5) for the dialect. Furthermore, the conventional word error rate was used for evaluating the MSA. The multi-reference word error rate (MR-WER) and averaged WER (AV-WER) were used to evaluate the dialect. The results were 12.5%, 27.5%, and 33.8% respectively, for the three datasets. These results were compared with those of native speakers, which remain outstanding. Finally, in another study (Azim et al. 2023), a novel ASR strategy was introduced for generating transcripts using a character-level sequence-to-sequence model. The model links the characters generated from the audio to their original text. The authors used two datasets, 1200 h of MGB2 and Modern Standard Arabic audio. The results were promising.

2.2 Speech-To-Text Tools

In this study, six commercial STT tools have been utilized to transcribe Arabic audio files generated by participants of varying ages and genders into text files. The tools are:

TurboScribe is an AI-powered transcription service that converts audio and video files into text for a monthly fee. It is powered by Whisper, a leading AI STT transcription technology, ensuring high accuracy and efficiency.

Notta specializes in real-time transcriptions for live meetings and webinars, boasting an accuracy of up to 98.86%. It integrates seamlessly with platforms like Zoom, Google Meet, Webex, and Microsoft Teams. Notta's official website provides high-level descriptions of its functionalities, such as real-time transcription, multi-language support, and AI-powered summarization. However, its backend architecture, algorithms and technical implementation details are not publicly available.

VEED is primarily a video editing platform, using STT algorithms. It enables users to add subtitles to videos automatically and provides tools for editing and customizing the transcription output. Specific details about its AI models are not disclosed, but it is known that VEED utilizes Gladia's transcription API to enhance its transcription capabilities. By integrating this API, VEED can provide users with high-quality transcriptions, boasting a reported accuracy of up to 98.5%.

Maestra is an AI-driven platform that automates transcription, captioning, and voiceover services for audio and video content across more than 125 languages. Most of the available references are not peer-reviewed and don't address technical details about Maestra, but mainly they provide insights into this tool's functionality and emphasize the use of cutting-edge AI and machine learning technologies to deliver high-quality transcription and subtitling services.

Clipto is an advanced STT platform that leverages artificial intelligence to transcribe audio and video content into text with high accuracy. The platform utilizes OpenAI's Whisper technology to deliver accurate transcriptions, boasting a transcription accuracy of up to 99%. Whisper is a state-of-the-art speech recognition model developed by OpenAI, known for its robustness and ability to handle diverse accents and background noises.

Sonix is an automated transcription platform that leverages advanced AI and NLP technologies to convert audio and video files into text. Sonix offers features such as automated translation, AI-powered summaries, and automated subtitles, enhancing the accessibility and utility of transcriptions. Additionally, it provides collaboration tools, enabling multiple users to work on transcripts simultaneously. While the company does not publicly disclose specific AI models, Sonix emphasizes the use of cutting-edge AI to ensure fast and accurate transcriptions.

3 Methodology

The primary aim of this study is to compare various commercial transcribing tools for transcribing Arabic audio and to advocate for future enhancements to better serve the Arabic community. This research methodology is structured into two main phases. The first phase consists of Tool Assessment, while the second phase involves Tool Enhancement. Figure 1 displays the methodology overview.

3.1 Phase 1: Tool Assessment

This phase was investigated using three key components, including data collection, preprocessing, and a comprehensive comparison study.

Data Collection: This study involved the participation of native Arabic speakers representing a diverse range of ages and genders. The recordings were categorized based on their duration into very long, long, medium, or short files. Then, the same recording was transcribed using six tools. Each transcription is saved in a separate text file.

Text Preprocessing: Each transcribed file is prepared for text analysis by eliminating noise and standardizing the data. Each generated text file underwent a series of meticulous steps to ensure its quality and suitability for further analysis. Tokenization was applied to remove punctuation, special characters, spaces, diacritics, and other elements

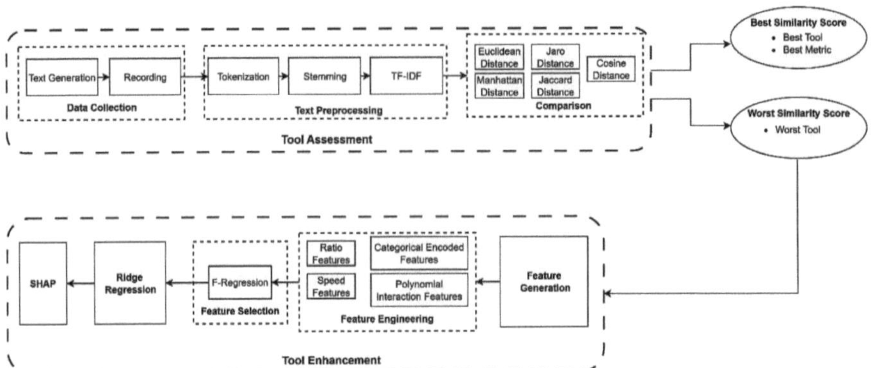

Fig. 1. The Research Methodology Overview

that could distort the results. Stemming is also employed to avoid different representations of the same token, for example, replacing " ة" with " ه" if these two letters are at the end of the word. Term frequency-inverse document frequency (TF-IDF) is used to analyze the occurrence and significance of each word in both text files (transcribed and original). Each word is represented by its frequency, indicating how often it appears in a document, or by zero if it is absent. For instance, if a word appears once in the original text but is missing from the converted text, it is recorded as 1 in the original file and 0 in the converted file.

Comprehensive Comparison: The original text is compared to the transcribed text for each recording using five distance measures: Cosine, Euclidean, Jaro-Winkler, Jaccard, and Manhattan metrics. These metrics aim to measure the difference between the original text and transcribed text. For more details, see (Larabi Marie-Sainte et al. 2021).

3.2 Tools Enhancement

Based on the comprehensive comparison, we focused on analyzing the weakest-performing tool using Explainable AI (xAI), with the intention of identifying the key factors influencing its low similarity scores. The investigation is carried out through the following steps:

- **Feature Generation:** extracting the relevant features based on the initial experimental setup, including recording duration and category, speaker age and gender, and other relevant factors. Full details of these features are provided in Sect. 4.3.
- **Feature Engineering:** Additional features were derived from the original ones to enhance the model's predictive accuracy. This step is detailed in the experiment.
- **Feature Selection:** The f_regression method was applied to identify features with strong linear relationships to the target variable. It uses an F-test to score each feature individually for relevance in regression tasks, helping reduce dimensionality while retaining predictive power (Pedregosa et al. 2011).

- **Ridge Regression:** is a linear model that applies L2 regularization to prevent over-fitting and enhance generalization. It was chosen in this study due to its robustness in handling multicollinearity and its suitability for datasets with continuous and moderately correlated features (Li, et al. 2020).
- **SHAP (SHapley Additive exPlanations):** explains individual predictions by attributing the impact of each feature. For Ridge Regression, it uses LinearExplainer, ensuring that the sum of SHAP values equals the difference between the prediction and the baseline, providing a precise explanation of the model's output (Lundberg, et al. 2017) and (Berriche, et al. 2025).

4 Experimental Results

4.1 Data Collection

In this study, 129 speech samples were collected from native Arabic speakers of different ages (15–54), genders, and nationalities to ensure diversity and reduce regional pronunciation bias, enhancing dataset generalizability. Informed consent was obtained from all participants before data collection. Scripted readings were generated using ChatGPT-4o to produce clear, neutral texts in White and Modern Arabic **known as العربية البيضاء**. This minimizes dialectal influence and ensures broad accessibility. Topics included Quranic verses, religion, transportation, politics, and social issues, providing linguistic diversity and rich phonetic coverage. The texts were saved in TXT format as reference files. Recordings were manually captured using a voice recorder to preserve natural pronunciation without the use of noise filtering. To ensure dataset balance, recordings were categorized by duration: very long (10–30 min), long (5–6 min), medium (2.5–3.5 min), and short (1.5–2 min). The recordings were stored in M4A and MP4 formats and subsequently transcribed using six different STT tools. Several preprocessing steps were applied to enhance the quality of the text and ensure consistency across transcriptions.

4.2 Comparison Study

The performance of six STT transcription tools was evaluated using five distance metrics. Then, the average and standard deviation were calculated for the scores obtained in each category, including the very long, long, medium, and short files. Table 1 displays the results of the two best and the worst scores obtained by each metric for the four categories. Figure 2 presents the similarity scores along with the standard deviation produced by the five metrics for the six tools using the long recordings category. The figures corresponding to other categories have been omitted for brevity, as the trends observed in the long recordings are similar to the overall behavior across categories.

Cosine and Euclidean Similarities: Turboscribe and Clipto achieved an average similarity of 0.99, the highest among the very long file category. This result indicates a strong alignment with the reference text. Maestra and Veed followed with Cosine scores of 0.94 and 0.93, respectively, and Euclidean scores of 0.97 for both tools. Notta had the lowest performance using both metrics. The standard deviation across all tools and categories ranges from 0.08 to 0.15 for Cosine and from 0.013 to 0.024 for Euclidean similarity.

Unlike the Euclidean distance, the Cosine metric exhibits greater variability in similarity scores, suggesting a stronger ability to distinguish between the tools or categories. This highlights the Cosine metric's enhanced sensitivity in capturing meaningful differences in textual representations.

Table 1. Best & Worst score obtained by the five metrics using the four categories

		Short	Medium	Long	Very Long
Cosine	Two Best	Veed (0,9158)	Veed (0,9317)	Maestra (0,9415)	Clipto (0.9881)
		Clipto (0,9108)	Maestra (0,9290)	Clipto (0,9409)	Turboscribe (0.9878)
	Worst	Notta (0.8021)	Notta (0.8381)	Notta (0.8557)	Veed (0.9478)
Euclidean	Two Best	Veed (0.9641)	Veed (0.9711)	Clipto (0.9752)	Turboscribe (0.9890)
		Clipto (0.9626)	Maestra (0.9703)	Maestra (0.9751)	Clipto (0.9889)
	Worst	Notta (0.9387)	Notta (0.9507)	Notta (0.9572)	Veed (0.9772)
Jaccard	Two Best	Sonix (0.7922)	Sonix (0.7943)	Sonix (0.8002)	Clipto (0.8212)
		Veed (0.7721)	Veed (0.7720)	Maestra (0.7577)	Turboscribe (0.8092)
	Worst	Notta (0.5331)	Notta (0.5346)	Notta (0.5199)	Notta (0.6216)
Jaro	Two Best	Sonix (0.9125)	Sonix (0.9111)	Maestra (0.9107)	Turboscribe (0.8920)
		Maestra (0.9118)	Maestra (0.9109)	Turboscribe (0.9060)	Sonix (0.8905)
	Worst	Notta (0.9004)	Notta (0.9017)	Veed (0.8968)	Veed (0.8685)
Manhattan	Two Best	Sonix (0.9734)	Sonix (0.8896)	Sonix (0.9227)	Clipto (0.7629)
		Veed (0.9724)	Veed (0.8894)	Clipto (0.9167)	Turboscribe (0.7549)
	Worst	Notta (0.9403)	Notta (0.7575)	Notta (0.8227)	Notta (0.4815)

Jaccard Similarity: Clipto and Turboscribe outperformed the other tools with scores of 0.82 and 0.81, respectively. This result is achieved using the very long category. The result shows a high degree of consistency with the lengthy reference text. Notta had the lowest Jaccard similarity, 0.52, showing more inconsistencies in word usage. The standard deviation ranges from 0.13 to 0.23 across all tools and categories. This demonstrates that the distance measure was moderately effective in capturing distinctions either between the tools themselves or among the different categories.

Jaro-Winkler Similarity: Maestra and Sonix are the top tools, with a score of 0.91, in the short file category. Turboscribe achieved 0.90 and 0.89 for the long and very long categories, respectively. Notta and Veed recorded the lowest similarity for the short/medium and the long/very long categories, respectively. This result indicates minor deviations in word arrangement. The standard deviation ranged between 0.011 and 0.04, indicating minimal variation across tools and categories. So, the metric lacks the sensitivity needed to differentiate between categories and tools effectively.

Manhattan Distances: Sonix and Veed achieved the highest scores of 0.97 using the short category. Clipto and Turboscribe achieved 0.75 and 0.76, respectively, using the very long file category. Notta achieved the worst score. The standard deviation varied between 0.12 and 0.21, reflecting a moderate level of variability in similarity scores across tools and categories.

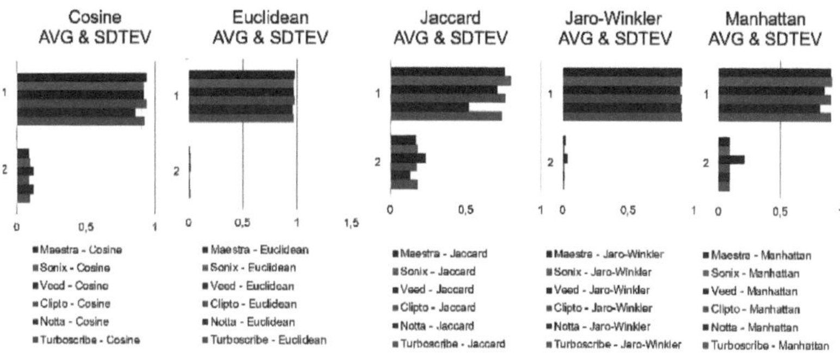

Fig. 2. The similarity scores and the standard deviation yielded by the five metrics for the four tools using the long recordings category.

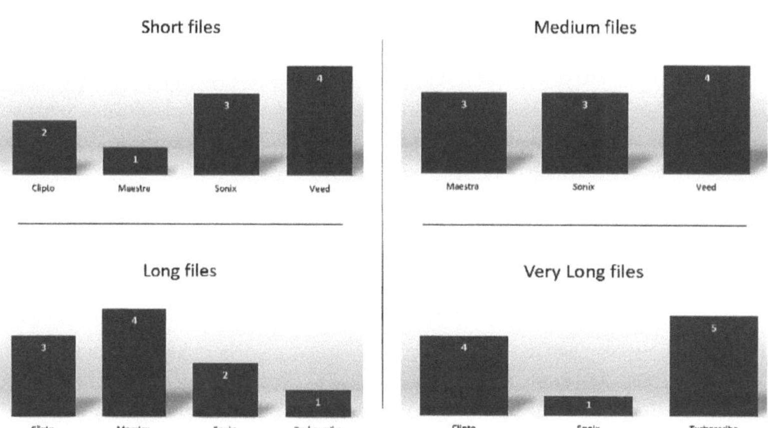

Fig. 3. The frequency with which each tool is identified as the top performer across the various categories, as determined by the five-similarity metrics.

Figure 3 summarizes the frequency with which each tool is recognized as the top performer across various categories, as determined by the five different similarity metrics. Veed performs well for short and medium recordings but shows clear limitations with long and very long files. Similarly, Sonix struggles beyond the medium category. On the other hand, Maestra demonstrates strong performance for long recordings and, to some extent, for medium files; however, it falls short when handling extremely long content. Turboscribe emerges as the top performer for very long recordings, although it is unexpectedly not ranked best for long recordings—an inconsistency that could be attributed to the limitations of certain metrics, such as Manhattan, Euclidean, and Jaro-Winkler, which may not effectively capture the nuances of similarity in long files. Notably, Clipto consistently ranks as the second-best performer for both long and very long files, indicating a stable and reliable performance in handling larger audio inputs.

To sum up, Notta exhibited the weakest performance across all measures. Moreover, Jaro-Winkler and Manhattan distances yielded consistently high similarity scores for the first three categories (see Table 2). However, for the very long recordings, it indicates a noticeable decline in performance (0.89 and 0.76, resp.) as the text length increases. This suggests that both metrics may be less robust when handling significantly longer textual inputs, potentially due to their sensitivity to string length and structure. Furthermore, Cosine Similarity outperforms the other metrics by showing the highest standard deviation, which indicates that it is the most effective in detecting meaningful variations between tools and categories in the text data. It is therefore the most suitable metric for this comparative task. This metric consistently identified Clipto as the best-performing tool across all categories, highlighting the overall reliability and adaptability of Clipto in handling diverse transcription scenarios.

4.3 Explainable AI-Based Analysis of the Lowest Tool Performance

Based on the results above, Notta consistently produced the lowest similarity scores across all metrics, particularly with the Jaccard distance. This experiment aims to investigate the reasons behind this outcome by analyzing the initial experimental setup, given that Notta operates as a black box with unknown models and features. To investigate this outcome, the xAI technique was applied using the SHAP model. For this purpose, a new dataset was constructed, incorporating all previously analyzed recordings and 11 well-defined features, detailed as follows:

Feature 1 represents the four categories, referred to as Length.
Feature 2 refers to the Topic used when generating the dataset.
Features 3 & 4 represent the Word count generated for each recording before and after preprocessing, respectively.
Features 5 & 6 represent the speaker's Gender and group age.
Feature 7 is the duration of the recording in seconds.
Feature 8 is the Speaking speed, divided into three categories: slow, normal, and fast.
Features 9 & 10 represent the Word count from Notta's transcription before and after preprocessing, respectively.
Feature 11 is the score yielded by Notta using Jaccard distance (the model Target).

To enrich the dataset and enhance the performance of the proposed model, 12 additional features were generated through Feature Engineering, as detailed in Table 2.

Table 2. New generated Features using Feature Engineering

Feature Type	Feature Description	Calculation
Ratio Features	These features capture compression during preprocessing (how many words were reduced)	**preprocessing_ratio** → # words after preprocessing / (# words before preprocessing + 1) **notta_preprocessing_ratio** → notta (# of words after pre-processing) / (notta (# of words before pre-processing) + 1)
Speed Feature	The # of words per second (Words_per_second)	→ # words after preprocessing / (Duration + 1)
Categorical Encodings	For each of the 4 categorical features (explained above), 2 new features were generated based on Label & Target Encoding: Topic_encoded, Topic_target_encoded, Length_encoded, Length_target_encoded, Recorded By (Male or Female)_encoded, Recorded By (Male or Female)_target_encoded, Age Group of Recorder_encoded, Age Group of Recorder_target_encoded	**Label Encoding** (converts category to integer) **Target Encoding** (average target value per category)
Polynomial Interaction	The product of the top correlated features (Talk Speed from the original feature and Words_per_second from the Speed feature) with the target was computed to capture interaction effects.	→ Talk Speed * Words_per_second

A total of 18 features were used, including 12 encoded and 6 original numerical features. After applying feature selection with f_regression and SelectKBest (from scikit-learn), the top 10 most relevant features were retained. The resulting dataset (129×10) was then split into 80% training and 20% testing sets for Ridge Regression modeling.

Table 3 presents the evaluation results of the proposed Ridge Regression model. The model achieved R-squared of 76%, demonstrating its effectiveness in predicting the similarity score. Moreover, the low RMSE and MAE values further validate the model's robustness, indicating that the predicted scores closely match the actual values. The model was also tested using 5-fold cross-validation to confirm the initial results.

Table 3. Ridge Regression model performance results

Performance Metrics	Score
Training R^2	0.8530
Test R^2	0.7605
Cross-validation R^2	mean $\pm$ std: 0.7637 ± 0.0727
Test RMSE	0.0435
Test MAE	0.0342

The final step involves applying SHAP to identify the most important and impactful features. Figure 4 illustrates the importance of features and their impact on the model's predictions. This visualization helps identify which features most strongly influence the predicted similarity scores, offering interpretability and transparency into the model's decision-making process. As shown, the Topic feature (encoded using target encoding) has a substantial impact on the similarity score. This is likely because certain topics contain complex or uncommon vocabulary, which can make pronunciation more challenging and, in turn, prevent the transcription tool from accurately recognizing words. Additionally, the Talk Speed feature (Feature 2) has a significant impact on tool performance, as fast speech is often less accurately transcribed than slow speech. Speaker age and recording length also influence transcription accuracy, highlighting their role in the tool's limitations.

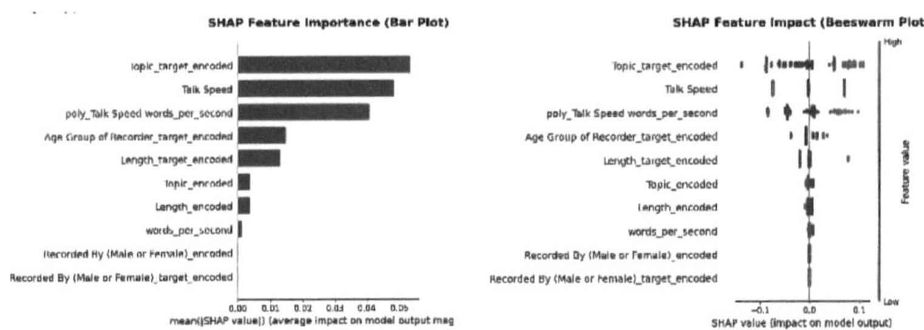

Fig. 4. Feature importance and feature impact provided by the proposed SHAP model.

5 Discussion and Conclusion

This research tackles the assessment and enhancement of some Arabic STT transcription tools. Two main experiments were conducted to fulfill the research objectives. A new MSA speech recordings dataset was developed and made publicly available to support future research (https://figshare.com/s/ccc33015d2e9abe8f4d8), fulfilling the study's first objective. Another research objective was to evaluate the accuracy and effectiveness of various STT transcription tools in transcribing Arabic recordings by measuring the similarity and distance measures to obtain a comprehensive assessment. The selected metrics provided prominent results. The Cosine Similarity is the most suitable metric for this comparative task in detecting meaningful variations between tools and categories in the text data. It reliably identified Clipto as the best-performing tool across all categories. To fulfill the third objective, xAI was applied to analyze the lowest similarity score obtained using the Notta tool with the Jaccard metric. The analysis revealed that factors such as Topic, Talk Speed, Speaker's Age, and Recording Length significantly influence Notta's transcription accuracy. The final objective, given that transcription tools operate as black boxes with unknown underlying models, is to consider the identified factors to help maintain transcription accuracy. Future work could focus on developing an efficient STT model that addresses the limitations identified in this study, with particular emphasis on handling MSA and its diverse dialects.

Acknowledgment. The authors acknowledge the support of Prince Sultan University, Saudi Arabia, for paying the article processing charges and conference fees related to this publication.

References

Alharbi, S., et al.: Automatic speech recognition: systematic literature review. IEEE Access **9**, 131858–131876 (2021). https://doi.org/10.1109/ACCESS.2021.3112535

Liu, K.K., Thurlow, M.L., Press, A.M., Dosedel, M.J.: A Review of the Literature on Computerized Speech-to-Text Accommodations (NCEO Report #414) (2008). http://www.nceo.info

Ghoul, E., et al.: Arabic Automatic Speech Recognition: A Systematic Literature Review (2022). https://doi.org/10.3390/app12178898

Besdouri, F.Z., Zribi, I., Belguith, L.H.: Arabic automatic speech recognition: challenges and progress. Speech Commun. **163**, 103110 (2024). https://doi.org/10.1016/J.SPECOM.2024.103110

Nasef, M.M., Elshall, A.A., Sauber, A.M.: ArabRecognizer: modern standard Arabic speech recognition inspired by DeepSpeech2 utilizing Franco-Arabic. Int. J. Speech Technol. **27**(3), 673–686 (2024). https://doi.org/10.1007/S10772-024-10130-8/METRICS

Toyin, H.O., Djanibekov, A., Kulkarni, A., Aldarmaki, H.: ArTST: Arabic text and speech transformer. In: ArabicNLP 2023 - 1st Arabic Natural Language Processing Conference, Proceedings, pp. 41–51 (2023). https://doi.org/10.18653/V1/2023.ARABICNLP-1.5

Rahman, A., Kabir, M.M., Mridha, M.F., Alatiyyah, M., Alhasson, H.F., Alharbi, S.S.: Arabic speech recognition: advancement and challenges. IEEE Access **12**, 39689–39716 (2024). https://doi.org/10.1109/ACCESS.2024.3376237

Azim, M.A., Hussein, W., Badr, N.L.: Using character-level sequence-to-sequence model for word level text generation to enhance Arabic speech recognition. IEEE Access **11**, 91173–91183 (2023). https://doi.org/10.1109/ACCESS.2023.3302257

Alqadasi, A.M.A., Abdulghafor, R., Sunar, M.S., Salam, M.S.B.H.J.: Modern standard Arabic speech corpora: a systematic review. IEEE Access **11**, 55771–55796 (2023). https://doi.org/10.1109/ACCESS.2023.3282259

Dhouib, A., Othman, A., El Ghoul, O., Khribi, M.K., Al Sinani, A.: Arabic automatic speech recognition: a systematic literature review. Appl. Sci. (Switzerland). MDPI (2022). https://doi.org/10.3390/app12178898

Al-Anzi, F.S., AbuZeina, D.: Synopsis on Arabic speech recognition. Ain Shams Eng. J. **13**, 2 (2022). https://doi.org/10.1016/j.asej.2021.06.020

Hussein, A., Watanabe, S., Ali, A.: Arabic speech recognition by end-to-end, modular systems and human. Comput. Speech Lang. **71**, 101272 (2022)

Larabi Marie-Sainte, S., Alnamlah, B.S., Alkassim, N.F., Alshathry, S.Y.: A new system for Arabic recitation using speech recognition and Jaro Winkler algorithm. Kuwait J. Sci. **49**(1) (2021). https://doi.org/10.48129/kjs.v49i1.11231

Pedregosa, F., et al.: Scikit-learn: machine learning in Python. J. Mach. Learn. Res. **12**, 2825–2830 (2011)

Li, T., Zhang, C., Liu, H.: Ridge regression and its extensions: a review. Stat. Sci. **35**(4), 518–542 (2020). https://doi.org/10.1214/19-STS749

Lundberg, S.M., Lee, S.-I.: A unified approach to interpreting model predictions. In: Advances in Neural Information Processing Systems (NeurIPS), vol. 30 (2017). https://proceedings.neurips.cc/paper_files/paper/2017/hash/8a20a8621978632d76c43dfd28b67767-Abstract.html

Berriche, L., Loulizi, A.: Prediction of failures in the project management knowledge areas using optimized ensemble models in software companies. Discov. Appl. Sci. **7**, 718 (2025). https://doi.org/10.1007/s42452-025-07337-y

A Practical Framework for Auditing Fairness in Medical AI

Andreea M. Oprescu[2(✉)], Jorge Vindel-Alfageme[1], Erik Campos-Espinosa[1], Marta Caro-Martínez[1], Belén Díaz-Agudo[1], M. Carmen Romero-Ternero[2], and Juan A. Recio-García[1]

[1] Department of Software Engineering and Artificial Intelligence, Instituto de Tecnologías del Conocimiento, Universidad Complutense de Madrid, Madrid, Spain
{jorgevin,erikcamp,martcaro,belend,jareciog}@ucm.es
[2] Departamento de Tecnología Electrónica, Instituto de Ingeniería Informática (I3US), Universidad de Sevilla, Sevilla, Spain
{aoprescu,mcromerot}@us.es

Abstract. The increasing deployment of Artificial Intelligence (AI) systems in healthcare has raised significant concerns about bias, fairness, and ethical implications of automated decision-making. While medical AI systems offer capabilities for diagnosis, treatment planning, and patient care, they also inherit and potentially amplify biases present in training data, particularly affecting underrepresented populations. This paper presents a comprehensive framework for auditing AI systems in medical domains through functional audit processes that systematically evaluate bias and fairness. Our approach integrates three key components: data quality assessment, fairness analysis, and application of explainable AI (XAI) techniques. We demonstrate the practical application through the auditing of a machine learning model to predict COVID-19 patient mortality. The results reveal disparities in model performance across different demographic groups.

Keywords: Fairness · Bias auditing · Explainable AI · Healthcare ethics

1 Introduction

AI technologies in healthcare enable diagnostic tools, personalized treatment recommendations, and predictive analytics for patient outcomes. However, AI models present critical challenges related to algorithmic fairness, bias detection, and ethical compliance in medical decision-making systems.

Medical AI systems are particularly susceptible to bias due to historical inequities in healthcare data collection and treatment practices. These biases manifest from underrepresentation of demographic groups in training datasets, systematic exclusion from clinical trials, and perpetuation of discriminatory practices embedded in medical records. Such biases can lead to suboptimal or

L. Martínez et al. (Eds.): IDEAL 2025, LNCS 16239, pp. 269–280, 2026.
https://doi.org/10.1007/978-3-032-10489-2_23

harmful outcomes for minority populations, women, elderly patients, and other vulnerable groups. AI auditing can be characterized by its purpose and methodological characteristics. According to Mokander [1], AI auditing is a governance mechanism utilized by different actors: (i) regulators assessing legal compliance, (ii) technology providers mitigating risks, and (iii) stakeholders making informed engagement decisions [2].

The literature on auditing AI systems includes academic articles and books, auditing tools developed by private companies, industry standards, and policy guidance documents. AI system audits can be approached from different perspectives: *functional audits* verify if systems perform tasks accurately and reliably; *code audits* review source code; and *impact audits* investigate outcome effects [3,4]. Audit strategies vary from generic governance-focused approaches to specific bias evidence collection methods based on domain regulations. This paper follows the latter approach.

Recent advances in eXplainable AI (XAI) have developed methods providing transparency and interpretability for AI systems. We propose our functional audit processes based on the synergy between explainability and auditing concepts. Research demonstrates XAI's utility for auditing: [5] in insurance, [6] using SHAP [7] in finance, [8] in health governance, and others developing "Accountable eXplainable AI" (AXAI) frameworks [9].

While our functional audit framework provides quantitative bias evidence through technical metrics, result interpretation requires interdisciplinary collaboration between AI experts, healthcare professionals, and social scientists. Identified disparities reflect complex social realities shaping healthcare delivery, not merely statistical anomalies. Effective healthcare AI auditing must address underlying social dynamics influencing data collection and model deployment. Furthermore, AI bias interpretation requires considering the social context of data collection and deployment. The COVID-19 pandemic exemplifies how social inequalities can be reflected and amplified through healthcare AI systems. Socioeconomic factors, healthcare access, and historical treatment patterns influence both data collection and model predictions. Therefore, effective AI auditing must incorporate sociotechnological perspectives considering how social realities shape technical artifacts and their interpretation [10].

This paper runs as follows: Sect. 2 describes our three-stage functional audit process for medical domains, including data quality assessment, fairness analysis using Aequitas, and XAI-enhanced auditing. Section 3 presents a COVID-19 mortality prediction case study demonstrating our framework's practical application. Section 4 discusses results from data quality assessment, fairness analysis, and explainable AI methods. Finally, Sect. 5 concludes with key contributions and outlines future work directions.

2   Functional Audit Process for Medical Domains

The medical domains presents unique challenges for AI fairness due to the complex interplay between biological, social, and historical factors that influence

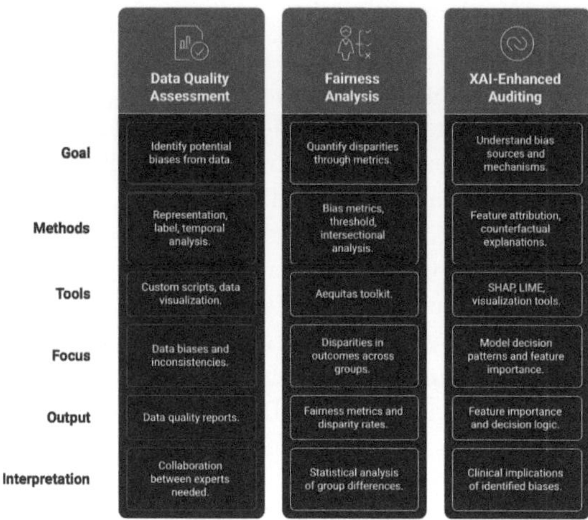

Fig. 1. Description of the proposed auditing process.

health outcomes and data collection practices. Understanding these biases is needed for developing effective auditing methodologies. Several key sources of data-centric ethical bias are presented: *Sampling or data representation bias*: biomedical datasets reflect the limitations of the healthcare systems they are extracted from, as, for example, systematic exclusion or undersampling of certain demographic groups. *Labeling bias*: Inconsistent or biased annotation practices that reflect historical prejudices. *Outcome availability bias*: adverse outcomes (e.g., death) can have a higher prevalence in specific demographics, limiting the diversity of positive cases available for model learning. *Temporal bias*: Changes in medical practices and diagnostic criteria over time that are not properly accounted for in historical datasets.

To address these biases, we propose a three-stage audit process: *(1) data quality assessment, (2) fairness analysis,* and *(3) eXplainable AI (XAI) integration* (see Fig. 1). This holistic approach identifies potential biases from data to model evaluation, requiring collaboration between domain experts, AI specialists, and social scientists for proper contextual interpretation.

2.1   Data Quality Assessment

Comprehensive data quality assessment forms a crucial component of our auditing procedures: *Representation Analysis*: Systematic evaluation of demographic representation throughout the different splits of the dataset (i.e. training and testing), identifying underrepresented groups and potential sampling biases. *Label Quality Assessment*: Analysis of annotation consistency and potential bias in labeling practices, particularly important in medical contexts where diagnostic labels may reflect historical biases. *Temporal Consistency*: Evaluation of how medical practices and diagnostic criteria have evolved over time and their impact on model training and performance.

2.2 Fairness Analysis

There are several toolkits for fairness analysis that can be applied to perform this stage. In our case, we have chosen Aequitas, an open-source bias and fairness assessment toolkit to perform functional bias audit processes [11]. Aequitas offers a comprehensive bias audit for medical AI, using fairness metrics such as demographic parity, equalized odds, calibration, and false positive/negative rate balance, with particular emphasis on metrics capturing disparities in diagnostic accuracy across demographic groups.

The tool receives as input the test set containing true outcome labels, model predictions, and sensitive, multivalued demographic attributes that serve as reference variables for organizing sample subgroups. While our current implementation focuses on basic bias metrics assessment, the full Aequitas framework enables more advanced analyses including threshold optimization and intersectional bias detection for comprehensive fairness evaluation.

The following steps describe a fairness audit with Aequitas: *(1) Define subgroups* using multivalue attributes to investigate for bias. *(2) Select fairness metrics*, maintaining consistency with model optimization metrics. *(3) Calculate group-specific metrics* based on selected fairness metrics. *(4) Establish reference group* using majority criterion (selecting the subgroup with the highest number of patients for each multivalue attribute). *(5) Quantify disparity* by comparing subgroup metrics to reference group through disparity rates.

2.3 XAI-Enhanced Auditing

Our framework integrates XAI techniques as a complementary component to Aequitas fairness metrics, creating a comprehensive approach to bias detection in medical AI systems. While Aequitas quantifies disparities through statistical metrics, XAI methods provide insights into underlying mechanisms causing biased outcomes.

XAI integration serves multiple purposes: Feature attribution methods like SHAP [7] and LIME [12] analyze how features contribute to decisions across demographic groups, revealing inappropriate reliance on sensitive attributes or proxies [13]. Beyond individual analysis, XAI identifies systematic bias patterns not apparent from aggregate metrics [14], such as visualization of feature importance distributions showing different decision logic across populations [15]. Counterfactual explanations [16] demonstrate how sensitive attribute changes affect predictions, distinguishing legitimate medical factors from discriminatory patterns. This synergistic approach enables auditors to detect bias presence and understand its sources, facilitating targeted mitigation strategies [17]. XAI interpretability is valuable for communicating results to healthcare professionals who need to understand clinical implications of identified biases [18].

3   Use Case: Audit of a COVID-19 Mortality Machine Learning (ML) Model

To demonstrate our functional audit process, we present a biomedical case study using a real-world COVID-19 clinical dataset to train a ML model to predict patient mortality.

3.1   Dataset

The dataset used is *COVID Data for Shared Learning (CDSL)*, a multimodal biomedical dataset available on Physionet [19], containing 4,479 hospitalized patients in Spanish hospitals HM Hospitales from 2019-12-26 to 2021-02-13. We focus exclusively on tabular data for binary mortality prediction to investigate sex and age biases. We adopt the data preprocessing methodology described in [20]. After preprocessing, features included laboratory values and vital signs from the first 48 h, admission duration, comorbidities, mechanical ventilation, ICU admission, age and sex. Exclusion criteria was applied: unknown patient destination after discharge (n=159), patients with a hospitalization stay of more than 30 day (n=159), patients without laboratory data within the first 48 h (n=2016). Following exclusion criteria, 2,145 patients were included with a class imbalance ratio of 5.96 (14.36% mortality class); 1,273 male and 872 female patients. Missing values were handled using Multiple Imputation by Chained Equations for numerical features and zero-filling for categorical ones. Categorical features were one-hot encoded and numerical features were Z-score normalized after stratified train-test split (80%-20%) to prevent data leakage. The split was performed using stratified sampling to preserve the proportion of the target classes (mortality vs. survival) in both subsets.

3.2   Classification Model

We trained and optimized an Extreme Gradient Boosting (XGBoost) model implementing cost sensitive learning to address class imbalance by assigning a higher weight to the positive class (mortality). To evaluate the model's ability to generalize, Repeated Stratified K-Fold cross-validation scheme with 10 splits and 5 repeats (n_splits = 10, n_repeats = 5) was employed. The metric used to optimize the model is F1-Score, to stress the importance of reducing the number of false negatives (FN) and trying to increase the number of true positives (TP).

The confusion matrix values for the test set were: 353 true negatives (TN), 15 false positives (FP), 16 FN, and 46 TP. The classifier achieved a precision of 0.757 (95% CI: 0.638–0.871), recall of 0.739 (95% CI: 0.627–0.844), and an F1-Score of 0.748 (95% CI: 0.648–0.829).

3.3   Results of the Data Quality Asessment

Our comprehensive data quality assessment revealed important characteristics of the data subsets that inform the subsequent bias analysis.

Table 1. Number and (%) of patients by hospital outcome for train and test sets

Set	Group	Category	Frequency		Percentage (%)	
			Survival	Death	Survival	Death
Train	Age	Adult (27–59 years old)	474	8	98.3	1.7
		Aged adult (60–79 years old)	666	78	89.5	10.5
		Oldest adults (80–98 years old)	312	156	66.7	33.3
	Sex	Male	858	166	83.8	16.2
		Female	612	80	88.4	11.6
Test	Age	Adult (27–59 years old)	144	0	100.0	0.0
		Aged adult (60–79 years old)	155	22	87.6	12.4
		Oldest adults (80–98 years old)	67	39	63.2	36.8
	Sex	Male	210	39	84.3	15.7
		Female	157	23	87.2	12.8

Representation Analysis: To identify sampling bias from underrepresented groups, we analyzed the training and testing sets, which were created by stratifying patients based on their outcome. The sensitive variables available in the dataset are "sex" and "age". Table 1 details the patient demographics for the training and testing sets, which were stratified by outcome.

In both the training and testing sets, survival rates decline significantly with increasing age. Regarding sex, the dataset includes a larger number of male patients than female patients across both splits. **Label Quality Assessment**: patient outcome (mortality or survival) is derived from the administrative variables. Patients were labeled with a mortality outcome (1) if their discharge status was recorded as "Death" within 30 d of admission. All other patients were labeled with a survival outcome (0). For this retrospective dataset, no further assessment of the discharge status can be performed. **Temporal Consistency**: Given that all data were collected during a specific pandemic period, temporal bias analysis was not relevant for this dataset.

3.4   Results of the Fairness Analysis

After evaluating the model's performance, we assessed for disparities in predictive outcomes across the available multivalue attributes: sex and age.

For the analysis, we have considered samples from the "Male" subgroup (253 individuals) and from the "Female" subgroup (177 individuals) regarding the variable sex. Also, we have created the following categories to stratify patients by age span with respect to the variable age: "Adult (27–59 years old)" (138 individuals), "Aged adult (60–79 years old)" (189 individuals), and "Oldest adults (80–98 years old)" (103 individuals). Table 2 shows how samples from the test set were classified regarding the different subgroups they belong to.

Although Aequitas does not use the F1 score, it does support its core components, sensitivity and precision. Therefore, we can analyze disparities in the F1 score by examining the fairness metrics for these two underlying performance measures. In Aequitas, sensitivity is equivalent to the True Positive Rate (TPR), and its corresponding fairness goal is Equal Opportunity. In parallel, Precision measures the accuracy of the model's positive predictions and its fairness metric is Predictive Parity.

Table 2. Classification of the samples in the test set

Multivalue attribute	Subgroup tag	FP	FN	TN	TP
"sex" (Sex)	"Female"	7	5	147	18
	"Male"	8	11	206	28
"age_cat" (Age category)	"Adult (27–59 years old)"	0	0	138	0
	"Aged adult (60–79 years old)"	4	8	162	15
	"Oldest adults (80–98 years old)"	11	8	53	31

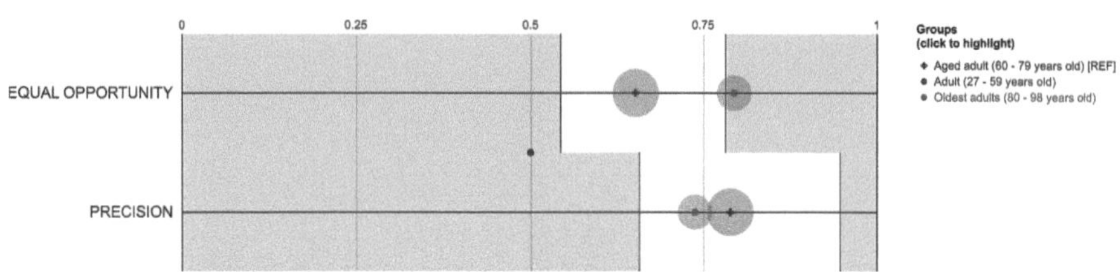

Fig. 2. Equal Opportunity and Precision results, grouped by age category.

Figure 2 shows Equal Opportunity and Precision metrics for age groups on a 0–1 scale. Aged adults score approximately 0.68 (Equal Opportunity) and 0.8 (Precision), while Adults show lower Equal Opportunity (0.6) but higher Precision (0.85). Oldest adults achieve the highest Equal Opportunity (0.75) with moderate Precision (0.78). Figure 3 presents the same analysis by sex. Females show higher Equal Opportunity than Males but lower Precision, resulting in more FP and reduced Precision.

Figure 4 shows the absolute performance of a model using two different metrics: TPR and Precision. The model's performance varies across the different age groups. The Aged Adults group achieved a TPR of 0.65 and a Precision of 0.79, which indicates moderate sensitivity and relatively high accuracy in positive predictions. For the Adult group, although the model correctly identified all 138 non-positive cases, the TPR and Precision metrics are undefined, as there were no positive instances in the test set for this subgroup. The Oldest Adults group yielded the highest TPR of 0.79, with a Precision of 0.74. The model is moderately precise in detecting TP within this older population.

Overall, these findings suggest that the model performs best in the Oldest Adults group and moderately in the Aged Adults group. For the Adult group, the model correctly classified all negative cases, altough the performance on positive cases could not be evaluated, as the test set for the Adult group contained no positive instances.

3.5 Results of the Explainable AI Methods

Following our three-stage audit methodology, after data quality assessment and fairness evaluation with Aequitas, we implemented the third component: XAI techniques (ALE for global explanations, LIME, SHAP, and DiCE for local explanations) to complement bias detection results and understand underlying model decision patterns.

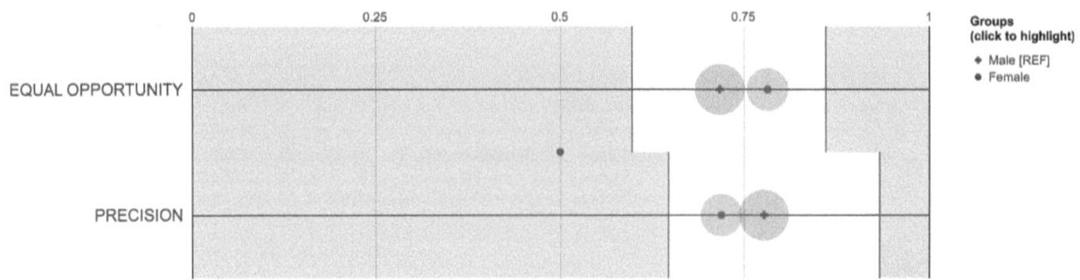

Fig. 3. Equal Opportunity and Precision results, grouped by sex.

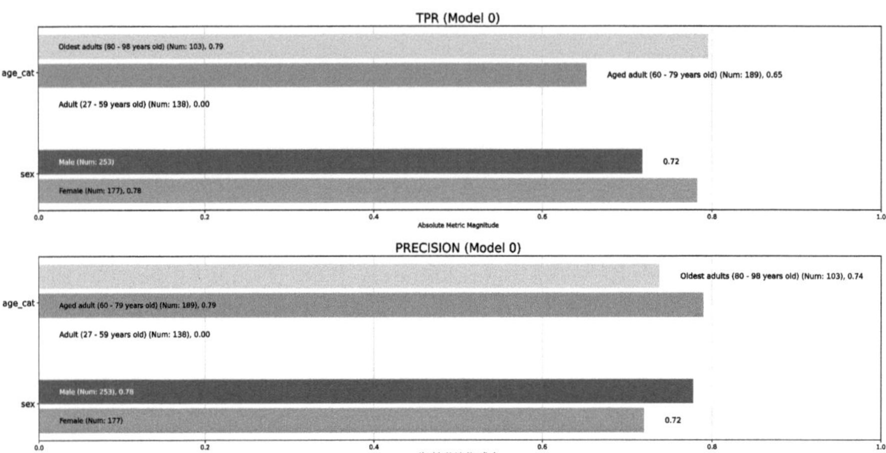

Fig. 4. Model performance (TPR and Precision) across demographic groups.

ALE analysis identified four key mortality predictors: *lab_creatinine*, *lab_til*, *sex*, and *mechvent*. Figure 5 shows serum creatinine increases mortality risk above 2.0, total bilirubin exhibits a U-shaped relationship (decreasing until 0.5, then increasing), while female sex and mechanical ventilation associate with lower mortality predictions.

Analyzing the local explanations obtained with LIME, SHAP, and DiCE, we found that the same four attributes are the most important ones when explaining a single instance (chosen randomly) from the testset. Therefore, global and local explanations supported each other. Figure 6 provides a SHAP local explanation, showing that the four most influential features are the same as those identified by ALE (Fig. 5) and LIME.

According to the SHAP analysis, sex has little influence, while the other three attributes have a similar impact. Only total bilirubin (*lab_til*) affects hospital outcome positively (as it increases, *hospital_outcome* also increases), while the rest have a negative effect. In the case of the result of DiCE, we obtain a slightly different explanation. DiCE obtains a counterfactual, which means an example of a different instance with a different outcome where there is at least one attribute value that has a wide variation comparing it with the instance to explain. Here,

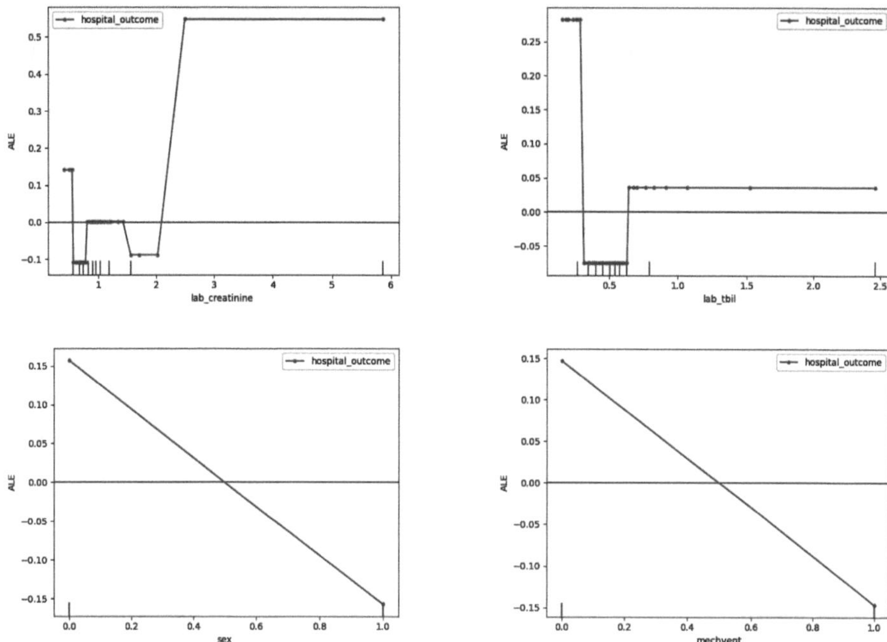

Fig. 5. Global explanation obtained with ALE.

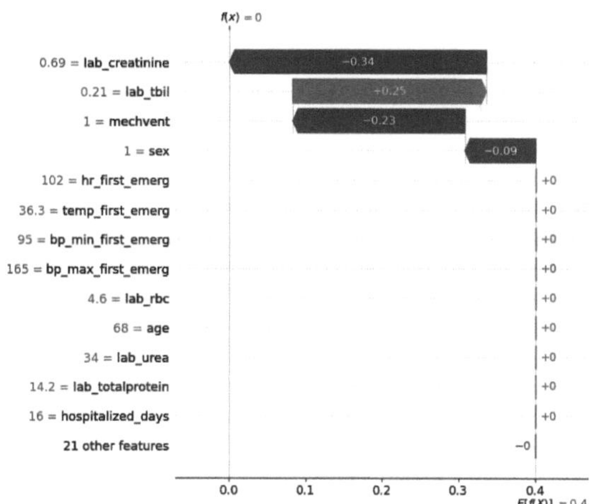

Fig. 6. Local explanation obtained with SHAP.

the only attribute that showed a variation was *lab_creatinine*, meaning that for this specific instance, changing only the *lab_creatinine* value is enough to return a different *hospital_outcome*.

This XAI analysis complements the Aequitas results regarding age and sex biases. While Aequitas detected disparities for both attributes, XAI reveals that age has minimal influence on model predictions, suggesting the detected bias stems from data imbalance rather than learned behavior. However, sex significantly affects *hospital_outcome* predictions, with the model predicting higher mortality for male patients. This confirms genuine algorithmic bias that requires mitigation strategies beyond simple data rebalancing.

4   Conclusions and Future Work

This paper presents a comprehensive framework for auditing AI systems in medical domains through functional audit processes that systematically evaluate bias and fairness. Our approach integrates three key components: data quality assessment, application of fairness toolkits such as Aequitas, and implementation of explainable AI techniques. The practical application through a COVID-19 case study demonstrates the framework's effectiveness in identifying significant disparities in model performance across demographic groups. The key contributions of our approach include: (1) Integration of various forms of discrimination, including intersectional bias that may affect vulnerable populations; (2) A systematic methodology that combines quantitative bias assessment with qualitative XAI insights; (3) Practical tools for healthcare practitioners and AI developers to identify and address ethical concerns in medical AI systems.

The technical analysis presented through our functional audit framework provides quantitative evidence of bias patterns, but understanding their full implications requires incorporating sociological interpretation. The disparities identified through Aequitas metrics and XAI explanations are not merely statistical anomalies but reflect complex social realities that shape healthcare outcomes. Our COVID-19 case study exemplifies this complexity: the choice of variables (sex and age) was constrained by data availability, yet missing socio-economic or ethnicity variables could provide crucial insights into health disparities. Traditional ethics models based on general principles are insufficient for healthcare AI; domain-specific ethical frameworks must consider how social determinants of health influence both data collection and model interpretation [21]. The intersection of technical bias detection with social context analysis is essential for developing comprehensive understanding of how AI systems may perpetuate or amplify existing healthcare inequalities.

Looking ahead, we plan to implement a Case-Based Reasoning (CBR) methodology based on our theoretical framework. CBR provides a problem-solving approach that learns from past experiences, well-suited for complex, context-dependent auditing scenarios where universal rules are insufficient. The CBR approach will transform AI auditing from ad-hoc practices into a structured, reusable methodology that captures collaborative insights from domain experts, AI specialists, and social scientists about bias patterns, appropriate fairness metrics, and effective XAI techniques for specific medical contexts.

Furthermore, our framework aligns directly with the principles of the new European Health Data Space (EHDS) regulation. The EHDS creates a legal and infrastructural environment for the trustworthy use of health data, mandating the traceability, transparency, and interoperability that are foundational to auditable AI. The intersection of the EHDS and AI regulation (particularly the AI Act) creates a dual compliance environment where both the ethical design of ML models and the lawful handling of biomedical data must be auditable and demonstrable. Additionally, our framework will incorporate regulatory and ethical compliance assessment, including GDPR data protection requirements, medical device regulations for AI diagnostics, and established medical ethics principles. Success requires ongoing collaboration between AI researchers, medical professionals, ethicists, and regulatory bodies to maintain relevant and effective auditing procedures aligned with evolving medical AI standards.

Acknowledgements. Supported by AUDITIA-X project PID2023-150566OB-I00 (MCIN/AEI/10.13039/501100011033), SHOW-X project (Comunidad de Madrid/ UCM), and ARTIFACTS project PID2022-141045OB-C4X (MCIN/AEI/10.13039/ 501100011033/FEDER).

References

1. Mökander, J.: Auditing of AI: legal, ethical and technical approaches. Digit. Soc. **2**(3), 49 (2023)
2. Brown, S., Davidovic, J., Hasan, A.: The algorithm audit: Scoring the algorithms that score us. Big Data Soc. **8**(1), 2053951720983865 (2021)
3. Berghout, E., Fijneman, R., Hendriks, L., de Boer, M., Butijn, B.J.: Advanced Digital Auditing: Theory and Practice of Auditing Complex Information Systems and Technologies. Springer (2023). https://doi.org/10.1007/978-3-031-11089-4
4. Mittelstadt, B.D.: Auditing for transparency in content personalization systems. Int. J. Commun. **10**, 4991–5002 (2016)
5. Zhang, C.A., Cho, S., Vasarhelyi, M.: Explainable artificial intelligence (XAI) in auditing. Int. J. Account. Inf. Syst. **46**, 100572 (2022)
6. Bussmann, N., Giudici, P., Marinelli, D., Papenbrock, J.: Explainable AI in fintech risk management. Front. Artif. Intell. **3**, 26 (2020)
7. Lundberg, S.M., Lee, S.: A unified approach to interpreting model predictions. arXiv preprint arXiv:1705.07874 (2017)
8. Khanna, S., Srivastava, S.: Ai governance in healthcare: explainability standards, safety protocols, and human-AI interactions dynamics in contemporary medical ai systems. Empirical Quests Manage. Essences **1**(1), 130–143 (2021)
9. Khan, M.M., Vice, J.: Toward accountable and explainable AI part 1: theory and examples. IEEE Access **10**, 99686–99701 (2022)
10. Cortina Orts, A.: ¿Ética o ideología de la IA? : el eclipse de la razón comunicativa en una sociedad tecnologizada. Paidós, Barcelona (2024)
11. Saleiro, P., et al.: Aequitas: a bias and fairness audit toolkit (2019). arXiv:1811.05577
12. Ribeiro, M.T., Singh, S., Guestrin, C.: why should i trust you? explaining the predictions of any classifier. In: Proceedings of the 22nd ACM SIGKDD Conference, pp. 1135–1144 (2016)

13. Dai, J., Upadhyay, S., Aivodji, U., Bach, S.H., Lakkaraju, H.: Fairness via explanation quality: evaluating disparities in the quality of post hoc explanations. In: Proceedings of the 2022 AAAI/ACM Conference on AI, Ethics, and Society, pp. 203–214 (2022)
14. Zhou, J., Gandomi, A.H., Chen, F., Holzinger, A.: Evaluating the quality of machine learning explanations: a survey on methods and metrics. Electronics **10**(5), 593 (2021)
15. Samek, W., Binder, A., Montavon, G., Lapuschkin, S., Müller, K.R.: Evaluating the visualization of what a deep neural network has learned. IEEE Trans. Neural Netw. Learn. Syst. **28**(11), 2660–2673 (2016)
16. Riccardo Guidotti. Counterfactual explanations and how to find them: literature review and benchmarking. Data Min. Knowl. Discov. **38**, 1–55 (2022). https://doi. org/10.1007/s10618-022-00831-6
17. Caro-Martínez, M., Recio-García, J.A., Díaz-Agudo, B., Darias, J.M., et al.: ISEE: a case-based reasoning platform for the design of explanation experiences. Knowl. Based Syst. **302**, 112305 (2024)
18. Silva, A., Schrum, M., Hedlund-Botti, E., Gopalan, N., Gombolay, M.: Explainable artificial intelligence: Evaluating the objective and subjective impacts of XAI on human-agent interaction. Int. J. Hum. Comput. Interact. **39**(7), 1390–1404 (2023)
19. Ritoré, Á., Oprescu, A.M., Bronchalo, A.E., de la Hoz, M.Á.A.: COVID data for shared learning (CDSL): a comprehensive, multimodal COVID-19 dataset from HM hospitales (2024)
20. Wu, J.Ty., de la Hoz, M.Á.A., Kuo, P.C., et al.: Developing and validating multimodal models for mortality prediction in COVID-19 patients: a multi-center retrospective study. J. Digit. Imaging, **35**(6), 1514–1529 (2022). https://doi.org/10. 1007/s10278-022-00674-z
21. Ratti, E., Morrison, M., Jakab, I.: Ethical and social considerations of applying artificial intelligence in healthcare—a two-pronged scoping review. BMC Med. Ethics **26**(68) (2025). https://doi.org/10.1186/s12910-025-01198-1

Explainable Multi-fault Predictive Maintenance Through Analysis of Wrong Predictions

Aurora Esteban[(✉)][iD], Aurora Ramírez[iD], Carlos García-Martínez[iD],
and Amelia Zafra[iD]

Department of Computer Science and Artificial Intelligence. Andalusian Research
Institute in Data Science and Computational Intelligence (DaSCI),
University of Cordoba, Cordoba, Spain
{aestebant,aramirez,cgarcia,azafra}@uco.es

Abstract. Predictive Maintenance (PdM) leverages data-driven models to anticipate equipment failures and optimize industrial operations. Explainability becomes a critical requirement for experts, as it helps identify the main issues leading to failures. This work proposes an explainable artificial intelligence (XAI) method suitable for online multi-label learning (MLL) in PdM. The approach is model-agnostic, although we base it on a recent and high-performing method by extending the MLHAT model to the PdM domain. To enhance user acceptance, a clustering method is incorporated to group misclassifications within each combined label, allowing the identification of representative instances to support the explanation of the model's behavior at those moments. The incremental version of SAGE provides the method for online model inspection, evaluating how feature contributions vary around selected predictions. The proposal is tested on one public, multi-label PdM problem. Our experiment shows that the approach is able to extract representative insights into failure patterns in an online setting, identify the most challenging fault combinations to detect, and support understanding of the main factors contributing to prediction errors.

Keywords: Explainable Artificial Intelligence · Predictive
Maintenance · Multi-label Learning

1 Introduction

Predictive Maintenance (PdM) leverages data-driven models to anticipate equipment failures, enabling timely maintenance that reduces downtime and operational costs [5]. Traditional batch learning methods, which rely on periodic retraining, are unsuitable for real-time data streams. To address this, adaptive online learning models have recently emerged for fault detection and diagnosis [13]. Despite these advances, most online PdM models still adopt a single-label classification setup, predicting only one fault at a time. This simplification does

L. Martínez et al. (Eds.): IDEAL 2025, LNCS 16239, pp. 281–293, 2026.
https://doi.org/10.1007/978-3-032-10489-2_24

not reflect the complexity of industrial systems where multiple, simultaneous faults frequently occur across interconnected components. Multi-Label Learning (MLL) offers a more suitable framework for modeling such scenarios, as it allows the prediction of concurrent faults. Nevertheless, applying MLL in online settings introduces specific challenges, including label co-occurrence, severe class imbalance, and concept drift. Recent approaches such as MLHAT [4], MLSVM [14] or MLCML [21], aim to improve prediction accuracy and responsiveness to system changes in generic MLL contexts.

The demand for explainable models in PdM is growing rapidly, as transparent and interpretable predictions are essential for building trust and facilitating adoption in industrial environments [9]. Operators and engineers need to understand why a failure is predicted, not just that it will happen, to take informed maintenance actions. eXplainable Artificial Intelligence (XAI) helps bridge the gap between complex algorithmic decisions and practical operational knowledge [3]. Understanding the causes of incorrect predictions is equally important, as it reveals whether errors are due to data characteristics or model limitations. However, few XAI methods are designed to explain such errors, and existing approaches are often limited to specific domains, regression tasks, or individual instances. Achieving explainability is even more challenging in online learning settings, where models must adapt to continuously evolving data without access to full historical information or retraining [7,17].

Our proposal introduces an explainable online MLL approach for PdM of multi-component systems based on Multi-Label Hoeffding Adaptive Tree (MLHAT), which is trained incrementally over a data stream of simultaneous failures. After the initial training, we perform a clustering of misclassified MLL instances, allowing us to identify temporal weaknesses in the model's predictions. Finally, to provide a detailed understanding of the model's behavior, we apply incremental Shapley Additive Global importancE (iSAGE) [17] adapted to the MLL setting, offering real-time, interpretable insights into the contribution of each feature to the prediction. The experimental study on one open dataset shows how our approach enables us to focus on and explain prototypical wrong predictions, in order to understand the most challenging fault combinations for the model.

The rest of the paper is structured as follows. Section 2 summarizes related work. Section 3 describes our proposed explainable online MLL model for PdM. The experimental setup is specified in Sect. 4, followed by the analysis of results in Sect. 5. Conclusions and future work are outlined in Sect. 6.

2   Related Work

As PdM systems become more complex and automated, the need for explainable methods increases, especially in high-risk industrial environments [3]. Most PdM studies apply standard XAI methods: global methods to explain the model behavior and local methods, such as SHAP or LIME for individual predictions [10,20]. However, these studies often simplify the problem by using binary classification, ignoring the multi-label nature of simultaneous fault combinations.

MLL problems introduce additional complexity to the XAI field, due to label co-occurrence, class imbalance, and the need for explanations that consider dependencies among fault types. Thus, the number of XAI approaches for MLL PdM remains limited [9,11,18]. Most of these approaches are designed for offline settings and are not readily applicable to online environments.

On the other hand, several recent works have integrated explainability into online PdM systems, although not using a MLL approach. The main XAI methods are incremental Permutation Feature Importance (iPFI) [7] and iSAGE [17], both model-agnostic and designed for incremental operation in streaming environments. iPFI assesses feature importance by measuring performance drops when feature values are permuted, while iSAGE extends SHapley Additive exPlanations (SHAP) to provide incremental global explanations. These methods provide a solid foundation for understanding models in real time, being particularly effective when historical data cannot be stored.

A reduced number of publications are specialized in explaining classification errors, with examples that analyze misidentified regions for image classification [2], excessive testing errors in sales applications [15], frequent features in erroneous predictions [19], and the generation of counterexamples [12].

3   Proposed Approach

This section provides an overview of the proposed approach based on four distinct phases that are detailed in the following sections. The process starts with incremental learning using the MLHAT algorithm to predict multiple faults over time (Sect. 3.1). During learning, model performance is evaluated prequentially, and misclassified instances are identified (Sect. 3.1). Due to the large data volume and concept drift, misclassified instances are clustered to simplify analysis and reduce cognitive load (Sect. 3.2). Finally, model behavior is explained by selecting representative prototypes from clusters and applying iSAGE to extract feature importance before and after faults (Sect. 3.3).

3.1   Online Learning

Online MLL is a machine learning paradigm where instances, arriving sequentially from a potentially infinite data stream, must be classified into multiple simultaneous labels in real-time. Let $S = \{(X_1, Y_1), ..., (X_n, Y_n), ...\}$ represent a stream of instances, where each $X_i \in R^d$ is a feature vector and each $Y_i \in \{0, 1\}^L$ is a binary vector across L possible labels. The objective is to learn a predictive model $h : R^d \rightarrow \{0, 1\}^L$ that can output the relevant subset of labels for each new instance X_i as it arrives, without accessing the full dataset.

Online MLL poses unique challenges, including concept drift, label imbalance and co-occurrence, and strict memory constraints that prevent revisiting past data. These challenges align closely with the demands of PdM, where systems must detect multiple, evolving faults in real time. To address this, we adopt MLHAT [4], an incremental decision tree tailored for multi-label streams. MLHAT enhances the Hoeffding Adaptive Tree by introducing a multi-label splitting criterion, adaptive leaf classifiers, and localized drift detection to maintain performance in dynamic environments. For a detailed description of MLHAT's inner mechanisms, readers are referred to [4].

MLHAT processes data incrementally as it arrives, making it ideal for industrial settings with limited resources. Using prequential evaluation, we identify misclassified instances through two strategies: strict and flexible labeling. The former considers an instance to be correctly classified if all positive labels are correctly identified and none of the other labels are activated. The latter relaxes the second condition, assuming that misclassification only occurs when one or more positive labels are not activated.

3.2 Clustering of Misclassified Instances

We perform a clustering analysis to detect areas of the data distribution where the classifier tends to fail. Depending on the temporal distribution of misclassified instances, we can also better understand at what points in the data stream the classifier experiences a performance degradation and for how long. A second goal is to reduce the number of instances for the explanation phase.

Based on the two objectives stated, we adopt Affinity Propagation (AP) [6]. AP has two characteristics that make it appropriate for our objectives: 1) the number of clusters does not need to be specified as parameter and, 2) it locates, for each cluster, the most representative instance, the so-called *exemplar*. AP is based on a "message passing" strategy between the instances to choose which of them is the best *exemplar* for other instances, and to determine which *exemplar* each of the remaining instances should be associated with.

AP has quadratic complexity with respect to the number of instances. However, the subset of misclassified instances will be small compared to the total number of instances, making the algorithm suitable for our purpose. This phase results in the separation of misclassified instances from each labelset, whether according to strict or flexible failure modes, into k clusters, which does not have to be the same for each labelset, nor for both failure modes of the same labelset.

3.3 Explaining Model Behavior

Understanding prediction failures in streaming contexts requires examining the model's behavior around the error, not just at a single point. Unlike traditional XAI methods, which analyze static models, we need a method able to analyze the model as it evolves, given a particular time interval for the explanation. iSAGE fulfills this purpose by extracting feature importance scores in the decision model for a specific instance (local approach), but adapting those importance values as the model learns (global approach). iSAGE has demonstrated better performance than iPFI because it inherits the properties of SHAP. Moreover, iSAGE values are more intuitive than in iPFI since their sum equals to the improvement in loss experienced by the model relative to using the mean prediction.

iSAGE incrementally estimates feature importance by approximating marginal contributions to prediction loss through sampling masked subsets, avoiding the computational cost of evaluating all possible feature combinations. To incorporate the time dependency, a reservoir sampling strategy is used to maintain a small representative set of past instances. We have extended it to MLL by setting *hamming loss* as the loss metric, which allows us to maintain a natively multi-label registry that takes into account the correlation between labels to obtain the importance of each feature.

In this way, iSAGE values reflect the evolving contribution of each feature over time. A rising SAGE value indicates a feature is becoming more influentialâĂŞpossibly due to a concept drift or change in operating conditions. Conversely, a declining score means redundancy between features or shifting relevance. This way, we can study the time interval centered in each prototype identified by the previous clustering phase to understand the cause of the wrong prediction.

4 Experimental Setup

First, the dataset characteristics are described in Sect. 4.1. Section 4.2 explains the parameter settings for each algorithm involved. The performance metrics used to evaluate the results are specified in Sect. 4.3.

4.1 Dataset

The AI4I dataset [16] captures realistic operational conditions and component interdependencies. It is a synthetic dataset designed to simulate industrial motor behavior, containing 10,000 working cycles described by six features: process type (ty), air temperature (at), process temperature (pt), rotational speed (rs), torque (to), and tool wear (tw). There are five possible failure modes: Tool Wear Failure (TWF), Heat Dissipation Failure (HDF), Power Failure (PWF), Overstrain Failure (OSF), and Random Failure (RNF). The dataset is highly imbalanced, with only about 3.39% of cycles exhibiting at least one failure.

4.2 Parameter Configuration

MLHAT has been optimized via Bayesian hyperparameter search [1] on the first half of the dataset stream, exploring growth, drift, and leaf classifier parameters. After the optimization, the most influential parameters were: *grace_period* $= 140$, *delta* $= 1.195e^{-4}$, *cardinality_threshold* $= 770$ and *entropy_threshold* $= 0.65$ for controlling the tree growth; $k = 2$ and *window_size* $= 140$ for k Nearest Neighbors (kNN) and *n_models* $= 10$ for the bagging of logistic regressions for the leaf classifiers; and for the concept drift adaptation are *drift_window_threshold* $= 270$ and *switch_significance* $= 0.35$.

The AP clustering algorithm is sensitive to: *damping*, a value between 0 and 1 that regulates message passing; *max_iter*, which sets the maximum number of iterations; and *convergence_iter*, which determines the maximum number of iterations before readjusting the number of groups. In our case, no convergence issues were observed, so AP was applied using scikit-learn's defaults: *damping* = 0.5, *max_iter* = 200, *convergence_iter* = 15, and Euclidean distance. Before applying clustering, the data features are scaled to $[0, 1]$ with *MinMaxScaler*.

The implementation of the iSAGE algorithm is publicly available in *iXAI* library[1]. The configuration employed in this work follows the options suggested by the authors to maximize reliability: *smoothing_alpha* = 0.001, *n_inner_samples* = 5, a geometric reservoir storage with *size* = 100, and the marginal *imputer*.

4.3 Performance Metrics

To evaluate the performance of the online learning model, we use a prequential evaluation scheme [8], with forgetting factor of $\alpha = 0.995$, and four well-known metrics from the MLL domain: *Subset Accuracy* (SAcc), *Hamming Loss* (HL), *Micro-averaged F1 score* (Micro-F1) and *Macro-averaged F1 score* (Macro-F1). These metrics provide insight into the evolution of performance at both the instance and label levels through the data stream.

For the cluster analysis, the following metrics are considered: number of clusters, *Silhouette Coefficient* (SC), *Calinski-Harabasz Index* (CHI), and *Davies-Bouldin Index* (DBI). These metrics analyze the dispersion of the created groups, not requiring the existence of a reference cluster to compare with.

The computation of the iSAGE explainability method is carried out incrementally and in parallel with the prequential evaluation, allowing it to capture the evolving contribution of each feature over time as the model adapts to potential concept drifts. To account for label dependencies, the feature importance is estimated using a joint loss across all labels based on the running HL.

5 Experimental Results

This section presents the results in three parts: the analysis of model performance (Sect. 5.1), the analysis of the clustering procedure (Sect. 5.2), and the generation of explanations (Sect. 5.3).

5.1 Performance Analysis

The AI4I dataset is highly imbalanced and contains sparse failure occurrences. At the end of the prequential evaluation, MLHAT achieves high subset accuracy (96.04%), indicating precise predictions at the instance level. However, the Micro-F1 (43.1%) and Macro-F1 (29.49%) scores reflect the challenge of correctly predicting the less frequent failure labels—an expected issue in imbalanced MLL settings. Moreover, performance varies across failure types. Figure 1 shows

---

[1] https://github.com/mmschlk/iXAI.

a detailed breakdown of how the count of misclassified instances evolves along the stream, in this case counting separately the False Positives (FPs) and False Negatives (FNs) that accumulate for each fault present in the problem.

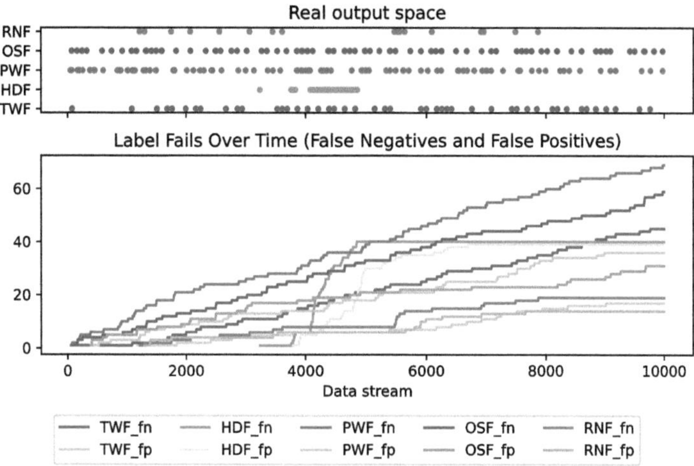

Fig. 1. Performance of MLHAT over AI4I dataset

The AI4I dataset is characterized by a high degree of failure dispersion, which causes MLHAT to often predict that there is no failure (i.e., that all labels are false). This maximizes the metrics measuring instance-level hits, since most instances are indeed no failures. However, at the label level, this causes many failures to go uncovered, as evidenced by the increase in FNs for all labels. This issue is exacerbated by sudden or concentrated time-dependent malfunctions, such as RNR, which represents a random failure that is difficult to predict, and HDF, which is related to *at* and *pt* peaks.

Next we focus on studying the misclassified instances and their distribution across the *labelsets*. The *labelsets* represent all the combinations of simultaneous faults that occur in each problem, and from the MLL point of view they help us to understand the most closely related faults. Table 1 shows the number of instances per *labeset*, followed by two columns expressing the percentages of misclassified instances according to the failure approach (strict or flexible). For both failure approaches, the lowest percentage corresponds to *labelset* 0, which predominates in the dataset as it is the absence of failures. For the remaining *labelsets*, the percentages vary between 34% (*labelset* 6) and 100%. Typically, high percentages correspond to *labelsets* with very few instances (5, 7, 10, and 11). Specifically, for the *labelsets* containing the random failure (RNR), 5 and 7, it is impossible to anticipate and therefore 100% misclassifications are obtained. The other ones correspond to the TWF fault in combination with other faults. In the same line, *labelset* 3 (TWF alone) is also difficult with 97% of failures. PWF and OSF are relatively easy to predict when they appears alone or together (*labelsets* 1, 4 and 2), and more difficult in combination with other labels, indicating a correlation between these two faults as intuited in the previous analysis. Finally, HDF (*labelsets* 6, 8 and 9) is relatively easy to predict probably because of its temporal clustering. The choice of failure mode does not yield any significant differences for 9 of the 12 label combinations. For those that undergo some changes, the flexible approach slightly reduces the percentages, as expected. However, only in *labelset* 0 the model makes no errors under the flexible approach, due to the absence of active labels.

Table 1. Percentage of misclassified instances per *labeset* in AI4I dataset, considering strict and flexible failure approaches.

Labelset	TWF-HDF-PWF-OSF-RNF	Instances	Strict	Flexible
0	0-0-0-0-0	9651	1.24%	0.00%
1	0-0-1-0-0	80	73.75%	73.75%
2	0-0-1-1-0	11	72.73%	72.73%
3	1-0-0-0-0	42	97.62%	97.62%
4	0-0-0-1-0	78	70.51%	65.38%
5	0-0-0-0-1	18	100.00%	100.00%
6	0-1-0-0-0	106	34.91%	33.96%
7	1-0-0-0-1	1	100.00%	100.00%
8	0-1-1-0-0	3	66.67%	66.67%
9	0-1-0-1-0	6	83.33%	83.33%
10	1-0-0-1-0	2	100.00%	100.00%
11	1-0-1-1-0	1	100.00%	100.00%

5.2   Cluster Analysis

Once the *labelsets* relevant to the problem have been identified, the clustering procedure is carried out to know if there is a temporal evolution and adaptation by MLHAT in the causes of these malfunctions. The results are detailed in Table 2. For each failure mode (M) and *labelset*, the table shows the number of clusters returned by the AP algorithm, and the three internal validation metrics (SC, CHI and DBI). Only the *labelsets* with more than one misclassified instance are shown since this is the minimum necessary to apply clustering.

We observe that for most *labelsets* AP returns between 2 and 5 clusters. A higher number of clusters is obtained for *labelset* 0 and 4, although the reasons are different. Despite the low failure rate of *labelset* 0, the absolute number of misclassified instances (120) is twice as high as the next-highest failure *labelset*, as this is the mayority *labelset* (0-0-0-0-0) in this problem with scarcity of faults. Each cluster contains between 5 and 30 instances. This diversity also influences the SC value, which is lower than the values obtained for other *labelsets*. For *labelset* 4, the AP algorithm also returned 7 clusters, but had to group 55 instances. After inspecting the clusters, we observed that 3 of them contain only one or two samples. In fact, clusters comprised of only one instance also appear in other *labelsets*, such as 2, 8 and 10. In the last two cases, both clusters have only one instance, so the metrics cannot assess the quality of the clusters. In general, the best clustering metrics are obtained for 3-cluster partitions, such as those obtained for the *labelsets* 3 and 5. Furthermore, the fact that AP usually groups a small number of instances into more than 2 clusters seems to suggest that the model's misclassifications could respond to different causes. Comparing the 2 failure approaches for those *labelsets* influenced by the mode (0, 4 and 6), not great differences are perceived. For *labelset* 4, the number of clusters remains stable, although some instances are grouped differently, and consequently, the number of instances also varies. For *labelset* 6, the number of clusters decreases to 3. Inspecting the clusters, we observe that 2 clusters obtained with the strict approach merged when adopting the flexible approach.

Table 2. Results of the clustering process for AI4I dataset.

Label set	Strict				Flexible			
	Clusters	SC (↑)	CHI (↑)	DBI (↓)	Clusters	SC (↑)	CHI (↑)	DBI (↓)
0	7	0.352	115.069	1.124	-	-	-	-
1	5	0.555	62.918	0.665	5	0.555	62.918	0.665
2	4	0.434	63.511	0.276	4	0.434	63.511	0.276
3	3	0.710	102.998	0.418	3	0.710	102.998	0.418
4	7	0.338	91.171	0.646	7	0.341	90.049	0.627
5	3	0.664	32.278	0.464	3	0.664	32.278	0.464
6	4	0.572	124.488	0.532	3	0.753	117.192	0.341
8	2	-	-	-	2	-	-	-
9	2	0.518	6.475	0.520	2	0.518	6.475	0.520
10	2	-	-	-	2	-	-	-

5.3   Explainability Analysis

The analysis in the previous section enabled the identification of prototypes that summarize distinct input data distributions associated with the misclassification made by MLHAT. Due to space limitations, the current section examines an illustrative case. iSAGE is employed to assess the contribution of individual features at the time of the wrong prediction, providing insight into how feature importance evolves across the stream.

We focus on *labelset* 4, which corresponds to the occurrence of the OSF failure. This *labelset* was selected due to its relatively higher number of instances

Table 3. iSAGE values for the prototypes of the clusters identified for *labelset* 4 in AI4I dataset

Instance	rs	pt	ty	to	at	tw
926	0.0387	0.0339	0.0319	0.0310	0.0271	0.0318
3356	0.1250	0.1204	0.1244	0.1282	0.1070	0.1236
5399	0.1984	0.1372	0.1510	0.1496	0.1456	0.1579
5488	0.1945	0.1349	0.1498	0.1537	0.1503	0.1609
7591	0.1646	0.1554	0.1646	0.1644	0.1644	0.1741
9414	0.1636	0.1506	0.1698	0.1713	0.1694	0.1699
9493	0.1628	0.1515	0.1711	0.1708	0.1707	0.1679

and clusters (as shown in Table 2), suggesting greater variability in the conditions leading MLHAT to misclassify this failure. Table 3 lists the prototypes identified as cluster centroids—7 in total—spanning from the 926th to the 9493rd instance in the stream. Alongside these, the table presents the corresponding iSAGE values for each input feature. Notably, the early instances in the stream generally exhibit lower iSAGE values across features. This is because, at the beginning of the stream, the model is still in its initial learning phase and has not yet adapted sufficiently to the data distribution. As the model is incrementally updated, its structure stabilizes and the importance of relevant features becomes more pronounced, particularly around prototype instances that mark significant changes. From instance 3356 onward, the importance of all features increases, but some attributes begin to stand out. The variable rs consistently shows the highest iSAGE values. This suggests that fluctuations in rotational speed are a key indicator for OSF failure. In contrast, features related to temperature (pt, at) tend to have lower relative importance, though they gradually catch up in later instances (9414 and 9493). This indicates their contribution may emerge later as the model refines its understanding of the failure mechanism. Other attributes (ty, to) show a consistent increase in importance, which suggests these variables may be associated with longer-term degradation patterns or cumulative effects tied to the OSF failure.

Figure 2 analyzes one of the prototypes in detailâĂŞthe instance 7591. To gain insight into this misclassification, the analysis considers a temporal window of 15 instances before and after the prototype. The top panel displays the evolution of the HL, while the bottom panel shows the corresponding evolution of the iSAGE values for each input feature. A noticeable peak in HL occurs precisely at the prototype instance, confirming that this is a point of classification error. On the other hand, the HL remains steady before and after the peak, suggesting that MLHAT quickly incorporates feedback and avoids repeating the same mistake. The lower panel reveals dynamic shifts in feature importance around the misclassification event. In particular, the importance of *torque* and *rotational speed* increases significantly following the failure, while *type*, which had previously been more influential, declines in relevance. This behavior suggests that

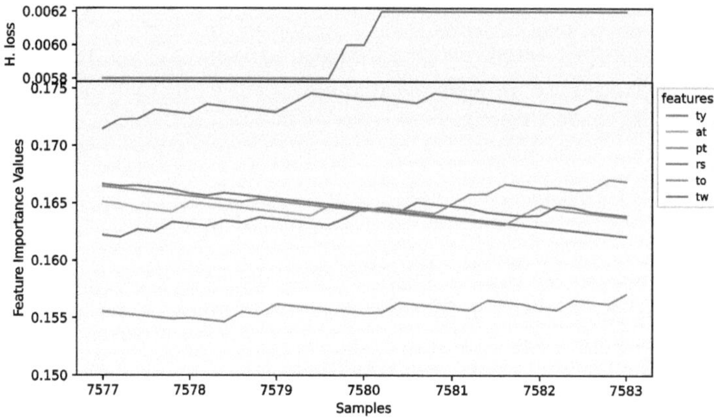

Fig. 2. iSAGE evolution for prototype 7591±15 instances in AI4I

the model may have initially underestimated the importance of these rotational attributes in predicting the OSF failure, possibly leading to the misclassification. The subsequent rise in their iSAGE values implies that these features were more informative than the model had initially recognized.

6   Conclusion and Future Work

Online learning, combined with a MLL approach, represents a suitable way to address PdM problems. However, the resulting models can be complex and highly variable over time, making it difficult to understand the operational failures experienced by the systems under analysis. We present an approach to apply clustering and incremental explainability techniques to inspect the behavior of MLHAT, an online MLL model, in the context of PdM.

Although MLHAT shows adequate performance, some label combinations are more difficult to predict. With our proposal, classification errors are isolated for each label combination. The efficient use of the AP clustering algorithm allows us to locate representative time points where it is most interesting to inspect the model's behavior. Measuring the evolution of feature importance allows us to identify trends that explain changes in the data stream. As the results show, MLHAT is capable of recovering performance by adapting the importance of the features most closely related to combined failure modes.

This work can be extended by applying the approach to more PdM problems to confirm the generalizability of the results. Regarding explainability, we could add contextual information about the circumstances of the classification errors, using it to prevent the model from making similar errors in the future. For example, trend changes in the values of some features could be monitored, focusing on those identified as more relevant to specific *labelsets* by iSAGE.

Acknowledgments. This research was supported in part by grant PID2023-148396NB-I00 funded by MICIN/AEI/10.13039/501100011033 and by the ProyExcel-0069 project of University, Research and Innovation Department of the Andalusian Board and the European Regional Development Fund.

References

1. Akiba, T., Sano, S., Yanase, T., Ohta, T., Koyama, M.: Optuna: a next-generation hyperparameter optimization framework. In: ACM SIGKDD International Conference on Knowledge Discovery and Data Mining, pp. 2623–2631 (2019)
2. Alirezaie, M., Längkvist, M., Sioutis, M., Loutfi, A.: A symbolic approach for explaining errors in image classification tasks. In: IJCAI Workshop on Learning and Reasoning (2018)
3. Cummins, L., et al.: Explainable predictive maintenance: a survey of current methods, challenges and opportunities. IEEE Access **12**, 57574–57602 (2024)
4. Esteban, A., Cano, A., Zafra, A., Ventura, S.: Hoeffding adaptive trees for multi-label classification on data streams. Knowl. Based Syst. **304**, 1–21 (10 2024)
5. Esteban, A., Zafra, A., Ventura, S.: Data mining in predictive maintenance systems: a taxonomy and systematic review. Wiley Interdiscip. Rev. Data Min. Knowl. Discov. **12**, 1–45 (2022)
6. Frey, B.J., Dueck, D.: Clustering by passing messages between data points. Science **315**(5814), 972–976 (2007)
7. Fumagalli, F., Muschalik, M., Hüllermeier, E., Hammer, B.: Incremental permutation feature importance (IPFI): towards online explanations on data streams. Mach. Learn. **112**(12), 4863–4903 (2023)
8. Gama, J., Rodrigues, P.P., Sebastião, R.: Evaluating algorithms that learn from data streams. In: 2009 ACM Symposium on Applied Computing, pp. 1496–1500 (2009)
9. Gawde, S., Patil, S., Kumar, S., Kamat, P., Kotecha, K.: An explainable predictive maintenance strategy for multi-fault diagnosis of rotating machines using multi-sensor data fusion. Decis. Anal. J. **10**, 100425 (2024)
10. Jiang, M., Chen, H., Yang, C.: A metro train air conditioning system fault diagnosis method based on explainable artificial intelligence: considering interpretability and generalization. Int. J. Refrig **174**, 47–59 (2025)
11. Kakogeorgiou, I., Karantzalos, K.: Evaluating explainable artificial intelligence methods for multi-label deep learning classification tasks in remote sensing. Int. J. Appl. Earth Obs. Geoinf. **103**, 102520 (2021)
12. Kim, B., Khanna, R., Koyejo, O.: Examples are not enough, learn to criticize! criticism for interpretability. In: 30th International Conference on Neural Information Processing Systems, pp. 2288–2296 (2016)
13. Le-Nguyen, M.H., Turgis, F., Fayemi, P.E., Bifet, A.: Real-time learning for real-time data: online machine learning for predictive maintenance of railway systems. Transport. Res. Procedia **72**, 171–178 (2023)
14. Li, S., Cao, H., Yang, Y.: Data-driven simultaneous fault diagnosis for solid oxide fuel cell system using multi-label pattern identification. J. Power Sources **378**, 646–659 (2018)
15. Lucic, A., Haned, H., de Rijke, M.: Why does my model fail? Contrastive local explanations for retail forecasting. In: Conference on Fairness, Accountability, and Transparency, pp. 90–98 (2020)

16. Matzka, S.: Explainable artificial intelligence for predictive maintenance applications. In: 3rd the International Conference on Artificial Intelligence for Industries, pp. 69–74 (2020)
17. Muschalik, M., Fumagalli, F., Hammer, B., Hüllermeier, E.: iSAGE: an incremental version of sage for online explanation on data streams. In: Koutra, D., Plant, C., Gomez Rodriguez, M., Baralis, E., Bonchi, F. (eds.) Machine Learning and Knowledge Discovery in Databases, vol. 14171 LNAI, pp. 428–445. Springer Science and Business Media Deutschland GmbH (2023). https://doi.org/10.1007/978-3-031-43418-1_26
18. Mylonas, N., Mollas, I., Bassiliades, N., Tsoumakas, G.: Local multi-label explanations for random forest. In: Joint European Conference on Machine Learning and Knowledge Discovery in Databases, pp. 369–384. Springer (2022). https://doi.org/10.1007/978-3-031-23618-1_25
19. Rizzi, W., Di Francescomarino, C., Maggi, F.M.: Explainability in predictive process monitoring: When understanding helps improving. In: International Conference on Business Process Management, pp. 141–158 (2020)
20. Tang, H., Wang, H., Li, C.: Explainable fault diagnosis method based on statistical features for gearboxes. Eng. Appl. Artif. Intell. **148**, 110503 (2025)
21. Yu, C., Ning, Y., Qin, Y., Su, W., Zhao, X.: Multi-label fault diagnosis of rolling bearing based on meta-learning. Neural Comput. Appl. **33**, 5393–5407 (2021)

Uncovering and Correcting XAI's Misconceptions Logic to the Rescue

Joao Marques-Silva(✉) [iD]

ICREA and University of Lleida, Lleida, Spain
`jpms@icrea.cat`

Abstract. The lack of proper formalization in non-symbolic explainable artificial intelligence (XAI) has resulted in a growing number of provable flaws in widely adopted methods of explainability. This paper reviews some of the most visible misconceptions of non-symbolic XAI, and shows how logic-based XAI has been applied to uncover and correct those misconceptions. The paper also summarizes ongoing research to develop a sound, scalable and human-understandable framework for rigorously defined XAI.

Keywords: Explainable AI · Symbolic AI · Logic

1 Introduction

The remarkable successes of machine learning (ML) are not without shortcomings. Among these, and due to the ever increasing complexity of ML models, human decision makers are most often unable to understand the rationale for the predictions made. Explainable Artificial Intelligence (XAI) [25, 26] aims to help human decision makers to fathom the predictions of ML models.

The explanation of ML models can be broadly organized into explanations by feature selection and those by feature attribution [62]. Feature selection identifies subsets of features (and their assignments) which are then used to produce rules that answer a *"Why (the prediction)?"* question. In contrast, feature attribution aims to assign relative importance to the features regarding some prediction. It is well-known that the best-known and most widely used methods of XAI are based on non-symbolic methods of AI [10, 24, 48, 64, 65], or focused on interpretability [27, 66, 67].[1] Unfortunately, non-symbolic methods of AI offers no formal guarantees of rigor when explaining ML models. As a consequence, a growing number of critical flaws has been identified with these methods. This paper provides a brief overview of the most significant misconceptions uncovered in non-symbolic XAI, and also glimpses over the ongoing work on rigorous explainability that offers solutions to overcome those misconceptions. The paper aggregates and extends results presented in earlier work by the author [41, 42, 50, 51, 57].

---

[1] These will be broadly defined as methods without formal guarantees of rigor. Among these, we analyze in detail the ones with unambiguous definitions.

L. Martínez et al. (Eds.): IDEAL 2025, LNCS 16239, pp. 294–306, 2026.
https://doi.org/10.1007/978-3-032-10489-2_25

The paper is organized as follows. Section 2 briefly introduces the notation and definitions used throughout the paper. Section 3 revisits the many issues with the claims about interpretability and interpretable models. Section 4 shows that non-symbolic methods for explaining with rules will most often provide incorrect rules. Section 5 shows that the application of Shapley values in XAI is flawed. Essentially, tools like SHAP attempt to approximate values that will mislead human decision makers. After presenting several misconceptions of non-symbolic XAI, in Sects. 6 and 7 we briefly glimpse over the emerging field of logic-based (and so symbolic) XAI. This emerging field not only offers strong guarantees of rigor, but also enables establishing important connections between different methods of XAI. The paper concludes in Sect. 8.

2   Preliminaries

The paper considers classification problems in ML. The extension to regression problems is straightforward but would complicate unnecessarily the notation given the goals of the paper.

An ML models is defined on a set of feature $\mathcal{F} = \{1, \ldots, m\}$. A domain $\mathbb{D}_i$ is associated with each feature $i \in \mathcal{F}$. Features can be categorical or ordinal. If ordinal, features can be real- or integer-valued. Feature space is the cartesian product (in order) of the domains of the features, i.e. $\mathbb{F} = \mathbb{D}_1 \times \cdots \times \mathbb{D}_m$. (For simplicity, throughout the paper, we assume all features to be discrete or that some sort of discretization has been adopted.)

A classifier implements a classification function κ that maps feature space to a set of classes $\mathcal{K} = \{c_1, \ldots, c_K\}$, i.e. $\kappa : \mathbb{F} \to \mathcal{K}$.

With each feature $i \in \mathcal{F}$ we associate a variable x_i that takes values from $\mathbb{D}_i$. Furthermore, points in feature space are represented by $\mathbf{v} = (v_1, \ldots, v_m) \in \mathbb{F}$, where each v_i is a constant from $\mathbb{D}_i$.

A logic rule is the form:

$$\text{IF cond THEN pred}$$

where cond denotes the condition for the rule to fire and predict pred. cond is most often a conjunction of literals, where a literal is represented as a tuple $(x_i, R_i, \mathsf{op}_i)$, where R_i is a subset of $\mathbb{D}_i$, and op_i is a relational operator, e.g. one of $\{\in, \notin, =, \neq, <, \leq, \geq, >\}$. A literal corresponds to a predicate $x_i \mathsf{op}_i R_i$, which can take values from $\{\bot, \top\}$.

A Running Example. Throughout the paper, we will consider the decision tree (DT) shown in Fig. 1, adapted from [32].

3   Most Interpretable Models Aren't

There is wide consensus that decision trees epitomize interpretable models.[2]

---

[2] Claims about the interpretability of DTs and related models can be traced back more than 2 decades. For example, quoting from [14, page 206]: *"On interpretability, trees rate an A+"*.

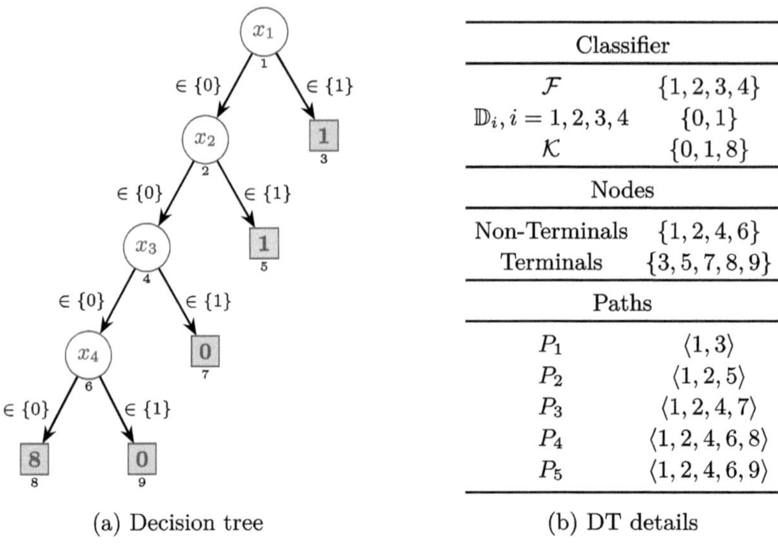

	Classifier	
$\mathcal{F}$	$\{1,2,3,4\}$	
$\mathbb{D}_i, i = 1,2,3,4$	$\{0,1\}$	
$\mathcal{K}$	$\{0,1,8\}$	
	Nodes	
Non-Terminals	$\{1,2,4,6\}$	
Terminals	$\{3,5,7,8,9\}$	
	Paths	
P_1	$\langle 1,3 \rangle$	
P_2	$\langle 1,2,5 \rangle$	
P_3	$\langle 1,2,4,7 \rangle$	
P_4	$\langle 1,2,4,6,8 \rangle$	
P_5	$\langle 1,2,4,6,9 \rangle$	

(a) Decision tree   (b) DT details

Fig. 1. Running example, adapted from [32].

On the one hand, it is simple to associate some sort of explanation with a decision tree; simply pick the literals in the path consistent with the input, and explain the prediction with a rule; this is referred to as a *path explanation*. On the other hand, it is rather intuitive to infer explanations by inspection, again by looking at path explanations. The appeal of interpretable models motivated calls for replacing complex models with interpretable models in high-risk applications [66,67], but also justified efforts towards the learning of optimal decision trees [1,12,22,27,74].

However, as pointed out in recent years, the concept of interpretability is subjective [47]. In addition, over the last few years, a number of works investigated explanations of DTs [4,5,30,41,42,57,60]. Furthermore, some works demonstrated that redundancy of path explanations is pervasive in many DTs [41,42], either used in earlier publications, but also produced by state of the art tree learners.

Example 1. For the DT shown in Fig. 1, let $\mathbf{v} = (0,1,0,0)$. Clearly, the prediction is 1. Invoking interpretability of DTs, one could explain the prediction with the following rule:

$$\text{IF } (x_1 = 0) \wedge (x_2 = 1) \text{ THEN } \kappa(x_1, x_2, x_3, x_4) = 1$$

However, it is plain that there is redundant information in the explanation, i.e. the fact that $x_1 = 0$ is irrelevant, and we could write the rule as follows:

$$\text{IF } (x_2 = 1) \text{ THEN } \kappa(x_1, x_2, x_3, x_4) = 1$$

Table 1. Results for Anchor [65].

Dataset	% Incorrect	% Redundant	% Correct
adult	80.5%	1.6%	17.9%
lending	3.0%	0.0%	97.0%
rcdv	99.4%	0.4%	0.2%
compas	84.4%	1.7%	13.9%
german	99.7%	0.2%	0.1%

For this example DT, there are more examples of path explanation redundancy. For example, for $\mathbf{v} = (0, 0, 0, 1)$, the prediction is 0. As a result, we can write,

$$\text{IF } (x_1 = 0) \wedge (x_2 = 0) \wedge (x_3 = 0) \wedge (x_4 = 1) \text{ THEN } \kappa(x_1, x_2, x_3, x_4) = 0$$

As before, there is redundant information in the explanation, because $x_3 = 0$ is unnecessary. Hence, we can write the simplified rule:

$$\text{IF } (x_1 = 0) \wedge (x_2 = 0) \wedge (x_4 = 1) \text{ THEN } \kappa(x_1, x_2, x_3, x_4) = 0$$

Although the analysis of Example 1 might hint otherwise, it would be difficult in general for a human decision maker to spot explanations that exhibit no redundancy. A number of polynomial-time algorithms to find such explanations is studied in [41, 42].

Experimental evidence demonstrated that path redundancy is most often significant [42], and can be arbitrarily large in the number of features. Furthermore, more recent work uncovered even more problematic difficulties with claiming interpretability in the case of decision lists [57], but also decision sets [71].

4   Feature Selection with Rules that Aren't

Most often, feature selection aims to infer a rule as the answer to a *"Why (the prediction)?"* question. A well-known example is Anchor [65].

Given an instance $(\mathbf{v}, c)$, Anchor identifies a subset $\mathcal{H}$ of $\mathcal{F}$ of the features from which it constructs the body of a rule that predicts the class c. Moreover, Anchor aims to infer a rule that predicts the expected prediction with high probability. In practice one would expect Anchor to produce rules similar to those shown in Example 1 for the DT running example.

Validating Anchor's Rules. The explanations produced by Anchor can be validated against the ML model using a rigorous model-based explainer (see Sect. 6), as follows [38]: The proposed explanation is assessed for sufficiency. This corresponds to deciding whether the literals in the explanation represent a logically valid rule. If that is not the case, the explanation is reported as incorrect. Otherwise, the rule is assessed for redundancy of some of its literals. It that is the case, the explanation is reported as redundant. Otherwise, it is reported as correct.

Practical Evidence. Earlier work showed that Anchor's rules are not, in most cases, logically valid rules given the ML model. Table 1 summarizes some of the results obtained in the recent past [33,38], obtained on well-known datasets. Boosted trees [17] were used as the model to explain. The approach outlined above was used to categorize the explanations for different instances.

As can be concluded, for most datasets, the vast majority of the rules computed by Anchor are not rules given the ML model. More recent work accounted for constraints on the inputs [75]. Although the quality of results improved, it still the case that Anchor's rules are anything but.

5   Feature Attribution with Misleading Values

One alternative to XAI by feature selection is feature attribution, as epitomized by the widely used tools LIME [64] and SHAP [48]. Among these two approaches, SHAP is based on the rigorous foundation of Shapley values [68]. Furthermore, SHAP is based on earlier work on applying Shapley values in ML [72,73], and has become ubiquitous in XAI by feature attribution [61–63], being massively popular at present. This section shows that it is actually simple to construct classifiers (but also regression models) where SHAP scores are misleading in terms of ranking feature importance [32,54].

Cooperative Games. A cooperative game [16] is a tuple $G = (N, v)$, where N is a set of players and v is a characteristic function, mapping subsets of N to the reals, $v : 2^N \to \mathbb{R}$. One important special case of cooperative games is often considered. For example, if v is monotonically increasing and maps to $\{0, 1\}$, then (N, v) is a *simple* game.

Shapley Values. For a cooperative game $G = (N, v)$, the Shapley values are defined as follows:

$$\mathsf{Sc}(i) = \sum_{S \subseteq N \setminus \{i\}} \frac{|S|!(|N| - |S| - 1)!}{|N|!} [v(S \cup \{i\}) - v(S)] \tag{1}$$

Shapley values respect a number of important properties, making them the *unique* values that respect those properties for the given game.

The set of players can represent different sets of entities. For example, Shapley values have been applied to cooperative games, voting power, measures of inconsistency and explainability, among many other domains. Recent work dissects uses of Shapley values [45,59].

SHAP Scores. SHAP scores are an instantiation of Shapley values to the case of ML, where the characteristic function is selected to be the expected value of the ML model given a target instance $(\mathbf{v}, c)$. Let $\mathsf{dom}(S) = \{\mathbf{x} \in \mathbb{F} \mid \wedge_{i \in S} x_i = v_i\}$, denote the set of points of $\mathbb{F}$ exhibiting the feature values dictated by $\mathbf{v}$ in the features contained in S. Then, the characteristic function becomes,

$$v_e(S) = {}^1\!/_{|\mathsf{dom}(S)|} \sum_{\mathbf{x} \in \mathsf{dom}(S)} [\kappa(\mathbf{x})] \tag{2}$$

Table 2. Exact SHAP scores for DT of Fig. 1 and instance $(1,1,1,1)$.

Classifier	Sc(1)	Sc(2)	Sc(3)	Sc(4)	Rank
DT in Fig. 1	0.000	0.000	−0.125	−0.125	$\langle 3:4, 1:2 \rangle$

(For real-valued features, either integration or domain discretization would be considered.)

Misleading Feature Importance. A number of recent works uncovered examples of classifiers for which the theoretical SHAP scores are misleading [32,54].

Example 2. Table 2 shows the SHAP scores computed with (1) with the v_e characteristic function. Although it is apparent that for the instance $((1,1,1,1),1)$, the prediction can only change if feature 1 and feature 2 change their value, the SHAP scores are 0, denoting **no** importance for the prediction. In contrast, features 2 and 3 have a non-zero Shapley value. However, these features are absolutely irrelevant for changing the value of the prediction, i.e. if both features 1 and 2 change their value, then the prediction will change independently of the value of features 3 and 4. And if neither feature 1 nor feature 2 change their value, then the prediction remains 1, again independently of the value assigned to features 3 and 4. Perhaps more importantly, features 3 and 4 are irrelevant [28,30,31], i.e. they do not occur in any AXp (and so CXp) for the instance $((1,1,1,1),1)$. This makes the assignment of non-zero Shapley value to these features even more unsettling.

The consequences of proving that theoretical SHAP scores can be misleading [32,54] are significant. First, the tool SHAP [48] is guaranteed to heuristically approximate misleading values. Second, misleading SHAP scores will induce human decision makers in error; this will foster distrust instead of promoting trust in ML models.

6   Rigorous Explainability

Logic-based XAI represents a rigorous alternative to the methods of non-symbolic XAI described in the previous sections. Although logic-based XAI has been most often formulated as a model-based symbolic approach to explaining ML models [49], more recent work has studied logic-based AI in model-agnostic (and so non-model-based) settings [2]. The origins of logic-based XAI can be traced back to two independent works [37,70], with one work [70] proposing rigorous explainability by compilation to canonical representations, and the other [37] proposing the use of modern automated reasoners [13]. In practice, there is ample evidence that the use of automated reasoners allows explaining significantly more complex ML models [34,35,43] than compilation-based methods.

As this and the following section observe, the flaws detailed in the previous sections are non-existent in the case of logic-based explanations.

Abductive and Contrastive Explanations. Logic-based explanations consider a specific instance $(\mathbf{v}, c)$, with $\mathbf{v} \in \mathbb{F}$, $c \in \mathcal{K}$ and $c = \kappa(\mathbf{v})$. As a result, logic-based explanations are defined *locally* to a given instance but are guaranteed to hold *globally*. Surveys of logic-based explainability include [20,21,49,51,56].

Abductive explanations aim to answer a *"Why (the prediction)?"* question, and capture sufficiency for the prediction, i.e. a set of features and respective value assignments that suffice for the prediction. Given an instance $(\mathbf{v}, c)$, a set of features $\mathcal{X} \subseteq \mathcal{F}$ is a weak abductive explanation (WAXp) if the following predicate holds true,

$$\mathsf{WAXp}(\mathcal{X}) \quad := \quad \forall(\mathbf{x} \in \mathbb{F}).\left(\bigwedge_{i \in \mathcal{X}} x_i = v_i\right) \rightarrow (\kappa(\mathbf{x}) = c) \tag{3}$$

Thus, a WAXp is a set of features which, if fixed to the values dictated by $\mathbf{v}$, guarantee that the prediction is c. An abductive explanation (AXp) is any WAXp that is minimal with respect to (3), i.e. (3) holds for $\mathcal{X}$, but not for any of its proper subsets.

In contrast to (W)AXps, contrastive explanations aim to answer a *"Why not (some other prediction)?"* question. Given an instance $(\mathbf{v}, c)$, a set of features $\mathcal{X} \subseteq \mathcal{F}$ is a weak contrastive explanation (WCXp) if the following predicate holds true,

$$\mathsf{WCXp}(\mathcal{Y}) \quad := \quad \exists(\mathbf{x} \in \mathbb{F}).\left(\bigwedge_{i \in \mathcal{F} \setminus \mathcal{X}} x_i = v_i\right) \wedge (\kappa(\mathbf{x}) \neq c) \tag{4}$$

Thus, a WCXp is a set of features which, if allowed to change their values with respect to those dictated by $\mathbf{v}$, suffices in obtaining a prediction other that c. A contrastive explanation (CXp) is any WCXp that is minimal with respect to (4). It should be noted that both predicates, WAXp and WCXp, are monotonically increasing, and so minimal sets can be computed as follows:

$$\mathsf{MinP}(\mathcal{Z}) \quad := \quad \mathsf{P}(\mathcal{Z}) \wedge \forall(t \in \mathcal{Z}).\neg\mathsf{P}(\mathcal{Z} \setminus \{t\}) \tag{5}$$

(The alternative to the definition above would be to quantify over all possible subsets, but that would require an exponential number of cases to be considered, and it is unnecessary.) (5) allows defining AXps and CXps as subset-minimal sets. Furthermore, it is known that AXps are minimal hitting sets of the CXps and vice-versa [36]. This property is referred to as minimal hitting set duality, and it is crucial for the efficient enumeration of AXps in practice.

For several families of classifiers, computing one AXp/CXp is in polynomial time. This is the case for example with Naive Bayes Classifiers (NBCs) [52], decision trees and graphs [30,41,42], monotone classifiers [53], d-DNNF classifiers [29], restricted cases of multi-variate DTs [15] and, more recently, kNN classifiers [11]. General conditions for polynomial-time explainability are studied in [19].

Example 3. In the case of DTs, there are polynomial-time algorithms for computing one abductive explanation. We consider the example DT of Fig. 1, and the instance $((1,1,1,1),1)$. Initially, the WAXp consists of all features: $\{1,2,3,4\}$. The features are analyzed in the order $\langle 1,2,3,4 \rangle$ (but any other order would do). We allow feature 1 to change its value, and then check whether any path from a prediction other than 1, can be made consistent. Clearly, because $x_2 = 1$, then the prediction is guaranteed not to change. This means that feature 1 can be dropped from the WAXp, which becomes $\{2,3,4\}$. Next, we allow feature 2 (and also feature 1) to change their values. In this case, one of the paths with a prediction other than 1 can be made consistent. Indeed, by setting $x_1 = x_2 = 0$, path $P_3 = \langle 1,2,4,7 \rangle$ becomes consistent, and so the prediction changes. As a result, we must keep feature 2 in the WAXp, and keep its value fixed to 1. In the next iterations, we would consider features 3 and 4. However, since feature 2 is fixed, and we have $x_2 = 1$, then the prediction is guaranteed to be 1, independently of the values assigned to features 3 and 4. As a result, the final AXp is $\{2\}$. Overall, for the instance $((1,1,1,1),1)$, the set of AXps is $\{\{1\},\{2\}\}$ and the set of CXps is $\{\{1,2\}\}$.

Logic-based explanations were developed for answering the aforementioned explanation questions, and so focus on explainability by feature selection. Besides polynomial-time cases, logic-based explanations have been applied to families of ML models for which finding on explanation is computationally hard [34,35,37, 40,43]. Additional problems and variants have also been studied [4,6–9,28,31, 39,75]. More importantly,, and as shown in the next section, recent work [46] showed that logic-based explanations provide a strong connection between XAI by feature selection and XAI by feature attribution.

Sample-Based Explanations. Despite the progress observed in logic-based XAI, a recurrent limitation is the scalability of logic reasoning. For example, recent work [40] reports explanations for neural networks (NNs) with up to 100K activation units. Although this is by any account a significant achievement, it is also the case that in many domains, NNs should be expected to be much larger than that.

One alternative line of research, which also builds on the framework detailed earlier in this section, are sample-based explanations [3,18,58]. Sample-based explanations instantiate model-agnostic explanations with a formal foundation [2]. Instead of reasoning about an ML model, which requires some sort of logic encoding of the model in order to compute explanations, sample-based explanations are computed with respect to a sample of the behavior of the ML model. When the sample matches feature space, sample-based explanations match plain (model-based) explanations. More recent work studied both sample-based AXps and CXps, proved hitting set duality, and developed more efficient algorithms [58]. The use of constraint-based approaches for model-agnostic explainability has also been studied [44].

7   Towards Safe Gambling

Section 5 exposed critical flaws with the best-known definition of SHAP scores [48]. The bottom line is that SHAP scores can mislead human decision makers into analyzing the non-important features, and ignoring the most important ones.

We now briefly summarize the ongoing efforts into developing definitions and that computation of SHAP scores that overcome their earlier critical flaws.

A first observation is that any flaws with SHAP scores are not related in any way with the definition of Shapley values. The key issue is the characteristic function that has been adopted by SHAP and by earlier works, i.e. the use of the expected value of the ML model. A detailed analysis of this fact is provided in fairly recent work [45, 46], that also proposes properties that characteristic function should respect. Besides pinpointing the reasons for the flaws with SHAP scores, this work also proposed characteristic functions that are guaranteed not to exhibit the same flaws. Here, we briefly overview one of these characteristic functions.

Instead of a game $G_e = (\mathcal{F}, v_e)$ such that the characteristic function computes an expected value, Léttofé et al. [46] propose a different game $G_a = (\mathcal{F}, v_a)$ where the characteristic function is defined by,

$$v_a(S) \quad := \quad \begin{cases} 1 & \text{if WAXp}(S) \\ 0 & \text{otherwise} \end{cases} \tag{6}$$

The characteristic function v_a mimics the use of Shapley values in several domains, including the measurement of a priori voting power [23, 69]. In addition, v_a shows a tight connection between feature attribution by computation of Shapley values and feature selection with logic-based explanations. Recent work [55] proposes a new algorithm, nuSHAP, that is based on v_a. However, for the analysis of highly complex ML models, the use of sample-based explanations [18, 58] is proposed as an alternative to plain model-based explanations.

8   Conclusions and Research Directions

The field of XAI is unfortunately littered with practical and theoretical misconceptions [48, 65–67]. This paper summarizes some of the most appalling examples of misconceptions that have found widespread use in recent years. The examples serve to underscore the importance of a rigorous foundation for explainability, one that logic-based XAI offers. Despite the guarantees of rigor of logic-based XAI, many challenges remain. Scalability and the size of explanations are well-known challenges [51]. Given the pace at which the most widely used ML models are growing, scalability is expected to remain one of the most important challenges in the foreseeable future. Furthermore, there is a clear gap between rule-based explanations based on conjunctions of input literals and more sophisticated approaches to explainability that involve analogies, e.g. by relating with similar concepts. These are open topics of research.

Acknowledgment. This work was supported in part by the Spanish Government under grant PID 2023-152814OB-I00, and by ICREA starting funds.

References

1. Aglin, G., Nijssen, S., Schaus, P.: Learning optimal decision trees using caching branch-and-bound search. In: AAAI, pp. 3146–3153 (2020)
2. Amgoud, L.: Explaining black-box classifiers: properties and functions. Int. J. Approx. Reason. **155**, 40–65 (2023)
3. Amgoud, L., Cooper, M.C., Debbaoui, S.: Axiomatic characterisations of sample-based explainers. In: ECAI, pp. 770–777 (2024)
4. Audemard, G., Bellart, S., Bounia, L., Koriche, F., Lagniez, J., Marquis, P.: On the computational intelligibility of Boolean classifiers. In: KR, pp. 74–86 (2021)
5. Audemard, G., Bellart, S., Bounia, L., Koriche, F., Lagniez, J., Marquis, P.: On the explanatory power of Boolean decision trees. Data Knowl. Eng. **142**, 102088 (2022)
6. Audemard, G., Bellart, S., Lagniez, J., Marquis, P.: Computing abductive explanations for boosted regression trees. In: IJCAI, pp. 3432–3441 (2023)
7. Audemard, G., Koriche, F., Marquis, P.: On tractable XAI queries based on compiled representations. In: KR, pp. 838–849 (2020)
8. Audemard, G., Lagniez, J., Marquis, P., Szczepanski, N.: Computing abductive explanations for boosted trees. In: AISTATS, pp. 4699–4711 (2023)
9. Audemard, G., Lagniez, J., Marquis, P., Szczepanski, N.: On contrastive explanations for tree-based classifiers. In: ECAI, pp. 117–124 (2023)
10. Bach, S., Binder, A., Montavon, G., Klauschen, F., Müller, K.R., Samek, W.: On pixel-wise explanations for non-linear classifier decisions by layer-wise relevance propagation. PLoS ONE **10**(7), e0130140 (2015)
11. Barceló, P., Kozachinskiy, A., Orth, M.A.R., Subercaseaux, B., Verschae, J.: Explaining k-nearest neighbors: abductive and counterfactual explanations. CoRR abs/2501.06078 (2025). https://doi.org/10.48550/arXiv.2501.06078
12. Bertsimas, D., Dunn, J.: Optimal classification trees. Mach. Learn. **106**(7), 1039–1082 (2017)
13. Biere, A., Heule, M., van Maaren, H., Walsh, T. (eds.): Handbook of Satisfiability - Second Edition, Frontiers in Artificial Intelligence and Applications, vol. 336. IOS Press (2021)
14. Breiman, L.: Statistical modeling: the two cultures. Stat. Sci. **16**(3), 199–231 (2001)
15. Carbonnel, C., Cooper, M.C., Marques-Silva, J.: Tractable explaining of multivariate decision trees. In: KR, pp. 127–135 (2023)
16. Chalkiadakis, G., Elkind, E., Wooldridge, M.J.: Computational Aspects of Cooperative Game Theory. Morgan & Claypool Publishers, Synthesis Lectures on Artificial Intelligence and Machine Learning (2012)
17. Chen, T., Guestrin, C.: XGBoost: a scalable tree boosting system. In: KDD, pp. 785–794 (2016)
18. Cooper, M.C., Amgoud, L.: Abductive explanations of classifiers under constraints: complexity and properties. In: ECAI, pp. 469–476 (2023)
19. Cooper, M.C., Marques-Silva, J.: Tractability of explaining classifier decisions. Artif. Intell. **316**, 103841 (2023)
20. Darwiche, A.: Logic for explainable AI. In: LICS, pp. 1–11 (2023)

21. Darwiche, A., Hirth, A.: On the (complete) reasons behind decisions. J. Log. Lang. Inf. **32**(1), 63–88 (2023)
22. Demirovic, E., et al.: MurTree: optimal decision trees via dynamic programming and search. J. Mach. Learn. Res. **23**, 26:1–26:47 (2022)
23. Felsenthal, D.S., Machover, M.: The Measurement of Voting Power. Edward Elgar Publishing (1998)
24. Guidotti, R., Monreale, A., Ruggieri, S., Turini, F., Giannotti, F., Pedreschi, D.: A survey of methods for explaining black box models. ACM Comput. Surv. **51**(5), 93:1–93:42 (2019)
25. Gunning, D., Aha, D.W.: DARPA's explainable artificial intelligence (XAI) program. AI Mag. **40**(2), 44–58 (2019)
26. Gunning, D., Stefik, M., Choi, J., Miller, T., Stumpf, S., Yang, G.: XAI - explainable artificial intelligence. Sci. Robot. **4**(37), eaay7120 (2019)
27. Hu, X., Rudin, C., Seltzer, M.I.: Optimal sparse decision trees. In: NeurIPS, pp. 7265–7273 (2019)
28. Huang, X., Cooper, M.C., Morgado, A., Planes, J., Marques-Silva, J.: Feature necessity & relevancy in ML classifier explanations. In: TACAS, pp. 167–186 (2023)
29. Huang, X., Izza, Y., Ignatiev, A., Cooper, M.C., Asher, N., Marques-Silva, J.: Tractable explanations for d-DNNF classifiers. In: AAAI, pp. 5719–5728 (2022)
30. Huang, X., Izza, Y., Ignatiev, A., Marques-Silva, J.: On efficiently explaining graph-based classifiers. In: KR, pp. 356–367 (2021)
31. Huang, X., Izza, Y., Marques-Silva, J.: Solving explainability queries with quantification: The case of feature relevancy. In: AAAI, pp. 3996–4006 (2023)
32. Huang, X., Marques-Silva, J.: On the failings of Shapley values for explainability. Int. J. Approx. Reason. **171**, 109112 (2024)
33. Ignatiev, A.: Towards trustable explainable AI. In: IJCAI, pp. 5154–5158 (2020)
34. Ignatiev, A., Izza, Y., Stuckey, P.J., Marques-Silva, J.: Using MAXSAT for efficient explanations of tree ensembles. In: AAAI, pp. 3776–3785 (2022)
35. Ignatiev, A., Marques-Silva, J.: SAT-based rigorous explanations for decision lists. In: SAT, pp. 251–269 (2021)
36. Ignatiev, A., Narodytska, N., Asher, N., Marques-Silva, J.: From contrastive to abductive explanations and back again. In: AIxIA, pp. 335–355 (2020)
37. Ignatiev, A., Narodytska, N., Marques-Silva, J.: Abduction-based explanations for machine learning models. In: AAAI, pp. 1511–1519 (2019)
38. Ignatiev, A., Narodytska, N., Marques-Silva, J.: On validating, repairing and refining heuristic ML explanations. CoRR abs/1907.02509 (2019). http://arxiv.org/abs/1907.02509
39. Izza, Y., Huang, X., Ignatiev, A., Narodytska, N., Cooper, M.C., Marques-Silva, J.: On computing probabilistic abductive explanations. Int. J. Approx. Reason. **159**, 108939 (2023)
40. Izza, Y., Huang, X., Morgado, A., Planes, J., Ignatiev, A., Marques-Silva, J.: Distance-restricted explanations: theoretical underpinnings & efficient implementation. In: KR (2024)
41. Izza, Y., Ignatiev, A., Marques-Silva, J.: On explaining decision trees. CoRR abs/2010.11034 (2020). https://arxiv.org/abs/2010.11034
42. Izza, Y., Ignatiev, A., Marques-Silva, J.: On tackling explanation redundancy in decision trees. J. Artif. Intell. Res. **75**, 261–321 (2022)
43. Izza, Y., Marques-Silva, J.: On explaining random forests with SAT. In: IJCAI, pp. 2584–2591 (2021)
44. Koriche, F., Lagniez, J., Mengel, S., Tran, C.: Learning model agnostic explanations via constraint programming. In: ECML, pp. 437–453 (2024)

45. Létoffé, O., Huang, X., Asher, N., Marques-Silva, J.: From SHAP scores to feature importance scores. CoRR abs/2405.11766 (2024). https://doi.org/10.48550/arXiv.2405.11766
46. Létoffé, O., Huang, X., Marques-Silva, J.: Towards trustable SHAP scores. In: AAAI, pp. 18198–18208 (2025)
47. Lipton, Z.C.: The mythos of model interpretability. Commun. ACM **61**(10), 36–43 (2018)
48. Lundberg, S.M., Lee, S.: A unified approach to interpreting model predictions. In: NeurIPS, pp. 4765–4774 (2017)
49. Marques-Silva, J.: Logic-based explainability in machine learning. In: Reasoning Web, pp. 24–104 (2022)
50. Marques-Silva, J.: Disproving XAI myths with formal methods - initial results. In: ICECCS, pp. 12–21 (2023)
51. Marques-Silva, J.: Logic-based explainability: past, present and future. In: ISoLA, pp. 181–204 (2024)
52. Marques-Silva, J., Gerspacher, T., Cooper, M.C., Ignatiev, A., Narodytska, N.: Explaining Naive Bayes and other linear classifiers with polynomial time and delay. In: NeurIPS (2020)
53. Marques-Silva, J., Gerspacher, T., Cooper, M.C., Ignatiev, A., Narodytska, N.: Explanations for monotonic classifiers. In: ICML, pp. 7469–7479 (2021)
54. Marques-Silva, J., Huang, X.: Explainability is Not a game. Commun. ACM **67**(7), 66–75 (2024)
55. Marques-Silva, J., Huang, X., Létoffé, O.: The explanation game - rekindled (extended version). CoRR abs/2501.11429 (2025). https://doi.org/10.48550/ARXIV.2501.11429
56. Marques-Silva, J., Ignatiev, A.: Delivering trustworthy AI through formal XAI. In: AAAI, pp. 12342–12350 (2022)
57. Marques-Silva, J., Ignatiev, A.: No silver bullet: interpretable ML models must be explained. Front. Artif. Intell. **6**, 1128212 (2023)
58. Marques-Silva, J., Lefebvre-Lobaina, J., Martinez, V.: Efficient and rigorous model-agnostic explanations. In: IJCAI (2025)
59. Marques-Silva, J., Mencía, C., Mencía, R.: The sets of power. CoRR abs/2410.07867 (2024). https://doi.org/10.48550/arXiv.2410.07867
60. McTavish, H., Boner, Z., Donnelly, J., Seltzer, M., Rudin, C.: Leveraging predictive equivalence in decision trees. CoRR abs/2506.14143 (2025). https://doi.org/10.48550/ARXIV.2506.14143
61. Mishra, P.: Explainable AI Recipes. Apress (2023)
62. Molnar, C.: Interpretable machine learning. Lulu.com (2020)
63. Molnar, C.: Interpreting Machine Learning Models With SHAP. Lulu.com (2023)
64. Ribeiro, M.T., Singh, S., Guestrin, C.: "why should I trust you?": explaining the predictions of any classifier. In: KDD, pp. 1135–1144 (2016)
65. Ribeiro, M.T., Singh, S., Guestrin, C.: Anchors: high-precision model-agnostic explanations. In: AAAI, pp. 1527–1535 (2018)
66. Rudin, C.: Stop explaining black box machine learning models for high stakes decisions and use interpretable models instead. Nat. Mach. Intell. **1**(5), 206–215 (2019)
67. Rudin, C., Chen, C., Chen, Z., Huang, H., Semenova, L., Zhong, C.: Interpretable machine learning: fundamental principles and 10 grand challenges. Stat. Surv. **16**, 1–85 (2022)
68. Shapley, L.S.: A value for n-person games. Contrib. Theory Games **2**(28), 307–317 (1953)

69. Shapley, L.S., Shubik, M.: A method for evaluating the distribution of power in a committee system. Am. Polit. Sci. Rev. **48**(3), 787–792 (1954)
70. Shih, A., Choi, A., Darwiche, A.: A symbolic approach to explaining Bayesian network classifiers. In: IJCAI, pp. 5103–5111 (2018)
71. Siala, M., Planes, J., Marques-Silva, J.: On trustworthy rule-based models and explanations. CoRR abs/2507.07576 (2025). https://doi.org/10.48550/arXiv.2507.07576
72. Strumbelj, E., Kononenko, I.: An efficient explanation of individual classifications using game theory. J. Mach. Learn. Res. **11**, 1–18 (2010)
73. Strumbelj, E., Kononenko, I.: Explaining prediction models and individual predictions with feature contributions. Knowl. Inf. Syst. **41**(3), 647–665 (2014)
74. Verwer, S., Zhang, Y.: Learning optimal classification trees using a binary linear program formulation. In: AAAI, pp. 1625–1632 (2019)
75. Yu, J., Ignatiev, A., Stuckey, P.J., Narodytska, N., Marques-Silva, J.: Eliminating the impossible, whatever remains must be true: on extracting and applying background knowledge in the context of formal explanations. In: AAAI, pp. 4123–4131 (2023)

Reducing Experimentation Training Overhead in Time Series Forecasting through LLM Adaptation

M. Germán-Morales[(✉)], A. J. Rivera, M. J. del Jesus, and C. J. Carmona

Andalusian Research Institute in Data Science and Computational Intelligence,
Department of Computer Science, University of Jaén, E-23071 Jaén, Spain
`{mgerman,arivera,mjjesus,ccarmona}@ujaen.es`

Abstract. The paradigm shift in the field of Deep Learning, due to the emergence of Foundation Models and, in particular, Large Language Models has transformed how models operate. These models handle multiple tasks by leveraging their inherent transfer learning capabilities. While powerful, these models raise concerns about computational and environmental costs. In response, we demonstrate that our lightweight methodology for adapting pre-trained Large Language Models to Time Series Forecasting using Parameter-Efficient Fine-Tuning techniques such as Low-Rank Adaptations, called LLIAM, is capable of significantly reducing total training time and resource consumption while maintaining accuracy comparable to other state-of-the-art Deep Learning models. This work contributes to the advancement of Green AI by demonstrating a more sustainable and accessible utilization of these large models for specialized tasks.

Keywords: Time Series Forecasting · Foundational Models · Transfer Learning

1 Introduction

Foundation Models (FMs) emerged as a new branch of models that has led to a paradigm shift in several areas of Machine Learning (ML), and in particular Deep Learning (DL) which has evolved greatly due to the good performance of these models. Specifically, in tasks related to the generation of new and meaningful content from training data, known as Generative AI [11]. In addition, their large generalization capacities allow them to be adapted easily and quickly to tasks from unknown domains. This is known as Transfer Learning (TL) [35]. These models undergo extensive pre-training, enabling the development of new, sustainable models aligned with the goals of Green AI [4], thanks to their inherent TL capabilities. Reducing training time and resource consumption without compromising performance has become an important objective in the field, especially given the high energy and computational costs associated with training large experimentations or even single models.

L. Martínez et al. (Eds.): IDEAL 2025, LNCS 16239, pp. 307–318, 2026.
https://doi.org/10.1007/978-3-032-10489-2_26

A time-series is a type of data consisting of a set of estimations ordered over time. This infers the presence of temporal dependencies between the different moments observed. Their analysis makes it possible to extract these relationships in order to estimate its future value at any time, which is called Time-Series Forecasting (TSF). The likeness between processing a numerical and a textual series allows intuiting that utilizing FMs dedicated to Language Modelling, such as Large Language Models (LLMs), can be adjusted to TSF. All things considered, a few proposals have argued and demonstrated that LLMs have the potential to perform differing Time-Series Analysis tasks [20], including the capability of being adequate predictors without providing additional time-series particular information to them [14]. In our previous work [13], we asked whether FMs, with their large knowledge base and especially their flexibility, could also be useful for TSF. Reusing a single model that learns from many series across different domains would allow us to move away from the traditional DL approach, where a model is usually trained on just one dataset.

In response to the growing concern over the computational demands of FMs, we hypothesize that efficient adaptation techniques such as LLIAM can substantially reduce total training time while maintaining predictive accuracy. Our focus is on the high computational cost typically associated with conventional DL approaches, where separate models are trained for each domain and exhaustive hyperparameter optimization is often required. To test this hypothesis, we compare our method against standard baselines in terms of both total training time and forecasting performance. Our aim is to demonstrate that LLIAM provides a more sustainable and computationally efficient alternative for TSF. Furthermore, we expand upon the key findings of our previous work, showing that LLIAM achieves equal or superior predictive quality compared to other DL models while requiring significantly less total training time.

This question is addressed in the article, which is structured as follows: Section 2 briefly describes related work, covering DL TSF methods, TL, FMs, and a brief description of the LLIAM methodology [13]. Section 3 presents the experimental framework, including the setup and an empirical comparison between LLIAM and traditional deep learning approaches including the total training times. Finally, Section 4 summarizes the main findings and conclusions from the study.

2   Related Work

In recent years, numerous methods have been developed for time-series analysis tasks, including TSF. Section 2.1 provides a description of the TSF task and the main DL methods, such as RNNs, TCNs, and Transformers. Section 2.2 discusses the definitions of TL, FM, and LLM, emphasizing their paradigm-shifting impact on TSF. Finally, Section 2.3 presents the LLIAM methodology, which adapts LLMs for the TSF task.

2.1 Deep Time-Series Forecasting methods

A wide range of domains presents temporal data that can be exploited to predict the behavior of different processes and enhance the decision-making process of individuals and organizations in order to maximize a benefit. TSF is a task related to the discover of the temporal dependencies of time-series in order to achieve a reliable forecast and support that process.

At its core, this task involves learning a function f that, given n past observations $S = \{x_0, x_1, x_2, ..., x_n\}$, can produce h future predictions $\hat{Y} = \{\hat{y}_{t+1}, \hat{y}_{t+2}, ..., \hat{y}_{t+h}\}$ while minimizing the error with respect to the actual future values $Y = \{y_{t+1}, y_{t+2}, ..., y_{t+h}\}$. In this context, the number of future time steps to predict is referred to as the *forecast horizon*, and the past values used as input are known as *lags*.

Machine Learning (ML) [29] is a subset of AI whose objective is bringing the capacity of learning from experience to computers without being explicitly programmed to do so. Inside this, another subset is found, known as DL [22]. Unlike ML methods, DL ones integrate the learning of representation into them, enabling the generation of representations with different levels of abstraction by stacking several processing or hidden layers. The following paragraphs introduce the main models used for sequential data.

Recurrent Neural Networks. Recurrent neural networks (RNNs) detect patterns in sequential data such as text, time series or genomes [30]. Sequences are processed step-by-step, with a hidden state that summarizes past information being updated. RNNs can act as both encoders and decoders [16], but they can struggle with long sequences due to issues such as vanishing/exploding gradients and the loss of long-range dependencies. Long Short-Term Memory (LSTM) [17] networks address this by using a long-term memory state and gated mechanisms, while Gated Recurrent Units (GRU) [7] employ two gates to control hidden state updates. Both of these improvements enhance the ability of RNNs to capture long-term dependencies and mitigate the previously mentioned issues.

Temporal Convolutional Networks. Temporal Convolutional Networks (TCNs) [3] combine some of the best practices of modern convolutional [21,37] architectures for sequence modeling, claiming to outperform RNNs in a broad collection of related tasks called Residual Block. A modification of the baseline convolutional operator originally used for audio generation [25] is employed, known as causal dilated convolution, inside this main component of these networks. The temporal order of the data is respected by the model due to the causality of this operator, while the receptive field is maximized for detecting longer time dependencies without the computational cost being increased or the number of trainable model parameters being increased by the dilatation technique.

Transformers. Transformers [34] emerged as state-of-the-art architectures for sequence translation. They feature multiple stacked encoders and decoders capable of processing the input sequence in parallel and achieving superior performance compared to traditional models. This architecture is composed of Multi-Head Attention mechanisms and Feed Forward Networks combined with residual connections and layer normalization. It is important to note that while the encoder stack processes all elements of the input sequence in a single forward pass, the decoder generates the output sequence in an autoregressive manner. The base Transformer architecture has been adapted for numerous tasks, including TSF, with these modifications remaining highly relevant due to their strong performance across various applications. Adaptations range from simple changes to individual components to entirely new architectures [6]. Notable Transformer models used in TSF are Informer [40], Autoformer [36], DSFormer [39], PatchTST [24] and the Temporal Fusion Transformer [23].

2.2 Transfer Learning, Foundational Models, and Large Language Models

DL methods require abundant amounts of training data from a specific domain to resolve tasks. In real-world scenarios, obtaining such data can be challenging or extremely expensive. Transfer Learning (TL) [27] addresses this limitation by enabling the reuse of knowledge acquired in one domain or task to improve performance in a different one. This allows models trained for a specific objective to be adapted to new tasks, reducing the need for large datasets and lowering computational costs.

Transfer Learning. This approach focuses on improving the predictive function f_T for a specific task T_T within a given domain D_T by using related information from a different task T_S in another domain D_S. Essentially, it means borrowing knowledge from a related but different area to enhance the performance of the target task. It is crucial to understand that the source and target domains and tasks may not be the same ($D_S \neq D_T$, $T_S \neq T_T$), and each one have its own unique set of features and labels [35].

Foundational Models. The combination of TL capabilities, advances in computing power and the availability of diverse and large datasets has enabled the emergence of FMs [1]. These models typically follow a two step training process, where they are first pretrained on a wide set of different data and then finetuned to adapt to specific downstream tasks. Thanks to their strong generalization abilities, a single model can often be applied to a wide range of tasks. Moreover, these models can perform well without additional training examples, which is referred to as zero-shot learning [38], or with only a few examples, known as few-shot learning [5]. This is possible because they are capable of understanding and learning from the context provided in their input, a property called in-context learning [9]. Several FMs have been proposed for Time Series Analysis tasks, such as TimesFM [8] or TimeGPT [12].

Large Language Models. LLMs are a subset of FM trained on large-scale textual corpora. Unlike traditional pretrained models, they perform well on a variety of tasks without fine-tuning, thanks to their strong general-purpose language understanding. Most modern LLMs are based on the Transformer architecture [34]. The LLaMA series by Meta, starting in 2023 [32], has evolved through LLaMA-2 [33] and LLaMA-3 [10], mainly improving in data scale, quality, and context window size. GPT models made by OpenAI, originally introduced as decoder-only Transformers [28], have progressed into multimodal systems with GPT-4 and GPT-4o [26]. Similarly, Gemini models developed by Google began as multimodal Transformer architectures designed for tasks ranging from complex reasoning to on-device usage [31]. Based on the similarities between a text sequence and a time series, some proposals adapt LLMs to the TSF taks such as TimeLLM [19], Chronos [2] or our proposal, LLIAM [13] which is briefly explained in the next section.

2.3   LLIAM: Lora-Llama Integrated Autorregresive Model

LLMs are trained using a two-phase approach. In the first phase, large and diverse datasets from sources such as ArXiv, Wikipedia, and social media are used to train the model, with the objective of learning the fundamental structure and use of language. The result of this step is referred to as the *pre-trained model*, which contains broad and general knowledge. The second phase, called *fine-tuning*, enables the model to adapt to a specific task by updating its knowledge using a task-specific dataset.

Parameter-Efficient Fine-Tuning [15] methods, such as Low-Rank Adaptations (LoRA) [18], provide a way to quickly and efficiently adapt large models to specific tasks by leveraging their existing knowledge without tuning all model parameters. In [13], we adapted and evaluated the LLaMA LLM for TSF by applying the LoRA technique. Additionally, we proposed a prompting scheme for time series data, allowing the model to perform TSF using only the input data typically employed by conventional deep learning algorithms, such as RNNs and TCNs. The resulting approach demonstrates the effectiveness of LLMs in forecasting tasks. Since this article is intended to complement the paper proposing LLIAM with an analysis of the total training times from different approaches, please consider referring to [13] for a detailed explanation of the method.

3   Experimentation Framework

In this section the design of the experiments done to evaluate our proposal is introduced. The datasets, pre-processing pipeline and metrics employed on the experimentation are described in Subsection 3.1. Section 3.2 presents a comparative study between our proposal, LLIAM, and a set of traditional DL techniques used in TSF, such as RNNs and TCNs.

3.1   Experimental setup

The experimentation made is an extension of the original one made for the proposal of LLIAM [13]. A subset of 7 datasets with different sampling frequency has been selected. The description of the datasets is provided below:

- **Electricity** (1 Dataset): Contains the electricity consumption of 370 customers in kilowatts from 2011 to 2014. All series have the same length. The version whose sampling frequency is weekly has been selected.
- **M1** (2 Datasets): This dataset is composed of a selection of series from one of the first competitions dedicated to its prediction, called M competitions (M, because its creator is called Makridakis). It consists of 617 monthly time series, related to economics, industry, and demography. Not all series have the same number of measurements. From it, he versions with monthly and quarterly sampling frequency have been selected.
- **M3** (2 Datasets): Like the M1 dataset, it contains 1428 series from the third edition of the M competitions. They are again not of the same length and add finance-related phenomena to their domains. From it, the versions that present a monthly and quarterly sampling frequency have been selected.
- **NN5** (2 Datasets): Dataset from another competition, with 111 series collecting transactions from ATMs located in Great Britain. In this case, the series have the same length.From it, the versions with a daily and weekly sampling frequency have been selected.

Table 1. Basic description of the data sets used. The columns #Series, H, Same Length and Input Size indicate the number of series in the data set, the size of the prediction horizon, the size of the input windows and whether all series have the same length.

Dataset	#Series	Frequency	H	Input size	Same Length?
electricity	321	weekly	8	65	✓
m1	617	monthly	18	15	✗
m1	203	quarterly	8	5	✗
m3	1428	monthly	18	15	✗
m3	756	quarterly	8	5	✗
nn5	111	daily	56	9	✓
nn5	111	weekly	8	65	✓

Table 1 presents the input window size and forecast horizon used to generate the sliding windows for each dataset. It also includes other relevant characteristics and identifies the studies in which each dataset is used. A sliding window strategy is employed to create instances for training the various DL models under evaluation. Preprocessing is applied to the LSTM, GRU, and TCN models. In contrast, LLIAM does not use any preprocessing or standardization. Instead, a textual prompt is constructed for each generated window, maintaining the raw format of the data. The hyperparameters, detailed preprocessing, datasets partition methodology and evaluation, including SMAPE, RMSE, and Missing Rate, used for each model follow the specifications described in [13]. Additionally, this study includes the training times recorded for each set of experiments conducted with each model.

3.2 Empirical comparison

In this study, our proposed model, LLIAM, is compared with various conventional DL algorithms, such as LSTM, GRU and TCN. The objective is to demonstrate that the general training approach adopted by LLIAM outperforms the traditional dataset-specific training paradigm, reducing the time expending in training while improving prediction quality. Unlike existing methods that are typically limited to individual datasets, LLIAM is designed to capture broader patterns across multiple datasets, enabling greater generalization.

Methodology. We have followed the same methodology described in the comparative study by [13]. As a further contribution, we report the training times for each of the models evaluated. Below, we briefly outline the methodology applied to each type of model.

- Traditional deep learning algorithms, such as RNNs and TCNs, are trained separately on each dataset using all combinations of hyperparameters specified in [13]. The metrics presented in the tables for these algorithms represent the averages across all tested configurations. This approach aims to provide a performance measure that reflects the typical or average behavior of these algorithms.
- LLIAM is trained only once, using all training partitions combined into a single dataset. Since it does not include components that depend on the length of the series to produce the forecast (such as a last linear layer with a fixed set of neurons equal to the forecast horizon), it results in a more general model compared to the deep learning approaches. Hyperparameter optimization is not performed for LLIAM. Instead, we use the same configuration specified in [13].

Table 2. RMSE metric per model and dataset. LSTM, GRU and TCN RMSE are averages for all the configurations tested, meanwhile LLIAM is the average RMSE obtained after aggregating the test instances.

		RMSE (Lower is better)					
	Electricity	M1 Monthly	M1 Quarterly	M3 Monthly	M3 Quarterly	NN5 Daily	NN5 Weekly
LLIAM T = 10	45120,043	2464,424	**1913,682**	**807,877**	**629,930**	5,918	**18,342**
LLIAM T = 20	**42338,992**	**2348,506**	2655,757	829,951	631,046	**5,906**	19,085
GRU	436904,097	27405,683	29961,260	1335,165	1028,065	6,887	24,109
LSTM	287414,428	24225,724	21713,865	1265,023	964,428	6,677	22,682
TCN D = 0	670686,818	34761,041	13997,682	1326,214	973,804	6,633	22,704
TCN D = 0.1	663362,613	34764,894	15835,914	1347,495	1003,729	6,712	22,721
TCN D = 0.25	683029,861	32666,035	17290,540	1368,059	1001,894	6,863	23,111

Results. As can be seen in Table 2, the differences between LLIAM and the other models are significant, with lower RMSE at both temperatures. For example, in the Electricity dataset, LLIAM T = 10 and T = 20 achieve RMSE values of 45120.043 and 42338.992 respectively, significantly outperforming the second-best model, LSTM, which achieves an RMSE of 287414.428. For the M1 monthly and quarterly data sets, LLIAM T = 10 also leads with RMSE values of 2464.424 and 1913.682, outperforming the conventional DL models. The same occurs with the rest of the datasets, with our proposal achieving the lowest of the results 4 times with T = 10 and three times with T = 20. While the RMSE values in some datasets might seem high, they reflect the larger scale and range of the data in those specific cases. Also, it is important to note that RMSE is sensitive to anomalies. The consistently lower RMSE values of LLIAM across different datasets indicate its robustness and superior performance in the TSF task.

Table 3. SMAPE metric per model and dataset. LSTM, GRU and TCN SMAPE are averages for all the configurations tested, meanwhile LLIAM is the average SMAPE obtained after aggregating the test instances.

	SMAPE (Lower is better)							
	Electricity	M1 Monthly	M1 Quarterly	M3 Monthly	M3 Quarterly	NN5 Daily	NN5 Weekly	Avg. of Avg.
LLIAM T = 10	0,099	**0,163**	**0,171**	**0,150**	0,102	**0,232**	**0,114**	**0,147**
LLIAM T = 20	**0,097**	0,166	0,175	0,153	0,103	0,237	0,119	0,150
GRU	0,241	0,604	0,616	0,165	0,113	0,259	0,135	0,305
LSTM	0,239	0,624	0,579	0,153	**0,102**	0,247	0,127	0,296
TCN D = 0	0,242	0,436	0,481	0,160	**0,102**	0,246	0,129	0,257
TCN D = 0.1	0,232	0,471	0,485	0,166	0,111	0,249	0,129	0,263
TCN D = 0.25	0,247	0,478	0,533	0,168	0,108	0,257	0,132	0,275

Unlike RMSE, SMAPE takes into account the scale of the data and determines whether a model can make predictions within the desired range. The results reported in Table 3 inform us that our proposal, LLIAM, achieves the best performance against conventional algorithms for each dataset and globally (as shown in the Avg. of Avg. column). As can be seen, compared to the best non-foundational DL model, LLIAM is able to obtain a SMAPE of 0.997 with LLIAM T = 20, respecting a SMAPE of 0.232 given by the TCN D = 0.1. In M1 Monthly, this performance improvement is pronounced, going from 0.436 of the TCN D = 0 to 0.163 with LLIAM T = 10. The M1 Quarterly results show the same tendency, obtaining a reduction of the error from 0.481 to 0.171 with LLIAM T = 10. However, on the M3 and NN5 datasets the error commitment is similar between LLIAM and the other methods, achieving a small error reduction on the M3 Monthly and NN5 datasets and matching it on the M3 Quarterly with LSTM and TCN D = 0 models. Overall, the performance of LLIAM T = 10 and T = 20 is similar, with an average SMAPE of 0.147 and 0.150 respectively, significantly outperforming all other methods.

Table 4. Percentage of test instances unable to decode into the expected forecast horizon. GRU, LSTM and TCN models have been omitted because their missing rate is always 0%.

	Electricity	Missing Rate % (Lower is better)						Average
		M1 Monthly	M1 Quarterly	M3 Monthly	M3 Quarterly	NN5 Daily	NN5 Weekly	
LLIAM T = 10	**0,000**	**0,000**	**0,000**	**0,000**	**0,000**	**3,604**	**0,000**	**0,5148**
LLIAM T = 20	**0,000**	0,486	**0,000**	**0,000**	**0,000**	6,306	**0,000**	0,9704

The missing rate for each dataset and LLIAM configuration is reported in Table 4. Hallucinations were observed only on NN5 Daily for LLIAM T = 10 and also on M1 Monthly for LLIAM T = 20. The most frequent anomaly is due to the model stopping early, failing to reach the expected forecast horizon [13]. Overall, the hallucinations are not significant with respect to the results obtained.

Table 5. Total training time in seconds. GRU, LSTM and TCNs training times are the sum of all the experiments made for all the configurations, meanwhile only one instance of LLIAM has been trained over all the datasets. A conversion from seconds to days (d), hours (h), minutes (m) and seconds (s) is also supplied. Training ran on a NVIDIA V100 32GB VRAM GPU.

Total training time (Lower is better)		
LLIAM	**266563,780 s**	**3d 02h 02m 43s**
LSTM	1612037,888 s	18d 15h 47m 18s
GRU	1648122,031 s	19d 01h 48m 42s
TCN D = 0	1819626,881 s	21d 01h 27m 07s
TCN D = 0.1	1899311,350 s	21d 23h 35m 11s
TCN D = 0.25	1883758,756 s	21d 19h 15m 59s

Finally, it is necessary to consider the training time of LLIAM compared to the other models, as detailed in Table 5. Since temperature is only an inference parameter, our proposal only required one training phase. Additionally, kindly note that this only instance was trained over all the datasets with its default hyperparameters unlike LSTM, GRU and TCNs. The total time for these three methods required an exhaustive set of experiments for each configuration and dataset, resulting in a total time that exceeds the time required to train LLIAM (about 3 days compared to 18 days in the best case). Its performance suggests that fine-tuning pre-trained FMs may lead to the creation of more sustainable proposals compared to training a specific model from scratch.

4   Conclusions

Unlike RNNs or TCNs, LLIAM allows a single model to be used across multiple domains in the TSF task thanks to the strong generalization capabilities of the underlying LLM. This allows for easy adaptation to TSF with minimal task-specific adjustments. The proposed prompting scheme aligns the numerical representation of time series data with the textual format used by LLMs. This enables the model to effectively recognize and process TSF tasks. Additionally, using LoRA for PEFT reduces computational costs by updating only a subset of parameters instead of the entire model. Consequently, combining minimal parameter tuning with the elimination of initial hyperparameter optimization results in a significant reduction in total training time and computational cost compared to classical methodologies based on conventional networks.

The results obtained in the experiments conducted let us verify the initial hypothesis established. Methodologies such as LLIAM can achieve a more sustainable and efficient approach in comparison of a conventional experimental methodology where one model is used for each dataset. The experiments conducted support the findings presented in [13], assuring the effectiveness of FMs for TSF and demonstrating how the combination of LLMs with the LoRA technique significantly improves forecasting performance while reducing overall training time.

Given the large amount of computational resources needed to pre-train LLMs, it is important to maximize their utility by efficiently reusing and adapting them. Techniques such as LLIAM offer a promising way to make LLMs more accessible and environmentally responsible while making them easier to adapt to specialized tasks like TSF.

Acknowledgments. The research carried out in this study is part of the project "Advances in the development of trustworthy AI models to contribute to the adoption and use of responsible AI in healthcare (TAIH)" with code PID2023-149511OB-I00 and supported via the FPU (Formación de Profesorado Universitario) fellowship program, both funded by the Spanish Ministry of Science, Innovation and Universities.

References

1. Bommasani, R., et al.: On the opportunities and risks of foundation models (2022)
2. Ansari, A.F., Stella, L., et al.: Chronos: learning the language of time series (2024). https://arxiv.org/abs/2403.07815
3. Bai, S., Kolter, J.Z., Koltun, V.: An empirical evaluation of generic convolutional and recurrent networks for sequence modeling. CoRR abs/1803.01271 (2018). https://doi.org/10.48550/arXiv.1803.01271
4. Bolón-Canedo, V., Morán-Fernández, L., Cancela, B., Alonso-Betanzos, A.: A review of green artificial intelligence: towards a more sustainable future. Neurocomputing **599**, 128096 (2024). https://doi.org/10.1016/j.neucom.2024.128096
5. Brown, T.B., et al.: Language models are few-shot learners (2020). https://arxiv.org/abs/2005.14165

6. Cabrera-Bermejo, M.I., Del Jesus, M.J., Rivera, A.J., Elizondo, D., Charte, F., Pérez-Godoy, M.D.: Analysis of transformer model applications. In: Hybrid Artificial Intelligent Systems. pp. 231–243. Springer Nature Switzerland, Cham (2023). https://doi.org/10.1007/978-3-031-40725-3_20

7. Chung, J., Gülçehre, Ç., Cho, K., Bengio, Y.: Empirical evaluation of gated recurrent neural networks on sequence modeling. CoRR abs/1412.3555 (2014). http://arxiv.org/abs/1412.3555

8. Das, A., Kong, W., Sen, R., Zhou, Y.: A decoder-only foundation model for time-series forecasting (2024). https://arxiv.org/abs/2310.10688

9. Dong, Q., et al.: A survey on in-context learning (2024). https://arxiv.org/abs/2301.00234

10. Dubey, A., Jauhri, A., et al.: The llama 3 herd of models (2024). https://arxiv.org/abs/2407.21783

11. Feuerriegel, S., Hartmann, J., Janiesch, C., Zschech, P.: Generative AI. Bus. Inf. Syst. Eng. **66**(1), 111–126 (2024)

12. Garza, A., Challu, C., Mergenthaler-Canseco, M.: TimeGPT-1 (2024). https://arxiv.org/abs/2310.03589

13. Germán-Morales, M., Rivera-Rivas, A., del Jesus Díaz, M., Carmona, C.: Transfer learning with foundational models for time series forecasting using low-rank adaptations. Inf. Fus. **123**, 103247 (2025). https://doi.org/10.1016/j.inffus.2025.103247

14. Gruver, N., Finzi, M., Qiu, S., Wilson, A.G.: Large language models are zero-shot time series forecasters (2023)

15. Han, Z., Gao, C., Liu, J., Zhang, J., Zhang, S.Q.: Parameter-efficient fine-tuning for large models: a comprehensive survey (2024). https://arxiv.org/abs/2403.14608

16. Hewamalage, H., Bergmeir, C., Bandara, K.: Recurrent neural networks for time series forecasting: current status and future directions. Int. J. Forecast. **37**(1), 388–427 (2021). https://doi.org/10.1016/j.ijforecast.2020.06.008

17. Hochreiter, S., Schmidhuber, J.: Long short-term memory. Neural Comput. **9**, 1735–80 (1997). https://doi.org/10.1162/neco.1997.9.8.1735

18. Hu, E.J., et al.: LoRA: low-rank adaptation of large language models (2021). https://arxiv.org/abs/2106.09685

19. Jin, M., et al.: Time-LLM: time series forecasting by reprogramming large language models (2024). https://doi.org/10.48550/arXiv.2310.01728, arXiv:2310.01728

20. Jin, M., et al.: Position: what can large language models tell us about time series analysis (2024). https://arxiv.org/abs/2402.02713

21. LeCun, Y., et al.: Backpropagation applied to handwritten zip code recognition. Neural Comput. **1**(4), 541–551 (1989). https://doi.org/10.1162/neco.1989.1.4.541

22. LeCun, Y., Bengio, Y., Hinton, G.: Deep learning. Nature **521**(7553), 436–444 (2015). https://doi.org/10.1038/nature14539, publisher: Nature Publishing Group

23. Lim, B., Arik, S.O., Loeff, N., Pfister, T.: Temporal fusion transformers for interpretable multi-horizon time series forecasting (2020). https://arxiv.org/abs/1912.09363

24. Nie, Y., Nguyen, N.H., Sinthong, P., Kalagnanam, J.: A time series is worth 64 words: long-term forecasting with transformers (2023). https://doi.org/10.48550/arXiv.2211.14730, arXiv:2211.14730

25. van den Oord, A., et al.: WaveNet: a generative model for raw audio. CoRR abs/1609.03499 (2016). http://arxiv.org/abs/1609.03499

26. OpenAI, Achiam, J., Adler, S., et al., S.A.: GPT-4 technical report (2024). https://arxiv.org/abs/2303.08774

27. Pan, S.J., Yang, Q.: A survey on transfer learning. IEEE Trans. Knowl. Data Eng. **22**(10), 1345–1359 (2009)
28. Radford, A., Narasimhan, K., Salimans, T., Sutskever, I.: Improving language understanding by generative pre-training (2018)
29. Sarker, I.H.: Machine learning: algorithms, real-world applications and research directions. SN Comput. Sci. **2**(3), 160 (2021). https://doi.org/10.1007/s42979-021-00592-x
30. Schmidt, R.M.: Recurrent Neural Networks (RNNs): a gentle Introduction and Overview (2019). https://doi.org/10.48550/arXiv.1912.05911, arXiv:1912.05911
31. Team, G., Anil, R., Borgeaud, S., et al., J.B.A.: Gemini: a family of highly capable multimodal models (2024). https://arxiv.org/abs/2312.11805
32. Touvron, H., et al.: Llama: open and efficient foundation language models (2023)
33. Touvron, H., Martin, L., et al., K.S.: Llama 2: open foundation and fine-tuned chat models (2023)
34. Vaswani, A., et al.: Attention is all you need. Proc. of NIPS **30**, 5999–6009 (2017)
35. Weiss, K., Khoshgoftaar, T.M., Wang, D.: A survey of transfer learning. J. Big Data **3**(1), 9 (2016). https://doi.org/10.1186/s40537-016-0043-6
36. Wu, H., Xu, J., Wang, J., Long, M.: AutoFormer: decomposition transformers with auto-correlation for long-term series forecasting. CoRR abs/2106.13008 (2021). https://arxiv.org/abs/2106.13008
37. Wu, J.: Introduction to convolutional neural networks. Nat. Key Lab Novel Softw. Technol. **5**(23), 495 (2017)
38. Xian, Y., Lampert, C.H., Schiele, B., Akata, Z.: Zero-shot learning – a comprehensive evaluation of the good, the bad and the ugly (2020). https://arxiv.org/abs/1707.00600
39. Yu, C., Wang, F., Shao, Z., et al.: DSformer: a double sampling transformer for multivariate time series long-term prediction (2023). https://doi.org/10.48550/arXiv.2308.03274, arXiv:2308.03274
40. Zhou, H., Zhang, S., et al.: Informer: beyond efficient transformer for long sequence time-series forecasting. CoRR abs/2012.07436 (2020). https://arxiv.org/abs/2012.07436

Taxonomy and Evaluation of XAI Tools for Explainable Machine Learning

Paola Montenegro-Cantos[1], Aurora Ramírez[2(✉)],
Carlos García-Martínez[2], and José Raúl Romero[2]

[1] University of Azuay, Cuenca, Ecuador
`pmontenegro@uazuay.edu.ec`
[2] Department Computer Science and Artificial Intelligence, University of Córdoba,
and DaSCI Research Institute, Cordoba, Spain
`{aramirez,cgarcia,jrromero}@uco.es`

Abstract. A persistent problem in machine learning (ML) lies in the lack of transparency of the resulting predictive models. The principles of explainable artificial intelligence (XAI) are especially needed to foster trust in ML applications by opening the "black box" to understand the inner workings of ML systems. XAI methods are effective for inspecting ML models, understand how data affect their predictions, or assessing their fairness. Tools facilitate access to XAI methods, providing graphical environments for generating and analysing different types of explanations. In this work, we present a taxonomy of XAI tools, offering a detailed analysis of the methods they implement and their user-oriented features. Based on three dimensions and 31 categories, we compare 15 XAI tools, highlighting their strengths and limitations. The goals of our study are to analyse the current landscape of XAI tools, expand the use of lesser-known XAI techniques and highlight the usefulness of these tools in analysing ML models beyond performance evaluation.

Keywords: Explainable Artificial Intelligence · Machine Learning · Taxonomy · Tool

1 Introduction

The large volumes of data currently generated and stored in computational systems represent a valuable resource for data-driven research. Machine learning (ML) plays a fundamental role in extracting relevant information from these systems, allowing the identification of patterns and trends. With the rise of ML research, predictive models have reached high levels of accuracy and their use has expanded to multiple application domains, such as healthcare, education, engineering and economy. However, these systems are ultimately aimed at professionals, who lack the in-depth knowledge to understand how these models work and how their outputs should be interpreted. This, among other technical, ethical and business barriers currently limits the widespread adoption of ML solutions in critical domains [3,15].

© The Author(s), under exclusive license to Springer Nature Switzerland AG 2026
L. Martínez et al. (Eds.): IDEAL 2025, LNCS 16239, pp. 319–330, 2026.
https://doi.org/10.1007/978-3-032-10489-2_27

The opacity of ML models, often described as "black boxes", is arguably the main cause of lack of user trust. Explainable artificial intelligence (XAI) has emerged to address the lack of transparency in ML, seeking trustworthy artificial intelligence (AI) systems [19]. The primary goal is to enhance user trust in the predictions of AI systems. Additional benefits of XAI include supporting informed decision-making, identifying vulnerabilities and biases, ensuring compliance with data protection policies, and raising industry standards for AI-based products. Interest in XAI has been growing due to its general applicability and focus on domain experts. Currently, the inclusion of explainable methods in fields such as healthcare [21], education [20], agriculture [2], engineering [25], or economy [22] is increasingly prevalent.

XAI methods can inspect the internal decision structures of ML models, analyse how the model behaves for particular data samples, and propose actionable changes to the data samples to achieve different predictions. The nature of explanations, and so the methods that generate them, can be categorised by different dimensions [14]. First, explanations may have a global scope, aiming to provide a comprehensive understanding of a ML model's internal mechanisms, or a local scope, focusing on predictions about individual instances and providing insights into the reasons behind specific decisions. On the other hand, methods can be classified as agnostic or model-specific. The former are capable of generating explanations for any ML model, as they only require analysing its outputs. The latter are dependent on the type of algorithm, as they analyse the internal structure such as weights, layers, or decision nodes.

As more XAI methods have emerged, their implementations have begun to be made available through libraries, packages, and software tools. Although some are more specialised than others, access to state-of-the-art implementations is essential to promote wider adoption. Given the expanding landscape of methods and tools, it is timely to offer a detailed assessment of the alternatives. Specifically, it would be interesting to analyse which open-access XAI tools are currently available and what features differentiate them. Aspects such as the number of methods implemented, the availability of a graphical user interface, the degree of customisation of explanations, and the type of data supported are all important considerations when choosing an XAI tool.

In this paper, we aim to identify, evaluate, and analyse existing XAI tools. To achieve this, we propose a taxonomy encompassing three dimensions: (1) *Metadata*, describing technical characteristics; (2) *Scope and general purpose*, including supported tasks, data types, and explanation types; and (3) *Explainability methods*, categorising the implemented XAI methods. We define and evaluate 31 characteristics across 15 tools, offering a comprehensive overview of the XAI tool landscape. Our taxonomy formalises the current landscape of general-purpose XAI tools, serving as a practical guide to identify and select the most suitable tool.

The rest of the paper is structured as follows. Section 2 presents background and related work, focusing on previously proposed taxonomies. Section 3 introduces our taxonomy of XAI tools, defining the dimensions and categories. This section also presents and analyses the evaluation of the tools for each dimension. Section 4 presents our conclusions and outlines future research directions.

2 Background and Related Work

Explainable AI encompasses many aspects aimed at improving the credibility, trust, and transparency of ML systems. For this reason, various authors have attempted to define its objectives and scope. From a more technical perspective, researchers have proposed classifications of XAI methods. Here, we summarise previous literature reviews that have sought to formalise these elements into taxonomies. As in other fields, taxonomies help identify appropriate methods for different problems and user needs. In each case, we highlight their limitations, which justify the need for a new, more comprehensive taxonomy of XAI tools.

The taxonomy proposed by Barredo Arrieta et al. [6] is based on two key questions: (1) What is explained? and (2) How is it explained? To answer these questions, the authors define two components: one for data and one for models. The first classifies explanations based on characteristics of the data used to train ML models. The second addresses the type of explainability that allows for the analysis of the model's behaviour. Within this component, the classification distinguishes between transparent models and post hoc explainability. Transparent models are those that are inherently interpretable, such as decision trees. In contrast, post hoc techniques are applied to complex models after training, such as deep neural networks and other "black box" models. While this taxonomy is useful for differentiating explainability approaches, its scope is limited to classifying XAI techniques at a high level, without considering the tools that implement them.

Ali et al. [5] propose a hierarchical taxonomy that categorises XAI techniques into four axes: (1) Data explainability, whose techniques focus on understanding the characteristics of training data; (2) Model explainability, which includes techniques aimed at making AI model structures more understandable; (3) Post hoc explainability, which involves techniques applied after model training; and (4) Explanation evaluation, i.e. methods for assessing the effectiveness of explanations generated by XAI techniques. Although this taxonomy provides a clear and more comprehensive structure to analyse XAI techniques, it does not incorporate any specific dimension to evaluate tools. Like the previous taxonomy, it focuses on existing literature rather than practical aspects of tool identification, selection, and evaluation.

The "XAI Toolsheet", developed by Karunagaran et al. [18], identifies 22 tool characteristics derived from a qualitative analysis of 152 tools. These characteristics are classified into three dimensions: metadata, utility, and usability. The metadata dimension includes basic descriptive information about the tools, such as type (visual tool, package, library), licence, and available documentation. The utility dimension evaluates the effective integration of XAI methods into users' projects, including supported data types, ML paradigms and models, and scope of explanations. The usability dimension refers to characteristics that facilitate the use of the tool, such as explanation formats and level of user's customisation. While this framework provides a specialised approach to describing XAI tools, it does not explore in depth the implementation of their explainability methods, despite this being their primary function.

3   Taxonomy of XAI Tools

This section outlines the methodology used to construct the taxonomy, including the search procedure for identifying XAI tools. Next, we present an overview of the taxonomy, structured into three dimensions: *metadata, scope and general purpose*, and *explainability methods*. The following sections define each dimension in detail and evaluate the selected tools according to their categories.

3.1   Methodology for Taxonomy Construction

Taxonomies are developed using either a top-down or a bottom-up approach [7]. In the top-down approach, the taxonomy and its dimensions are defined based on previous classifications and formal concepts. In the bottom-up approach, a literature search is conducted to identify existing works, and the dimensions of the taxonomy are determined during analysis. We adopt a mixed approach to developing our taxonomy. The taxonomies compiled in Sect. 2 were scrutinised to gather the relevant dimensions and categories. However, the rapid emergence and development of new tools necessitate an iterative review of the dimensions to encompass all observed functionalities.

We performed an online search of code repositories and academic literature to identify open-access XAI tools. Specifically, we searched GitHub using the keywords 'XAI' and 'Explainable AI', focusing on repositories with recent activity. We then established and applied exclusion criteria, removing tools that met any of these conditions. If a tool did not meet any exclusion criteria, we assessed whether it fulfilled the inclusion criteria. They ensure that a tool meets minimum requirements that not only allow its analysis with the taxonomy, but also serve to review and expand the definition of the taxonomy itself.

Tools were excluded if (1) they were proprietary or had licences restricting open access, (2) lacked verifiable implementations of XAI techniques, (3) had insufficient documentation or usage examples in their repository, or (4) had not been developed in the last five years. Conversely, tools were included if (1) they were open access or had licences permitting academic use, (2) were implemented in Python, (3) supported explainability methods either at the model level (intrinsic explainability) or at the post hoc level (post-hoc explainability), (4) had accessible documentation and activity on GitHub, assessed through popularity, number of forks, and recent updates, and (5) had been referenced in academic literature.

After applying these criteria, we confirmed that 15 tools met the requirements and included them in our analysis. The selected XAI tools are: Arena [11], ExplAIner [12], ModelStudio [23], What-if Tool [27], AIX360 [4], ALIBI [1],

An eXplainability toolbox for machine learning [28], CARLA [8], CEML [9], DALEX [10], ExplainerDashboard [13], Intel XAI tools [16], InterpretML [17], OmniXAI [24] and SHAPASH [26].

3.2   Overview of the Taxonomy Dimensions

Based on existing taxonomies in the literature (see Sect. 2), we have unified and expanded classification categories to address identified limitations. Additionally, a detailed analysis of each tool has enabled the iterative refinement of the taxonomy categories. As a result, the proposed taxonomy comprises three dimensions:

1. *Metadata*: Classification of tools based on technical attributes and availability.
2. *Scope and general purpose*: Evaluation of tool functionalities, including supported data types, ML models, and explanation forms.
3. *Explainability methods*: Detailed categorisation of implemented explainability techniques.

Figure 1 summarises the three dimensions and their corresponding categories. Overall, our taxonomy defines 31 categories. Categories marked with asterisk have not been included in previous taxonomies. Due to space constraints, the following sections provide a brief explanation of the categories within each dimension. Where applicable, tables summarise the presence of defined categories in the 15 selected tools, or their degree of compliance for qualitative characteristics. Our data analysis highlights the most relevant aspects within each dimension. The full taxonomy assessment of the 15 tools is available as supplementary material.

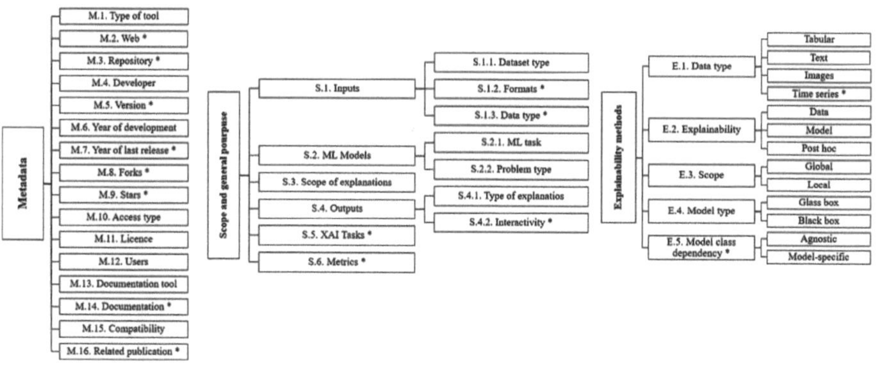

Fig. 1. Dimensions and categories of the proposed taxonomy of XAI tools.

3.3   Dimension 1: Metadata

This dimension encompasses categories related to tool development, distribution, and documentation. It includes categories that define the tool type, version, developer, access type, licence, and available documentation. These categories originate from the XAI Toolsheet. Additional categories have been introduced, including a rubric to assess documentation quality,[1] update frequency, and GitHub engagement metrics such as forks and stars. These measures indicate tool adoption and usage levels. Including these new categories offers a more comprehensive view of the status and evolution of each tool from a software perspective. To validate this dimension, we conducted a thorough review of the documentation and distribution repositories. A summary of the metadata dimension evaluation is shown in Table 1.

Table 1. Metadata dimension: Software development and distribution

Tool	M.5: Version	M.6: Year of Development	M.7: Last Release	M.8 Forks	M.9: Stars	M.11: License
Arena	0.2.2	2020	N/A	1	22	GPL-3.0
ExplAIner	N/A	2020	N/A	14	30	N/A
Model Studio	3.1.2	2019	2023	32	332	GPL-3.0
What-if tool	1.8.1	2018	2021	174	958	Apache 2.0
AIX360	0.3.0	2019	2023	313	1,700	Apache 2.0
ALIBI	0.9.6	2019	2024	257	2,500	BSL 1.1
eXplainability toolbox	0.1.0	2019	2021	179	1,200	MIT
CARLA	0.0.1	2021	N/A	63	291	MIT
CEML	0.7.0	2019	2023	11	44	MIT
DALEX	1.7.2	2018	N/A	168	1,400	GPL-3.0
Explainer Dashboard	0.5.0	2020	2025	342	2,400	MIT
Intel XAI tools	1.3.0	2020	2025	7	39	Apache 2.0
Interpret ML	0.6.13	2019	2025	751	6,500	MIT
OmniXAI	1.3.2	2021	2024	930	101	BSD-3-Clause
SHAPASH	2.7.9	2020	2025	2,900	347	BSD-3-Clause

Most tools have published new releases in the last two years, indicating active maintenance and ongoing development. Several tools have more than 1,000 stars on Github, suggesting widespread adoption and use. SHAPASH has the highest number of forks (2,900). While most tools provide documentation rated as complete or average, we have detected cases with insufficient documentation. Documentation platforms vary, ranging from GitHub to more structured systems such as ReadTheDocs.

---

[1] The rubric is available in the supplementary material.

Table 2. Metadata dimension: Compatibility with ML libraries (M.15)

Tool	caret	catboost	h20	keras/tensorflow	lightgbm	mlr	parsnip	pytorch	sklearn	skorch	xgboost
Arena	✓		✓	✓	✓	✓	✓		✓		✓
ExplAIner				✓							
Model Studio	✓		✓	✓	✓	✓	✓		✓		✓
What-if tool				✓					✓		✓
AIX360				✓				✓	✓		
ALIBI				✓				✓	✓		
eXplainability toolbox				✓				✓	✓		
CARLA				✓				✓	✓		✓
CEML				✓				✓	✓		
DALEX	✓		✓	✓	✓	✓	✓		✓		✓
Explainer Dashboard		✓			✓				✓	✓	✓
Intel XAI tools				✓				✓	✓		
Interpret ML				✓				✓	✓		
OmniXAI				✓				✓	✓		
SHAPASH		✓			✓				✓		✓

Regarding compatibility (M.15), most tools support widely used ML libraries such as TensorFlow/Keras, scikit-learn, pytorch, and XGBoost for model explainability (see Table 2). This enables users to integrate explainability aspects into ML pipelines across various development environments. Some tools specialise in specific ML frameworks, which may influence their selection based on the predominant development environment. Many tools are accompanied by publications that detail their approaches and validate their methodologies.

3.4 Dimension 2: Scope and General Purpose

This dimension classifies tools based on functional aspects, including the types of data they process (structured or unstructured), supported ML models, scope of explanations, and output formats (visual, numerical, or textual). These categories originate from the XAI Toolsheet but have been revised and expanded to encompass all currently available options. Our taxonomy also considers the level of interactivity offered by the tools, categorised as basic or advanced. We also assess whether the tasks supported by these tools relate to diagnosis, fairness, trust, and compliance. Additionally, a category classifies the metrics used to evaluate explanation effectiveness. While Ali et al. [5] treats metrics as an independent dimension, our taxonomy integrates them here, as evaluation is an essential functionality. Metrics are categorised based on the XAI tasks they support: diagnostic, fairness, trust, and compliance.

We validated this dimension by analysing tool functionalities, reviewing documentation, and testing provided examples. As shown in Table 3 for S.1 (Inputs), most tools handle structured data, particularly tabular data. Additionally, a significant number of tools support unstructured data, especially images and text. We observe less interest in supporting time series. Regarding ML models (S.2),

XAI tools widely support supervised learning tasks, including binary and multiclass classification, as well as regression (see Table 4). With the exception of CEML and CARLA, which provide only local explanations (counterfactuals), all other tools support both local and global explanations (S.3).

Table 3. Scope and general purpose dimension: Inputs (S.1)

Tool	S.1.1 Dataset Type		S.1.2 Data Type			
	Structured	Unstructured	Tabular	Images	Text	Time Series
Arena	✓		✓			
ExplAIner	✓	✓	✓	✓		
Model Studio	✓		✓			
What-if tool	✓	✓	✓	✓	✓	
AIX360	✓	✓	✓	✓	✓	✓
ALIBI	✓	✓	✓	✓	✓	
eXplainability toolbox	✓		✓			
CARLA	✓		✓			
CEML	✓		✓			
DALEX	✓		✓			
Explainer Dashboard	✓		✓			
Intel XAI tools	✓	✓	✓	✓	✓	
Interpret ML	✓		✓			
OmniXAI	✓	✓	✓	✓	✓	✓
SHAPASH	✓		✓			

As summarised in Table 5, the types of explanations (S.4.1) are predominantly visual and numerical. As for user interactivity (S.4.2), most offer basic functionalities, while only *What-if tool* has advanced interactivity. Diagnosis, fairness, and trust are common XAI tasks (S.5), though not all tools support all three. This reflects variations in tool purpose, with only a few extending beyond diagnosis to address ethical considerations and trustworthiness. Regarding metrics, while some fairness and trust indicators, such as *Demographic Parity* and *Trust Scores*, are mentioned, their adoption lacks uniformity. Identified indicators assess monotonicity, demographic parity, equality of opportunity, equitable precision, and trust factors such as fidelity and stability.

3.5   Dimension 3: Explainaibility Methods

This dimension builds on the taxonomy of XAI methods proposed by Ali et al. [5]. Additional categories have been introduced, including model class dependency and data type, with specific consideration for time series. Methods are classified based on the data type they operate on (tabular, images, text, time

Table 4. Scope and general purpose dimension: ML models (S.2) and Scope of Explanations (S.3)

Tool	S.2. ML Models			S.3. Scope of Explanations	
	Binary Classification	Multiclass Classification	Regression	Local	Global
Arena	✓	✓	✓	✓	✓
ExplAIner	✓	✓	✓	✓	✓
Model Studio	✓		✓	✓	✓
What-if tool	✓	✓	✓	✓	✓
AIX360	✓	✓	✓	✓	✓
ALIBI	✓	✓	✓	✓	✓
eXplainability toolbox	✓	✓	✓	✓	✓
CARLA	✓	✓		✓	
CEML	✓	✓	✓	✓	
DALEX	✓	✓	✓	✓	✓
Explainer Dashboard	✓	✓	✓	✓	✓
Intel XAI tools	✓	✓	✓	✓	✓
Interpret ML	✓	✓	✓	✓	✓
OmniXAI	✓	✓	✓	✓	✓
SHAPASH	✓	✓	✓	✓	✓

Table 5. Scope and general purpose dimension: Outputs (S.4) and XAI Tasks (S.5)

Tool	S.4.1. Types of Explanations			S.4.2 Interactivity			S.5. XAI Tasks			
	Visual	Textual	Numerical	Static	Basic	Advanced	Diagnosis	Fairness	Trust	Compliance
Arena	✓		✓				✓	✓		
ExplAIner	✓		✓							
Model Studio	✓		✓		✓		✓			
What-if tool	✓		✓			✓	✓	✓		
AIX360	✓		✓	✓			✓		✓	
ALIBI	✓		✓	✓			✓		✓	
eXplainability toolbox	✓		✓	✓			✓			
CARLA	✓		✓				✓			
CEML	✓		✓				✓			
DALEX	✓		✓	✓			✓	✓		
Explainer Dashboard	✓		✓		✓		✓			
Intel XAI tools	✓		✓	✓			✓	✓		
Interpret ML	✓		✓		✓		✓			
OmniXAI	✓	✓	✓		✓		✓			
SHAPASH	✓		✓	✓			✓		✓	

series), explainability class (data, model, post hoc), and method scope (global or local). This dimension also differentiates between intrinsically interpretable models ("glass box") and non-interpretable models ("black box"). Another category determines whether explainability methods are model-agnostic or model-specific. This dimension enables the taxonomy to offer a comprehensive classification, assisting researchers in identifying the most suitable XAI method implementation for their needs.

We identified 127 methods across the tools. Table 6 summarises the number of methods falling into each explainability category. The supplementary material

Table 6. Dimension of explainability methods: Summary of methods implemented in each tool (E.1-E.5)

Tool	E.1. Data Type				E.2. Explainability			E.3. Scope		E.4. Model Type		E.5. Dependency	
	Tabular	Images	Text	Time Series	Data	Model	Post hoc	Global	Local	Glass box	Black box	Agnostic	Specific
AIX360	15	11	6	3	9	12	13	6	12	6	13	10	9
ALIBI	12	7	3			1	10	5	9	1	10	9	3
eXplainability toolbox	4				1	3		4			4	4	
CARLA	13						13		13		13	13	
CEML	1						1		1		1	1	
DALEX	12		5		7	4	12	7	8		12	12	
ExplAIner	5	23	1		2	22	3	6	18		24	3	21
Explainer Dashboard	2					1	1	2			2	2	
Intel XAI tools	1	2	1				2		2		2		2
Interpret ML	9	2	2			9	4	9	9	5	4	4	5
OmniXAI	14	13	5	2	3	12	18	9	16	1	20	11	11
SHAPASH	4	2	2		1	2	4	2	3		4	4	
What-if tool	3	3	3				3		3		3	3	
Total	95	63	28	5	23	66	84	50	94	13	112	76	51

provides detailed information on the methods available in each tool. Arena and Model Studio are excluded from this table, as they use DALEX internally to provide XAI methods. The three tools are complementary components of the DrWhy.AI project,[2] each serving a different purpose: DALEX provides the core explanatory methods, Model Studio facilitates the generation of detailed reports for individual models, and Arena enables interactive comparison between multiple models.

Regarding data types (E.1), the tools predominantly support explaining models trained on tabular data (95 methods). Additionally, many methods focus on image processing (63) and text analysis (28). However, time series analysis methods are scarce, with only five identified. Post hoc methods are more prevalent, used by 84 methods (E.2). Similarly, methods to explaining "black box" models (112 methods) are more common than those for "glass box" (13) in E.3. Model-agnostic methods (76) also outnumber model-specific ones (51) in E.5.

Regarding method scope (E.3), we identified 50 methods that provide a global view of model behaviour. In contrast, 94 methods provide local-level explanations. This suggests a preference for detailed insights into individual model decisions, likely driven by demand for practical applications requiring such granularity. Additionally, 19 methods are classified as both global and local, as they allow aggregating local explanations to generate them at a global level.

Some tools, such as CARLA and CEML, specialise in counterfactual explanations. Counterfactual methods are available in 5 tools, while CEM, a method for contrastive explanations, is implemented by 4 tools. Among the most frequently implemented methods, SHAP appears in 8 of the analysed tools, while LIME is implemented in 6 tools. Partial dependence plots are present in 5 tools. Other methods present in at least 3 tools are: accumulated local effects, decomposition graphs, permutation importance, ceteris-paribus profiles, and gradient-

---

[2] https://modeloriented.github.io/DrWhy/ (Last accessed: 18 September 2025).

based methods. Recently, Intel XAI Tools and OmniXAI have introduced the use of large language models to process outputs generated by SHAP-based methods.

4   Conclusions

Explainable artificial intelligence offers methods and approaches of significant value to many application domains, enabling a deeper understanding the inner operation of ML techniques. To support the selection of publicly accessible tools, this study proposes a comprehensive taxonomy to characterise and evaluate them. Our taxonomy comprises three dimensions and 31 categories, classifying tools based on their support for various XAI tasks, data types, compatible ML models, and explanation scope. This work serves as a structured and up-to-date reference for XAI stakeholders, offering a clear overview of general-purpose XAI tools currently available. Notably, we provide additional valuable information about each tool in our supplementary material.

Our evaluation of XAI tools reveals insights into their current strengths, but also into some limitations. Support for tabular data is widespread, while support for images, text, and time series is less common. The vast majority of tools include XAI methods for generating both local and global explanations, usually presented in visual formats. However, the tools are primarily focused on diagnostics, and the level of user interactivity is still low. The full catalogue of algorithms is quite extensive, with over 100 implemented methods. AIX360, ALIBI, ExplAIner, and OmniXAI stand out in this regard, offering over 20 different methods for a wider variety of data types.

For future work, we plan to assess the suitability of XAI tools in addressing the specific needs and expectations of stakeholders in the agriculture, healthcare and education sectors. This evaluation will offer insights into how these tools can be adapted or enhanced to support transparency, fairness, and practical usability in specific application domains.

Acknowledgments. Work partially supported by the Spanish Ministry of Science, Innovation and Universities, grants PID2023-148396NB-I00 and PID2024-155284NA-I00 funded by MICIU/AEI/10.13039/501100011033; and project ProyExcel-0069, funded by the Regional Government of Andalucía and the European Regional Development Fund. Paola Montenegro-Cantos is supported by a doctoral scholarship funded by Carolina Foundation.

Data Availability Statement. Supplementary material is available in a public repository at https://zenodo.org/ records/15856737. The shared data includes the documentation checklist and the exhaustive evaluation of the 15 selected XAI tools using the proposed taxonomy.

Disclosure of Interests. The authors have no competing interests to declare that are relevant to the content of this article.

References

1. ALIBI (2025). https://github.com/SeldonIO/alibi
2. Ahmed, M., Ahmed, M.W., Kamruzzaman, M.: A systematic review of explainable artificial intelligence for spectroscopic agricultural quality assessment. Comput. Electron. Agric. **235**, 110354 (2025)
3. Ahmed, M.I., Spooner, B., Isherwood, J., Lane, M., Orrock, E., Dennison, A.: A systematic review of the barriers to the implementation of artificial intelligence in healthcare. Cureus **15**(10), e46454 (2023)
4. AIX360 (2025). https://github.com/Trusted-AI/AIX360
5. Ali, S., et al.: Explainable Artificial Intelligence (XAI): what we know and what is left to attain trustworthy artificial intelligence. Inf. Fus. **99**, 101805 (2023)
6. Barredo Arrieta, A., et al.: Explainable Artificial Intelligence (XAI): concepts, taxonomies, opportunities and challenges toward responsible AI. Inf. Fus. **57**, 82–115 (2019)
7. Broughton, V.: Essential Classification. Facet Publishing (2015)
8. CARLA (2025). https://github.com/carla-recourse/CARLA
9. CEML (2025). https://github.com/andreArtelt/ceml
10. DALEX (2025). https://dalex.drwhy.ai/
11. Arena DrWhy (2024). https://arena.drwhy.ai/docs/
12. Explainer AI (2024). https://explainer.ai/
13. ExplainerDashboard (2025). https://explainerdashboard.readthedocs.io
14. Guidotti, R., Monreale, A., Ruggieri, S., Turini, F., Giannotti, F., Pedreschi, D.: A survey of methods for explaining black box models. ACM Comput. Surv. **51**(5), 1–42 (2018)
15. Hall, A., Agarwal, V.: Barriers to adopting artificial intelligence and machine learning technologies in nuclear power. Prog. Nucl. Energy **175**, 105295 (2024)
16. Intel XAI tools (2025). https://github.com/IntelAI/intel-xai-tools
17. InterpretML (2025). https://interpret.ml/docs/
18. Karunagaran, S., Lucic, A., Custis, C.: XAI ToolSheet: towards a documentation framework for XAI Tools. In: Workshop on Explainable Artificial Intelligence (IJCAI) (2022)
19. Kaur, D., Uslu, S., Rittichier, K.J., Durresi, A.: Trustworthy artificial intelligence: a review. ACM Comput. Surv. **55**(2), 1–38 (2022)
20. Khosravi, H., et al.: Explainable artificial intelligence in education. Comput. Educ. Artif. Intell. **3**, 100074 (2022)
21. Korica, P., Gayar, N.E., Pang, W.: Explainable artificial intelligence in healthcare: opportunities, gaps and challenges and a novel way to look at the problem space. In: International Conference on Intelligent Data Engineering and Automated Learning (IDEAL), pp. 333–342 (2021)
22. Lyu, Q., Wu, S.: Explainable artificial intelligence for business and economics: methods applications and challenges. Expert Syst. **42**(4), e70017 (2025)
23. ModelStudio (2025). https://modelstudio.drwhy.ai/
24. OmniXAI (2025). https://github.com/salesforce/OmniXAI
25. Shadi, M.R., Mirshekali, H., Shaker, H.R.: Explainable artificial intelligence for energy systems maintenance: a review on concepts, current techniques, challenges, and prospects. Renew. Sustain. Energy Rev. **216**, 115668 (2025)
26. SHAPASH (2025). https://shapash.readthedocs.io/en/latest/
27. The What-if tool (2025). https://pair-code.github.io/what-if-tool/
28. An eXplainability toolbox for machine learning. https://github.com/EthicalML/xai (2025)

Preliminary Analysis of Loss Functions in Generative Model Inversion Attacks

F. Moreno$^{(\boxtimes)}$, A. Rivera, M. J. del Jesus, and M. D. Pérez-Godoy

Department of Computer Science, University of Jaén, Jaén, Spain
`{fmjimene,arivera,mjjesus,lperez}ujaen.es`

Abstract. Nowadays, the number of cyberattacks that affect machine learning and deep learning models is increasing. These models may be vulnerable to different attacks, including adversarial attacks that can compromise information at various stages of data processing. For example, the model inversion attack is an exploratory attack that focuses on compromising the information privacy, recovering the training data, or inferring sensitive information from these data. In this type of attack, multiple loss functions have been employed in the literature, including Cross-Entropy loss, Max-Margin loss, and Poincaré loss. The objective of this paper is to study the stability and effectiveness of these loss functions in the training process of the selected Model Inversion attack method. In addition, the number of iterations required to achieve high attack performance is analyzed, taking into account the loss function. To this end, target models with varying degrees of complexity and datasets of different sizes are utilized.

Keywords: Model inversion attack · Loss function · Face recognition

1 Introduction

In recent years, the number of cyberattacks has increased, affecting a variety of entities and producing significant economic, social, and legal impacts on society [14]. This increase is due to greater use of digital technology and the adoption of various tools that facilitate the execution of cyberattacks, some of them developed with Artificial Intelligence (AI). In this context, Machine Learning and Deep Learning models may be vulnerable to various types of attacks, including adversarial attacks.

There are different types of adversarial attacks according to their attack surface, that is, the stages of data processing that are vulnerable [1,11]. The main attack scenarios include evasion attacks, poisoning attacks, and exploratory attacks. The objective of exploratory attacks is to extract information about the target model without altering the dataset in the training phase. To this end, adversarial instances that the model correctly classifies during the testing phase are generated.

In exploratory attacks, three types can be found: model inversion, membership inference, and model extraction. Model inversion (MI) [3,11] attempts to

© The Author(s), under exclusive license to Springer Nature Switzerland AG 2026
L. Martínez et al. (Eds.): IDEAL 2025, LNCS 16239, pp. 331–342, 2026.
https://doi.org/10.1007/978-3-032-10489-2_28

recover training data or infer sensitive information in these data from the target model. This type of attack focuses on compromising the information privacy of the target models. In the context of facial recognition models, which are employed in this study, a privacy information leak may result in severe consequences, such as, identity theft, target model replication, etc.

In the last years, multiple MI attacks have been proposed to compromise Deep Neural Networks (DNNs). Attacks on image classification models, particularly those used for facial recognition, as analyzed in this study, have become a primary focus of MI attacks. These methods can be classified as black-box attacks or white-box attacks. [4], depending on whether the attacker has access to the weights and outputs of the target model, respectively.

In white-box attacks, the adversary has full access to the weights and the output of the target model [4]. The GMI method [19] is considered the first MI attack on DNNs. This method employs a Wasserstein GAN to facilitate knowledge extraction for guiding the inversion process. The KEDMI [2] and VMI [17] methods recover the private data distribution, instead of performing the instance-level data reconstruction. All these methods use the Cross-Entropy loss. However, other methods employ alternative loss functions. The Plug & Play Attacks [16] and IF-GMI [13] approaches utilize the Poincaré loss function. Pseudo Label-Guided MI (PLG-MI) [18] is a multi-stage method that introduces a top-n selection strategy for pseudo-label generation and incorporates the Max-Margin function loss.

After presenting the white-box attack methods, it can be observed that these methods utilize different loss functions for the training of the selected GANs, including Cross-Entropy, Max-Margin and Poincaré. Indeed, different tests with various loss functions were conducted in the PLG-MI [18] and the Plug & Play Attacks [16] methods. Given this scenario, in this paper, we study the impact of these loss functions on the employed white-box MI attack method, with the aim of identifying those that provide greater stability to the MI method training process and improve the effectiveness of this attack. Thus, the number of iterations required to achieve high performance in MI attack is analyzed, with the objective of reducing overfitting and computational cost. For this purpose, target models that exhibit variation in complexity and depth are employed. Specifically, the VGGNet-16 and ResNet-50 target models are utilized. These models are trained on CelebA dataset, while the FFHQ, LFW and CFP datasets are used for the attack execution.

The paper is structured as follows: In Sect. 2, the selected MI method and the target models are described. Furthermore, the dataset and loss functions employed are presented. Subsequently, in Sect. 3, the results of the MI method with different loss functions are presented. After the presentation of the results, a comprehensive analysis of the results is provided in Sect. 4. Finally, conclusions are presented in Sect. 5.

2   Material and Methods

In this section, the selected MI attack method and the target models are presented. Additionally, a brief description is provided regarding the public and private datasets, along with the loss functions that have been utilized.

2.1   Model Inversion Attack Method

In this study, the PLG-MI method [18] was utilized to execute the MI attack. This method is considered a white-box attack, as the adversary has complete access to the target model's parameters. In Fig. 1, phases of the PLG-MI method are shown: pseudo-label generation (A), and the Pseudo Label-Guided MI Attack (B and C), which is composed of two stages. In the first phase (B), a conditional GAN is trained using the public dataset with pseudo-labels. In the second stage (C), the trained conditional GAN is employed to reconstruct private images from the original training dataset.

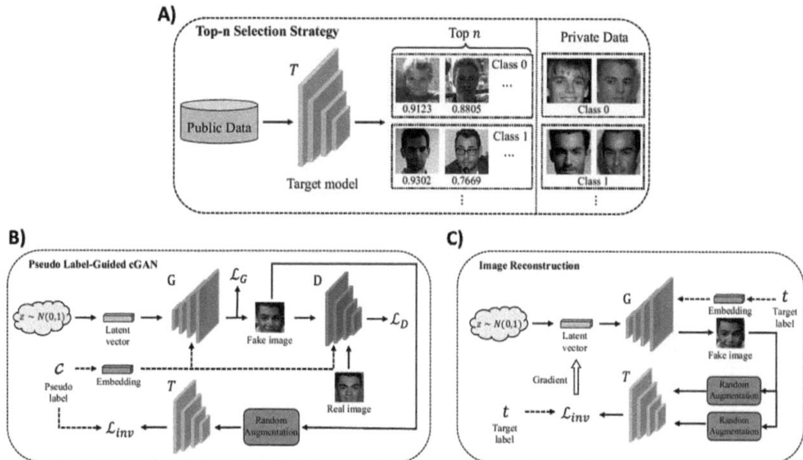

Fig. 1. Stages of the PLG-MI method: pseudo-label generation (A), training of conditional GAN (B), and reconstruction of private images (C). Source: [18].

Initially, this method applies a top-n selection strategy for pseudo-label generation. In this strategy, the n images with the highest confidence scores provided by the target model are selected for each pseudo-label in the public dataset. The public dataset with pseudo-labels is employed to guide the training of the conditional GAN. Consequently, the distribution of features for each class can be modeled. Furthermore, a set of random transformations (cropping, color jittering, resizing, etc.) is applied to the generated images to achieve more stable convergence to realistic images. Finally, the trained conditional GAN is employed to reconstruct the private images, that is, given a target class, the objective is to find the appropriate latent vectors, so that the generated images are progressively more proximate to the original images of the target class. In this phase, random transformations are also used to ensure that the generated images contain the key features of the target class.

2.2 Target Models

PLG-MI has demonstrated to be an effective attack method against various image classification models. In this study, the facial recognition models known as VGGNet-16 and ResNet-50 were selected as target models. The VGGNet-16 network [12] comprises 11 blocks, where the first 8 blocks are convolutional, and the remaining 3 are called fully connected. These fully connected blocks are implemented using convolutional layers, where the size of the filters matches the dimensions of the input data. Finally, the softmax layer is used to classify facial images. The ResNet-50 model [5] employs residual connections, which facilitate the learning of residual functions and help mitigate the vanishing gradient problem. This network is composed of convolutional layers, fully connected layers, as well as identity and convolutional blocks, which consist of multiple convolutional layers.

Fine-tuning was applied to the VGGNet-16 and ResNet-50 models to adapt these networks to the selected datasets and improve their performance. In this case, the weights of the convolutional layers were unmodified, while the fully connected layers were replaced. Specifically, a flatten layer, a batch normalization layer, and a dense layer were introduced after the convolutional layers (see Fig. 2).

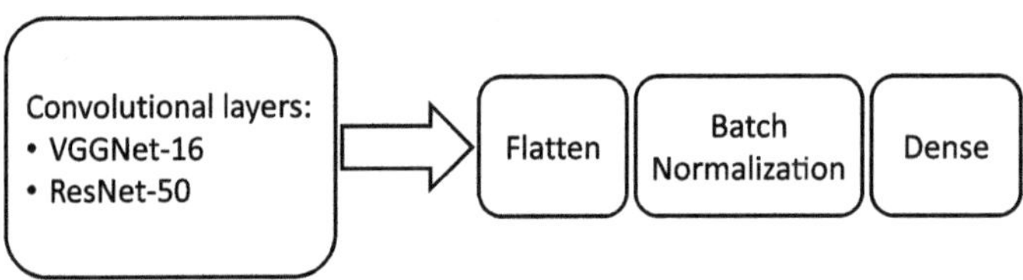

Fig. 2. Fine-tuning process applied to the VGGNet-16 and ResNet-50 architectures.

2.3 Datasets

The execution of the PLG-MI method requires at least two datasets: a private dataset and a public dataset. The private dataset is utilized to exclusively train the target model, while the public dataset is employed to generate pseudo-labels using the top-n selection strategy and to subsequently train the conditional GAN to recover the training facial images.

In this case, the selected private dataset used for the target model training was CelebFaces Attributes (CelebA). CelebA [10] is a dataset that was created by labeling images from the CelebFaces dataset. It comprises 10,000 individuals, with each individual represented by 20 images. Thus, this dataset contains a total of 200,000 facial images. PLG-MI uses the CelebA dataset split defined by the KEDMI method [2] to obtain the private and public datasets. In particular, the private dataset consists of 30,027 facial images of 1,000 identities.

The Labeled Faces in the Wild (LFW), Celebrities in Frontal-Profile (CFP) and FlickrFaces-HQ (FFHQ) datasets were selected as the public datasets used to execute the PLG-MI method. LFW [7] is an unconstrained dataset that contains 13,233 facial images of 5,749 unique individuals, where the facial images exhibit a wide range of variations seen in everyday life (age, pose, gender, facial expressions, illumination, etc.). CFP [15] combines unconstrained and constrained settings, as pose variation is studied in a controlled way, while the remaining variations are unconstrained. Thus, this dataset only includes frontal and profile faces. CFP consists of 10 frontal and 4 profile facial images for each of the 500 individuals, resulting in a total of 7,000 instances. FFHQ [8] comprises 70,000 high-quality facial images that present extensive variations in age, accessories, background, and ethnicity. Despite the presence of labels in the LFW and CFP datasets, these labels were not used, because the PLG-MI attack generates pseudo-labels from the selected public dataset.

2.4   Loss Functions

In this study, three loss functions were employed as inversion losses to assess their impact on the training process. Specifically, Cross-Entropy, Max-Margin, and Poincaré losses were used.

MI methods commonly apply the Cross-Entropy as inversion loss when dealing with a classification problem involving n classes. This loss function takes as input (x) the unnormalized logits for each class that have been extracted from the last layer of the target model, without applying the softmax function, along with their corresponding real pseudo-labels (y). It measures the difference between predicted probabilities and the true class labels. Equation 1 illustrates the calculation of the Cross-Entropy loss.

$$l_{ce} = \sum_{n=1}^{N} \frac{1}{\sum_{n=1}^{N} w_{y_n}} (-w_{y_n} log \frac{e^{x_{n,y_n}}}{\sum_{c=1}^{C} e^{x_{n,c}}}) \tag{1}$$

where w represents the weight, C is the number of classes and N denotes the batch size. It is important to note that this loss function was implemented using the Torch package[1].

The Max-Margin loss was introduced by Xiaojian Yuan et al. [18] to address the gradient vanishing problem and produce stronger attacks. This function loss focuses on identifying the most representative instance in the target class, while distinguishing it from the remaining classes. Equation 2 shows the computation of the Max-Margin loss.

---

[1] https://docs.pytorch.org/docs/stable/.

$$l_{mm} = -l_c(x) + \max_{j \neq c} l_j(x) \tag{2}$$

where c denotes the target class, and l_c represents the logits of this target class.

The Poincaré loss [9] leverages the Poincaré space, where the distances on the surface of the Poincaré ball increase exponentially when you move towards the surface of the ball, thereby solving the noise curing problem. The noise curing occurs when the gradient decreases during the iterative attack, causing two successive noises to be highly similar due to the accumulation of momentum. The formula of Poincaré loss is included in Eq. 3.

$$l_p = d(u, v) = arcosh(1 + \frac{2\,\|u - v\|_2^2}{(1 - \|u\|_2^2)(1 - \|v\|_2^2)}) \tag{3}$$

where u denotes the logits which are normalized using the l_1 distance, and v is the target vector encoded as a one-hot vector with respect to the target class.

3   Experimentation and Results

In this section, the results of all experiments that employ the VGGNet-16 and ResNet-50 target models and the PLG-MI method are presented. In addition, the experimental framework is included, where hyperparameters, metrics, and execution of the experiments are indicated.

3.1   Experimental Framework

The values of the main hyperparameters are summarized in Table 1. In this case, the VGGNet-16 and ResNet-50 target models have the same hyperparameters. For the PLG-MI method, the main hyperparameters for each stage are also provided. The remaining hyperparameters can be found in the original article [18]. The value of the top-n parameter for pseudo-label generation is set to 30 by default, which means that the 30 images with the highest confidence scores for each class are selected.

In this study, the Fréchet Inception Distance (FID) metric [6] was employed to assess the quality of the generated images. The objective of this metric is to evaluate the distance between two sets of images by comparing their statistical properties through the pretrained Inception network. The Eq. 4 illustrates the calculation of the FID metric.

$$FID = \|\mu_r - \mu_g\|^2 + Tr(\Sigma_r + \Sigma_g - 2(\Sigma_r \Sigma_g)^{1/2}) \tag{4}$$

where μ_r and Σ_r represent the mean and the covariance of the real images and μ_g and Σ_g denote the mean and the covariance of the generated images.

Figure 3 shows the executed experiments using the VGGNet-16 and ResNet-50 target models. The training of these models was performed utilizing the CelebA private dataset. For the PLG-MI method, the LFW, CFP, and FFHQ

Table 1. Selected hyperparameters of the target models and the PLG-MI method.

Models	MI stages	Hyperparameters	Values
Target model	-	Optimizer	Adam
		Learning rate	0.0001
		Epochs	10
		Batch size	32
		Loss function	Cross-Entropy
PLG-MI attack	Pseudo-label generation	Top-n	30
	Training of conditional GAN	Batch size	64
		Learning rate	0.0002
		Generator and discriminator optimizer	Adam
	Reconstruction of private images	Learning rate	0.1
		Iterations	300
		Number of images per class	5

public datasets were employed to generate the pseudo-labels and train the conditional GAN. The Cross-Entropy, Max-Margin, and Poincaré loss functions were applied to each public dataset. It is important to note that the number of iterations ranges from 1 to 25,000 to study the inversion loss and FID values.

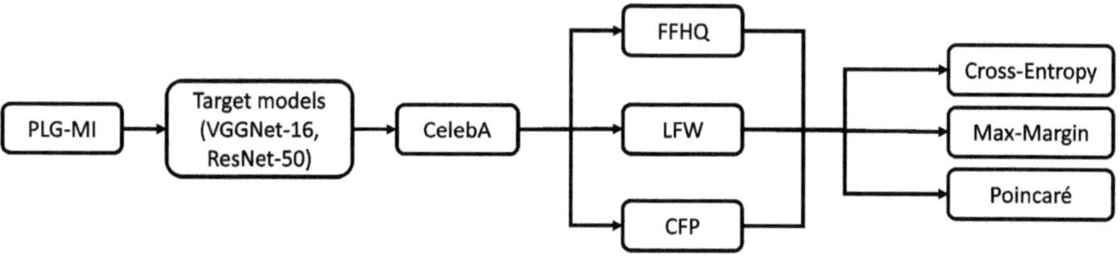

Fig. 3. Experiments execution using the VGGNet-16 and ResNet-50 as the target models and the PLG-MI attack.

The subsequent subsections present the results for the VGGNet-16 (see Fig. 4) and ResNet-50 (see Fig. 5) target models, respectively. For each target model, a figure is included that exhibits the FID (the smaller, the better) and the inversion loss values extracted from the respective architecture. These results were obtained from the training data during the conditional GAN training phase of the PLG-MI attack.

3.2   VGGNet-16 Results

According to the FID values (left column of Fig. 4), it can be observed that using the Cross-Entropy and Max-Margin losses, the optimization exhibits a relatively standard behavior, as the FID values decrease after some initial iterations and reach a minimum, whereas the Poincaré loss demonstrates erratic behavior throughout the entire training process. For the Cross-Entropy loss, the FID values do not vary significantly beyond the 10,000th iteration, suggesting that a reduced value of FID can be achieved within the first 10,000 iterations. After this point, the FID values attain a stabilization phase. In this loss function, minimum FID values of approximately 100 are reached for the CFP and LFW datasets, and 40 for the FFHQ dataset. The FID values obtained with the Max-Margin loss decrease during the initial iterations but increase when the minimum is reached. Depending on the minimum FID value, a lower FID is achieved with the Max-Margin loss compared to that obtained with the Cross-Entropy loss. In this particular loss function, minimum FID values ranging from 50 for the FFHQ and LFW datasets, and 70 for the CFP dataset are attained. The FID values using the Poincaré loss (4) exhibit an erratic behavior, since no clear optimization of the FID metric is achieved. Even, the LFW and CFP datasets exhibit an increase in this measure during the training. In this specific loss function, the lowest FID values observed are around 200 for the CFP dataset, 130 for the LFW dataset, and 50 for the FFHQ dataset.

Regarding the inversion loss values (right column of Fig. 4), it can be observed that the Max-Margin loss exhibits considerable instability, as it produces significantly reduced values, reaching a value of approximately -900 in the 23,000th iteration. This reduction in the loss function when its values decrease below 0 coincides with the increase in FID values, which suggests the presence of overfitting, particularly beyond the 10,000th iteration for the CFP dataset and the 5,000th and 7,000th iterations for the FFHQ and LFW datasets, respectively. The Poincaré loss provides similar values across all iterations, just like the FID values obtained with this function. This observation indicates that the conditional GAN may not be learning effectively when using this loss function. Conversely, the Cross-Entropy loss shows greater stability in comparison to the other loss functions. This loss results in a more gradual reduction, reaching a value close to 0 from the 10,000th iteration. Beyond this point, the Cross-Entropy loss stabilizes, and this stabilization coincides with the iteration where the minimum FID value is achieved.

3.3 ResNet-50 Results

The ResNet-50 model provides lower FID values (left column of Fig. 5) in all loss functions in comparison to the VGGNet-16 target model. For the Cross-Entropy loss, FID exhibits a gradual decrease until the 7,000th iteration. After this point, the FID values are maintained. In this loss function, minimum FID values of approximately 80 are reached for the CFP dataset, 40 for the LFW dataset, and 20 for the FFHQ dataset. Conversely, the Max-Margin loss shows an analogous behavior concerning the VGGNet-16 network. In this case, the FID values initially diminish until the 10,000th iteration for the LFW and FFHQ datasets and, approximately, the 19,000th iteration for the CFP datasets. Subsequently, these values increased. In this particular loss function, a minimum FID values ranging from 60 for the CFP dataset, 40 for the LFW dataset and 20 for the FFHQ dataset are attained. Similar to the VGGNet-16 model, the Poincaré loss

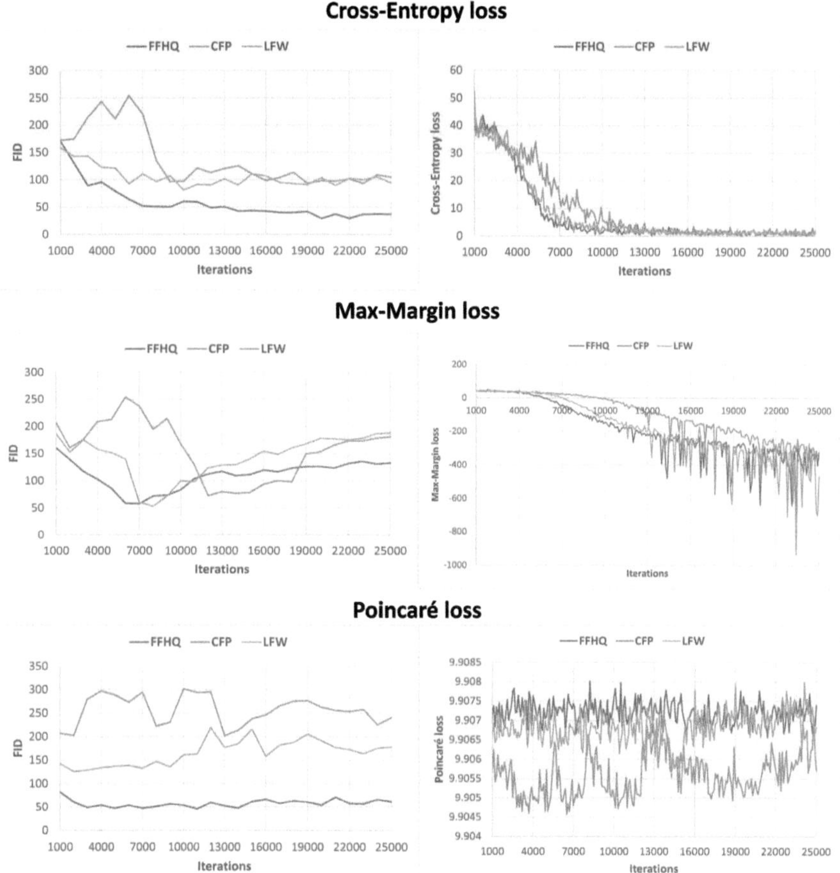

Fig. 4. FID and inversion loss values using the VGGNet-16 as the target model. The Cross-Entropy, Max-Margin, and Poincaré losses were used.

in the ResNet-50 model exhibits analogous values across all iterations, except for the LFW and FFHQ datasets, which display a pronounced decrease in the initial iterations after which the values remain unchanged or even increase. This loss function attains minimum values of approximately 110 for the CFP dataset, 40 for the LFW dataset, and 20 for the FFHQ dataset.

With respect to the inversion loss values (right column of Fig. 5), the Cross-Entropy and Max-Margin losses demonstrate a relatively standard behavior, yielding a more gradual reduction over the training stage, compared to the aforementioned network. Minimum FID values are achieved when the Cross-Entropy loss stabilizes its optimization. The decrease in the Max-Margin values below 0 is observed in later iterations compared to the VGGNet-16 model, corresponding to an increase in the FID values, similar to the VGGNet-16 model. The Poincaré loss using the ResNet-50 target model also exhibits erratic behavior similar to that of the VGGNet-16 model.

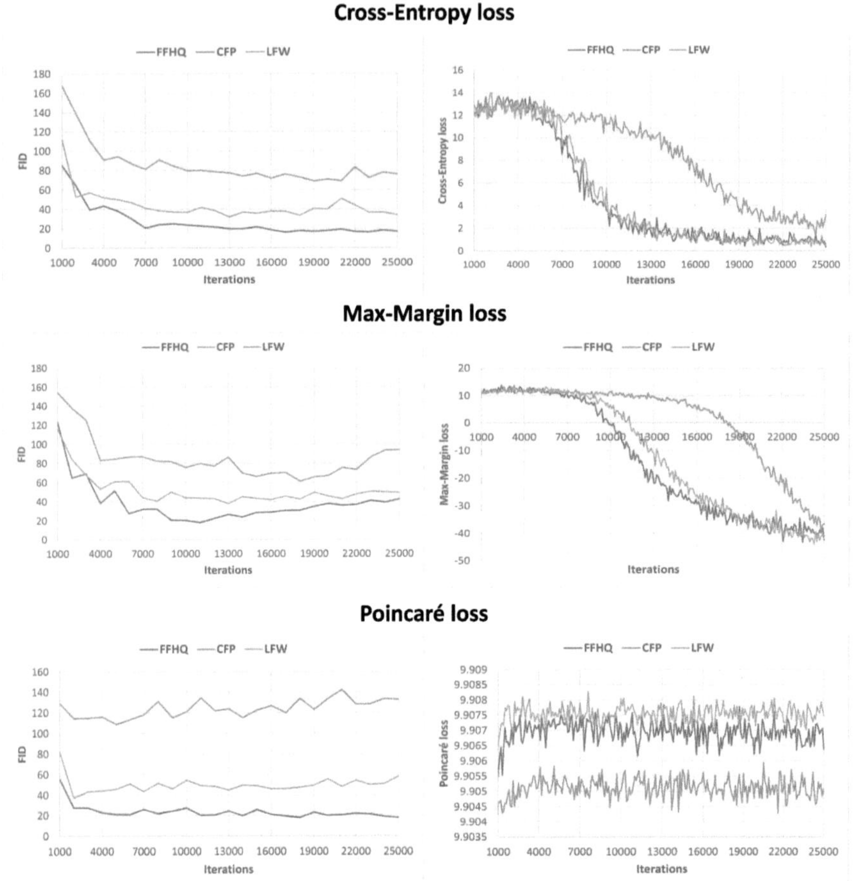

Fig. 5. FID and inversion loss values using the ResNet-50 as the target model. The Cross-Entropy, Max-Margin, and Poincaré losses were used.

4   Analysis of the Results

Considering the previously mentioned results, it can be deduced that the optimization of the FID metric using Cross-Entropy loss shows a standard behavior, decreasing after some initial training iterations, depending on the selected model and the public dataset. Afterward, the FID values achieve only slight improvements beyond a certain point (the well-known elbow). In general, this loss function reaches satisfactory FID values, provided that a sufficient number of iterations are executed. The Cross-Entropy loss determines when the training process should be stopped. Specifically, this stopping point is produced when these values stabilize (after the optimization elbow).

Regarding the Max-Margin loss, the FID values exhibit a V-shaped pattern, that is, they decrease after some initial iterations, depending on the public dataset and the target model, followed by an increase when the minimum values are achieved. However, this loss function provides similar or even lower minimum FID values than those obtained with the Cross-Entropy loss. The Max-Margin loss values indicate that the training process should be terminated when these values become negative, as beyond this point the FID values degrade.

According to the Poincaré loss, it can be observed that the FID metric exhibits an erratic pattern from the start of training, making it difficult to draw meaningful conclusions. Furthermore, the FID values are higher compared to those obtained with the remaining loss functions.

In addition, it can be observed that the CFP dataset exhibits greater FID values compared to the remaining datasets and optimization shapes with an initial upward trend. This may be attributed to the reduced number of samples that makes learning difficult, with this particular dataset, as it contains only 7,000 samples.

FID values obtained with ResNet-50 are lower than those in VGGNet-16. This may be attributed to the greater depth of the ResNet-50 model compared to the VGGNet-16 network. Thus, the ResNet-50 architecture suggests an improved ability to capture facial features and extract more sensitive information. As a result, the PLG-MI attack may reconstruct images that reveal highly sensitive information from private data.

5   Conclusions

In this paper, we examine the impact of the Cross-Entropy, Max-Margin and Poincaré loss functions on the training of the PLG-MI white-box attack, for identifying those that provide greater stability and effectiveness for this MI attack. In addition, we investigate the effect of these loss functions on the number of iterations required to achieve high performance in this attack. The results show that, although the Max-Margin loss achieves satisfactory outcomes, it is necessary to analyze its values, as the FID scores tend to deteriorate from a certain iteration. In contrast, the Cross-Entropy loss achieves greater stability improvements in the FID metric. Finally, the Poincaré loss results in erratic behavior and produces higher FID values. For future work, this study can be extended by incorporating additional MI attack methods and loss functions. Furthermore, various defense strategies against such attacks could be utilized to evaluate their impact on the training and effectiveness of MI attacks.

Acknowledgments. The research carried out in this study is part of the project "Advances in the development of trustworthy AI models to contribute to the adoption and use of responsible AI in healthcare (TAIH)" with the code PID2023-149511OB-I00, funded by MICIU/AEI /10.13039/50110001 1033 and by ERDF, EU.

References

1. Chakraborty, A., et al.: A survey on adversarial attacks and defences. CAAI Trans. Intell. Technol. **6**, 25–45 (2021)
2. Chen, S., et al.: Knowledge-enriched distributional model inversion attacks. In: IEEE/CVF International Conference on Computer Vision, pp. 16178–16187 (2021)
3. Dibbo, S.V.: SoK: model inversion attack landscape: taxonomy, challenges, and future roadmap. In: 2023 IEEE 36th Computer Security Foundations Symposium, pp. 439–456 (2023)
4. Fang, H., et al.: Privacy leakage on DNNs: a survey of model inversion attacks and defenses. ArXiv abs/2402.04013 (2024)
5. He, K., et al.: Deep residual learning for image recognition. In: IEEE Conference on Computer Vision and Pattern Recognition, pp. 770–778 (2016)
6. Heusel, M., et al.: GANs trained by a two time-scale update rule converge to a local nash equilibrium. In: Neural Information Processing Systems, vol. 30 (2017)
7. Huang, G.B., et al.: Labeled faces in the wild: a database for studying face recognition in unconstrained environments. In: Faces in 'Real-Life' Images: Detection, Alignment, and Recognition (2008)
8. Karras, T., et al.: A style-based generator architecture for generative adversarial networks. In: IEEE/CVF Conference on Computer Vision and Pattern Recognition, pp. 4396–4405 (2018)
9. Li, M., et al.: Towards transferable targeted attack. In: IEEE/CVF Conference on Computer Vision and Pattern Recognition, pp. 641–649 (2020)
10. Liu, Z., et al.: Deep learning face attributes in the wild. In: IEEE International Conference on Computer Vision, pp. 3730–3738 (2015)
11. Paracha, A., et al.: Machine learning security and privacy: a review of threats and countermeasures. EURASIP J. Inf. Secur. **2024**, 10 (2024)
12. Parkhi, O., et al.: Deep face recognition. In: British Machine Vision Conference, pp. 1–12 (2015)
13. Qiu, Y., et al.: A closer look at GAN priors: exploiting intermediate features for enhanced model inversion attacks. In: European Conference on Computer Vision, pp. 109–126 (2024)
14. Roumani, Y., et al.: Examining the factors that impact the severity of cyberattacks on critical infrastructures. Comput. Secur. **148**, 104074 (2024)
15. Sengupta, S., et al.: Frontal to profile face verification in the wild. In: 2016 IEEE Winter Conference on Applications of Computer Vision, pp. 1–9 (2016)
16. Struppek, L., et al.: Plug & play attacks: towards robust and flexible model inversion attacks. ArXiv abs/2201.12179 (2022)
17. Wang, K.C., et al.: Variational model inversion attacks. ArXiv abs/2201.10787 (2022)
18. Yuan, X., et al.: Pseudo label-guided model inversion attack via conditional generative adversarial network. In: AAAI Conference on Artificial Intelligence, vol. 37, no. 3, pp. 3349–3357 (2023)
19. Zhang, Y., et al.: The secret revealer: generative model-inversion attacks against deep neural networks. In: IEEE/CVF Conference on Computer Vision and Pattern Recognition, pp. 253–261 (2020)

Personalized Counterfactual Explanations via Cluster-Based Fine-Tuning of GANs

Ahmed A. Fares[1,2]([✉]) and João Mendes-Moreira[1,2]

[1] Faculty of Engineering, University of Porto, Porto, Portugal
ahmed.a.fares@inesctec.pt
[2] INESC Tec, Porto, Portugal

Abstract. Counterfactual explanations (CFs) help users understand and act on black-box machine learning decisions by suggesting minimal changes to achieve a desired outcome. However, existing methods often ignore individual feasibility, leading to unrealistic or unactionable recommendations. We propose a personalized CF generation method based on cluster-specific fine-tuning of Generative Adversarial Networks (GANs). By grouping users with similar behavior and constraints, we adapt immutable features and cost weights per cluster, allowing GANs to generate more actionable and user-aligned counterfactuals. Experiments on the German Credit dataset show that our approach achieves a 6× improvement in prediction gain and a 30% reduction in sparsity compared to a baseline CounterGAN, while maintaining plausibility and acceptable latency for online use.

Keywords: Counterfactual Explanations · GANs · Personalization · Actionability

1 Introduction

Machine learning models are increasingly being used in high-stakes domains such as credit scoring, hiring, and healthcare. Despite their predictive power, many of these models act as black boxes, offering little transparency in how decisions are made, raising concerns on trust, fairness, and accountability[1].

Counterfactual explanations (CFs) have emerged as a popular method for interpreting black-box predictions [6,7]. A CF highlights the minimal changes needed to an input instance to achieve a desired outcome, e.g., "Had your income been higher, your loan would have been approved." Despite existing CF methods, including GAN-based techniques like CounterGAN [5], they lack personalization and often propose unrealistic or unactionable changes, limiting practical usability. They typically ignore user-specific constraints such as immutable features (e.g., age) or effort-based costs (e.g., changing occupation), resulting in explanations that are unhelpful or infeasible in practice. In reality, what is feasible for one

---

[1] The full codebase is publicly available on https://github.com/ahmedfaresfeup/PersonalizedCounterfactualExplanations

L. Martínez et al. (Eds.): IDEAL 2025, LNCS 16239, pp. 343–348, 2026.
https://doi.org/10.1007/978-3-032-10489-2_29

individual may be difficult or impossible for another, making both immutability and cost highly subjective. To overcome these limitations, we propose a method that generates personalized counterfactual explanations by fine-tuning GANs on user clusters. Each cluster encodes shared behavioral and constraint patterns, enabling context-sensitive and user-aligned CFs.

Our main contributions are threefold: (i) we introduce a cluster-based fine-tuning approach to personalize CFs using GANs, (ii) we derive actionable and immutable features from user clusters based on intra-cluster stability and domain knowledge, and (iii) we demonstrate a 6× increase in prediction gain and a 30% reduction in sparsity over a baseline CounterGAN.

2   Background and Related Work

Counterfactual explanations (CFs) have gained traction as a method to interpret the decisions of complex black-box models by identifying minimal changes to input features that would alter a prediction [7]. In contrast to feature attribution methods, CFs provide actionable "what-if" scenarios, making them particularly relevant in decision-critical domains like credit, hiring, and healthcare [3].

Several generative approaches have been proposed to improve the realism and diversity of CFs, notably CounterGAN [5]. It combines a residual GAN with proximity and classification losses to ensure plausibility and label flip. However, these methods are typically trained once on the full dataset, ignoring user variability in constraints and feasibility. This one-size-fits-all paradigm may lead to CFs that are technically correct but practically unhelpful.

To improve practicality, other works have explored constraint-aware or user-sensitive CFs. Actionable recourse [6] incorporates domain constraints like immutable features or cost bounds. Mahajan et al. [4] propose preserving causal structures to avoid invalid recommendations. Still, many of these approaches impose constraints post hoc, or lack a generative component that ensures realism.

Despite these advances, personalization is often either superficial or omitted entirely. In practice, feasibility is user-specific: changing jobs may be easy for one person but infeasible for another. Such differences should be encoded during CF generation, not after. To address this, our method clusters users based on raw feature similarity and fine-tunes GANs per cluster using custom cost weights and immutable masks. This integration ensures that personalization, actionability, and plausibility are embedded directly into the generation process.

3   Proposed Methodology

Our method builds on **CounterGAN's architecture**, which uses a generator G and discriminator D to produce realistic counterfactuals through adversarial training. We extend this framework by integrating a user-specific personalization component directly into the training process. Figure 1 illustrates the overall structure of our personalized counterfactual generation pipeline.

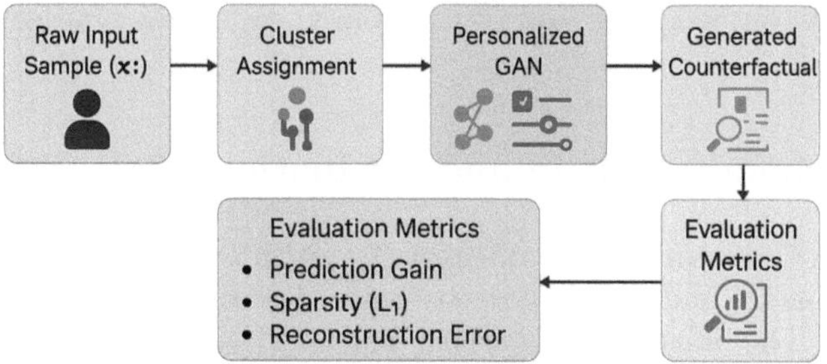

Fig. 1. Overall structure of the personalized counterfactual generation framework. The model combines user clustering, constraint-aware fine-tuning of GANs, and evaluation based on sparsity, prediction gain, and plausibility.

Given an input sample x, its ground-truth label y, and a desired counterfactual label y^*, the generator aims to produce a counterfactual x' such that $f(x') = y^*$, while maintaining realism and proximity to x. To account for user-specific constraints, we incorporate a personalized constraint loss L_{user} based on cost weights and immutable masks derived from the user's cluster.

The total loss function is:

$$L_{\text{total}} = L_{\text{GAN}} + \lambda_{\text{clf}}L_{\text{clf}} + \lambda_{\text{prox}}L_{\text{prox}} + \lambda_{\text{user}}L_{\text{user}} \tag{1}$$

where L_{clf} promotes label change, L_{prox} ensures minimal deviation from x, and L_{user} penalizes changes to immutable features and high-cost modifications.

To **inject personalization at scale, we cluster** training samples in the raw input space using Gower distance [2], which supports mixed feature types. Using hierarchical clustering and silhouette analysis, we selected $k = 4$ clusters. Each cluster is treated as a behavioral profile, and a separate CounterGAN model is fine-tuned for each. This enables learning localized distributions and constraint patterns, improving the plausibility and actionability of generated CFs.

We define **immutable features** within each cluster by selecting attributes with near-zero variance (numeric std < 0.05 or categorical mode share $> 90\%$), along with domain-specific constraints (e.g., Age is always immutable). **Actionable features** are the complement set.

We assign cost weights to actionable features based on inverse variability—specifically, $1/\text{std}$ for numeric features and $1/(1 - \text{mode share})$ for categorical ones.

Weights are normalized to mean 1 per cluster to balance learning. These weights guide the L_{user} loss to prioritize low-cost, realistic changes during counterfactual generation.

To evaluate counterfactual quality, we follow the original CounterGAN metrics [5]: (i) **Prediction Gain**, measuring the increase in predicted probability for the desired class; (ii) **Sparsity (L1)**, the total feature change magnitude, serving as a proxy for actionability; (iii) **Reconstruction Error**, assessing plausibility via an autoencoder; and (iv) **Latency**, capturing runtime per instance and per batch.

4 Experimental Setup and Evaluation

We used the **German Credit dataset** [1], which contains 1000 instances with a mix of numeric and categorical attributes describing financial and personal information. The binary target variable indicates credit risk (good or bad).

This dataset is well-suited for counterfactual recourse, as its features correspond to real-world, human-understandable concepts such as age, employment status, savings, and credit amount—making it meaningful for evaluating actionable recommendations.

Preprocessing involved mapping raw codes to interpretable categories, label-encoding four ordinal variables (e.g., credit history), and one-hot encoding nine nominal features (e.g., job type, housing). The final feature space consisted of 37 encoded attributes. We performed an 80/20 train-test split and applied standard scaling to all numeric fields.

To enable personalized counterfactual generation, we clustered training instances into user profiles that reflect similar constraints and behavioral patterns. Clustering was performed using hierarchical agglomerative clustering with Gower distance [2], which handles mixed numerical and categorical data. Complete linkage was used, and silhouette analysis indicated an optimal number of clusters $k = 4$, with sizes 294, 230, 156, and 119.

Each cluster was treated as a distinct user profile for personalization. Within each cluster, immutable features were identified based on low variability (numeric standard deviation < 0.05 or categorical mode share $> 90\%$), along with domain-enforced constraints (e.g., Age). The remaining features were considered actionable. Cost weights were assigned to actionable features based on inverse variability and normalized to have a mean of 1, guiding the personalization loss during GAN fine-tuning.

To balance sparsity and prediction gain, we performed a per-cluster grid search over the sparsity penalty coefficient λ in the loss function. This parameter controls the strength of the proximity term L_{prox}, which encourages minimal changes between the input x and the generated counterfactual x'. For each cluster, we selected the smallest λ that achieved at least the baseline prediction gain. Among these, we chose the λ value that minimized L1 sparsity. To further improve actionability, we applied a refinement strategy we call the "cost-weight bump." For each cluster, we identified the actionable features with the highest average change magnitude across generated CFs. These features (top 20%) had their cost weights increased by 3050%, discouraging over-reliance. Retraining with updated profiles reduced sparsity while preserving high prediction gain.

5   Results and Discussion

We evaluated our personalized GAN models against a baseline CounterGAN trained on the full dataset without any user-specific tuning. All evaluation metrics follow the original CounterGAN setup [5] for comparability. The baseline results are based on our own reimplementation of CounterGAN applied to the German Credit dataset; while we follow the original architecture and evaluation procedure, numerical results are not directly comparable to those in the original paper due to dataset differences.

Table 1 summarizes the overall performance. Personalization resulted in a substantial improvement in prediction gain (from 0.029 to 0.177) and a 30% reduction in sparsity (L1 distance). Lower sparsity directly reflects improved actionability—fewer, more targeted feature changes make the counterfactuals easier for users to apply in practice, which is central to our objective of user-aligned explanations. Reconstruction error increased slightly but remained within plausible bounds. Latency rose due to the added overhead of selecting and running a cluster-specific model; however, it remains well below one second per instance, which is acceptable for real-time feedback in online decision-support scenarios.

Table 1. Baseline vs Personalized GAN results (averaged across all test samples)

Metric	Baseline	Personalized
Prediction Gain	0.029	**0.177**
Reconstruction Error	2.679	2.882
Sparsity (L1)	6.969	**4.893**
Latency (ms)	0.000	32.722
Batch Latency (ms)	136.206	6544.48

To better understand the impact of personalization across user groups, Fig. 2 illustrates per-cluster performance for prediction gain, reconstruction error, and sparsity. These results highlight that different clusters respond differently to fine-tuning. For example, Cluster 2 achieves the highest prediction gain with moderate sparsity, while Cluster 1 produces the most conservative CFs, favoring minimal changes. This diversity confirms that tailoring counterfactuals to user-group-specific constraints leads to more practical and context-sensitive recommendations than a single global model could provide.

Despite these improvements, several limitations remain. Smaller clusters may introduce instability during training and tuning due to limited data. Although latency remains below one second per instance—acceptable for online use—it is higher than the baseline and may require architectural optimization in latency-sensitive environments. Finally, the need to train and maintain multiple GANs introduces added complexity in deployment pipelines, which could be addressed in future work through end-to-end dynamic personalization.

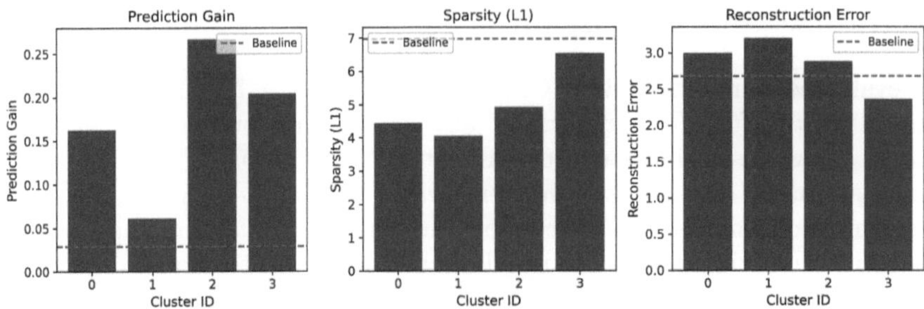

Fig. 2. Per-cluster evaluation of personalized GANs across three metrics. Red dashed lines indicate baseline performance (CounterGAN). (Color figure online)

Future work will explore dynamically learning user-specific constraints at inference time, removing the need for predefined clustering. We also plan to extend our framework to support multimodal inputs and evaluate the generated counterfactuals through human-in-the-loop experiments to assess perceived actionability and trust. Additionally, reducing training and inference overhead through model-sharing strategies remains an open challenge.

Acknowledgements. This work is funded by national funds through FCT Fundação para a Ciência e a Tecnologia, I.P., under the support UID/50014/2023 (https://doi.org/10.54499/UID/50014/2023). Ahmed Adel Fares is also supported by FCT (Fundação para a Ciência e a Tecnologia, Portugal), through the Ph.D. grant with reference number 2023.02282.BD.

References

1. Dua, D., Graff, C.: UCI machine learning repository: (German credit data) (2019). https://archive.ics.uci.edu/ml/datasets/statlog+(german+credit+data)
2. Gower, J.: A general coefficient of similarity and some of its properties. Biometrics **27**(4), 857–871 (1971)
3. Karimi, A.H.: A survey of algorithmic recourse: definitions, formulations, solutions, and prospects. CoRR abs/2010.04050 (2020)
4. Mahajan, D.: Preserving causal constraints in counterfactual explanations for machine learning classifiers. In: Proceedings of the NeurIPS 2020 Workshop on Causal ML (2020)
5. Nemirovsky, P.: CounterGAN: generating counterfactuals for real-time recourse and interpretability using residual GANs. In: Proceedings of the Conference on Uncertainty in Artificial Intelligence, pp. 1488–1497 (2022)
6. Ustun, B.: Actionable recourse in linear classification. In: Proceedings of the Conference Fairness, Accountability, and Transparency, pp. 10–19 (2019)
7. Wachter, S.: Counterfactual explanations without opening the black box: Automated decisions and the GDPR. Harvard J. Law Technol. **31**(2), 1–47 (2018)

A Multi-layer Trust and Privacy Architecture for Decentralized AI Systems

Jiban Kumar Ray, Sheikh Sharfuddin Mim$^{(\boxtimes)}$, and Doina Logofatu

Department of Computer Science and Engineering, Frankfurt University of Applied Sciences, Frankfurt am Main, Germany
`jiban.ray@stud.fra-uas.de`, `{sheikh.mim,logofatu}@fb2.fra-uas.de`

Abstract. As decentralized AI systems become more prevalent, ensuring privacy, security, and trust is no longer optional—it is essential. Existing approaches often address these concerns in isolation, leaving critical gaps in accountability and resilience. This paper proposes a multi-layer architecture that integrates privacy, security, and trust as interconnected components of responsible AI design. The Privacy Layer applies differential privacy and data minimization techniques, the Security Layer ensures encrypted communication and authenticated computation, and the Trust Layer supports auditability through dynamic reputation scoring and blockchain-based logging. Together, these layers form a cohesive and adaptable framework suitable for sectors such as healthcare, finance, and smart cities. To validate these concepts, we present a conceptual simulation illustrating trust–privacy–utility trade-offs, supported by a qualitative comparison with existing frameworks and a security analysis. While not a substitute for large-scale implementation, this conceptual validation highlights the feasibility of our approach and identifies key engineering challenges. By aligning technical mechanisms with legal and ethical requirements, the paper offers a foundational blueprint for building decentralized AI systems that are not only secure and private but also accountable.

Keywords: Decentralized AI System · Trustworthy AI · Privacy Preserving Machine Learning · Federated Learning · Blockchain · Secure Aggregation · Differential Privacy · Trusted Execution Environments · Explainability

1 Introduction

Artificial intelligence (AI) is becoming a cornerstone of innovation across sectors such as healthcare, finance, and public services, helping organizations extract insights from distributed and often sensitive data [13]. To comply with privacy regulations and reduce reliance on centralized data collection, many modern systems are turning to decentralized AI—an approach that distributes data and computation across devices, institutions, or regions [14]. Federated learning (FL), edge computing, and collaborative analytics are prominent examples of this emerging paradigm [9].

L. Martínez et al. (Eds.): IDEAL 2025, LNCS 16239, pp. 349–360, 2026.
https://doi.org/10.1007/978-3-032-10489-2_30

Yet decentralization introduces new challenges. When data and computation are spread across heterogeneous environments, risks do not vanish—they become harder to detect and manage. The breach of psychotherapy records in Finland [2,7] or the reconstruction of sensitive training data from model outputs [6] underscores the severity of these risks. As AI increasingly powers critical infrastructure, its trustworthiness must be deliberately designed rather than passively assumed.

This paper contributes a novel perspective by proposing a *multi-layer architecture* that integrates privacy, security, and trust into a unified framework. Unlike existing works that treat these dimensions in isolation, our design emphasizes cross-layer interactions and explicit alignment with legal and ethical requirements. Beyond protecting data, the architecture introduces mechanisms for accountability, transparency, and governance—making it a foundational step toward responsible and explainable decentralized AI.

1.1 Problem Statement

While decentralized AI systems often use techniques like data anonymization, differential privacy (DP), and encrypted communication to protect sensitive information [9], these measures typically fall short when it comes to establishing trust. Users and institutions are rarely able to verify who accessed the data, how decisions were made, or whether actions align with ethical or regulatory standards [12].

Security tools may defend against external threats but offer little protection against insider misuse or governance failures. Likewise, privacy controls help limit exposure but don't guarantee accountable or ethical use of information [14]. In practice, decentralized AI remains vulnerable to well-known risks such as model poisoning, inference attacks, or Sybil-style client manipulation. In this context, trust requires more than technical safeguards—it demands transparency, traceability, and shared responsibility. Without an architecture that weaves privacy, security, and trust into a unified foundation, decentralized AI risks becoming opaque, fragile, and unfit for responsible deployment [12].

1.2 Research Questions and Contributions

To address the challenges outlined above, this paper is guided by the following research questions:

- How can privacy, security, and trust be integrated into a unified framework for decentralized AI?
- What are the functional roles of each layer, and how do they interact?
- How can such an architecture enhance resilience, transparency, explainability, and regulatory compliance?

To answer these questions, we propose a modular, multi-layer architecture that treats privacy, security, and trust as co-equal design goals. Unlike existing approaches that address these aspects in isolation, our framework emphasizes *cross-layer integration* and explicit alignment with legal requirements. The *Privacy Layer* handles data minimization and anonymization; the *Security Layer* ensures secure communication and authentication; and the *Trust Layer* supports transparency, reputation management, and traceability. This structure is adaptable across domains with varying risks and regulatory constraints.

We demonstrate the architecture's applicability through use cases in healthcare, finance, and smart cities. A qualitative evaluation and a lightweight conceptual simulation illustrate the trade-offs of our design, while a security analysis highlights its robustness against known vulnerabilities. Although not a substitute for full implementation, these results validate the feasibility of the approach and clarify the engineering challenges ahead. By bridging ethical, legal, and technical perspectives, this paper offers a foundational blueprint for building responsible AI systems in decentralized environments.

2   Related Work

This section reviews developments in privacy, security, and trust in decentralized AI. While there has been progress in each area individually, integrated architectural solutions remain rare. Below, we highlight key methods and identify where connections are still missing.

2.1   Privacy-Preserving Techniques in Decentralized AI

Privacy is fundamental in decentralized AI, especially in sensitive fields like healthcare and finance. FL allows local data to remain on-device, sharing only model updates [9]. However, FL alone is not fully privacy-preserving—gradient leakage attacks can expose user data, especially when clients are few or unevenly distributed [14].

To address this, many systems implement *DP* [5], which injects noise to obscure individual contributions. Still, DP can degrade accuracy and is difficult to tune at scale. More advanced tools like *homomorphic encryption (HE)* and *secure multiparty computation (SMC)* [15] enable encrypted computation but are often too resource-intensive for real-time or edge use.

2.2   Security in Distributed and Federated Systems

Security measures in decentralized AI aim to protect data in motion and during computation. Yet, systems remain vulnerable to threats like *model poisoning*, where malicious clients manipulate training updates [3].

Techniques like *secure aggregation* [4] prevent intermediaries from seeing individual contributions. Additionally, *Trusted Execution Environments (TEEs)* offer memory-level protection for sensitive operations. However, even with these tools, security alone doesn't guarantee trustworthy behavior—there's still a need for validating model sources and decisions.

2.3 Trust Mechanisms in AI Collaboration

In federated and decentralized AI, trust is crucial but often underdeveloped. Traditional models use static credentials or reputation scores, which fall short in dynamic, multi-party settings [11]. Newer approaches evaluate trust based on behavior—such as accuracy contributions or policy compliance [10]—and record actions on blockchains to ensure traceability [8].

However, these methods are often isolated from privacy and security mechanisms. Without integration, trust remains shallow, and accountability is limited—especially in cross-organizational contexts where transparency and explainability are also essential.

2.4 Architectural Gaps Across Domains

Many domain-specific frameworks (e.g., in IoT or healthcare) address one dimension—like encryption or access control—but neglect others such as auditability or dynamic trust [1]. A common flaw is assuming trust in central aggregators, which becomes unrealistic in cross-border or multi-institutional AI systems. Recent calls for "Trustworthy AI" emphasize the need for architectures that combine technical, legal, and ethical safeguards [12].

Existing literature provides strong tools in isolation—privacy techniques to protect confidentiality, security mechanisms to guard data flows, and trust models to verify participants—yet these layers remain fragmented. Without coordinated integration, decentralized AI risks opacity and fragility. A holistic, multi-layer framework is needed—one that is modular, scalable, and aligned with evolving regulations such as the GDPR and AI Act. To our knowledge, no prior work has proposed such an integrated approach that explicitly combines privacy, security, and trust into a single architecture. The next section introduces such an architecture.

3 Proposed Architecture

We propose a multi-layer architecture for decentralized AI that treats privacy, security, and trust as interdependent pillars rather than isolated components. Unlike prior frameworks that bolt on these features individually, our design emphasizes cross-layer integration—so that privacy controls are auditable, secure channels are accountable, and trust scores reflect verified behaviors.

3.1 Architecture Overview

As shown in Fig. 1, client nodes (e.g., hospitals, banks, or edge devices) submit data or model updates. These pass through three coordinated layers: the **Security Layer** handles encryption and authentication; the **Privacy Layer** applies data minimization techniques like DP; and the **Trust Layer** logs metadata, manages reputation scores, and maintains auditability via blockchain. At the center of this pipeline is a **Central Aggregator**, which collects updates from clients, verifies integrity, and combines them into a global model. This component serves as the coordination hub for learning, ensuring that updates are applied securely and in accordance with trust and privacy policies. The aggregator maintains required communication with both the Security and Privacy Layers, enabling both secure update handling and adaptive privacy control.

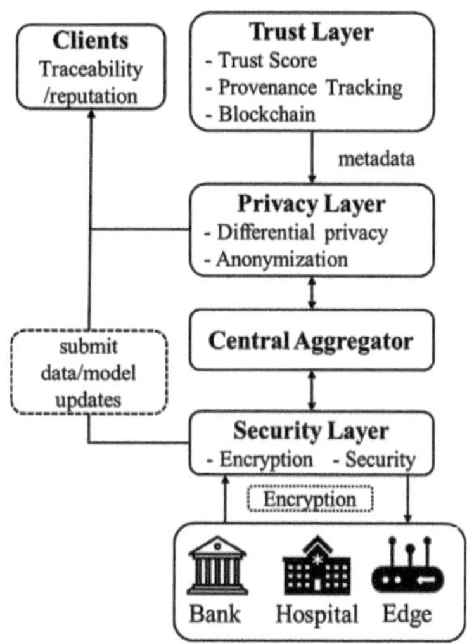

Fig. 1. Proposed multi-layer architecture linking security, privacy, and trust layers with feedback loops for accountability.

A typical interaction involves a client submitting a locally trained update, which is first differentially privatized, then encrypted, verified by the aggregator, and finally logged with a trust score, clarifying how the three layers operate together in practice. See Fig. 2 for a compact sequence view.

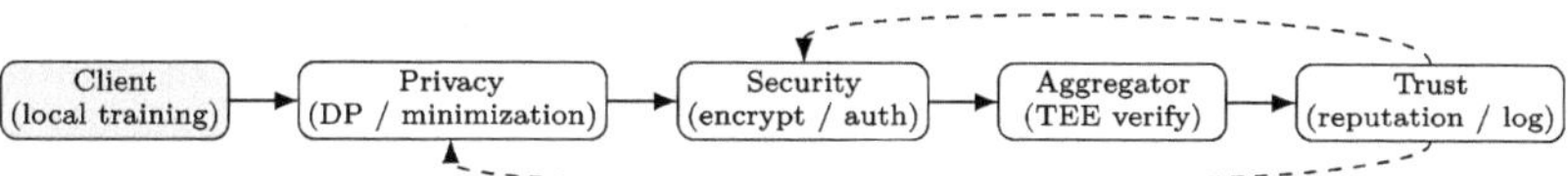

Fig. 2. Sequence of a client update: training → Privacy (DP) → Security (encryption/authentication) → Aggregator (TEE) → Trust (reputation/log), with feedback to Privacy and Security layers.

3.2   Privacy Layer: Personalized Data Control

This layer protects raw data through local DP, applied at the client level. Each participant perturbs data or gradients before sharing, offering mathematically bounded privacy guarantees. Importantly, the architecture supports *personalized privacy budgets*, allowing organizations to tune sensitivity levels based on policy or domain needs—unlike traditional DP schemes that use a fixed global parameter [5].

3.3 Security Layer: End-to-End Integrity

The Security Layer enforces encrypted communication and secure computation. Techniques such as *TLS/SSL, secure aggregation* [4], and *Trusted Execution Environments (TEEs)* help ensure that updates remain protected even in hostile environments. Participation is gated through verified credentials and trust scores, minimizing exposure to poisoned or malicious clients.

3.4 Trust Layer: Transparent and Auditable AI

What sets our architecture apart is the Trust Layer, which introduces accountability through *decentralized provenance*. Using blockchain ledgers [8], the system records pseudonymized metadata such as update times, trust scores, and privacy usage. A *dynamic reputation system* evaluates client reliability based on behavior and peer feedback [10,11], influencing both access control and privacy enforcement. Nodes with low trust may be subject to stricter privacy controls or be excluded from training.

Unlike prior approaches that treat trust as an isolated add-on, our design tightly couples it with privacy and security layers, ensuring that reputation metrics directly shape authentication, privacy budgets, and auditability. This cross-layer integration is a central novelty of the proposed framework.

3.5 Layer Interactions and Enforcement

Each layer reinforces the others: the Privacy Layer generates metadata used by the Trust Layer; the Trust Layer influences authentication in the Security Layer; and the Security Layer ensures that integrity is preserved across privacy and trust operations. This tight coupling reduces attack surfaces and improves system robustness.

3.6 Adaptability Across Domains

The architecture is modular and adaptable to different sectors. In healthcare, strict privacy budgets and auditable logs support regulatory compliance. In smart cities, dynamic trust scores offer a lightweight alternative to static credentials. This flexibility makes the framework suitable for real-world FL deployments in legal, medical, and financial domains.

4 Use Case Scenarios

To demonstrate the applicability and versatility of the proposed architecture, we present three representative use cases: healthcare data collaboration, smart city infrastructure, and interbank fraud detection. These scenarios illustrate how the multi-layer framework can be tailored to domain-specific risks, regulatory constraints, and operational requirements.

4.1 Healthcare Data Collaboration

Healthcare organizations increasingly rely on AI to support diagnosis, predictive analytics, and resource planning. However, patient data is sensitive and highly regulated under laws like HIPAA[1] and GDPR. In this setting, the Privacy Layer ensures compliance by applying DP at the hospital level, enabling medical institutions to contribute to shared models without exposing individual records [13].

The Security Layer protects data transmission between hospitals and aggregators using encrypted channels and secure aggregation protocols. TEEs further safeguard computation during aggregation. Trust mechanisms play a vital role in healthcare collaborations. Each participating hospital is assigned a dynamic trust score based on previous compliance, update quality, and contribution history. Blockchain-based logging ensures auditability, enabling legal accountability in the event of disputes or breaches [8].

4.2 Smart City Infrastructure

Smart cities rely on distributed sensors and edge devices to manage services like traffic, public safety, and utilities. These devices often operate in untrusted, resource-limited environments, requiring privacy and security mechanisms that are both lightweight and robust. In this setting, the *Privacy Layer* anonymizes real-time sensor data using techniques such as local DP, which removes or masks identifying metadata like time and location [14]. The *Security Layer* ensures encrypted transmission over potentially unreliable networks, using lightweight public-key credentials and secure aggregation to prevent eavesdropping or tampering. Meanwhile, the *Trust Layer* tracks device reliability based on behavioral metrics—21 s

uch as update frequency or failure rates—and can flag or exclude devices with suspicious patterns. This dynamic trust scoring enables scalable governance in highly heterogeneous networks [1].

4.3 Interbank Fraud Detection

In finance, institutions are often hesitant to share transaction data due to regulatory and competitive concerns. Yet collaboration is key to identifying fraud patterns across distributed datasets. Using our framework, banks can jointly train models via FL, ensuring that sensitive data remains within institutional boundaries. The *Privacy Layer* enforces domain-specific anonymization and DP to prevent re-identification. The *Security Layer* encrypts model updates during transmission and enables secure aggregation, with TEEs protecting computations from internal or external threats [4]. The *Trust Layer* maintains tamper-resistant logs and computes reputation scores based on factors like protocol adherence or fraud detection accuracy. Institutions with stronger trust scores may have greater influence in aggregation, while lower-trust participants can

---

[1] HIPAA compliance, https://www.ncbi.nlm.nih.gov/books/NBK500019/.

be audited or excluded. This scenario highlights how the layered architecture adapts to sector-specific needs—enhancing financial transparency without compromising confidentiality.

5 Evaluation and Discussion

The primary aim of this section is to conceptually assess the effectiveness and relevance of the proposed multi-layer architecture in real-world scenarios. As the architecture emphasizes privacy, security, and trust simultaneously, we evaluate it across those three axes using qualitative metrics. Additionally, we compare it with existing frameworks and provide an illustrative simulation to highlight its distinct advantages, trade-offs, and practical limitations.

5.1 Evaluation Criteria

We assess the architecture based on the following core criteria:

- **Privacy Assurance:** The extent to which sensitive information remains protected from inference, leakage, or unauthorized access.
- **Security Robustness:** The system's resilience against attacks such as data poisoning, model inversion, and communication interception.
- **Trust Enablement:** The ability to enforce accountability, traceability, and transparency in multi-party collaborations.
- **Regulatory Alignment:** The framework's compatibility with legal standards such as GDPR, HIPAA, or domain-specific guidelines.
- **Operational Flexibility:** Ease of integration into federated or edge AI environments with varying resources and network topologies.
- **Explainability and Fairness:** The degree to which outcomes remain interpretable and equitable across participants.

5.2 Comparative and Quantitative Evaluation

To assess the distinct advantages of the proposed architecture, we conducted both a comparative analysis of related frameworks and a lightweight conceptual simulation. These two complementary evaluations help clarify where our approach stands in the landscape of decentralized AI solutions and how it performs under varying trust and privacy conditions.

Comparative Analysis with Existing Frameworks: Table 1 compares our architecture with three well-known frameworks. While Google's FL implementation emphasizes privacy through optional DP and secure aggregation, it lacks integrated trust mechanisms or auditability. Blockchain-based FL for IoT improves traceability but is limited by its trust model. TrustFed introduces score-based trust evaluation but does not enforce privacy or ensure auditability.

In contrast, our architecture integrates all three concerns through tightly coordinated layers, offering a unified foundation for building responsible AI systems, where privacy, security, and trust are explicitly integrated rather than addressed in isolation.

Table 1. Comparison of the proposed architecture with existing approaches.

Framework	Privacy	Security	Trust Mechanism	Auditability
Google FL [4]	Basic (DP optional)	Secure Aggregation	None	No
Blockchain-FL for IoT [8]	Moderate	Medium	Ledger-based reputation	Partial
TrustFed [11]	Weak	Moderate	Score-based trust	No
Proposed Architecture	**Strong (DP + anonymization)**	**High (TEEs + aggregation)**	**Dynamic trust + blockchain**	**Full audit trail**

Simulation of Trust-Privacy-Utility Trade-Offs: To complement the comparison, we implemented a small-scale simulation to explore how variations in trust and privacy settings affect model accuracy and system overhead. Ten client nodes were assigned random trust scores and DP budgets, and we estimated model accuracy and audit log sizes for each. This simulation serves as a conceptual illustration of trade-offs, rather than an empirical benchmark.

Nodes with higher trust and looser privacy budgets yield better accuracy but pose greater privacy risks, while strict privacy settings (low ϵ values) increase noise and reduce accuracy. Audit log sizes reflect recorded metadata—such as trust scores, privacy usage, and update activity—and introduce modest storage overhead that must be considered when scaling. Table 2 summarizes selected results.

Table 2. Simulation of trust, privacy budget, and model accuracy across nodes.

Node	Trust Score	Privacy Budget (ϵ)	DP Noise Std. Dev.	Estimated Accuracy	Audit Log Size (KB)
Node_1	0.69	0.14	7.14	0.13	76.72
Node_2	0.98	1.94	0.52	0.83	62.67
Node_3	0.87	1.68	0.60	0.73	68.92
Node_4	0.80	0.50	2.00	0.56	69.31
Node_5	0.58	0.45	2.22	0.39	55.45

For illustration, we modeled accuracy as a simple function of both the privacy budget (ϵ) and the trust score (t): a higher ϵ reduces differential privacy noise, while a higher t increases the reliability of contributions. This conceptual formulation, though simplified, highlights how the architecture can dynamically balance privacy, trust, and utility when weighting updates in the aggregation process.

5.3 Trade-Offs and Limitations

Although the simulation illustrates the interplay between privacy, trust, and accuracy, real-world deployment presents several trade-offs:

- **Overhead:** Blockchain-based logging and trust scoring can increase computation and storage costs.
- **Complexity:** Managing multiple layers may require orchestration tools, which could challenge smaller organizations.
- **Trust Initialization:** New participants need carefully seeded trust scores to prevent manipulation.

In addition, decentralized AI remains exposed to security threats such as data poisoning, inference attacks, or Sybil-style manipulation. The proposed design mitigates these risks through authenticated and encrypted communication, client-side privacy noise injection, and trust-weighted aggregation recorded on a tamper-resistant ledger. While these measures enhance robustness, they cannot eliminate all vulnerabilities, making continuous monitoring and adaptation to evolving threats essential. However, the modular architecture allows for practical adjustments. Lightweight alternatives like hash chains can replace blockchains where needed, and trust parameters can be tuned to fit domain-specific requirements.

5.4 Legal and Ethical Considerations

An important strength of the framework is its alignment with regulatory frameworks like GDPR. The architecture supports:

- **Data Minimization:** Implemented via anonymization and local DP mechanisms.
- **Purpose Limitation:** Expressed via consent parameters embedded at the Privacy Layer.
- **Accountability and Transparency:** Enforced through blockchain audit trails and trust metrics.

From an ethical standpoint, the framework advances transparency and user control. Clients are not merely data providers but participate in the governance of shared models. This reduces power asymmetries between central aggregators and distributed entities, aligning with principles of responsible AI [12]. While these features illustrate strong regulatory alignment, achieving full compliance and fairness in practice will depend on careful implementation, explainability of model outcomes, and ongoing oversight.

5.5 Scalability and Adaptability

Our architecture is adaptable to varying system scales. In small-scale deployments (e.g., 5–10 hospitals), all three layers can be activated without excessive overhead. In larger environments, such as national financial networks or urban IoT systems, optimization techniques such as federated trust scoring and batched audit logging can be applied to maintain performance. While these adaptations demonstrate scalability in principle, real-world deployments will face engineering challenges related to overhead, complexity, and interoperability that require further study. Overall, the proposed architecture offers a forward-looking solution to the growing demands for privacy, security, and trust in AI systems deployed across decentralized, sensitive, and dynamic environments.

6   Conclusion and Future Work

This paper proposed a multi-layer architecture that unifies privacy, security, and trust into a cohesive and auditable framework for decentralized AI. Guided by the research questions, we showed how each layer functions, how they interact, and how the design can enhance resilience, transparency, and regulatory compliance. Unlike prior approaches that treat these pillars independently, our framework emphasizes cross-layer integration, making privacy controls auditable, secure channels accountable, and trust scores verifiable. Its applicability was illustrated through use cases in healthcare, finance, and smart cities, and assessed via comparative analysis, a conceptual simulation, and a discussion of security threats.

While these evaluations highlight feasibility, they remain conceptual rather than empirical. A key limitation is the absence of large-scale implementation or benchmarking with established federated learning frameworks, as well as the need for deeper security testing against known vulnerabilities. Future work will therefore explore lightweight alternatives to blockchain, formalize trust computation models, and extend the framework with explainability and fairness. Deploying the architecture on real-world federated platforms and testing scalability under realistic workloads will be critical for validating its impact. In summary, this paper offers a blueprint for responsible decentralized AI in which privacy, security, and trust are not optional add-ons but core design principles.

References

1. Adam, M., Hammoudeh, M., Alrawashdeh, R., Alsulaimy, B.: A survey on security, privacy, trust, and architectural challenges in IoT systems. IEEE Access **12**, 57128–57149 (2024). https://doi.org/10.1109/ACCESS.2024.3382709
2. AFP in Helsinki: "Shocking" hack of psychotherapy records in Finland affects thousands (2020). https://www.theguardian.com/world/2020/oct/26/tens-of-thousands-psychotherapy-records-hacked-in-finland. leak of therapy records and patient extortion
3. Bhagoji, A.N., Chakraborty, S., Mittal, P., Calo, S.: Analyzing federated learning through an adversarial lens. In: Chaudhuri, K., Salakhutdinov, R. (eds.) Proceedings of the 36th International Conference on Machine Learning. Proceedings of Machine Learning Research, vol. 97, pp. 634–643. PMLR (09–15 Jun 2019) (2019). https://proceedings.mlr.press/v97/bhagoji19a.html

4. Bonawitz, K., Ivanov, V., Kreuter, B., et al.: Practical secure aggregation for privacy-preserving machine learning. In: Proceedings of the 2017 ACM SIGSAC Conference on Computer and Communications Security, pp. 1175–1191. CCS 2017, Association for Computing Machinery, New York, NY, USA (2017). https://doi.org/10.1145/3133956.3133982

5. Dwork, C., Roth, A.: The Algorithmic Foundations of Differential Privacy, Foundations and Trends in Theoretical Computer Science, vol. 9. Now Publishers Inc. (2014). https://doi.org/10.1561/0400000042

6. Fredrikson, M., Jha, S., Ristenpart, T.: Model inversion attacks that exploit confidence information and basic countermeasures. In: Proceedings of the 22nd ACM SIGSAC Conference on Computer and Communications Security, pp. 1322–1333. CCS 2015, Association for Computing Machinery, New York, NY, USA (2015). https://doi.org/10.1145/2810103.2813677

7. Ghanbari, H., Koskinen, K.: When data breach hits a psychotherapy clinic: the VASTAAMO case. J. Inf. Technol. Teach. Cases (2024). https://doi.org/10.1177/20438869241258235

8. Jiang, Y., et al.: BlockChained federated learning for internet of things: a comprehensive survey. ACM Comput. Surv. **56**(10) (2024). https://doi.org/10.1145/3659099

9. Li, T., Sahu, A.K., Talwalkar, A., Smith, V.: Federated learning: challenges, methods, and future directions. IEEE Signal Process. Mag. **37**(3), 50–60 (2020). https://doi.org/10.1109/MSP.2020.2975749

10. Ma, C., Li, J., Ding, M., Wei, K., Chen, W., Poor, H.V.: Federated learning with unreliable clients: performance analysis and mechanism design (2021). https://doi.org/10.48550/arXiv.2105.06256

11. Tariq, A., et al.: Trustworthy federated learning: a comprehensive review, architecture, key challenges, and future research prospects. IEEE Open J. Commun. Soc. **5**, 4920–4998 (2024). https://doi.org/10.1109/OJCOMS.2024.3438264

12. Tomašev, N., Cornebise, J., Hutter, F., et al.: AI for social good: unlocking the opportunity for positive impact. Nat. Commun. **11**(1), 2468 (2020). https://doi.org/10.1038/s41467-020-15871-z, guidelines and opportunities for applying AI to the UN Sustainable Development Goals (SDGs)

13. Voigt, P., Von dem Bussche, A.: The EU General Data Protection Regulation (GDPR): a practical guide. Springer (2024). https://doi.org/10.1007/978-3-031-62328-8, purposes and principles of GDPR

14. Yang, W., et al.: Deep learning model inversion attacks and defenses: a comprehensive survey (2025). https://doi.org/10.48550/arXiv.2501.18934

15. Yu, S., Cui, L.: Secure multi-party computation in federated learning, pp. 89–98. Springer Nature Singapore, Singapore (2023). https://doi.org/10.1007/978-981-19-8692-5_6

Special Session on Trustworthy and Intelligent Data Infrastructures: Distributed Ledger Technologies and Cybersecurity for automated Learning Systems

Guided and Federated RAG: Architectural Models for Trustworthy AI in Data Spaces

Carlos Mario Braga[(✉)] [iD], Manuel A. Serrano [iD],
and Eduardo Fernández-Medina [iD]

University of Castilla La-Mancha, Ciudad Real, Spain
`carlosmario.braga1@alu.uclm.es`

Abstract. The growing deployment of large language models (LLMs) in policy-sensitive and distributed environments raises concerns about hallucinations, semantic opacity, and lack of traceability, especially in regulated domains such as healthcare and finance. Retrieval-augmented generation (RAG) helps address these issues by grounding outputs in external sources, while data spaces (DSs) offer a governance-driven infrastructure for federated and sovereign data exchange. However, their integration remains underdeveloped, limiting the design of AI systems that are both explainable and compliant with regulatory demands.

This paper analyzes three RAG–DSs integration models from the literature and introduces two new architectures, Guided RAG and Federated RAG, that directly address their structural limitations by combining coordination, data sovereignty, and trust in generative systems. Guided RAG and Federated RAG are formally specified using Business Process Model and Notation (BPMN), and all five models, together with a non-retrieval baseline, are analyzed across operational dimensions including transparency, traceability, and control delegation. This work establishes an architectural foundation for trustworthy generative AI operating under distributed governance constraints, with explicit guarantees of data sovereignty. The proposed framework supports systematic evaluation in federated settings and serves as a foundation for methodological refinement and regulatory alignment.

Keywords: RAG · Data spaces · Trustworthy AI

1 Introduction

Despite impressive progress in natural language generation, large language models (LLMs) still suffer from hallucinations [4,19], opacity [7,15,17], and limited explainability [9,10]. These issues challenge the reliability, traceability, and ethical compliance of AI systems, particularly in regulated domains such as clinical decision-making or financial supervision [18].

Retrieval-augmented generation (RAG) has emerged as a promising paradigm to mitigate these limitations by grounding responses in external documents retrieved at query time, thereby enhancing factual accuracy, contextual relevance, and traceability in content generation [14,15,19].

L. Martínez et al. (Eds.): IDEAL 2025, LNCS 16239, pp. 363–374, 2026.
https://doi.org/10.1007/978-3-032-10489-2_31

Meanwhile, the concept of federated data spaces (DSs) has emerged as a governance-driven infrastructure for secure, sovereign, and policy-compliant data sharing across domains. DSs preserve data locality, enforce usage policies, and enable semantic and legal interoperability [11,12].

Although complementary in purpose, integration between RAG and DSs remains largely unexplored. As detailed in Sect. 2, only a few studies address this convergence, and none offer a formal architectural or trust-oriented perspective. RAG enhances relevance and verifiability but typically relies on centralized corpora that conflict with the principles of federated governance [19], whereas DSs ensure data sovereignty and access control [11] but lack native support for semantic retrieval or natural language interaction. Bridging these paradigms is essential to develop explainable AI aligned with emerging governance frameworks in distributed, policy-sensitive settings [3,12].

These findings motivate the formalization of RAG–DSs integration models and their evaluation through controlled simulation. This approach builds on a previously proposed trustworthiness taxonomy for AI systems [1], providing a concrete instantiation in distributed generative architectures. The operational dimensions selected for architectural comparison, such as traceability, control delegation, and semantic interoperability, are directly grounded in this taxonomy, which provides the conceptual backbone for both design and evaluation throughout the paper. This work addresses the lack of formal and trust-oriented RAG–DSs integration by introducing two novel architectures, Guided RAG and Federated RAG, explicitly designed to enhance semantic coordination, data sovereignty, and provenance in federated environments. The proposed architectures are formally specified in Business Process Model and Notation (BPMN) 2.0, and all five models are evaluated in controlled simulation with emulated roles of data providers (DPs), data consumers (DCs), and federators (Fs). The assessment combines quantitative metrics, such as latency, source diversity, and hallucination risk, with qualitative judgments on relevance, transparency, and traceability, aligning with key principles of trustworthy AI [1,3].

The paper is structured as follows. Section 2 provides background on RAG, DSs, and trust in AI. Section 3 introduces five integration models, including two novel architectures. Section 4 outlines the simulation framework and reports results. Finally, Sect. 5 discusses the implications for trustworthy AI and future directions, and concludes the paper.

2   Background and Related Work

This section reviews RAG and DSs, focusing on their principles, applications, and limitations in the context of trustworthy AI.

2.1   Retrieval-Augmented Generation (RAG)

The original RAG architecture, introduced by Lewis et al. [8], combines dense retrieval with sequence-to-sequence generation via a dual encoder–decoder pipeline.

RAG improves factual reliability and contextual accuracy by grounding responses in retrieved documents. Unlike traditional generative models that rely solely on training parameters, RAG dynamically retrieves external knowledge relevant to each query via an integrated retrieval component [14,15,19].

A RAG pipeline typically consists of three stages: indexing, retrieval, and conditional generation. Each query triggers retrieval of relevant documents, which are embedded into the LLM prompt. This design enables access to current external knowledge and improves traceability while reducing hallucination risk [14,15].

RAG has been successfully applied in domains such as biomedical question answering, clinical diagnostics, regulatory support, and industrial process design [2,4,7]. However, deploying RAG in distributed settings poses technical challenges, particularly around decentralized indexing, semantic heterogeneity, data privacy, and document provenance [10,19].

2.2 Data Spaces

DSs are decentralized ecosystems that enable secure, sovereign data exchange across institutions, sectors, or jurisdictions. They are based on the principle of data locality: DPs retain full control over their datasets and define usage policies without transferring ownership [11].

The typical architecture of DSs defines three roles: DPs, who publish data and manage access; DCs, who request and process that data; and Fs, who coordinate interoperability and enforce usage policies [11].

Key enabling components include secure connectors, semantic catalogs, and policy-based access control [11]. DSs are increasingly adopted in domains such as manufacturing, smart mobility, and digital health, and form a key pillar of Europe's digital data strategy [12].

However, DSs continue to face challenges in semantic alignment, auditability, and dynamic policy enforcement, especially when DCs require real-time access to heterogeneous and evolving knowledge [11].

2.3 Integration Potential and Research Gap

Although RAG and DSs address complementary dimensions of trustworthy AI, namely semantic relevance and governance, their integration remains largely unexplored. RAG typically relies on centralized knowledge corpora, which conflicts with the distributed, policy-constrained architecture of DSs. Conversely, DSs offer strong data control but lack native support for semantic retrieval and interactive AI capabilities.

To assess the current state of research on RAG–DSs integration, a systematic literature review was conducted using Petersen et al.'s methodology [13]. The research question was: *To what extent has the integration of RAG into DSs been formally studied and evaluated from a trustworthiness perspective?*

A search was conducted in January 2025, using the query (''Retrieval-Augmented Generation'' OR ''RAG'') AND

("Data Spaces"), restricted to publications from 2018 onward, in the ACM Digital Library (20 results), SCOPUS (23), and IEEE Xplore (1), yielding 44 papers in total. After removing duplicates, 41 unique results remained, of which three studies explicitly addressed this integration [5,6,16]. These works propose three integration variants but face critical limitations. They lack integration and traceability, requiring consumers to query multiple providers without a unified coordination layer, which exposes sensitive data and compromises semantic consistency. Additionally, concentrating both retrieval and generation in Fs lead to excessive centralization, reducing transparency and limiting DCs control. None of the studies offer a formal comparison of integration models or a systematic evaluation of trust-related aspects such as privacy, sovereignty, semantic alignment, or multi-DS coordination, where consumers query multiple DSs in parallel through distinct Fs.

To overcome these limitations, this article formalizes two new architectures, Guided RAG and Federated RAG, specified in BPMN and evaluated through a reproducible simulation framework. This work directly addresses the gaps of personalization and traceability identified in the literature and provides a systematic basis for integrating RAG into data spaces.

3   Architectural Integration Models

This section presents five architectural models for integrating RAG into DSs, three drawn from existing literature and two proposed contributions, Guided RAG and Federated RAG, which are specified in BPMN 2.0. The analysis focuses on how indexing, retrieval, and generation are distributed across roles and the resulting implications for sovereignty, latency, transparency, and traceability.

3.1   Models from Existing Literature

Three integration models are commonly described in the literature. In the Autonomous RAG approach, DCs download datasets from multiple DPs, build a local index, and independently execute both retrieval and generation, while Fs only provide metadata catalogs in the ingestion process [5,6]. Although this setup reduces latency and grants full consumer-side control, it cannot be considered a genuine RAG–DS integration. Once the ingestion is completed, neither the DPs nor the Fs play any role during query execution, meaning that the online interaction is detached from the federated infrastructure. As a consequence, traceability and provenance guarantees are severely weakened, since generated responses are based on a local copy rather than on federated, policy-governed retrieval.

In RAG at Source, each of the DPs maintains its own retrieval index. DCs send embedded queries directly to the DPs, which return ranked document fragments, while Fs are only responsible for metadata management and query routing [6,16]. After receiving the relevant fragments, the DC performs the augmentation and forwards the enriched query to its local LLM to generate the final response.

Finally, in the Custodian RAG model, Fs manage both indexing and response generation, possibly replicating data from the DPs. In this setup, DCs only receive the final response without visibility into how it was constructed. This architecture centralizes control in the Fs, who become the main actors responsible for retrieval and generation within the data space [6,16].

These three models illustrate contrasting trade-offs in terms of sovereignty, latency, and control, but they still lack formalization and cohesive support for trust-related properties. In particular, Autonomous RAG cannot be considered a genuine integration since DPs and Fs play no role during query execution, RAG at Source increases consumer exposure by requiring queries to be broadcast to multiple providers and complicates the integration of heterogeneous fragments in the absence of global coordination, and Custodian RAG centralizes both retrieval and generation in the Fs, thereby removing the DCs' ability to adapt outputs to their specific context and further reducing transparency, since intermediate steps remain opaque. Moreover, by concentrating all responsibilities in the Fs, the Custodian model lacks a coordination layer beyond this single actor, which limits its capacity to handle semantic normalization or provide robust provenance management. As a result, none of these approaches effectively address the challenges of personalization, traceability, semantic coordination across providers, or interoperability in multi–data-space environments.

3.2 Proposed Integration Models

Two new architectures are proposed to address key limitations in existing models. Guided RAG is a hybrid configuration in which Fs manage retrieval through a shared index within a single DS, while DCs remain responsible for generating the final response. This avoids the excessive centralization of the Custodian model by preserving consumer-side personalization, and ensures consistency and provenance inside the DS. Federated RAG extends this approach to scenarios involving multiple DSs, where coordination of retrieval takes place at the level of the DPs, and Fs consolidate the results before passing them to the DC for generation. This design resolves the fragmentation and weak traceability of Source-based models while maintaining personalization and data sovereignty.

Figure 1 presents the BPMN representation of the two proposed models. The roles of DPs, DCs, and Fs are placed in distinct pools, covering both the publication phase (left) and the query execution phase (right). This layout highlights how responsibilities are distributed and how semantic traceability is maintained

In Guided RAG, Fs manage a shared retrieval index within a single DS while DCs retain control of generation, avoiding over-centralization. In Federated RAG, DPs maintain their own infrastructures and Fs coordinate distributed retrieval, grouping fragments by source before passing them to DCs for generation. These models address key limitations of previous approaches by combining coordination, traceability, and consumer-side flexibility, and they naturally extend to cross-domain and multi-DS environments. BPMN 2.0 was selected because it makes roles and message flows explicit, which is essential for analyzing trust-related properties.

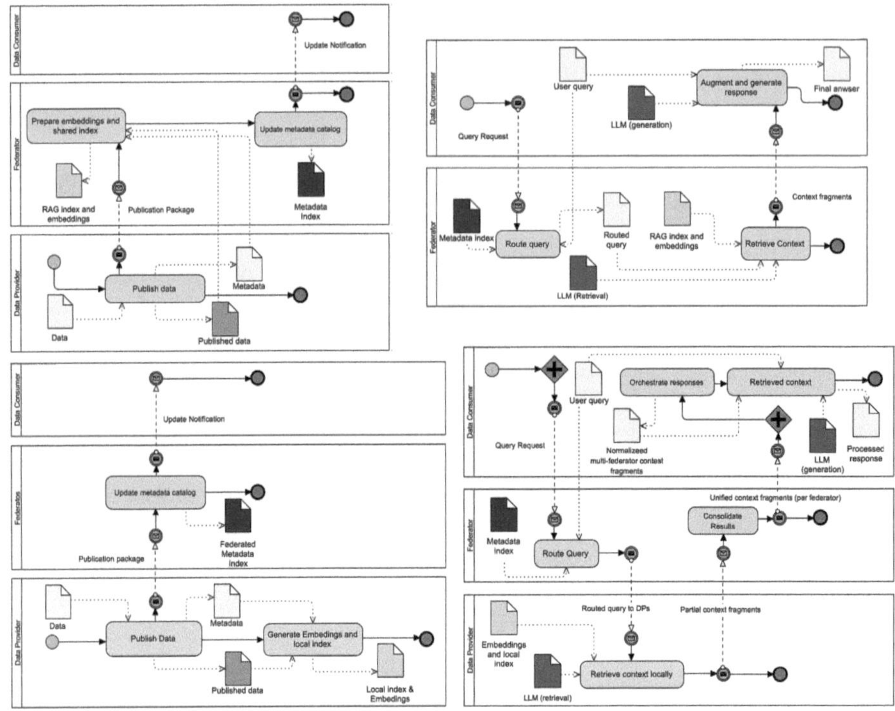

Fig. 1. BPMN 2.0 diagrams of the proposed models: Guided RAG (top) and Federated RAG (bottom), with DPs, DCs, and Fs represented in distinct pools covering publication (left) and query execution (right).

3.3 Theoretical Comparison of Integration Models

Table 1 summarizes the five integration models and a non-retrieval baseline. The comparison uses a normalized scale (1 = low, 2 = medium, 3 = high) across five trust-related dimensions. Scores reflect the role of each actor and the balance between consumer control, federator centralization, and provider exposure. Sovereignty captures how much control remains with the data consumer (low if authority shifts to the federator, high if generation stays with the consumer); latency measures expected response time (low for local execution, medium for coordinated retrieval, high for distributed settings); privacy indicates data exposure (low when datasets are replicated or queries broadcast, high when data remains local); transparency refers to source visibility (low if the consumer only sees a final answer, high with clear source attribution); and interoperability covers semantic and technical compatibility (low with isolated copies, high across multiple DSs). This comparison highlights trade-offs among the five dimensions and provides the conceptual foundation for the simulation-based evaluation in Sect. 4.

Table 1. Comparison of RAG–DS integration architectures across five dimensions (1 = low, 2 = medium, 3 = high). Guided and Federated RAG are highlighted in gray.

Model	Sovereignty	Latency	Privacy	Transparency	Interop.
Autonomous RAG	1	1	1	3	1
RAG at Source	2	2	2	2	2
Custodian RAG	1	2	2	1	1
Guided RAG	2	2	2	3	2
Federated RAG	3	3	3	3	3
Baseline (No RAG)*	N/A	3	N/A	1	N/A

* The baseline generates responses without retrieval, and does not involve data exchange, access policies, or federated roles.

4 Experimental Setup and Evaluation

The evaluation combines quantitative and qualitative analysis to assess the six architectures. Quantitative metrics include latency, number of retrieved fragments, number of contributing DPs and redundancy. Qualitative evaluation considers hallucination risk, relevance, semantic consistency, traceability and transparency. Together these dimensions provide complementary evidence on efficiency, coverage, sovereignty and trustworthiness, which are compared against the theoretical expectations from Sect. 3.

To emulate a realistic federated setup, a synthetic corpus was distributed across three DPs. DP1 was a hospital research institute, DP2 a European data protection authority and DP3 a nonprofit bioethics organization. Each provided two documents of 300–400 words on privacy, data sovereignty, bias and compliance. The corpus was intentionally small to allow full manual inspection and semantic control and was used consistently across all models.

The experimental framework was implemented in Python with six architectures (`autonomous.py`, `source.py`, `custodian.py`, `guided.py`, `federated.py` and `baseline.py`) that share a core module (`core.py`) for document loading, FAISS-based indexing and semantic retrieval using OpenAI embeddings. The setup abstracts away asynchronous API interactions and infrastructure variability to isolate architectural differences, and in `baseline.py` a non-retrieval setting is simulated where the LLM generates answers directly from the query without external documents, acting as a control to isolate the contribution of retrieval within DSs.

All models were evaluated on the same set of six queries defined in `queries.py`. Five addressed domain-specific concerns such as governance, privacy and ethics, and the sixth was out of scope to test hallucination by checking whether models produced unsupported content, expressed uncertainty or declined to answer. Although limited in size, the set balanced coverage and manual evaluability, enabling detailed qualitative inspection across architectures. All architectures were executed under identical conditions, with retrieval from multiple DPs performed sequentially to simulate worst-case latency and isolate architectural impact while avoiding concurrency effects.

In `federated.py`, retrieved fragments are grouped and explicitly labeled by source (e.g., `[Context from DP2]`), enabling origin-based semantic structuring. This step is essential in Federated RAG to ensure traceability across multiple DPs, while in Guided RAG provenance is already maintained through the shared index managed by the Fs.

All responses were stored in structured `.json` files, and a helper script generated a CSV template with one row per model–query pair for annotation. Multiple annotators scored each qualitative dimension, with agreement measured using Krippendorff's α. Code and evaluation data are available at the following github repository: https://github.com/GSYAtools/DSRAG.

Although methodologically rigorous, the evaluation relies on a controlled simulation, which may limit generalizability. Following guidelines for empirical software engineering [13], threats to validity and mitigation strategies are identified. The simulated environment abstracts away real-world conditions such as asynchronous APIs, network variability and infrastructure constraints, and retrieval is executed sequentially to isolate architectural effects at the cost of real-time realism. Operational robustness under deployment scenarios with containerized microservices or semantic brokers (e.g., Eclipse Dataspace Connector) is left to future work. The framework does not address fairness, bias or interpretability, and the use of OpenAI embeddings and LLMs may affect replicability across platforms. All models were tested on a fixed corpus and shared query set to control input variation, though output stochasticity remains, and the synthetic, domain-specific corpus may limit generalization to other settings.

Despite these constraints, the framework remains robust and reproducible, isolating architectural effects while combining quantitative and qualitative insights for trustworthy RAG–DS design.

4.1   Results and Analysis

Table 2 summarizes the average performance of five retrieval-based architectures and one baseline across four quantitative metrics. Custodian and Autonomous showed the lowest latency ($\approx$2.2 s), Source and Federated were slower ($\approx$6.5–6.8 s) due to distributed retrieval, and Guided was intermediate ($\approx$2.8 s) with high variability ($\sigma = 1.4$ s). Source and Federated consistently retrieved six fragments from all three DPs. The baseline, lacking retrieval, had the highest latency (15.7 s). Redundancy was zero in all cases, indicating effective deduplication. Although Custodian had been rated as medium latency in the theoretical comparison due to its centralized design, in practice it performed comparably to Autonomous under the controlled simulation.

Qualitative results are shown in Fig. 2 and detailed by dimension in Fig. 3. Twenty annotators rated responses on a 1–5 Likert scale using a structured rubric, with Krippendorff's $\alpha = 0.782$ confirming substantial agreement. All retrieval-based models scored low on hallucination risk, reflecting conservative generation or omissions. Federated and Custodian RAG achieved the highest

Table 2. Quantitative evaluation of the six architectures. Values are averaged over six queries and include standard deviation (σ).

Model	Latency (ms)		Frag. Ret.		DPs Contr.		Redundancy	
	Mean	σ	Mean	σ	Mean	σ	Mean	σ
Autonomous	2260.95	440.79	3.00	0.00	2.17	0.41	0.00	0.00
Source	6817.27	720.22	6.00	0.00	3.00	0.00	0.00	0.00
Custodian	2247.47	319.55	3.00	0.00	2.17	0.41	0.00	0.00
Guided	2836.80	1424.37	3.00	0.00	2.17	0.41	0.00	0.00
Federated	6499.81	315.60	6.00	0.00	3.00	0.00	0.00	0.00
Baseline	15760.85	7907.17	0.00	0.00	0.00	0.00	0.00	0.00

Note. Latency is measured under sequential retrieval. Asynchronous execution would reduce absolute values but introduce infrastructure variability.

relevance, semantic consistency, and traceability, with the latter benefiting from strong provenance logging at the Fs despite low consumer transparency. Baseline responses, while fluent and topically aligned, lacked source attribution and were penalized on traceability.

Overall, the empirical results broadly confirm the theoretical expectations in Table 1. Architectures predicted to have low latency (Autonomous and Custodian) performed fastest, while those with distributed retrieval (Source and Federated) were slower. Guided RAG, expected to be intermediate, showed latencies closer to the lower range but with higher variability, confirming its position between groups. The evaluation also clarified the distinction between transparency and traceability. Custodian RAG scored high in traceability due to federator-side logging yet low in transparency since intermediate steps remain hidden from the DC. This confirms that Custodian, although rated as medium latency in the theoretical framework, behaved as low-latency in the experimental setup. Privacy, like interoperability, was not directly measured and therefore remains part of the theoretical assessment.

5 Conclusions and Future Directions

This paper examined the integration of RAG with federated DSs as a foundation for trustworthy generative systems. Five architectures were analyzed, including two novel contributions, Guided RAG and Federated RAG, specified in BPMN and evaluated through controlled simulation.

The results confirm that architectural choices directly shape both performance and trust-related properties. Federated RAG achieved the highest qualitative scores, combining semantic coverage, provenance and consumer-side personalization. Custodian and Guided RAG also performed well, producing traceable outputs with slightly lower semantic integration. Autonomous and Source-based models lacked provenance cues and showed partial source coverage, which

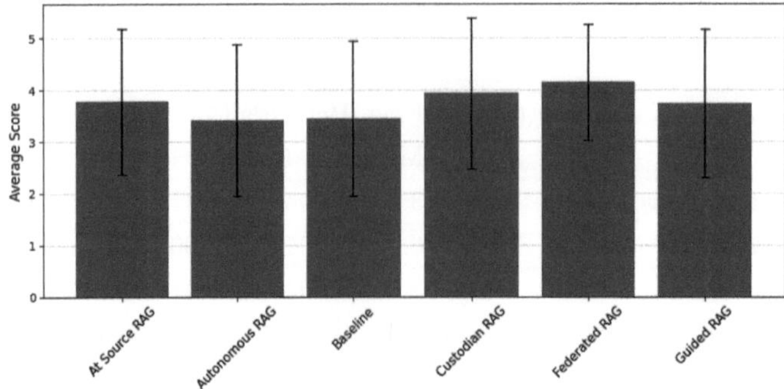

Fig. 2. Average qualitative scores per model (excluding hallucination risk). Error bars indicate standard deviation.

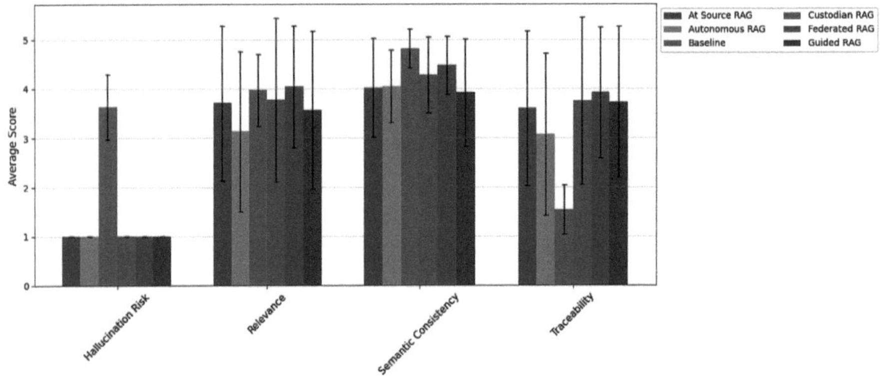

Fig. 3. Average scores per dimension and model (±STD). For hallucination risk, lower values are better.

reduced reliability despite moderate relevance. The baseline, although fluent, failed to provide grounded outputs and highlighted the trade-off between robustness and factual verifiability. Quantitative results further showed that Custodian and Autonomous minimized latency through localized execution, while Federated and Source achieved broader coverage at higher latency. These patterns illustrate that sovereignty, responsiveness and interpretability cannot be optimized simultaneously and that system design must adapt to contextual requirements. Overall the empirical findings are consistent with the theoretical expectations, with minor deviations such as Custodian behaving as a low-latency model despite being rated as medium. Privacy remained part of the theoretical assessment, with Federated RAG offering the strongest guarantees by design.

Guided and Federated RAG were proposed to overcome the structural limitations of Source and Custodian models by combining coordination and personalization without resorting to full centralization. Guided centralizes retrieval under the Fs while preserving consumer-side generation, and Federated coordinates distributed retrieval across DPs while ensuring provenance and leaving generation to the DCs.

Several challenges remain. None of the models performed semantic alignment or ontology mapping across heterogeneous vocabularies, which limits interoperability in cross-domain scenarios. In all architectures, provenance grouping available at retrieval time was not preserved during generation, highlighting the need for workflows that propagate source-level traceability into the final output. Response synthesis remains limited to fragment concatenation, requiring methods for semantic integration and reconciliation. Another open issue is interoperability across multiple DSs. While Federated RAG can be extended to such settings, further research is needed on meta-federation mechanisms and optional DCs-side coordinators to ensure coherence and provenance. Distributed ledger technologies could further enhance auditability through immutable provenance logs. Methodologically, results were supported by substantial inter-annotator agreement ($\alpha = 0.782$), but the synthetic corpus and small query set limit generalizability. Future evaluations should expand to multilingual corpora, token-level metrics, and adversarial prompts.

This study constitutes the first formal and trust-oriented comparative framework for RAG–DS integration. It provides a methodological foundation for the design and evaluation of trustworthy generative AI under federated governance constraints and paves the way for future developments aligned with emerging AI governance frameworks such as the EU AI Act and ISO/IEC 42001.

Acknowledgments. This publication would not have been possible without the support of the following projects: Di4SPDS (PCI2023145980-2), funded by MCIN/AEI/10.13039/501100011033 and by the European Union (Chist-Era Program); KOSMOS-UCLM (PID2024-155363OB-C44) funded by MCIN/AEI/10.13039/501100011033/FEDER, UE; AURORA (SBPLY/24/180225/000074), funded by the Consejería de Educacirón, Cultura y Deportes de la Junta de Comunidades de Castilla La Mancha, and Fondo Europeo de Desarrollo Regional FEDER (Programa Operativo de Castilla-La Mancha 2021–2027); RADAR (2025-GRIN-38447) funded by FEDER; and Grant RED2024-154240-T funded by MICIU/AEI/10.13039/501100011033.

Disclosure of Interests. The authors have no competing interests to declare that are relevant to the content of this article.

References

1. Braga, C.M., Serrano, M.A., Fernández-Medina, E.: Towards a methodology for ethical artificial intelligence system development: a necessary trustworthiness taxonomy. Expert Syst. Appl., 128034 (2025)
2. Dalal, S., Tilwani, D., Gaur, M., Jain, S., Shalin, V.L., Sheth, A.P.: A cross attention approach to diagnostic explainability using clinical practice guidelines for depression. IEEE J. Biomed. Health Inform. (2024)
3. Floridi, L., et al.: AI4People–an ethical framework for a good AI society: opportunities, risks, principles, and recommendations. Mind. Mach. **28**, 689–707 (2018)

4. Freyer, O., Wiest, I.C., Kather, J.N., Gilbert, S.: A future role for health applications of large language models depends on regulators enforcing safety standards. Lancet Digit. Health **6**(9), e662–e672 (2024)

5. Gamage, G., et al.: Multi-agent RAG chatbot architecture for decision support in net-zero emission energy systems. In: 2024 IEEE International Conference on Industrial Technology (ICIT), pp. 1–6. IEEE (2024)

6. Hermsen, F., et al.: On data spaces for retrieval augmented generation. In: INFORMATIK 2024, pp. 701–707. Gesellschaft für Informatik eV (2024)

7. Kim, J., et al.: PhenoFlow: a human-LLM driven visual analytics system for exploring large and complex stroke datasets. IEEE Trans. Vis. Comput. Graph. (2024)

8. Lewis, P., et al.: Retrieval-augmented generation for knowledge-intensive NLP tasks. Adv. Neural. Inf. Process. Syst. **33**, 9459–9474 (2020)

9. Lin, F., et al.: PE-GPT: a new paradigm for power electronics design. IEEE Trans. Ind. Electron. (2024)

10. Lu, S., Li, S., Guo, J.: Scalable, explainable, adaptive information extraction from structure-aware nearest neighbor. Neurocomputing **605**, 128261 (2024)

11. Möller, F., et al.: Industrial data ecosystems and data spaces. Electron. Mark. **34**(1), 41 (2024)

12. Penedo, A.C.: the regulation of data spaces under the EU data strategy: towards the 'actification' of the fifth European freedom for data? Eur. J. Law Technol. **15**(1) (2024)

13. Petersen, K., et al.: Systematic mapping studies in software engineering. In: 12th International Conference on Evaluation and Assessment in Software Engineering (EASE). BCS Learning & Development (2008)

14. Remy, F., Demuynck, K., Demeester, T.: BioLORD-2023: semantic textual representations fusing large language models and clinical knowledge graph insights. J. Am. Med. Inform. Assoc., ocae029 (2024)

15. Schneider, J.: Explainable generative AI (GenXAI): a survey, conceptualization, and research agenda. Artif. Intell. Rev. **57**(11), 289 (2024)

16. Shi, Z.F., et al.: Meta data retrieval for data infrastructure via RAG. In: 2024 IEEE International Conference on Web Services (ICWS), pp. 100–107. IEEE (2024)

17. de Silva, D., et al.: Opportunities and challenges of generative artificial intelligence: research, education, industry engagement, and social impact. IEEE Ind. Electron. Mag. (2024)

18. Wang, D., et al.: Large language models in medical and healthcare fields: applications, advances, and challenges. Artif. Intell. Rev. **57**(11), 299 (2024)

19. Yang, J., Shu, L., Duan, H., Li, H.: RDguru: a conversational intelligent agent for rare diseases. IEEE J. Biomed. Health Inform. (2024)

Governing Data Lakehouses – An Alternative Model

André Martins[1] (iD), Bruno Oliveira[2] (iD), and Orlando Belo[1](✉) (iD)

[1] ALGORITMI Research Centre / LASI, University of Minho, 4710-059 Braga, Portugal
pg50739@alunos.uminho.pt, obelo@di.uminho.pt
[2] CIICESI, School of Management and Technology, Porto Polytechnic, Felgueiras, Portugal
bmo@estg.ipp.pt

Abstract. Like any other data system, data lakehouses require the application of effective governance strategies and models for ensuring compliance, security and quality of the data they host. The absence of an effective data governance strategy can seriously compromise data analysis and decision-making processes in data lakehouses, reducing trust and compromising data usefulness. Despite the various initiatives carried out on the field, there are gaps in the definition and application of governance models in data lakehouse environments, particularly concerning data lakehouse system development methodologies. After studying the recent development models, architectures, and methodologies for data lakehousing, we designed a specific data governance model for data lakehouses, creating a roadmap for its application to the various stages of system development. In this paper we present and discuss such a model, as well as its concrete application to the development of a concrete data lakehouse application.

Keywords: Data Lakehousing Systems · Data Lakehouses · Data Governance · Data Governance Strategies and Models · Data Lakehouse Governance

1 Introduction

These days, organizations develop their activities by exchanging and manipulating enormous amounts of diverse data – structured, semi-structured or unstructured –, which often causes great constraints in traditional data storage and processing systems. Usually, these systems worked well with limited volumes and uniformly structured data, having very satisfactory levels of operability. However, today, they are not able anymore to manage high complex and diverse data formats in an adequate manner. In addition, there are some application scenarios in which data arrives at great speed to storage structures, provided by high-performance data ingestion processes [1]. This new reality of data storage systems has promoted the study and gestation of new storage paradigms and data storage systems, such as data lakes [2] and data lakehouses [3].

Data warehouses [4] were designed to organize the storage of structured data for analytical purposes and improve performance of analytical systems in satisfying results for very broad queries, involving different analysis perspectives over very diverse periods

L. Martínez et al. (Eds.): IDEAL 2025, LNCS 16239, pp. 375–386, 2026.
https://doi.org/10.1007/978-3-032-10489-2_32

of time. In turn, data lakes have emerged as a slightly more flexible solution, largely as a substantial reinforcement for data staging areas of data warehouses, allowing data of any type to be stored without the need to structure it, and receiving it quickly, even when it is provided by sources in (near) real time. When integrated into a data warehouse system, a data lake allows for getting solutions having high analytical performances in application scenarios. The idea of this combination of valence characteristics, functionalities and services, led to the emergence of a new system for decision support, the data lakehouses, which have the capacity to handle a large volume of heterogeneous data, with a very low supply latency – "very fast" data. Despite data lakehouses are still a relatively recent innovation, they are quickly gaining their place in the analytical systems domain. However, its adoption is still in a maturing process, and so far, it does not have the same level of recognition in the technological community as more consolidated approaches, such as data warehouses or even data lakes [5].

As with other data storage systems, companies that have adopted and implemented data lakehouse systems also reveal increasingly pronounced concerns about all issues related to data governance. The reasons are several. From the simple need to have to comply with what is required of them by the new operational regulations, such as the General Data Protection Regulation, which imposes inflexible rules on how to collect, process and exploit data, to the need to improve the quality of the data they have, often incorrect, duplicate or incomplete, which can originate a lack of confidence in their analysis and decision-making processes or reduce the quality of their way of acting. Thus, governance must be dealt with effectively and with well-defined application and regulatory standards [6]. The absence of a robust data governance model, particularly in companies with little experience (or even immature) can lead to inappropriate organizational decisions, as well as compromise compliance, security, and data quality [7].

With the growing use of data lakehouse systems, as a solution for decision-making processes, dealing with application scenarios involving large volumes of information, of diverse nature and with high levels of volatility, and the increase in sensitivity around the security and quality of business data, as a guarantee for the increase of their skills in the medium and long term, it is necessary to study and create new scientifically based approaches to develop data governance models for companies accordingly to their way of being and business culture. However, we note that there are still no methodologies for the development of data governance models specifically oriented to data lakehouse systems, including concrete guidelines to support the various stages of their implementation, addressing effective data governance practices to be applied throughout the system development lifecycle and, subsequently, in its application and exploitation. These absences of methodologies (and data governance models) are sometimes due to the recognition of corporate data as an essential asset, with value, the complexity that data presents, or the lack of recognition and attribution of skills and responsibility to individuals who apply and promote concrete data governance actions [8].

In this paper we present the result of a study we made aiming to establish a set of principles for the development of a governance model for data lakehouses, which allow for accommodating the various data domains usually involved in the implementation of a data lakehouse system, taking particular attention in the design and implementation of the staging area. The remaining part of this paper is organized as follows. Section 2 approach several aspects of data lakehouse systems, and Sect. 3, exposes and discusses the governance model designed for data lakehouse systems, and Sect. 4 describes a data lakehouse prototype we developed using the governance model designed. Finally, Sect. 5 presents some conclusions and future work.

2 Data Lakehousing

The term "data lakehouse" gained popularity about 2019, when the company Databricks adopted it to describe the approach its technicians usually do to combine data management structures and capabilities of a data warehouse into low-cost storage solutions using data lakes [3]. Today, data lakehouse is regularly used by many information and communication technology companies, such as AWS[1], Snowflake[2], or Google[3], among many others. Despite the strong use of the term, data lakehouses are still data storage systems of recent design that, application by application, welcomes new features and functionalities. So, it is evolving naturally. However, its rationale often appears in a rather informal way, such as the combination of a data lake and a data warehouse. Although they are a little more complex than that. We can use these two "ingredients" to introduce some important aspects in their field and, particularly for supporting some of the ideas we express and discuss in this the work.

Data warehouses [4, 9] are data storage structures designed especially to support data analysis and decision processes. In some scenarios, involving a great diversity of information sources, data warehouses cannot host the data provided by these sources, given the difficulty that their populating systems, commonly recognized as ETL (Extract-Transform-Load) systems, have in recognizing, collecting and processing data from them. To mitigate such incapacities, in some cases, data lakes are implemented to act as their staging areas. Given their characteristics, data lakes have no problem dealing with data from sources with characteristics such as those we have mentioned. Data lakes do not have to worry about adapting data that reaches their structures through data ingestion or extraction processes, according to the populating model established for the system. Simply, data is hosted in the data lake, in its original format. The great flexibility and scalability of a data lake make this process much easier. Later, when defined (and possible) data will begin to be worked on. In practice, in this type of architecture, only a part of the data received will be processed. The data contained in a data lake is gradually known, as well as its structure (if any), as queries are launched over the data lake – an action that is known as schema-on-read [10]. This feature gives us great advantages. It allows a natural adaptation of the system to any change in requirements, both in terms of the data available, and the way we intend to exploit them, which does not present any special requirement for its reading - the data can be consulted freely, as we wish, without restrictions, making it possible to explore very varied and alternative decision scenarios and analytical interpretations. In fact, data lakes provide a very flexible and scalable environment, with characteristics suitable for the implementation of a data

---

[1] https://docs.aws.amazon.com/whitepapers/latest/best-practices-building-data-lake-for-games/lake-house-architecture.html.

[2] https://www.snowflake.com/en/product/use-cases/data-lakehouse/.

[3] https://services.google.com/fh/files/misc/building-A-data-lakehouse.pdf.

staging area of a data warehouse, having structures and services suitable for carrying out data transformation and preparation tasks usually required by a conventional data warehouse populating process. However, we must not see a data lake as an *ad hoc* data system, which is used without any care or discernment in the application of criteria, norms or models to sustain data processing tasks.

Since their "adoption", data lakes have been the subject of study by several authors. Without breaking the beauty of its nature and intrinsic characteristics, several works have appeared suggesting architectural models and levels of functional abstraction for data lakes. One of these models has gained some relevance in recent times: the medallion architecture [11]. The medallion architecture was proposed and developed by Databricks [12] with the aim of providing a computational model to organize data processing tasks that usually occur in a data staging area and to improve the data flow established between the various stages of a populating process. The way this model is organized enhances better data quality, improves performance of data processing tasks, and allows establishing an effective data governance model. In a medallion architecture, data processing in a staging area is structured into three distinct layers, organized according to the state of the data throughout the data transformation process. The three layers are recognized as: "bronze", in which raw data are positioned, received directly by ingestion or extraction from the various sources of information in the system; "silver", which hosts data already transformed in some way, involving data cleaning, structuring, conversion or reconciliation tasks; and "gold", which receives prepared data, especially oriented to populate data warehouses, reporting and dashboarding platforms or even analytical sandboxes. Using this model to structure a data lake for supporting a data staging area, we are promoting a functional architecture that allows to carry out the various data processing operations of a data warehouse populating process in a gradual and sustained way, in which data goes through different quality states, changing according the requirements, layer by layer, while preserving the original data (raw data), which can be reused in its various forms in different transformation tasks (or even in intermediate analysis processes), providing robust means for high-performance query operations, with adequate means to track the evolution of data (data lineage) along the various layers, and, finally, capable of welcoming and supporting an appropriate governance model. All these aspects are crucial to successful implementation (and use) of a data lakehouse system in which the data lake is integrated. This makes the data lake a vital framework to sustain data lakehouses' populating process and, consequently, justify its existence.

Any data lakehouse (or data warehouse) that receives improperly prepared data will certainly not meet its basic objectives and fulfill its mission – a "single source of truth" for the organization in which it was implemented. Within the scope of this work, we focused essentially on the aspect of data governance, its importance, influence and usefulness in a data lakehouse system. These systems are, by nature, very flexible and versatile, with the ability to scale easily. However, they do not have, by default, an associated governance model, something that can be applied through a technical-administrative recipe for data systems. The existence of a robust data governance model (and strategy) is essential to any data storage system, particularly in a data lakehouse, to prevent it from becoming a data swamp [13]. In the next section we will look at this issue in more detail and present the governance model we have devised for a data lakehouse system.

3   Data Lakehouse Governance

3.1   An Overview

Over the years, organizations have collected enormous volumes of information from disparate and heterogeneous information sources, constructing large and complex data storage systems. To be valuable and useful, these systems need to be managed and maintained securely and reliably, which imposes defining and implementing effective data governance models and techniques. In the business world, data governance is often confused with data management. But they are different concepts. While data governance defines who has authority and decision-making power over data in an organization, defining data management policies or establishing business glossaries, data management is about structuring, controlling and protecting data, developing data modelling processes or defining quality control or security measures[4].

Data governance represents the planning and control of data management at a higher level in an organization [14]. The definition and implementation of a data governance model and a policy in an organization requires a structured approach that clearly sets out the objectives and principles of governance, involving the people and processes of the organization and considering the existing organizational culture. To achieve a successful governance model, it is necessary to define good practices of action, which may involve metadata management, data life cycle establishment, or data quality evaluation, among other things [15]. To define a model for data governance, in [6] were identified five interrelated domains that should guide an organization's data governance. These domains served as the foundation for the proposed governance model: data principles, data quality, metadata, data access, and data lifecycle. The first domain, *data principles*, sets the direction for all other decisions by defining the boundary requirements for the intended uses of the data. The *data quality* domain focuses on establishing clear and appropriate requirements to ensure the data meets the specific needs of the system. The third domain, *metadata*, aims to ensure that data is interpretable to users by documenting its semantics. *Data access* is a crucial domain that covers the specification of access requirements, involving the assessment of the data's business value and the implementation of continuous risk assessments. Finally, the *data lifecycle* determines the definition, production, retention, and deletion of data, which are crucial aspects for applying governance principles within the organization's infrastructure. Decision-making in each of these domains is guided by a set of pertinent questions that help establish a robust governance model for a data lakehouse.

3.2   A Data Lake and Staging Area

To understand and apply a practical governance model, first it is necessary to outline the key components of a data lakehouse. We focus on the staging area of a traditional data warehouse, which comprises both storage and the ETL processes that prepare data for downstream analytics (Kimball & Ross, 2002). In a lakehouse, this staging area serves as a strategic checkpoint where raw data is ingested, assessed, and – if needed – refined

---

[4] https://dama.org/.

before being integrated into the warehouse, thereby ensuring analytical data quality, integrity, and reliability.

The staging workflow begins with data collection: sources are tapped, and raw data is stored in the data lake "raw data" zone, preserving data in its native form [16]. Next, a quality assessment against predefined standards is performed. If it meets the criteria, it is promoted to the "selected zone" for subsequent transformation; otherwise, it is routed through preprocessing to correct any issues. Thus, the staging area comprises four core entities – data sources, raw data, selected data, and the data lake – augmented by two functional zones: the transformation area, where data is shaped to warehouse specifications, and the loading area, which transfers the processed data into the warehouse. The primary modeling decision for the staging area is how to integrate the selected data and the data lake itself. Since the warehouse will ultimately structure and organize the data, it is inefficient to re-structure it prematurely within the lake. Therefore, both raw and selected datasets reside in distinct lake zones: a raw zone and a selected zone. After storage, selected data enters the transformation zone, following further preprocessing determined by detailed quality analyses. Finally, the transformed data is loaded into the warehouse, completing the staging-area model while preserving data accessibility for users (Fig. 1).

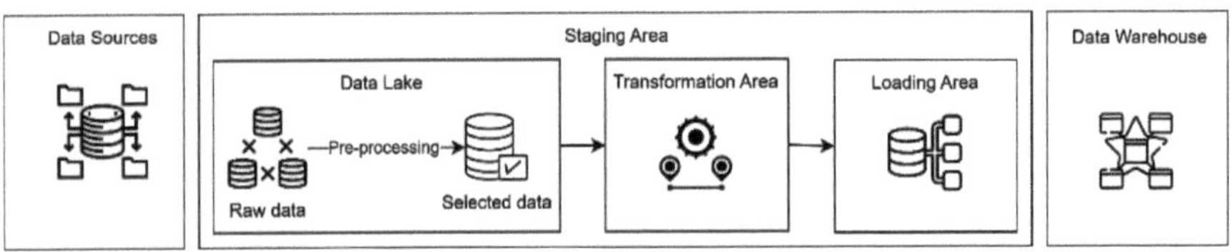

Fig. 1. Staging Area proposed model for data lakehouse

3.3 Data Governance Model

Once the data-governance model and staging-area architecture are defined, most decisions for building the lakehouse are in place. The next step is a process that lays out every stage of implementation and ensures alignment with governance practices. We adapted Kimball's lifecycle-diagram approach, detailed in [17], to the data lakehouse context. The process sequences nine parallelizable steps:

1. *Project planning*: define scope, objectives, justification, resources, roles, and budget to secure stakeholder approval.
2. *Primary data-source inventory*: catalog available internal systems, lakes, files, and APIs.
3. *User-requirements gathering*: clarify analytical use cases (e.g., reporting, machine learning applications), data formats, latency, versioning, and security policies.
4. *Final "working-data" specification*: reconcile source inventory with user needs, identify gaps, and specify exact datasets, schemas, and quality metrics.
5. *Data-and-metadata catalog creation*: build a central repository describing each dataset's origin, structure, semantics, update frequency, and governance rules.
6. *Tool selection*: choose ingestion, processing, transformation, and orchestration platforms.

7. *ETL/ELT process definition*: design extraction, quality checks, preprocessing, trans-formation, and loading workflows.
8. *Integration of tools and processes*: align chosen technologies with pipeline require-ments, iterating as needed.
9. *Implementation*: build the raw-zone data lake (per zoned-architecture principles), ingest sources, perform deep quality analysis [16], apply any preprocessing, and load clean data into the selected-zone – bypassing raw-zone analysis when prequalified – before executing full transformation into the warehouse.

For validating the methodology, a proof-of-concept data-lakehouse application was developed using Amazon unlocked-phone reviews dataset[5].

Table 1. An excerpt of the Amazon unlocked-phone reviews dataset

Product	Brand	Price	Rating	Review Text	Helpful Votes
Apple (…)	Apple	194.99	1	The phone (…)	1
Motorola (…)	Motorola	174.99	5	I was (…)	3
CNPGD (…)	CNPGD	49.99	1	an (…)	0
Nokia (…)	Nokia	95.00	5	I fell (…)	0
(…)	(…)	(…)	(…)	(…)	(…)

The dataset provides (Table 1) consumer feedback sourced from the Amazon e-commerce platform. Each entry corresponds to a single user review and is characterized by several distinct attributes. The "Product" provides the full name of the device as listed, which is complemented by the "Brand", a categorical field that explicitly identifies the manufacturer, such as Apple, Samsung, or Motorola. The commercial context is provided by the "Price", indicating the cost of the item at the time the review was submitted. The core of each entry consists of the consumer's evaluation, which is captured in two forms: the "Rating", representing the overall satisfaction level and the social validation of each review is quantified by "Helpful Votes", a discrete integer that counts the number of other users who found the review to be valuable. The dataset revealed two natural dimensional attributes for decision-support reports: "Brand" (to compare manufacturer performance) and "Product" (to compare individual models). With these dimensions, a set of reports were developed to support consumer trends, brand, product-level satisfaction, and price-to-rating correlations. The case study follows the architecture presented in Fig. 1, providing an additional visualization layer with reports using the data warehouse tables to present data.

---

[5] https://www.kaggle.com/datasets/PromptCloudHQ/amazon-reviews-unlocked-mobile-phones.

Table 2. Excerpt of questions with examples from the proposed governance model.

Principle	Question	Example Answer
Data Principles	What will be the uses of the data?	To support decisions in the presented problem domain according to the defined dimensions.
Motorola (…)	(…)	(…)
Data Quality	What are the data quality standards with respect to the established dimensions?	Appropriate quantity, completeness, ease of manipulation, absence of errors.
Nokia (…)	(…)	(…)
Metadata	What will be the program to document the semantics of the data?	Develop a data and metadata catalog.

4　Data Lakehouse Implementation

4.1　The Governance Model

Following the proposed governance model, we began by answering as many governance-framework questions as possible across its five domains. Although not every question applied in this proof-of-concept, completing the relevant ones clarified scope, data assets, and decision-points, eliminating ambiguities and revealing subsequent design choices. Considering the *Use and Value of Data*, the lakehouse support decision-making on unlocked-phone reviews from Amazon, analyzing price, numeric ratings, and "helpful" votes by brand and product. Because this application has a single use case, there are no broader sharing or reuse opportunities beyond this domain. For *Data Quality Dimensions*, a preliminary scan showed that three dimensions: credibility, concise/consistent representation, and interpretability, could be excluded without impairing our proof-of-concept. For example, product names were too inconsistent to normalize cost-effectively, and temporal currency is irrelevant in this retrospective analysis. We retained four dimensions (appropriate volume, completeness, ease of manipulation, and error-free), defined concrete metrics, and established a quality-assurance program:

1. *Baseline Assessment*: measure each dimension at project outset.
2. *Reporting*: publish a "quality snapshot" covering all four dimensions.
3. *Preprocessing*: apply corrections as needed.
4. *Validation*: re-report to confirm that quality thresholds are met.

Considering *Metadata Governance*, every dataset attribute is documented in a single, centralized data-and-metadata catalog. Because neither the data nor its usage will change post-deployment, no ongoing metadata-update process is required. For *Data Access and Security*, the data's commercial value lies in our chosen dimensions. Access is restricted to the dimensional warehouse – end users authenticate via login/password and query only preprocessed, structured data. Since the source data are static and publicly obtainable, no backup or recovery plan is necessary: running the ETL pipeline always produces the same result. Finally, for *Data Lifecycle*, metadata exist in the catalog, and because neither the underlying data nor their semantics will evolve, no production-stage tooling or update schedule is needed. Data removal poses no issue since any record can be purged and later restored by reloading the original source files. Table 2 presents an excerpt from this framework, detailing a subset of guiding questions and their respective answers. These responses form the basis of the data strategy, ensuring that data is treated as an asset, its quality is maintained, and its access is controlled throughout its lifecycle.

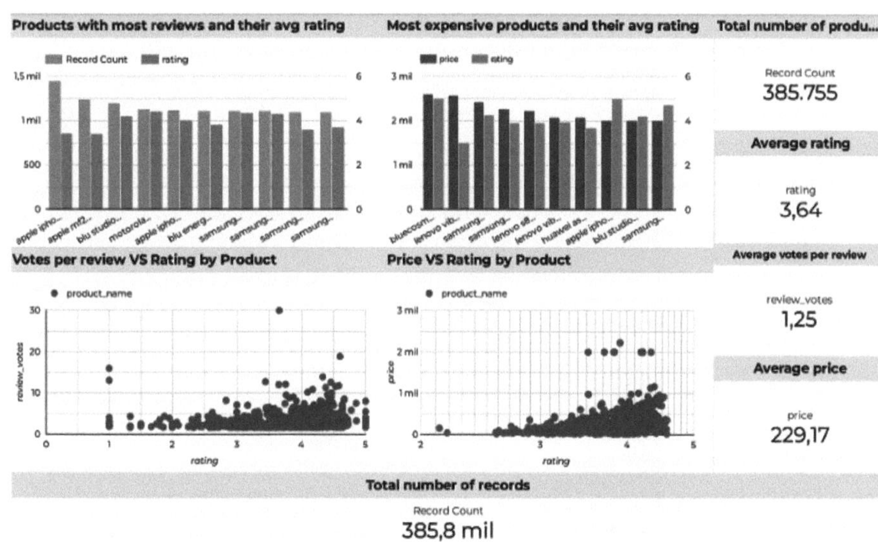

Fig. 2. An example of a report for product analysis.

After answering governance-framework questions to clarify scope, data assets, and quality requirements, it was confirmed that end users would access only cleansed, dimensional data via batch loads into the warehouse, focusing analyses on brand and product. Then, tools were selected to implement each stage. AWS S3[6] hosts both the raw and pre-processed zones of staging area, ensuring flexible, zone-based storage and future OLAP support [16]. PostgreSQL[7], deployed on Amazon RDS[8], serves as both the transformation engine and final warehouse, while dbt[9] handles SQL-based transformations into two-dimension tables and a fact table. Python scripts (using yData Profiling[10]) automate quality assessments (measuring completeness, volume, manipulability, and error rates), generate preprocessing reports and orchestrate data processing. Finally, Looker

---

[6] https://aws.amazon.com/pt/s3/.

[7] https://www.postgresql.org/.

[8] https://aws.amazon.com/pt/rds/.

[9] https://www.getdbt.com/.

[10] https://docs.profiling.ydata.ai/.

Studio[11] connects to the warehouse to produce interactive dashboards for strategic decision support.

4.2 Implementation

Data lakehouse was instantiated by creating an Amazon S3 bucket divided into "raw" and "selected" folders, then uploading the dataset is CSV (Comma Separated Values) to the raw zone. A data quality analysis was executed for assessing volume, completeness, manipulability and error rate, which revealed 413 840 total records, 64 079 exact duplicates, and numerous non-phone items (cases, SIM cards, headsets) that we filtered using keyword matching. Missing brand values (15.75% of records) were backfilled by detecting common manufacturer names within product titles, while price and vote-count gaps were imputed using the median and mode respectively [18]. Textual reviews left blank were tolerated, given their irrelevance to this proof-of-concept, and after preprocessing we retained 385 755 high-quality entries in the selected zone.

A PostgreSQL database was created on Amazon RDS with distinct transformation and loading schemas, bulk-loaded the cleansed CSV into a staging table, and applied dbt models to generate three dimensional tables – "dim_brand", "dim_product" – and a single "fact_review" table. Referential integrity was enforced via foreign keys, timestamps normalized to UTC, and the final tables atomically swapped into the production schema that feeds Looker Studio dashboards (Fig. 2). The dashboard delivers multi-view summary of the 385 755 record dataset. In the upper-left panel, a dual-axis bar chart ranks the ten most-reviewed products alongside their average ratings, revealing which devices attract the most feedback and how customer satisfaction varies among them. The top-right sidebar presents four key performance indicators: total reviews, overall average rating, mean "helpful" votes per review, and average price. Below these summaries, two scatterplots unpack deeper relationships. On the left, review-vote counts are plotted against rating scores, illustrating whether higher ratings tend to garner more peer endorsements. On the right, individual ratings are mapped against product price to show how satisfaction distributes across the entire price spectrum.

5 Conclusions and Future Work

The ever-increasing data volumes and diverse analytics requirements, data lakehouse has emerged as an architectural paradigm, remaining variably implemented across academia and industry. In this paper we propose a governance-driven framework to support data lakehouse implementation, validating it through a comprehensive proof-of-concept. The governance framework systematically addresses data scope, quality dimensions, metadata management, and access controls. These decisions are supported by a process that guides implementation, and improves clarity, minimizing ad-hoc decisions and streamlined development, which highlight the framework's practical value. The prototype system we developed integrates raw and selected-zone storage on Amazon S3, automated ELT pipelines implemented in Python (pandas) and dbt, dimensional-model loading into

---

[11] https://lookerstudio.google.com/.

PostgreSQL on Amazon RDS, and interactive BI dashboards built in Google Looker Studio. This end-to-end lakehouse implementation not only validates the methodological coherence and technical feasibility of our governance-driven approach [16] but also offers a fully reproducible template for future academic investigations and industrial pilot projects.

Future work should focus in incorporating heterogeneous and unstructured data sources such as application logs, images, and social-media feeds, to rigorously evaluate schema evolution and format-flexibility capabilities. In parallel, systematic performance benchmarking is essential: measuring ingestion throughput, transformation latency, and query response times will reveal scalability limits and inform optimizations. It would also be instructive to conduct a comparative analysis of multi-tier architectures versus metadata-oriented solutions (for example, Delta Lake) to clarify trade-offs in cost, complexity, and operational overhead.

Acknowledgements. This work has been supported by FCT – Fundação para a Ciência e Tecnologia within the R&D Unit Project Scope UID/00319/Centro ALGORITMI (ALGORITMI/UM).

References

1. Hariri, R.H., Fredericks, E.M., Bowers, K.M.: Uncertainty in big data analytics: survey, opportunities, and challenges. J Big Data. **6**, 44 (2019). https://doi.org/10.1186/s40537-019-0206-3
2. Inmon, B.: Data Lake Architecture: Designing the Data Lake and Avoiding the Garbage Dump, 1st edn. Technics Publications, Sedona (2016)
3. Armbrust, M., Ghodsi, A., Xin, R., Zaharia, M.: Lakehouse: a new generation of open platforms that unify data warehousing and advanced analytics. In: Conference on Innovative Data Systems Research (CIDR) (2021)
4. Kimball, R., Rosse, M.: The Kimball Group Reader - Relentlessly Practical Tools for Data Warehousing and Business Intelligence. Wiley, Hoboken (2016)
5. Orescanin, D., Hlupic, T.: Data lakehouse - a novel step in analytics architecture. In: 2021 44th International Convention on Information, Communication and Electronic Technology (MIPRO), pp. 1242–1246. IEEE (2021). https://doi.org/10.23919/MIPRO52101.2021.9597091
6. Khatri, V., Brown, C.V.: Designing data governance. Commun. ACM **53**, 148–152 (2010). https://doi.org/10.1145/1629175.1629210
7. Quinto, B.: Big data governance and management. In: Next-Generation Big Data, pp. 495–506. Apress, Berkeley, CA (2018). https://doi.org/10.1007/978-1-4842-3147-0_11
8. Nargesian, F., Zhu, E., Miller, R.J., Pu, K.Q., Arocena, P.C.: Data lake management. Proc. VLDB Endow. **12**, 1986–1989 (2019). https://doi.org/10.14778/3352063.3352116
9. Golfarelli, M., Rizzi, S.: Data Warehouse Design: Modern Principles and Methodologies. McGraw-Hill, New York (2009)
10. Hashem, I.A.T., Yaqoob, I., Anuar, N.B., Mokhtar, S., Gani, A., Ullah Khan, S.: The rise of "big data" on cloud computing: review and open research issues. Inf. Syst. **47**, 98–115 (2015). https://doi.org/10.1016/j.is.2014.07.006
11. Gupta, N., Yip, J.: Delta lake - deep dive. In: Databricks Data Intelligence Platform, pp. 61–88. Apress, Berkeley, CA (2024). https://doi.org/10.1007/979-8-8688-0444-1_4
12. Databricks: Medallion Architecture

13. Derakhshannia, M., Gervet, C., Hajj-Hassan, H., Laurent, A., Martin, A.: Data lake governance: towards a systemic and natural ecosystem analogy. Future Internet **12**, 126 (2020). https://doi.org/10.3390/fi12080126
14. Al-Ruithe, M., Benkhelifa, E., Hameed, K.: A systematic literature review of data governance and cloud data governance. Pers. Ubiquitous Comput. **23**, 839–859 (2019). https://doi.org/10.1007/s00779-017-1104-3
15. Seiner, R.S.: Non-Invasive Data Governance: The Path of Least Resistance and Greatest Success. Technics Publications, Sedona (2014)
16. Giebler, C., Gröger, C., Hoos, E., Eichler, R., Schwarz, H., Mitschang, B.: The data lake architecture framework: a foundation for building a comprehensive data lake architecture. In: Conference for Database Systems for Business, Technology and Web (BTW) (2021)
17. Kimball, R., Reeves, L., Ross, M., Thornthwaite, W.: The Data Warehouse Lifecycle Toolkit : Expert Methods for Designing, Developing, and Deploying Data Warehouses. Wiley, New York (2008)
18. Marino, M., Lucas, J., Latour, E., Heintzman, J.D.: Missing data in primary care research: importance, implications and approaches. Fam. Pract. **38**, 199–202 (2021). https://doi.org/10.1093/fampra/cmaa134

A Real-Time Traceability System Based on Blockchain and IoT Technologies

Juan A. Torres de la Chica, Bruno Ramos-Cruz[✉], Pedro José Sánchez, and Luis Martínez

Computers Science Department, University of Jaen, Las Lagunillas, 23071 Jaen, Jaen, Spain
`jatc0011@red.ujaen.es`, {`brcruz,pedroj,martin`}`@ujaen.es`

Abstract. Traceability is a critical component in supply chains to ensure transparency, quality control, and trust among stakeholders. However, current traceability systems face significant challenges, including the risk of data manipulation by unauthorised actors, a lack of system integration across participants, inefficient centralized processes, often manual in nature, and delays in accessing real-time information. In this work, we propose a novel real-time traceability system based on Blockchain and IoT technologies to address previous limitations. The proposed system uses IoT devices to automatically monitor environmental data and save it directly on a blockchain, ensuring immutability and eliminating the need for manual data handling. The use of blockchain guarantees decentralisation and trust among participants, while the IoT infrastructure provides real-time visibility throughout the supply chain. To enable efficient real-time data retrieval, the system employs an event-driven architecture, where data collected by IoT devices triggers events through smart contracts, allowing effective indexing and optimised querying of traceability data.

Keyword: Traceability, Blockchain, IoT, Supply chain, Security

1 Introduction

Traceability in supply chains is essential to ensure product quality, safety, and transparency throughout the distribution process [1]. However, traditional traceability systems present significant limitations, including the possibility of data manipulation by unauthorised actors, a lack of integration among heterogeneous systems used by supply chain participants, inefficient and often manual processes, and delays in obtaining real-time data [2]. Moreover, these systems usually suffer from limited scalability when the volume of traceability data increases, leading to performance bottlenecks during critical operations. There is also a dependency on centralised infrastructure, making the systems vulnerable to single points of failure and cybersecurity attacks that may compromise the integrity and availability of traceability data. Additionally, traditional systems often lack

L. Martínez et al. (Eds.): IDEAL 2025, LNCS 16239, pp. 387–398, 2026.
https://doi.org/10.1007/978-3-032-10489-2_33

interoperability with emerging digital technologies, which makes automated validation and traceability particularly challenging in globally distributed supply chains.

Blockchain and Internet of Things (IoT) technologies offer inherent properties that can address previous challenges effectively [3]. Blockchain ensures data immutability and decentralisation, reducing the risks of unauthorised modifications and eliminating the need for a central authority to maintain trust among participants [4]. IoT devices, on the other hand, enable real-time data collection from the environment, reducing manual intervention and ensuring the timeliness of the information captured across the supply chain [5].

In this contribution, we propose a real-time traceability system based on Blockchain and IoT technologies, with a focus on event-driven architecture using smart contracts to overcome the limitations of traditional supply chain systems. The proposed system automatically collects data through IoT devices and writes it to the blockchain through the event-emitting smart contract, the main feature of this system. This not only eliminates human error in data recording but also ensures efficient queries when obtaining real-time traceability, where each data set collected by IoT devices triggers an event via a smart contract deployed on the blockchain network. This strategy enables the indexing of data sets, optimising query and retrieval operations while maintaining the consistency and transparency required in traceability systems. Once recorded, the data is immutable, providing a trust foundation shared by all actors within the supply chain ecosystem.

Unlike other related projects, this system is developed in Hyperledger Besu, so it has both the advantages of being able to operate on a private network and the possibility of extending to a hybrid network as it is compatible with the Ethereum ecosystem.

The paper is structured as follows: Sect. 2 gives a background on blockchain and IoT technologies. Section 3 reviews related work. Section 4 introduces the proposed framework. Section 5 defines a particular use case in the food supply chain. Section 6 shows the implementation features and present the key results. Finally, Sect. 7 discuss the conclusions and future work, respectively.

2 Background

Blockchain and IoT represent complementary technological paradigms that, when integrated, can enable secure, transparent, and automated data collection and processing across distributed environments. This section provides a definition related two areas: blockchain and IoT.

Blockchain is a decentralised technology that enables the secure verification and recording of transactions without the need for trusted intermediaries. It achieves this through a data structure that addresses two of the primary challenges in decentralised networks: double spending and the Byzantine Generals' Problem [6,7].

As its name suggests, the blockchain is composed of blocks, each storing a set of validated transactions and linked to the previous block through a cryptographic hash. This chained structure guarantees data integrity, as any attempt to modify a block would affect the entire chain. Due to this property, only new transactions can be added, and those already recorded cannot be altered or deleted [7].

To fully understand its impact and practical utility, it is essential to analyse the main characteristics that define its nature and operation: decentralisation, immutability, transparency, cryptographic security, and traceability. Traceability, in particular, is a key feature that enables the detailed tracking of any stored item or transaction. The blockchain structure, combined with immutability, allows the historical journey of an asset to be followed step by step, from its origin to its current state, thereby facilitating accurate and effective audits [8]. Traceability, therefore, not only strengthens trust but also enables the rapid detection of errors or fraud, significantly improving operational efficiency [6].

An important element within blockchain systems is the smart contract, which can be defined as a self-executing program stored on the blockchain that automatically enforces and executes the terms of an agreement when predefined conditions are met, without the need for intermediaries [9].

The **Internet of Things (IoT)** can be defined as a system composed of embedded devices such as sensors, actuators, and microcontrollers connected to communication networks, which cooperate to collect, exchange, and process information from the physical world for monitoring, control, and analysis purposes [10–12].

The synergy between Blockchain and IoT create opportunities for the development of advanced systems capable of improving operational efficiency [13], traceability, and trust in domains such as healthcare [14], maritime industry [15], supply chain [16], and among others.

3   Related Work

Modern supply chain management (SCM), characterized by its complexity and global reach, has driven the need for sophisticated traceability systems to ensure quality, authenticity, security, and efficiency [1]. Saidu et al. [1] provide a comprehensive review on the convergence of Blockchain, IoT, and AI for enhanced traceability systems. Their systematic analysis identifies key trends, such as the increase in research in agricultural supply chains and SCM, with a prevalence of experimental and hybrid methodologies. Traditional approaches are often prone to data manipulation, fragmentation, and a lack of transparency, especially in environments involving multiple intermediaries. In response to these challenges, recent literature has explored emerging technologies such as blockchain and IoT. For instance, the author in [17] focus on food safety, traceability, and data integrity verification using private blockchains, specifically Hyperledger Fabric. They argue that, unlike public blockchains, private blockchains can overcome scalability, latency, and privacy issues for real-time food safety systems.

Another important problem arises when executing queries directly on blockchain systems in traceability contexts [18]. While blockchain offers immutable, tamper-evident records, efficiently querying and retrieving data for traceability purposes poses challenges due to the inherent design constraints of blockchain architectures. For instance, data retrieval often requires scanning multiple blocks, leading to high latency and computational overhead, especially when real-time traceability and analytics are required across large-scale supply chains [19]. Furthermore, the lack of native query functionalities in many blockchain platforms necessitates auxiliary structures, which introduces potential trust issues and partial centralisation [20]. Additionally, scalability issues surface when the volume of traceability queries increases, leading to congestion, higher gas fees (in public blockchains), or bottlenecks in smart contract-based querying logic. These limitations emphasise the need for architectures combining blockchain's integrity with querying and analytics frameworks to support advanced traceability requirements while maintaining security and performance.

The integration of blockchain and IoT not only improves efficiency but also addresses security and transparency issues. For instance, in the security domain, Belfqih and Abdellaoui in [21] address the significant security challenges in IoT device interconnectivity, arguing that traditional centralised approaches are insufficient. They present a decentralised authentication protocol that leverages blockchain technology (Ethereum) and the IPFS data management framework for secure, real-time communication between IoT devices.

Current research demonstrates that the integration of IoT and blockchain is critical for developing advanced traceability systems that address transparency, security, efficiency, and quality management in complex supply chains, particularly in the food and asset management sectors, although significant challenges related to implementation and standardisation remain [5].

4   Traceability System

A real-time traceability system based on Blockchain and IoT technologies is presented. The inherent properties of Blockchain and IoT technologies can address the limitations of traditional supply chain systems, providing them with enormous potential to inform new approaches to traceability across all sectors.

4.1   General Description

The system illustrated in Fig. 1 involves two main actors: TRACERS, who are responsible for recording data on the blockchain, and VIEWERS, who analyse the recorded data. **TRACERS** consist of IoT devices equipped with sensors that collect environmental data in real time. Each dataset captured by these devices triggers an event through a smart contract deployed on the blockchain, ensuring that the data is automatically recorded without the need for human intervention. This mechanism prevents data manipulation and enables the indexing of datasets, which facilitates efficient information retrieval. **VIEWERS**, on

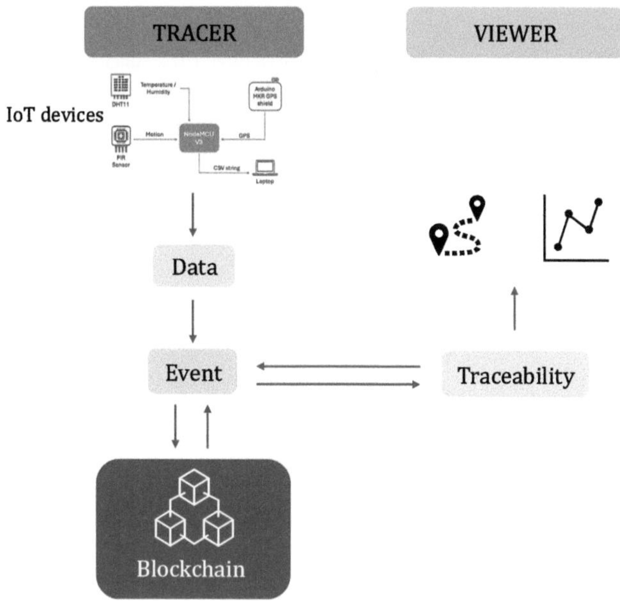

Fig. 1. Traceability system architecture.

the other hand, interact with the blockchain to query and analyse the stored data. The immutability of the blockchain ensures that the information accessed by VIEWERS is reliable and has not been altered, establishing a foundation of trust shared by all actors in the system. By adopting an event-driven architecture and leveraging the properties of the blockchain, the system supports real-time data collection and retrieval, ensuring transparency, traceability, and efficiency across all stages of data processing.

4.2 Design and Specifications

The proposed system is structured around two main actors: TRACERS for data collection and VIEWERS for data analysis. In addition, the design prioritises automation, immutability, and efficient querying, which are essential requirements for real-time traceability. This is achieved through a perspective of events issued by a smart contract, which will be the main axis of the system. Since we will be using smart contracts, it is necessary to use blockchain ecosystems that are compatible with them, such as Ethereum.

The tracing process begins with IoT devices equipped with environmental sensors (e.g., temperature, humidity, motion, GPS). These devices continuously collect data and generate CSV-formatted entries containing a timestamp, sensor readings, and location metadata; as illustrated in Fig. 2. Each line of data is transmitted to a designated blockchain node, where it is encapsulated into a transaction. This transaction invokes the smart contract, passing the data as the payload and using the product identifier as the indexing key. This smart contract not only ensures secure and immutable storage but also emits an event that allows efficient indexing and real-time retrieval of traceability records. Unlike traditional blockchain-based systems that rely solely on passive data storage, this approach introduces active logic and responsiveness at the blockchain layer— marking a significant innovation in traceability applications.

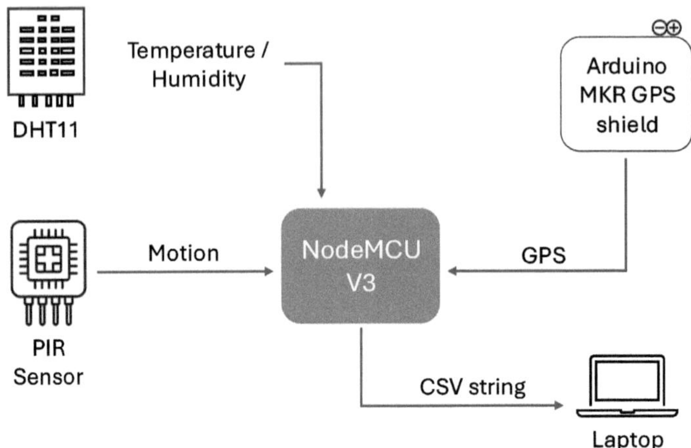

Fig. 2. IoT architecture for real-time data collection.

IoT devices play a foundational role in the system's autonomy and reliability. Their main function is to capture real-world phenomena with high temporal resolution and translate them into structured digital records. These devices must operate with low power consumption, resilience to environmental conditions, and support for secure data transmission. Their distributed nature reduces reliance on central infrastructure and enables scalable deployment across diverse locations in the supply chain.

At the core of the system lies a smart contract deployed on a permissioned blockchain network. This contract is responsible for emitting structured events that encapsulate traceability data. Each event is indexed by a unique identifier (id) corresponding to a specific product and includes a payload (data) containing sensor information. It is important to note that the smart contract implemented in this system does not follow common token standards such as ERC-20, as it is not intended to manage balances, user accounts, or inter-contract interactions. Instead, its functionality is limited to emitting indexed events upon receiving structured traceability data. To ensure modularity and flexibility, the smart contract is conceptually designed to be agnostic of data structure, accepting inputs in standardised formats that can accommodate multiple types of environmental parameters. It acts as a neutral registry layer, responsible solely for the integrity, indexing, and time-ordering of events. This design choice promotes scalability and adaptability. The Algorithm 1 summarises the data collection and submission.

Once the transaction is validated and added to the blockchain, the smart contract emits the event. Due to Ethereum's native event indexing mechanism, this data becomes efficiently searchable, enabling real-time access to historical records without the need for external databases or complex querying logic.

Algorithm 1. Traceability Data Submission

    Input   : Product identifier id, IoT sensor data $data$
    Output : Submission status
1: **function** SUBMITTRACEDATA(id, $data$)
2:    $csvData \leftarrow$ FormatData($data$, timestamp, location)   ▷ Format sensor data as CSV
3:    $tx \leftarrow$ CreateTransaction(id, $csvData$)   ▷ Create blockchain transaction
4:    $isValid \leftarrow$ ValidateTransaction(tx)   ▷ Validate via blockchain consensus
5:    **if** $isValid$ **then**
6:       $event \leftarrow$ SmartContract.EmitEvent(id, $csvData$)   ▷ Emit indexed event
7:       LogEventToBlockchain($event$)   ▷ Add to blockchain
8:       **return** $True$ ▷ Submission successful
9:    **else**
10:      **return** $False$ ▷ Submission failed
11:    **end if**
12: **end function**

On the retrieval side, VIEWERS execute queries to the blockchain for all events associated with a given product id. This mechanism allows them to reconstruct the full traceability path of the product, tracking environmental conditions and movements over time through a sequence of recorded data points. The integrity and ordering of these records are inherently guaranteed by the blockchain's consensus and timestamping mechanisms. This event-based approach ensures low gas costs, logical simplicity, and scalable performance—making it well suited for real-time applications in complex supply chains.

To facilitate human interaction with the system, the architecture includes a user-facing layer that enables access to traceability data. This layer allows VIEWERS to query, filter, and interpret product information associated with specific identifiers. It is conceptually designed to support the visualisation of temporal and spatial trends in an intuitive and accessible format. By separating user interaction from the underlying blockchain logic, this component enables non-technical stakeholders such as auditors, regulators, or consumers to benefit from the transparency and verifiability provided by the system. It also promotes informed decision-making through clear and timely visual insights. The Algorithm 2 illustrates the retrieval process.

Algorithm 2. Traceability Data Retrieval

    Input   : Product identifier id, Viewer query $query$
    Output : Traceability records
1: **function** RETRIEVETRACEDATA(id, $query$)
2:    $events \leftarrow$ QueryBlockchain(id)   ▷ Query events by product id
3:    $records \leftarrow$ ParseEvents($events$)   ▷ Extract data payloads
4:    $filteredRecords \leftarrow$ FilterRecords($records$, $query$)   ▷ Apply query filters
5:    $visualization \leftarrow$ GenerateVisualization($filteredRecords$)   ▷ Create temporal/spatial trends
6:    **return** $filteredRecords$, $visualization$ ▷ Return traceability data
7: **end function**

---

5 Use Case: Food Traceability

Food traceability represents a critical environment where product registration, control and verification processes must be accurate, secure and efficient. Furthermore, food supply chains often involve a wide range of heterogeneous actors such as producers, transporters, regulators, and retailers, each operating with different levels of technological maturity. This heterogeneity increases the risk of miscommunication, delays, and inconsistencies in data reporting.

The specific case chosen focuses on the monitoring of products that are sensitive to parameters such as temperature, humidity, physical handling, and geographical location. These factors not only affect product quality, but are also crucial for decision making in case of sanitary or logistical incidents. In this context, the proposed system ensures data integrity and traceability, facilitates real-time access to the data, and significantly reduces the possibility of manipulation by unauthorized third parties.

Another advantage of the proposed system lies in its ability to provide historical visibility and auditability. In the event of a contamination or quality issue, the full lifecycle of a product can be reconstructed and verified, allowing faster response and targeted recalls. This not only protects consumers but also enhances brand reliability and regulatory compliance.

The use case also reflects growing societal and regulatory demands for transparency in food production and distribution. Consumers increasingly expect accurate information about the origin, handling, and environmental impact of the food they consume. A system that ensures real-time, trustworthy traceability supports these expectations while offering companies a competitive edge in a market that prioritizes sustainability and digital trust.

6 Implementation

Motivated by the food traceability use case. The implementation was built around two fundamental components: the IoT devices and Blockchain network.

The IoT devices were implemented on a NodeMCU v3 board equipped with DHT11 sensors (temperature and humidity), PIR (tamper detection) and an

MKR GPS Shield (positioning). The Arduino IDE environment was used for programming. The data obtained is transmitted to a node, which integrates a script developed in Node.js that listens to the serial data, encapsulates it in a transaction, signs it using ethers.js v6 and sends it to the blockchain network via Hardhat.

Regarding blockchain, Hyperledger Besu was used as an Ethereum client in a private-permissioned environment. Interaction with the network was carried out through its RPC API, facilitating integration with the rest of the system. The smart contract was developed in Solidity, and deployed using the Hardhat framework, which simplified the testing, deployment, and maintenance cycle. The design was based on an event-driven model to record data, prioritising gas efficiency and logical simplicity. The use of parallel databases or complex structures within the contract was avoided, aiming for scalability and performance.

A web interface was also developed to facilitate user interaction with the system. The interface was built using React as the front-end framework and Vite as the build tool, with Tailwind CSS used for styling. It communicates with the blockchain via JSON-RPC calls to Besu clients, allowing users to query data associated with specific product identifiers. The interface features include a product traceability viewer, a block explorer, and a transaction log visualizer. Figure 3 provides a visual summary of the technologies and tools used throughout the system, categorized by domain and usage context.

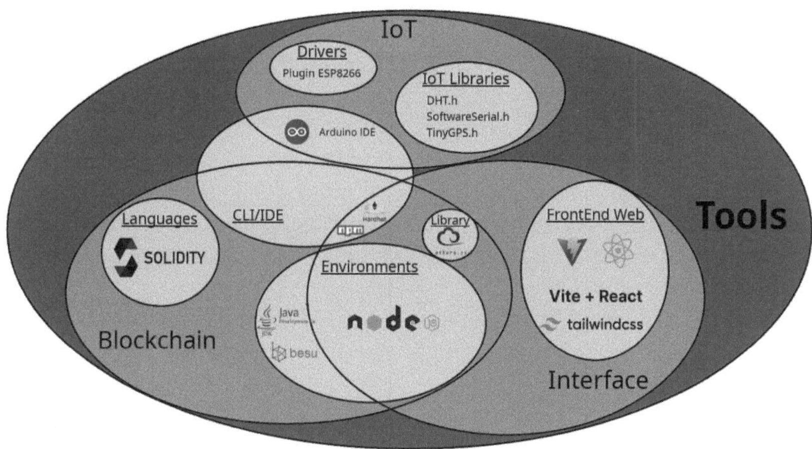

Fig. 3. Technological ecosystem used in the implementation.

All source files generated during the development and implementation of the system, including smart contracts, IoT device code, network configuration files, and integration scripts, are available at the following GitHub repository: https://github.com/Juanan151/Trazability. This repository ensures full reproducibility of the proposed solution.

7 Conclusion

The validation results confirm that integrating blockchain and IoT technologies within a traceability system is both feasible and operationally effective. The proposed architecture reliably saves and retrieves critical data in real time while maintaining security, efficiency, and automation. By eliminating human intervention in data handling, the system reduces errors and fraud, ensuring transparency and tamper-resistance across the supply chain. A key innovation of this work lies in the use of a purpose-built smart contract that emits structured, indexed events upon receiving traceability data from IoT devices. This event-driven mechanism enables real-time querying with minimal computational and storage overhead, offering a scalable alternative to traditional blockchain models focused solely on immutable storage. Unlike conventional smart contract designs that manage tokens or account logic, the contract here is streamlined and stateless, optimised for efficient data access and system modularity. The deployment of the smart contract required approximately 5,000,000 gas, while emitting a traceability event during contract execution averaged around 20,000 gas. Data retrieval, performed through off-chain log inspection, incurs no transaction costs. These results demonstrate that the event-driven design significantly reduces operational expenses compared to conventional on-chain storage approaches.

A current limitation of the system is the need for IoT devices to connect directly via cable to a blockchain node. While this setup offers reliability and low latency, it becomes costly and impractical at scale, as it would require deploying a dedicated node for each device. A more efficient alternative is to enable multiple IoT devices to connect to a single node through lightweight gateways or aggregators, significantly reducing infrastructure costs and enhancing scalability. However, this approach also introduces the need for additional security measures. This work lays the groundwork for scaling toward more complex systems that address data scalability and interoperability without compromising trust. Future developments will aim to extend the solution to hybrid blockchain networks and integrate predictive analytics and advanced data visualization to support decision-making in decentralized environments. The proposed system aligns with the broader digital transformation of supply chains, addressing emerging demands for accountability and resilience in global markets.

Acknowledgment. This work is partially supported by the Andalucía Excellence Research Program (ProyExcel-00257) and the Spanish Ministry of Science, Innovation, and Universities through the Knowledge Generation Projects grant (PID2024-161073NB-I00). Additionally, it is supported by the University of Jaén, through the Operational Reach Plan, Action 8a, granted to the second author.

Data Availibility. Data will be made available on request.

Declaration of Competing Interest. The authors declare that they have no known competing financial interests or personal relationships that could have appeared to influence the work reported in this paper.

References

1. Saidu, Y., Shuhidan, S.M., Aliyu, D.A., Abdul Aziz, I., Adamu, S.: Convergence of blockchain, IoT, and AI for enhanced traceability systems: a comprehensive review. IEEE Access **13**, 16838–16865 (2025). https://doi.org/10.1109/ACCESS. 2025.3528035
2. Apeh, O.O., Nwulu, N.I.: Improving traceability and sustainability in the agrifood industry through blockchain technology: a bibliometric approach, benefits and challenges. Energy Nexus **17**, 100388 (2025). https://doi.org/10.1016/j.nexus. 2025.100388
3. Li, L., Wang, W., Zhou, X., Qu, H., Ni, W., Jamalipour, A.: A trustworthy IoT-based supply chain traceability system with semantic multichain and preblockchain data verification. IEEE Internet Things J. **12**(14), 28245–28258 (2025). https:// doi.org/10.1109/JIOT.2025.3565740
4. Awasthy, P., Haldar, T., Ghosh, D.: Blockchain enabled traceability – an analysis of pricing and traceability effort decisions in supply chains. Eur. J. Oper. Res. **321**(3), 760–774 (2025). https://doi.org/10.1016/j.ejor.2024.10.019
5. Bendaouch, F., Zaydi, H., Merzouk, S., Assoul, S.: Securing IoT architectures with blockchain: opportunities, challenges, and future directions. In: 2025 5th International Conference on Innovative Research in Applied Science, Engineering and Technology (IRASET), pp. 1–10 (2025). https://doi.org/10.1109/IRASET64571. 2025.11008201
6. Bashir, I.: Mastering Blockchain. Packt Publishing, Germany (2017). https:// books.google.es/books?id=urkrDwAAQBAJ
7. Singhal, B., Dhameja, G., Panda, P.S.: Beginning Blockchain: A Beginner's Guide to Building Blockchain Solutions. Apress, Berkeley (2018). https://doi.org/10. 1007/978-1-4842-3444-0
8. Drescher, D.: Blockchain Basics: A Non-Technical Introduction in 25 Steps. Apress, Berkeley (2017). https://doi.org/10.1007/978-1-4842-2604-9
9. Antonopoulos, A.M., Wood, G.: Mastering Ethereum: Building Smart Contracts and DApps. O'Reilly Media (2018). https://github.com/ethereumbook/ ethereumbook
10. Serpanos, D., Wolf, M.: Industrial Internet of Things, pp. 37–54. Springer, Cham (2018). https://doi.org/10.1007/978-3-319-69715-4_5
11. Pathak, N., Bhandari, A.: IoT, AI, and Blockchain for .NET: Building a Next-Generation Application from the Ground Up. Apress, Berkeley (2018). https:// doi.org/10.1007/978-1-4842-3709-0
12. Kanagachidambaresan, G.R., Anand, R., Balasubramanian, E., Mahima, V. (eds.): Internet of Things for Industry 4.0. EICC, Springer, Cham (2020). https://doi.org/ 10.1007/978-3-030-32530-5
13. García, L., Cancimance, C., Asorey-Cacheda, R., Zúñiga-Cañón, C.-L., Garcia-Sanchez, A.-J., Garcia-Haro, J.: Lightweight blockchain for data integrity and traceability in IoT networks. IEEE Access **13**, 81105–81117 (2025). https://doi. org/10.1109/ACCESS.2025.3567773
14. Mazhar, T., et al.: Generative AI, IoT, and blockchain in healthcare: application, issues, and solutions. Discover Internet Things **5**(1), 5 (2025)
15. Tahsin, R., Rantu, S.B.A., Rahman, M., Salman, S., Karim, M.R.: Towards the adoption of AI, IoT, and blockchain technologies in Bangladesh's maritime industry: challenges and insights. Results Eng. **25**, 103825 (2025). https://doi.org/10. 1016/j.rineng.2024.103825

16. Kumar, N., Kumar, K., Aeron, A., Verre, F.: Blockchain technology in supply chain management: innovations, applications, and challenges. Telemat. Inform. Rep. **18**, 100204 (2025). https://doi.org/10.1016/j.teler.2025.100204
17. Oh, S.E., Kim, J.-H., Kim, J.-Y., Ahn, J.-H.: Food safety distribution systems using private blockchain: ensuring traceability and data integrity verification. Foods **14**(8) (2025). https://doi.org/10.3390/foods14081405
18. Sui, Y., Yang, X., Wang, B., Zhang, Y., Wang, W., Wang, L.: ScorpioBase: an efficient batched blockchain system for traceability applications in supply chain. IEEE Trans. Industr. Inf. **21**(5), 3575–3584 (2025). https://doi.org/10.1109/TII.2025.3528530
19. Asiamah, E.A., et al.: A storage-efficient learned indexing for blockchain systems using a sliding window search enhanced online gradient descent. J. Supercomput. **81**(1), 321 (2025)
20. Wu, H., Tang, Y., Shen, Z., Tao, J., Lin, C., Peng, Z.: TELEX: two-level learned index for rich queries on enclave-based blockchain systems. IEEE Trans. Knowl. Data Eng. **37**(7), 4299–4313 (2025). https://doi.org/10.1109/TKDE.2025.3564905
21. Belfqih, H., Abdellaoui, A.: Decentralized blockchain-based authentication and interplanetary file system-based data management protocol for internet of things using ascon. J. Cybersecur. Priv. **5**(2) (2025). https://doi.org/10.3390/jcp5020016

A Self-sovereign Identity Framework Using Blockchain for Privacy-Preserving Data Management

Manuel Garrido-Fúnez, Bruno Ramos-Cruz[✉], Jessica Zaqueros-Martinez[✉],
Manuel José Barranco, and Pedro José Sánchez

Computers Science Department, University of Jaen, Las Lagunillas, 23071 Jaen,
Jaen, Spain
`mgf00042@red.ujaen.es`, `{brcruz,zaqueros,barranco,pedroj}@ujaen.es`

Abstract. Digital identity management is a critical enabler for secure, privacy-respecting access to digital services, yet traditional systems often suffer from centralised control, data leaks, and limited user data management. To overcome these limitations, this contribution presents a novel Self-Sovereign Identity (SSI) framework that leverages blockchain technology to empower users with full control over their digital identities. The proposed system integrates Decentralised Identifiers (DIDs), Verifiable Credentials (VCs) encoded as JSON Web Tokens (JWTs), and Elliptic Curve Cryptography (ECC) to provide a lightweight, secure, and privacy-preserving identity management solution. Furthermore, the use of blockchain-based verifiable data registries ensures immutability, transparency, and trust in credential verification processes. Privacy is further reinforced through Zero-Knowledge Proofs (ZKPs) and secure peer-to-peer communication via DIDComm. A complete SSI architecture is implemented and validated through a functional prototype in the hotel reservation domain, demonstrating the feasibility of decentralised identity provisioning, selective disclosure, and user-centric data sovereignty. The framework supports scalability and interoperability across services and stakeholders while maintaining compliance with privacy regulations and World Wide Web Consortium (W3C) standards. This research contributes a practical and extensible SSI infrastructure that bridges blockchain and identity technologies, offering a promising foundation for next-generation intelligent digital trust ecosystems.

Keyword: Self-Sovereign Identity, Privacy data, JWTs, DIDs, Security

L. Martínez et al. (Eds.): IDEAL 2025, LNCS 16239, pp. 399–410, 2026.
https://doi.org/10.1007/978-3-032-10489-2_34

1   Introduction

Digital identity management has become a highly relevant area of research as digital services continue to expand in diverse areas, such as banking, healthcare, education, and public administration [1]. The way users authenticate and manage their identity in these environments is critical to ensuring both system security and individual privacy, directly impacting trust in digital interactions [2]. However, traditional identity management systems present multiple challenges, including dependence on centralised entities, exposure to security vulnerabilities, and lack of user control over their own data [3]. These challenges create single points of failure, risks of data breaches, difficulties in interoperability between systems, and limitations in complying with privacy and data protection principles in distributed environments.

One proposal to address these limitations is the use of a Self-Sovereign Identity (SSI). We can find specialised applications in different areas of SSI, including government, private, and emerging sectors. For example, in finance, SSI supports KYC (Know Your Customer), AML (anti-money laundering), and FATF (Financial Action Task Force) compliance [4]. It also allows online document verification during the loan process [5]. In the public sector, SSI helps decentralise licence and document management [6], improving transparency and security through targeted information disclosure. In healthcare, patients gain full control over access and sharing of health data, enabling interoperability between providers [7]. In education, SSI aids credential verification in cross-border infrastructure [8]. For vehicular networks and IoT, SSI offers decentralised identity management for smart vehicles [9] and IoT assets [10]. In offline environments, it ensures secure communication and access control [11]. SSI supports cross-domain authentication to register once and then use verifiable credentials (VCs) in other domains [12]. It also enables copyright-attributing, privacy-preserving content sharing [13].

This paper proposes the development of an SSI framework, a decentralised approach that seeks to eliminate centralised control and empower individuals to manage their own identities securely and privately. This system allows users to maintain sovereignty over their personal data, sharing it selectively and in a controlled manner based on their interactions with trusted entities. The proposal is based on the use of decentralised identifiers (DIDs) and VCs [14], encoded as JSON Web Tokens (JWTs), and blockchain technology to facilitate identity portability and verifiability across different domains and create a secure and privacy-preserving framework. Elliptic Curve Cryptography (ECC) is also used to ensure the security of interactions, maintain the lightweight capabilities necessary for resource-constrained devices, and ensure the integrity and confidentiality of transmitted data. To validate the proposed architecture, a framework has been developed and implemented to evaluate the practical capabilities of self-sovereign identity management in decentralised environments, ensuring scalability, interoperability, and compliance with security and privacy principles. The system integrates blockchain-based verifiable data registries for tamper-resistant record-keeping, Zero-Knowledge Proofs (ZKPs) for privacy-preserving verifications, and secure communication protocols such as DIDComm to ensure confidentiality and authenticity during credential exchanges.

The rest of the paper is structured as follows: Sect. 2 gives a background on technologies such as SSI, VCs, DIDs and JWTs. Section 3 introduces the proposed framework. Section 4 defines a particular use case in the hotel reservation domain. Section 5 shows the implementation features and presents the key results. Finally, Sect. 6 discusses the conclusions and future work, respectively.

2   Background

This section provides an overview of the key technologies underpinning the SSI ecosystem.

Self-Sovereign Identity (SSI) represents an innovative framework for digital identity management that emphasises individual control over personal data. SSI leverages technologies such as DIDs, VCs, and JWTs [15] where individuals are the primary custodians of their digital identities, distinguishing SSI from traditional systems where institutions delegate authority. SSI enhances privacy and security by enabling users to present verified credentials selectively, ensuring that only the necessary information is revealed during interactions, maintaining data protection [16]. SSI provides a scalable and interoperable solution for identity management across diverse digital ecosystems.

Verifiable Credentials (VCs) developed by the World Wide Web Consortium (W3C) as a standardised and interoperable framework for encoding claims in a format that is both cryptographically secure and resistant to tampering [17]. VCs can be used across diverse platforms, stored in digital wallets, and ensure their authenticity remains verifiable at all times. Key components include a Subject Uniform Resource Identifier (URI) for ownership verification, an issuer URI for origin confirmation, expiration conditions, and digital signatures for the integrity of the information over time and allowing verifiers to confirm their validity through cryptographic means [3]. VCs also support selective disclosure, enabling individuals to share only the specific attributes needed for a given interaction, such as proving age without revealing a full name, thereby preserving privacy while meeting verification requirements [14].

JSON Web Tokens (JWTs) provide a compact, URL-safe mechanism for transmitting claims, structured data representing user attributes or authorisation details, between parties in a JSON-based format that supports digital signatures and encryption to ensure integrity and confidentiality [18]. JWTs eliminate the need for repeated database queries and enhance system performance and scalability. A JWT consists of a header, a payload, and a signature, each encoded in base64url format. The header specifies the cryptographic algorithm (like HS256 or RS256) and token type, and the payload holds claims such as user identifiers as name-value pairs within a JSON object. The signature, generated by combining the header and payload with a secret or private key, ensures the token's authenticity and prevents tampering, with options for symmetric (HMAC) or asymmetric (RSA) schemes depending on the application context

[19]. This modular design allows for nested structures, where a JWT can be further signed or encrypted, accommodating complex security requirements.

Decentralised Identifiers (DIDs) defined by the W3C as an innovative of identifiers that provide globally unique and persistent identifiers for subjects such as individuals, organisations, or objects without reliance on centralised entities [3]. This allows the DID controller to demonstrate ownership through cryptographic means, independent of external authorisation, marking a significant departure from traditional federated identifiers. A DID is structured as a URI comprising three parts: 'did' scheme, identifier method, and a method-specific identifier unique to that system [14]. DIDs resolve to a DID document, a JSON-LD file that contains metadata about the subject, including cryptographic public keys, verification methods, and service endpoints facilitating trusted interactions [20]. While the DID is static, the DID document is dynamic and can be updated by the controller using a private key, ensuring both control and security [3].

3   SSI-Blockchain System Architecture

This paper presents a novel SSI-Blockchain system for digital identity management, addressing the limitations of centralised and federated systems, such as OAuth and OpenID, which are prone to single points of failure, data breaches, and limited user privacy due to third-party reliance [21,22]. By leveraging DIDs, VCs, JWTs, and a blockchain-based verifiable data registry, the system ensures tamper-proof, decentralised security and scalability. Moreover, it introduces selective disclosure via VCs, enhanced by ZKPs for privacy-preserving interactions, and secure peer-to-peer communication through DIDComm [11,23]. Supported by specific algorithms for VC issuance, management, and verification, this framework empowers users to maintain sovereignty over their personal data, share it selectively, and interact with trusted entities in a trustless environment.

3.1   Design and Specifications

This subsection delineates the technical design and specifications of the proposed SSI-Blockchain system, engineered to achieve full compliance with the SSI principles through a decentralised architecture. The system leverages blockchain-based verifiable data registries, ZKPs, and secure communication protocols to ensure sovereignty, privacy, and interoperability. In the verifiable credentials ecosystem, three primary entities interact to enable secure and privacy-preserving data exchange: the **issuer, holder**, and **verifier**. The issuer is responsible for creating and issuing the verifiable credential, ensuring its authenticity and integrity before transmitting it to the Holder [1]. The holder is an entity, typically an individual or organisation, that receives and manages a verifiable credential [2,24]. The holder has full control over the credential, selectively disclosing specific data attributes to third parties at their discretion [24,25]. The verifier, in turn, is the entity to whom the holder presents the credential. The verifier validates the integrity and authenticity of the credential using a trust relationship with the issuer, often through cryptographic mechanisms, to confirm the legitimacy of the data presented [1,25]. These interactions form the "trust triangle" shown in Fig. 1, establishing a decentralised, trust-based framework for secure digital identity management. These entities are detailed in the subsequent paragraphs.

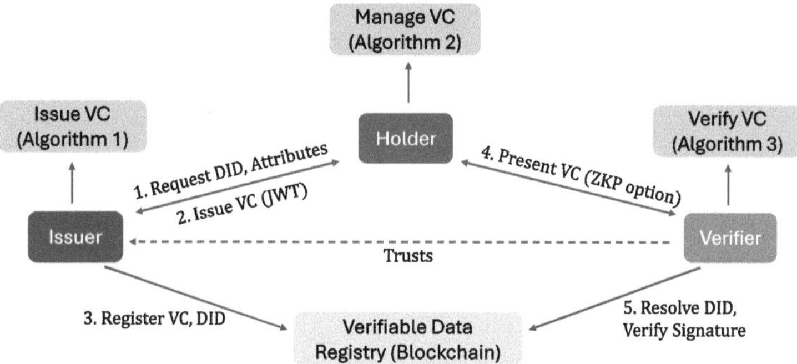

Fig. 1. Figure displays the system architecture and the main actors.

Issuer

The issuer, a trusted entity such as a governmental institution or corporate authority, is responsible for creating and issuing VCs. The issuance process begins when the issuer receives a user's request (holder), which serves as input. This request is a structured form containing the user's DID, a set of attributes (like name, date of birth, or employment status), and a timestamp for freshness. The output of the algorithm is the issued VC, ready for transmission to the user. The intermediate steps involve several actions: first, the issuer validates the user's DID by resolving it through the universal resolver to retrieve the corresponding DID document and verify the user's public key. Next, the issuer checks the submitted attributes against an internal policy or external data source, ensuring accuracy and eligibility. Once validated, the issuer constructs the VC by assembling the required elements 'id', 'type', 'issuer', 'issuanceDate', and 'credentialSubject' into a JWT format. The issuer then generates a cryptographic signature using its private key with the ECDSA algorithm, embedding it in the 'proof' object. Finally, the VC is registered on the blockchain registry, and the completed credential is sent to the user via DIDComm, completing the issuance cycle. This process is summarised in the flow diagram in Fig. 2.

Holder

The holder (user) receives the Verifiable Credential (VC) through a mobile wallet, which verifies the signature of the issuer using the issuer's public key resolved from the verifiable data registry. Once verified, the VC is encrypted (e.g., using AES-256) and stored locally, accessible only with the user's private key. The wallet enables selective disclosure, allowing the user to generate presentations with

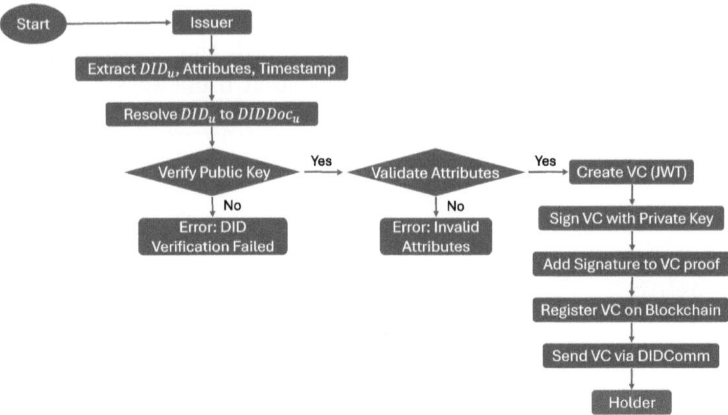

Fig. 2. Figure displays the flow diagram of the Issuer.

only the necessary claims, supporting privacy-preserving sharing aligned with SSI principles. When sharing with a verifier, the user can optionally apply ZKPs to hide sensitive data while proving required claims. The presentation, encrypted with the verifier's public key, is transmitted via DIDComm, ensuring privacy and authenticity. This process is summarised in the flow diagram in Fig. 3, where the VC is verified, securely stored, and selectively disclosed when requested by the user, with ZKPs applied as needed before encrypted transmission to the verifier.

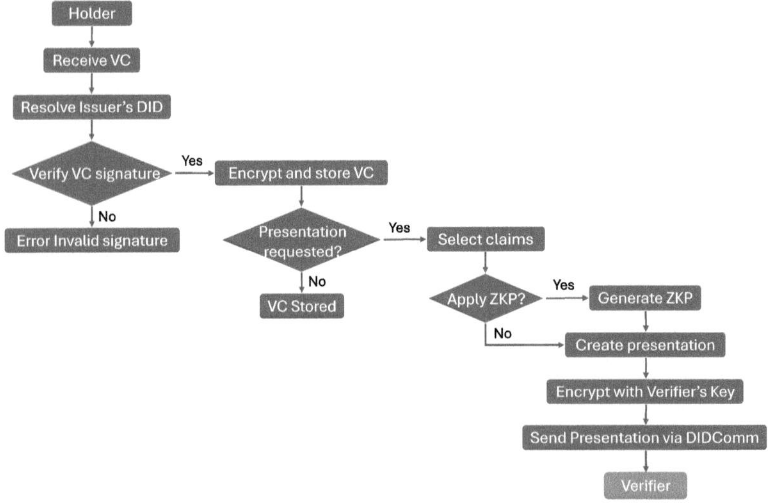

Fig. 3. Figure displays the flow diagram of the Holder.

Verifier

The verifier, such as a service provider, receives the user's VC presentation via DIDComm or QR code and validates it to authorise services or access. It extracts the JWT, resolves the issuer's DID using the universal resolver, and retrieves the issuer's public key from the blockchain registry to verify the ECDSA signature, ensuring the credential's integrity. If the presentation includes ZKPs, the verifier checks these to validate specific claims (e.g., age verification) without accessing full user data, enhancing privacy while maintaining trust. Verification occurs off-chain, using the blockchain registry solely as a trust anchor. Based on these checks, the verifier accepts or rejects the presentation, completing a secure, decentralised interaction aligned with SSI principles. DIDComm ensures encrypted, peer-to-peer communication throughout the process. This workflow is summarised in the flow diagram in Fig. 4, where the verifier processes the received presentation, verifies the signature and optional ZKPs, and outputs a boolean decision indicating acceptance or rejection.

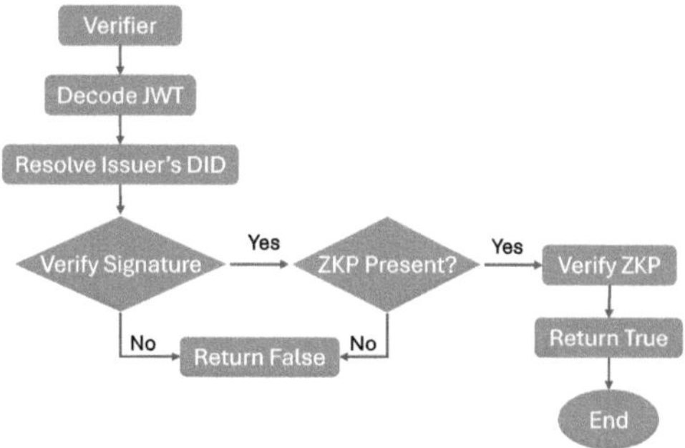

Fig. 4. Figure displays the flow diagram of the Verifier.

The proposed system architecture replaces conventional databases with a blockchain or verifiable data registry, providing a tamper-proof, decentralised infrastructure for storing DIDs, DID documents, and VCs. ZKPs enhance privacy by enabling selective disclosure without revealing raw data, while ECC ensures robust cryptographic security. DIDComm serves as the primary communication protocol, supporting encrypted, asynchronous interactions between entities. This design achieves full SSI compatibility, offering scalability, privacy, and user sovereignty across diverse applications.

4   Use Case: Hotel Reservation

The SSI framework, integrated with blockchain technology, transforms digital identity management in the hotel reservation domain by eliminating centralised control and empowering guests to manage their identities. This use case enables secure, privacy-preserving interactions during booking and check-in processes, ensuring guests maintain sovereignty over their personal data while hotels verify credentials efficiently without relying on centralised databases. In this framework, three key roles interact seamlessly, as illustrated in Fig. 1:

1. **Issuer**: Trusted entities, such as government agencies, health institutions, or loyalty program providers, issue VC to guests. These credentials include identity attributes (e.g., name, date of birth, contact information) or specialised claims (e.g., vaccination records, loyalty membership status), ensuring authenticity through cryptographic signatures.
2. **Holder**: The guest, acting as the Holder, receives and manages VCs. They can selectively share specific claims, such as proof of identity or membership status, without disclosing unnecessary personal details, aligning with the SSI privacy principles.
3. **Verifier**: The hotel reservation or check-in system, which acts as a Verifier, validates the VCs presented using cryptographic proofs and trust relationships with the Issuer, confirming the legitimacy of the data in a decentralised manner.

The process begins when a guest requests a hotel reservation or checks in, presenting VCs via a secure mechanism, such as a QR code. For example, a guest may share a credential proving they are over 18 or a loyalty member to access discounts, without revealing their full date of birth or other sensitive data. The hotel verifies these credentials against a blockchain-based verifiable data registry (e.g., Ethereum or Hyperledger), which stores Issuer public keys and VC metadata, ensuring tamper-proof validation without centralised intermediaries. This approach enhances privacy, reduces data breach risks, and streamlines trust establishment compared to traditional systems reliant on centralised databases. The use case demonstrates significant advantages, for instance, regarding user sovereignty: guests control their identity data, choosing what to share and with whom. Another is privacy through selective disclosure: guests share only necessary claims, protecting sensitive information. And decentralised trust: blockchain-based verification eliminates reliance on single points of failure, enhancing security and scalability. This SSI-Blockchain framework redefines hotel reservation processes by prioritising user control and privacy while maintaining robust verification, supporting the way for broader adoption in the hospitality sector.

5   Implementation

The implementation of the proposed SSI system began with the adaptation of the Veramo Node.js demo, as outlined in the official documentation [26], replacing 'yarn ts-node' with 'yarn tsx' to ensure compatibility and functionality. Initial testing confirmed the creation of identifiers and credentials, providing a baseline for further development. Subsequently, enhancements were introduced to align with the SSI framework and improve the user experience. These included dynamic creation of multiple DIDs and VCs based on user-provided data or system-recommended options (e.g., using timestamps as default credential names), with all credentials logged for record-keeping. To address the complexity of associating credentials with specific identifiers, a numbered list of existing DIDs was displayed during credential creation, allowing users to select the target identifier by number. Additionally, a form-based input system was implemented, prompting users to specify data to share and offering options for encryption (using a dynamically generated private key) or plain text, thereby supporting selective disclosure and enhancing privacy in line with SSI principles.

The initial terminal-based prototype facilitated rapid testing, consolidating issuer, holder, and verifier functionalities. For the transition to a mobile-integrated backend, the primitive backend was restructured into a RESTful API, during which the database was modified to include new tables for storing credentials, app users, and the mapping of identifiers to users, tailoring it to the system's specific requirements. This API supported user registration and login, with the issuer fully responsible for generating user DIDs and VCs, linked to user accounts via backend processes or QR codes. The mobile app, developed using the Dart language, the Flutter framework, and Android Studio, was custom-built to seamlessly integrate with the backend. Featuring multiple user-friendly screens, the app's control panel enabled credential management, allowing users to select specific claims for sharing encrypted or in plain text generating a QR code containing the resulting JWT for verifier scanning and decoding.

However, the current prototype relies on a centralised database, deviating from the proposed blockchain-based verifiable data registry, and lacks full integration of ZKPs, local storage, a user-initiated DID submission form, and the DIDComm protocol. These gaps reflect the project's early stage, with the implementation serving as a proof-of-concept to validate the innovative design outlined in Sect. 3. The source files created during the development and realisation of the first proof of concept can be found on the following GitHub repository: https://github.com/gfmanuel/SSI-app. Future work will focus on replacing the database with a blockchain solution, implementing ZKPs for privacy-preserving

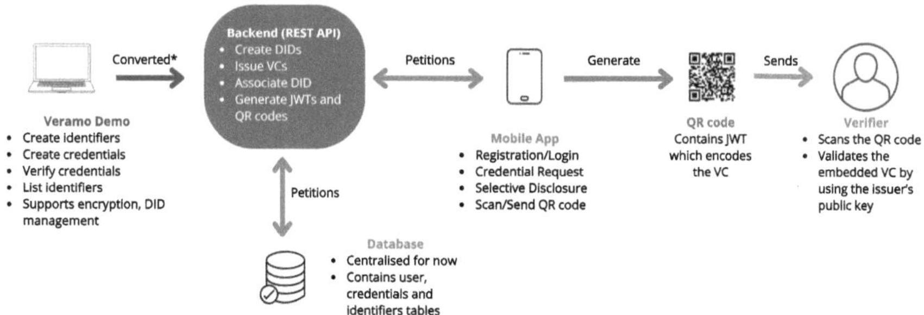

Fig. 5. Figure displays the implementation diagram of the system.

validation, and adopting DIDComm for secure communication, aligning the system fully with the SSI architecture (Fig. 5).

6   Conclusion

This paper has introduced and validated a comprehensive Self-Sovereign Identity (SSI) framework that integrates decentralised technologies to address critical limitations in traditional identity management systems. By leveraging blockchain-based verifiable data registries, decentralised identifiers (DIDs), verifiable credentials (VCs) encoded as JSON Web Tokens (JWTs), and elliptic curve cryptography (ECC), the proposed system enables secure, user-centric, and privacy-preserving identity management. The inclusion of Zero-Knowledge Proofs (ZKPs) and the DIDComm protocol further strengthens confidentiality, integrity, and trust in identity interactions across distributed digital ecosystems. The proposed architecture has been prototyped and tested in a practical use case within the hotel sector, demonstrating the feasibility and effectiveness of decentralised identity management in real-world scenarios. The detailed operational algorithms for credential issuance, management, and verification establish a solid foundation for the deployment of interoperable SSI solutions, aligned with World Wide Web Consortium (W3C) standards and privacy regulations. Future research will focus on extending the current framework by replacing centralised components with blockchain-native modules, fully integrating non-interactive ZKPs, and enabling advanced functionalities such as anonymous credential schemes and decentralised revocation mechanisms. Performance benchmarking across different blockchain platforms and domains will also be explored. By advancing these capabilities, the framework aims to support scalable, secure, and privacy-enhancing digital identity infrastructures suitable for cross-sectoral adoption in finance, healthcare, education, IoT, and public services.

Acknowledgment. This work is partially supported by the Andalucía Excellence Research Program (ProyExcel-00257) and the Spanish Ministry of Science, Innovation, and Universities through the Knowledge Generation Projects grant (PID2024-161073NB-I00). Additionally, it is supported by the University of Jaén, through the Operational Reach Plan, Action 8a, granted to the second author.

Data Availability Statement. Data will be made available on request.

Declaration of Competing Interest. The authors declare that they have no known competing financial interests.

References

1. Mühle, A., Grüner, A., Gayvoronskaya, T., Meinel, C.: A survey on essential components of a self-sovereign identity. Comput. Sci. Rev. **30**, 80–86 (2018). https://doi.org/10.1016/j.cosrev.2018.10.002

2. Preukschat, A., Reed, D.: Self-Sovereign Identity: Decentralized Digital Identity and Verifiable Credentials. Manning Publications, Shelter Island (2021). https://ieeexplore.ieee.org/document/10280453
3. Mazzocca, C., Acar, A., Uluagac, S., Montanari, R., Bellavista, P., Conti, M.: A survey on decentralized identifiers and verifiable credentials. IEEE Commun. Surv. Tutor., 1 (2025). https://doi.org/10.1109/COMST.2025.3543197
4. Giacobino, A., Grierson, D., Singh, H.P., McHale, P., Maggs, S.: Cosmos cash: public permissionless approach towards SSI and use cases. In: 2022 IEEE International Conference on Blockchain (Blockchain), pp. 462–467 (2022). https://doi.org/10.1109/Blockchain55522.2022.00071
5. Satybaldy, A., Subedi, A., Nowostawski, M.: A framework for online document verification using self-sovereign identity technology. Sensors **22**(21) (2022). https://doi.org/10.3390/s22218408
6. Lorico, H.D.Q., Zamuco, A.D.P., Caro, J.D.L., Juayong, R.A.B.: Decentralizing Philippine driver's license application and identity management for transparency, efficiency, and security: a design proposal. In: Novel and Intelligent Digital Systems: Proceedings of the 4th International Conference (NiDS 2024), pp. 30–41. Springer, Cham (2024). https://doi.org/10.1007/978-3-031-73344-4_2
7. Popa, M., Stoklossa, S.M., Mazumdar, S.: Chaindiscipline - towards a blockchain-IoT-based self-sovereign identity management framework. IEEE Trans. Serv. Comput. **16**(5), 3238–3251 (2023). https://doi.org/10.1109/TSC.2023.3279871
8. Tan, E., Lerouge, E., Du Caju, J., Du Seuil, D.: Verification of education credentials on European blockchain services infrastructure (EBSI): action research in a cross-border use case between Belgium and Italy. Big Data Cognit. Comput. **7**(2) (2023). https://doi.org/10.3390/bdcc7020079
9. Zeydan, E., Mangues, J., Arslan, S.S., Turk, Y.: Self-sovereign identity management for hierarchical federated learning in vehicular networks. In: 2023 IEEE 24th International Conference on High Performance Switching and Routing (HPSR), pp. 191–196 (2023). https://doi.org/10.1109/HPSR57248.2023.10147937
10. Pino, A., Margaria, D., Vesco, A.: On PQ/T hybrid verifiable credentials and presentations to build trust in IoT systems. In: 2024 9th International Conference on Smart and Sustainable Technologies (SpliTech), pp. 1–6 (2024). https://doi.org/10.23919/SpliTech61897.2024.10612295
11. Enge, A., Satybaldy, A., Nowostawski, M.: An offline mobile access control system based on self-sovereign identity standards. Comput. Netw. **219**, 109434 (2022). https://doi.org/10.1016/j.comnet.2022.109434
12. Zhang, Z., Xiong, R., Di, X., Ren, W.: CroAuth: a cross-domain authentication scheme based on blockchain and decentralized identity. In: 2024 27th International Conference on Computer Supported Cooperative Work in Design (CSCWD), pp. 2016–2021 (2024). https://doi.org/10.1109/CSCWD61410.2024.10580261
13. Fraser, A., Shehu, A.-S., Frymann, N., Haynes, P., Schneider, S.: Privacy-preserving photo sharing: an SSI use case. In: Patil, V.T., Krishnan, R., Shyamasundar, R.K. (eds.) Information Systems Security, pp. 320–329. Springer, Cham (2025). https://doi.org/10.1007/978-3-031-80020-7_18
14. Brunner, C., Gallersdörfer, U., Knirsch, F., Engel, D., Matthes, F.: DID and VC: untangling decentralized identifiers and verifiable credentials for the web of trust. In: Proceedings of the 2020 3rd. ICBTA 2020, pp. 61–66. Association for Computing Machinery, New York (2021). https://doi.org/10.1145/3446983.3446992
15. Mattei, L., Morpurgo, F., Occhipinti, C., Ratto Vaquer, L.M., Vasylieva, T.: Self-sovereign identity model: ethics and legal principles. Digit. Soc. **3**(2), 25 (2024). https://doi.org/10.1007/s44206-024-00113-2

16. Stokkink, Q., Ishmaev, G., Epema, D., Pouwelse, J.: A truly self-sovereign identity system. In: 46th Conference on LCN, pp. 1–8. IEEE Computer Society, Los Alamitos (2021). https://doi.org/10.1109/LCN52139.2021.9525011
17. World Wide Web Consortium (W3C): Verifiable Credentials Data Model v1.1. W3C Recommendation (2022). https://www.w3.org/TR/vc-data-model/
18. Jones, M., Bradley, J., Sakimura, N.: JSON Web Token (JWT) (2015)
19. Mahindraka, P.: Insights of JSON web token. Int. J. Recent Technol. Eng. (IJRTE) (2020). https://doi.org/10.35940/ijrte.F7689.038620. ISSN 2277–3878
20. Hoops, F., Mühle, A., Matthes, F., Meinel, C.: A taxonomy of decentralized identifier methods for practitioners. In: 2023 IEEE International Conference on Decentralized Applications and Infrastructures (DAPPS), pp. 57–65 (2023). https://doi.org/10.1109/DAPPS57946.2023.00017
21. Leiba, B.: Oauth web authorization protocol. IEEE Internet Comput. **16**(1), 74–77 (2012). https://doi.org/10.1109/MIC.2012.11
22. Recordon, D., Reed, D.: OpenID 2.0: a platform for user-centric identity management. In: Proceedings of the Second ACM Workshop on Digital Identity Management, DIM 2006, pp. 11–16. Association for Computing Machinery, New York (2006). https://doi.org/10.1145/1179529.1179532
23. Aad, I.: In: Mulder, V., Mermoud, A., Lenders, V., Tellenbach, B. (eds.) Zero-Knowledge Proof, pp. 25–30. Springer, Cham (2023). https://doi.org/10.1007/978-3-031-33386-6_6
24. Babel, M., et al.: Self-sovereign identity and digital wallets. Electron. Mark. **35**(1), 28 (2025). https://doi.org/10.1007/s12525-025-00772-0
25. Naik, N., Jenkins, P.: Your identity is yours: take back control of your identity using GDPR compatible self-sovereign identity. In: 2020 7th International Conference on Behavioural and Social Computing (BESC), pp. 1–6 (2020). https://doi.org/10.1109/BESC51023.2020.9348298
26. Veramo: Node.js Setup Identifiers Tutorial (2023). https://veramo.io/docs/node-tutorials/node_setup_identifiers. Accessed 14 July 2025

Trustworthy Emergency Management in Radiation Oncology: A Blockchain-Powered System with AI and IoT Integration

Cristina Muñoz-Higueras[1,2(✉)], Pilar Moreno-Colmenero[2],
Bruno Ramos-Cruz[1], Jessica Zaqueros-Martínez[1],
and Francisco José Quesada-Real[1]

[1] University of Jaén, Jaén, Spain
{cmh00031,brcruz,zaqueros,fqreal}@ujaen.es
[2] Hospital Universitario de Jaén, Jaén, Spain
mariap.moreno.colmenero.sspa@juntadeandalucia.es

Abstract. Rapid and reliable emergency response is essential in hospital environments, particularly in high-risk units such as radiation oncology. This paper presents the design of an intelligent alert system that integrates Artificial Intelligence (AI), Blockchain, and the Internet of Things (IoT) to enable real-time decision-making and secure event traceability. IoT sensors (e.g., smoke, temperature, and presence detectors) monitor environmental conditions, while a permissioned blockchain immutably records the captured data. An AI agent continuously analyzes this data to detect hazards—such as fire outbreaks—and autonomously triggers evacuation protocols, logging the decisions made on the blockchain.

The system computes safe evacuation routes based on the incident's location and real-time occupancy data, guiding patients and staff through visual and acoustic signals. The architecture is modular, secure, and compatible with hospital infrastructures, ensuring auditable records for compliance and post-incident analysis. A simulation within a radiation oncology department demonstrates its feasibility and potential for real-world deployment. This work lays the foundation for future implementations and highlights the promise of combining AI, blockchain, and IoT for smarter and more trustworthy emergency management in healthcare.

Keywords: Artificial Intelligence · Blockchain · IoT · Emergency Management · Smart Hospitals

1 Introduction

Ensuring real-time and reliable alert systems is essential in environments with high human occupancy, such as schools, shopping centers, or hospitals. In particular, within hospital settings, these systems are especially critical due to the

© The Author(s), under exclusive license to Springer Nature Switzerland AG 2026
L. Martínez et al. (Eds.): IDEAL 2025, LNCS 16239, pp. 411–422, 2026.
https://doi.org/10.1007/978-3-032-10489-2_35

high likelihood of patients with reduced mobility. This is especially true in units such as radiation oncology, where patient safety and clinical operation integrity are paramount. The rapid detection and response to hazardous events (e.g., fire, smoke, or occupancy anomalies) can save lives and prevent facility damage, yet remains a challenge in complex hospital infrastructures [5].

Despite advances, existing alarm systems in healthcare often rely on static thresholds or centralized architectures, suffering from delayed reactions, lack of traceability, and vulnerability to sensor tampering or data loss [1].

The integration of AI, IoT, and blockchain offers a compelling solution to these limitations. IoT sensors capture real-time data (e.g., smoke, temperature, presence) throughout the hospital environment; AI enables intelligent detection and dynamic decision-making based on this data; and blockchain ensures immutable, auditable logging of all captured information and decisions— enhancing transparency, integrity, and accountability [11]. Recent frameworks that combine edge AI with sharded proof-of-authority blockchains demonstrate feasible platforms for secure, real-time healthcare decision support [8]. Systematic reviews highlight how such integration improves emergency response accuracy, minimizes human error, and supports regulatory compliance in medical contexts [9].

In this paper, we propose a secure and intelligent alert system specifically tailored to a radiation oncology unit. The system architecture includes IoT sensors deployed in clinical areas, a permissioned blockchain for storing both the captured data and the decisions made by the intelligent agent, and the agent itself, which continuously monitors environmental conditions and triggers evacuation actions when critical thresholds (e.g., smoke levels) are exceeded. Evacuation routes are dynamically planned based on the origin of the incident and the location of available exits, with guidance provided via speakers and visual signals. In addition to enabling timely evacuation, the proposed system offers explainability by logging the rationale behind each agent decision in the blockchain. This feature enhances transparency, supports post-incident analysis, and reinforces institutional accountability.

The structure of the paper is as follows. Section 2 reviews the related work. Section 3 describes the system architecture and its components, including the IoT layer, blockchain layer, intelligent agent, and actuators. Section 4 presents an illustrative fire scenario that demonstrates the system's coordinated response through sensor detection, blockchain logging, and AI-driven evacuation planning. Section 5 discusses the system's strengths and limitations. Finally, Sect. 6 concludes the paper and outlines future work.

2   Related Work

The widespread adoption of smart environments based on IoT sensors has enabled new scenarios and significant advances in monitoring and alert systems within the healthcare domain. Wearable devices support real-time collection of physiological and environmental data in clinical settings, reducing response times and facilitating timely medical interventions [10,20]. In particular, IoT-based frameworks have enabled remote patient monitoring, contributing to reduced hospitalization periods and mortality rates by alerting clinical staff as soon as critical thresholds are exceeded [7].

Given the critical nature of healthcare, it is essential to ensure that monitoring processes and medical interventions are performed correctly and transparently. In this context, Distributed Ledger Technologies (DLTs), such as blockchain, have emerged as powerful tools for secure, immutable data logging and enhanced transparency. Numerous studies have demonstrated the benefits of integrating IoT with DLTs to certify patient health status [13], detect and record abnormal vital signs that require medical intervention [19], and ensure traceability in areas such as clinical trials, electronic health records, and insurance claims [2,3,12]. Despite these advantages, large-scale adoption remains constrained by regulatory challenges, interoperability issues [4], and the need for low-latency algorithms in real-time scenarios [15].

Intelligent agents constitute a key element in cyber-physical systems, autonomously perceiving their environment and making decisions to achieve predefined goals. Recent systematic reviews in disaster risk management and emergency response highlight their ability to detect hazards early and coordinate automated actions in critical scenarios [6]. Agent-based decision support systems have been developed to forecast incident likelihood, allocate emergency resources, and optimize response strategies—functions that are central to the requirements of cyber-physical infrastructures [17].

In the specific context of emergency detection and response, most existing work combining IoT, AI, and secure logging has focused on large-scale public settings or infrastructure-level deployments. For instance, AI-driven systems have improved situational awareness and decision-making in firefighting operations [5], while cyber-physical decision support platforms have been used for ambulance dispatching and multi-agency coordination [17]. However, relatively few studies address the unique challenges of hospital environments, where evacuation planning must consider patient immobility, infection control procedures, and zone-specific safety protocols.

Our proposed system advances the state of the art by integrating IoT, blockchain, and an intelligent agent tailored to the specific operational needs of a radiation oncology department. In this approach, the agent relies on trustworthy data securely stored on the blockchain, ensuring that its decisions are based on verifiable, tamper-proof information. Moreover, all decisions made by the agent are explicitly documented and logged on the blockchain, facilitating post-incident traceability and auditability.

3   System Architecture

The proposed system adopts a modular architecture composed of three main layers (see Fig. 1): (i) an IoT layer responsible for sensing and actuation, (ii) a blockchain layer for secure data logging and traceability, and (iii) an intelligent agent layer that performs real-time reasoning and decision-making.

The system is built on top of the Phonendo framework [14], which facilitates the integration of smart environments with DLTs. Data capture and transmission to the blockchain are managed through a unified gateway, which also hosts the execution of the intelligent agent.

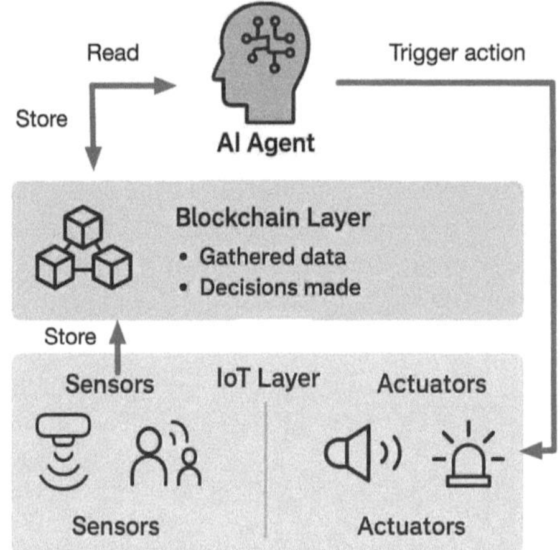

Fig. 1. Overview of the proposed system architecture.

3.1   IoT Layer

The IoT layer includes a network of sensors and actuators deployed across the radiation oncology service. The sensors capture environmental and occupancy data essential for risk detection and evacuation planning:

- *Smoke detectors*: installed in all rooms and corridors to identify fire outbreaks.
- *Temperature sensors*: complement smoke detection by capturing rapid increases in ambient temperature.
- *Presence sensors*: detect human presence in rooms and hallways to estimate occupancy in real time.

The actuation sublayer comprises:

- *Speakers*: deliver real-time audio instructions during evacuation.
- *Emergency lights*: activate visual signals to guide occupants along safe evacuation routes.

Sensor data is transmitted to an edge gateway, which also hosts the blockchain node and the AI agent, ensuring unified processing and decision-making at the edge.

3.2   Blockchain Layer

The blockchain layer functions as a secure, tamper-evident ledger that immutably records both environmental data and AI-generated decisions. Built on *Hyperledger Besu*[1], a permissioned blockchain framework, it stores timestamped sensor readings, trigger metadata, and evacuation commands issued by the agent. This architecture ensures system-wide auditability and supports post-incident analysis and regulatory compliance through transparent and verifiable logging.

3.3   AI Agent Layer

The AI agent, deployed within the edge gateway, continuously queries the blockchain to retrieve current environmental data. It maintains a spatial representation of all deployed sensors across the radiation oncology service, enabling the detection of abnormal conditions—such as fire scenarios—by correlating readings from multiple sources (e.g., elevated smoke and temperature levels). Based on this input and knowledge of the facility layout, the agent computes the safest evacuation plan, accounting for hazard location and real-time occupancy.

To preserve data privacy and ensure low-latency response, the agent uses the Llama 3.1 large language model (LLM) [21], running locally on the gateway. It is configured through prompt-engineered interactions, guided by the sensor layout, emergency exits, and predefined hospital safety protocols. This configuration enables the dynamic selection of evacuation routes and the generation of natural language instructions tailored to each situation.

Once a decision is made, the agent activates visual indicators—such as directional lights—and issues verbal alerts via the speaker system. All actions and their justifications are immutably recorded on the blockchain to ensure full traceability and post-event accountability. The following prompt (see Prompt 3.1) illustrates the structure used to configure the intelligent agent's reasoning process.

---

[1] https://besu.hyperledger.org.

Prompt 3.1: Evacuation Planning Prompt

You are an intelligent agent installed at the gateway of a hospital system. You have access to real-time sensor data, as well as a detailed map showing the distribution of sensors and emergency exits in the Radiation Oncology Service of the Hospital Universitario de Jaén.

Your task is to analyze the current data and generate an evacuation plan in case of an emergency. To do so, you must:

1. Identify high-risk zones (presence of fire or elevated temperature).
2. Determine which exits are blocked.
3. Redirect staff and patients to the nearest safe exit, prioritizing accessible routes with lower occupancy.

Sensor and Exit Layout:

Zones:

- Z0: Entry (sensors SMK001, SMK002)
- Z1: CorridorA (sensors SMK001, SMK002, SMK003)
- Z2: CorridorB (sensors SMK002, SMK005, SMK006, SMK007)
- Z3: CorridorC (sensors SMK008, SMK009)
- Z4: TreatmentWaitingRoom (sensors SMK003, SMK010, SMK011, SMK012, SMK013, SMK014)
- Z5: WaitingRoom (sensor SMK015)
  ...

Evacuation Exits:

- Exit A: Closest sensors are SMK001, SMK002; connected to zones Z0 and Z1.
- Exit B: Closest sensors are SMK013, SMK014; connected to zones Z3 and Z4.
- All routes are equipped with lights and speakers to guide the evacuation.

Rules and Priorities:

- If fire or smoke is detected in a zone, consider the associated exit blocked if nearby sensors report out-of-range values.
- Prioritize the shortest routes, avoiding risky areas.
- Prioritize evacuation from zones with higher occupancy (based on presence sensors).
- Instructions must be clear and zone-specific.

Expected Output:

1. Location of the fire.
2. Identification of blocked exits.
3. Evacuation plan indicating which zones should use which exits.
4. Verbal instructions to be broadcast.
5. Activation of light signals (zones → direction).

4   Illustrative Example

To contextualize the deployment of the proposed system, we consider the Radiation Oncology Service at the Hospital Universitario de Jaén, a high-risk clinical environment that combines sensitive radiological equipment with vulnerable patients undergoing treatment. The facility includes treatment bunkers, planning and control rooms, a patient waiting area, offices, and multiple corridors with at least two emergency exits per sector, compliant with national health and safety regulations. This architectural configuration presents a complex spatial topology where emergency evacuation planning must consider not only physical layout constraints but also the health status and mobility of the occupants.

4.1   Sensor and Actuator Deployment

Building upon the workflow described by Muñoz-Higueras et al. [16], where IoT presence sensors and a mobile beacon system were used to monitor patient flow and reduce wait times through DLT-based logging, our system extends this infrastructure to support real-time emergency detection and dynamic evacuation management.

The proposed alert system integrates additional sensing devices—specifically, smoke and temperature detectors—strategically installed in treatment rooms, equipment enclosures, electrical panels, and corridors. Presence sensors are also deployed near restricted access points and exit routes to provide real-time occupancy data. Actuation components include wall-mounted acoustic speakers and LED-based light panels capable of displaying dynamic visual guidance in the form of arrows or blinking signals, depending on the scenario.

All devices are uniquely identified and mapped to specific zones within the facility. This mapping is used not only for visualization and control but also for the registration of all sensor data and agent actions as signed, timestamped transactions on a permissioned blockchain. This design ensures that every change in environmental conditions, as well as any system response, is verifiable and audit-ready.

Figure 2 illustrates the spatial distribution of the deployed sensing and actuation infrastructure throughout the Radiation Oncology Service.

4.2   Fire Scenario Response

Figure 3 illustrates an example scenario that demonstrates the system's behavior in response to a critical emergency event. In this case, a fire originates near *Exit A*, possibly due to equipment overheating or an electrical fault. The adjacent smoke detector begins reporting elevated values, followed shortly by the temperature sensor registering an abnormal rise in ambient heat. Once the predefined safety thresholds are exceeded, both sensors trigger an alert condition.

These readings are immediately transmitted to the edge gateway, where they are immutably recorded on the blockchain. Simultaneously, the AI agent— deployed within the gateway—queries the most recent blockchain entries and

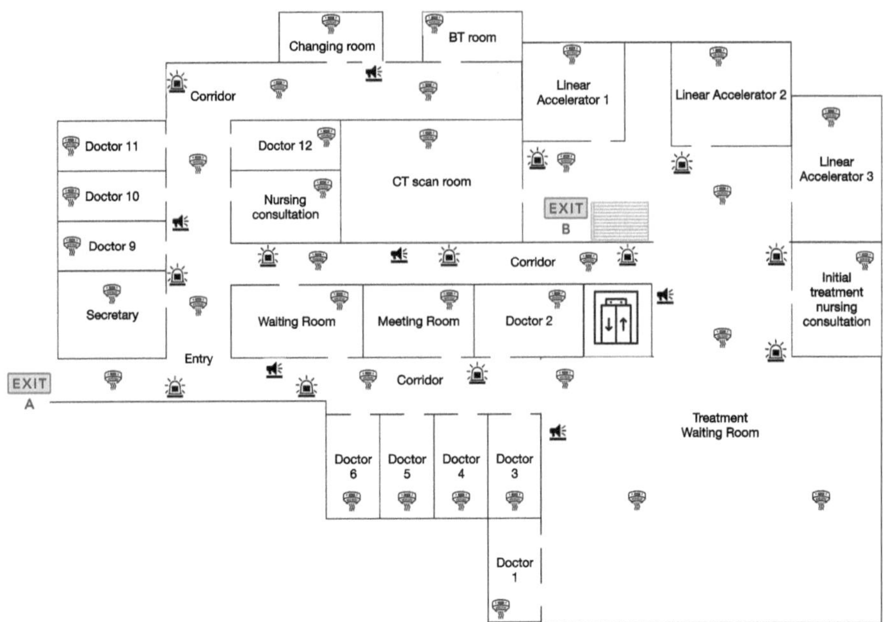

Fig. 2. Layout of the Radiation Oncology Service showing sensor and actuator placement.

processes the combined sensor data. Leveraging its internal spatial model and emergency reasoning logic (implemented via prompt-engineered interactions with an embedded LLM), the agent determines that *Exit A* is compromised and should be avoided.

The agent dynamically recalculates the evacuation plan, taking into account real-time occupancy levels, the spatial distribution of hazards, and the availability of emergency exits. It concludes that all occupants must be redirected toward *Exit B*, which remains operational and is located at a safe distance from the affected area. This decision—along with its justification and relevant environmental context—is encoded, signed, and recorded on the blockchain to guarantee traceability and forensic auditability.

Upon committing the evacuation plan, the agent activates the appropriate emergency light panels, which display directional arrows to guide staff and patients to the designated exit based on their current location. In parallel, verbal evacuation instructions are broadcast via the distributed speaker system, clearly announcing the nature and location of the emergency, the designated exit to be used, and behavioral guidance to maintain order and safety.

The following output, generated by the agent and stored on the blockchain, reflects the reasoning and actions taken:

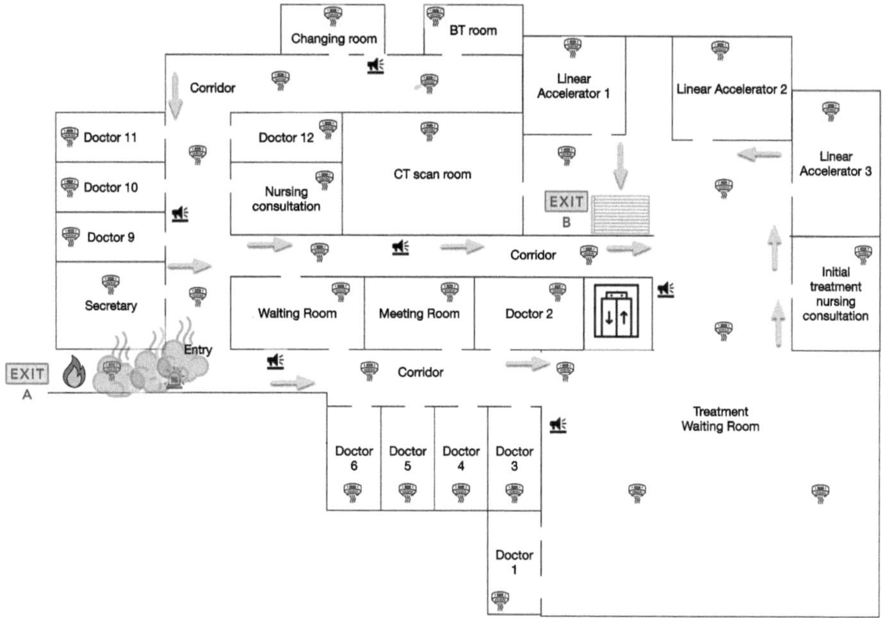

Fig. 3. Illustrative example of fire detection and visual guidance activation.

Agent Output Stored in the Blockchain

1. **Fire Location:** The fire is detected in *Zone Z0 (Entry)*. Sensor SMK001 reports a smoke level of *9.2* (threshold = 7.0) and a temperature of *65°C* (threshold = 50°C), indicating abnormal environmental conditions consistent with fire.
2. **Blocked Exit:** *Exit A* is considered blocked as it is connected to Z0 and Z1, and the nearby sensor SMK001 reports critical values. Therefore, Exit A must not be used for evacuation.
3. **Evacuation Routing:**
   - Zone Z1 (Corridor A) → evacuate via *Exit B*.
   - Zone Z2 (Corridor B) → evacuate via *Exit B*.
   - Zone Z3 (Corridor C) → evacuate via *Exit B*.
   - Zone Z4 (Treatment Waiting Room) → evacuate via *Exit B*.
   - Zone Z5 (Waiting Room) → evacuate via *Exit B*.
4. **Verbal Instructions (to be announced):**
   - "Attention: A fire has been detected near the main entry (Zone Z0). Exit A is currently inaccessible."
   - "All staff and patients in Zones Z1 to Z5 must evacuate through Exit B."
   - "Please follow the illuminated arrows and remain calm. Assistance will be available along the evacuation route."
5. **Activation of Light Signals:**
   - **Z1** → directional lights toward Corridor B (Z2), then Corridor C (Z3), and onward to Exit B.
   - **Z2** → directional lights toward Corridor C (Z3) and Exit B.
   - **Z3** → directional lights to Exit B.
   - **Z4** → directional lights to Exit B.
   - **Z5** → directional lights toward Z4 and then to Exit B.

Importantly, the system remains adaptive: if the situation evolves—for example, if smoke propagates to additional zones or occupancy levels change—the agent is capable of reassessing the context and issuing updated commands accordingly. This continuous feedback loop ensures that the system maintains responsiveness and safety throughout the duration of the emergency.

5   Discussion

The proposed system presents a technically feasible and modular architecture for intelligent emergency management in healthcare environments. Although this work centers on the Radiation Oncology Service, its layered design—built upon standardized interfaces and modular components—enables seamless adaptation to other critical hospital units, such as intensive care, surgical theaters, and emergency departments. Key strengths of the system include its real-time decision-making capabilities, achieved through the integration of IoT, blockchain, and AI technologies, as well as its support for explainability, auditability, and system accountability.

Nonetheless, several limitations and areas for enhancement should be addressed. First, the reliance on Llama 3.1 as the embedded LLM warrants further benchmarking. Future work should compare alternative open-source LLMs that can operate locally, analyzing trade-offs in response time, memory footprint, and reasoning quality—particularly under edge-computing constraints. Second, the current agent design involves frequent blockchain queries to retrieve sensor data. Performance and responsiveness could be significantly improved by using a locally replicated database that synchronizes periodically with the blockchain, preserving data integrity while reducing latency.

From a security and compliance perspective, the use of a permissioned blockchain ensures immutability, non-repudiation, and fine-grained access control. However, dealing with sensitive data such as occupancy and movement patterns necessitates full compliance with privacy regulations (e.g., GDPR, HIPAA). In a production environment, additional safeguards—including pseudonymization, encryption, and secure audit trails—would be essential to guarantee patient confidentiality.

Finally, integration with Hospital Information Systems (HIS) represents a key direction for future enhancement. Such integration would provide the agent with richer contextual data (e.g., staff presence, patient mobility, treatment status) and enable system-wide coordination. This would allow the emergency response framework to transcend individual units, facilitating interaction with clinical workflows, inventory systems, and disaster recovery protocols at the hospital level.

6   Conclusion

This work has introduced a modular architecture for intelligent emergency management in hospital environments, integrating IoT sensing, blockchain-based traceability, and AI-driven reasoning. The proposed system, deployed within the Radiation Oncology Service of the Hospital Universitario de Jaén, enables real-time detection of fire-related hazards, dynamic evacuation planning, and explainable decision-making, while ensuring auditability and data integrity. By combining a prompt-engineered LLM-based agent with a distributed sensor network and secure data storage, the solution enhances patient safety and institutional accountability in critical clinical scenarios.

Future work will focus on large-scale validation in real healthcare settings, benchmarking alternative LLMs for local inference, and optimizing communication workflows to reduce latency. Further developments include integration with HIS to incorporate patient-specific data—such as mobility constraints or vital signs from wearable sensors—to support personalized evacuation strategies. Additionally, the system will be extended to coordinate with external emergency response services (e.g., fire and police departments), enabling automated multi-agency alerts and a unified crisis response protocol [18].

References

1. Ajakwe, S.O., et al.: Medical IoT record security and blockchain: systematic review of milieu, milestones, and momentum. Big Data Cogn. Comput. **8**(9), 121 (2024)
2. Al-Khasawneh, M.A., Faheem, M., Alarood, A.A., Habibullah, S., Alzahrani, A.: A secure blockchain framework for healthcare records management systems. Healthc. Technol. Lett. **11**(6), 461–470 (2024)
3. Alamsyah, A., Setiawan, I.P.S.: Enhancing privacy and traceability of public health insurance claim system using blockchain technology. Front. Blockchain **8**, 1474434 (2025)
4. Attaran, M.: Blockchain technology in healthcare: challenges and opportunities. Int. J. Healthc. Manage. **15**(1), 70–83 (2022)
5. Bajwa, A.: AI-based emergency response systems: a systematic literature review on smart infrastructure safety. SSRN (2025)
6. Hamid, H., Abedlmajid, E.: Intelligent agents in disaster risk management: a systematic review of advances and challenges. Int. J. Adv. Comput. Sci. Appl. **16**(6) (2025)
7. Iqbal, F.M., Lam, K., Joshi, M., Khan, S., Ashrafian, H., Darzi, A.: Clinical outcomes of digital sensor alerting systems in remote monitoring: a systematic review and meta-analysis. NPJ Digit. Med. **4**(1), 7 (2021)
8. Jun, M.: Platform framework for blockchain-enhanced healthcare AIoT systems. Front. Commun. Netw. **6**, 1538965 (2025)
9. Kasralikar, P., Polu, O.R., Chamarthi, B., Rupavath, R., Patel, S., Tumati, R.: Blockchain for securing AI-driven healthcare systems: a systematic review and future research perspectives. Cureus **17**(4) (2025)
10. Li, C., Wang, J., Wang, S., Zhang, Y.: A review of IoT applications in healthcare. Neurocomputing **565**, 127017 (2024)
11. Manh, B.D., Nguyen, C., Hoang, D.T., Nguyen, D.N., Zeng, M., Pham, Q.: Privacy-Preserving cyberattack detection in Blockchain-Based IoT systems using AI and homomorphic encryption. IEEE Internet Things J. (2025)
12. Mohammadi, M., Javan, R., Beheshti-Atashgah, M., Aref, M.R.: Scalhealth: acalable blockchain integration for secure IoT healthcare systems. arXiv preprint arXiv:2403.08068 (2024)

13. Moya, F., Quesada, F.J., Martínez, L., Estrella, F.J.: CertifioT: an IoT and DLT-based solution for enhancing trust and transparency in data certification. In: International Conference on Ubiquitous Computing and Ambient Intelligence, pp. 127–138. Springer (2023)

14. Moya, F., Quesada, F.J., Martinez, L., Estrella, F.J.: Phonendo: a platform for publishing wearable data on distributed ledger technologies. Wireless Netw. **30**(7), 6507–6521 (2024)

15. Moya Perez, F., Quesada Real, F.J., Martínez López, L., Estrella Liebana, F.J.: Energy: reducing latency in IoT DLTs for AI-driven real-time solutions. Int. J. Web Inf. Syst. (2025)

16. Muñoz-Higueras, C., Serradilla-Gil, A.M., Moreno-Colmenero, P., Quesada-Real, F.J.: Integrating IoT and DLT to enhance patient wait time traceability in radiotherapy oncology. In: International Conference on Ubiquitous Computing and Ambient Intelligence, pp. 932–942. Springer (2024)

17. Pettet, G., et al.: Designing decision support systems for emergency response: challenges and opportunities. In: 2022 Workshop on Cyber Physical Systems for Emergency Response (CPS-ER), pp. 30–35. IEEE (2022)

18. Quesada, F.J., McNeill, F., Bella, G., Bundy, A.: Improving dynamic information exchange in emergency response scenarios. In: 14th International Conference on Information Systems for Crisis Response and Management, pp. 824–833. Information Systems for Crisis Response and Management, ISCRAM (2017)

19. Ramos-Cruz, B., Quesada-Real, F.J., Rodriguez-Garcia, M., Andreu-Pérez, J., Martínez, L.: Combining distributed ledger technologies and differentially private sketching techniques for securing health monitoring. In: International Conference on Ubiquitous Computing and Ambient Intelligence, pp. 920–931. Springer (2024)

20. Uddin, R., Koo, I.: Real-time remote patient monitoring: a review of biosensors integrated with multi-hop IoT systems via cloud connectivity. Appl. Sci. **14**(5), 1876 (2024)

21. Vavekanand, R., Sam, K.: Llama 3.1: an in-depth analysis of the next-generation large language model. Preprint (2024). https://doi.org/10.13140/RG.2.2.10628.74882

Evaluating AI Agents for Cyber Defense: A Comparison of Deep Reinforcement Learning and LLM Approaches

Hassan Chowdhry⬤, Jaume Manero$^{(\boxtimes)}$⬤, and Srinivas Sampalli⬤

Dalhousie University, Halifax, NS, Canada
`{hassan.chowdhry,jaume.manero}@dal.ca, srini@cs.dal.ca`

Abstract. Cyber-threats continue to grow in both scale and sophistication, increasingly exceeding the capacity of human analysts and traditional static defenses. Autonomous Cyber Defense (ACD) aims to develop intelligent agents that can monitor, analyze, and counter attacks in real time. In this paper, we present a controlled evaluation of two distinct ACD paradigms within the CAGE-2 environment: (i) a Deep Reinforcement Learning (DRL) defender trained using Proximal Policy Optimization (PPO) combined with an Intrinsic Curiosity Module (ICM), and (ii) a Large Language Model (LLM) defender that applies state-of-the-art models with prompt-based decision making to select and explain defensive actions. The agents were tested against a range of red-team strategies and assessed with standardized performance metrics as well as qualitative measures of interpretability. Results indicate that the DRL-based agent achieved more consistent and reward-efficient performance, whereas the LLM-based agent demonstrated stronger transparency and adaptability to previously unseen attack tactics, albeit with greater computational cost and latency. Together, these findings point to complementary advantages of the two approaches and provide guidance for the design of future autonomous defense systems that combine high performance with explainability and robustness in dynamic operational settings.

Keywords: Autonomous Cyber Defense · Artificial Intelligence · Deep Reinforcement Learning · Large Language Models · AI Agents

1 Introduction

Modern organizations and government institutions increasingly rely on digital systems for daily operations, and the cybersecurity landscape has become significantly more complex and vulnerable to threats. The integration of Information Technology (IT) and the Internet of Things (IoT) devices that are highly vulnerable to cyber threats, coupled with a shortage of qualified cybersecurity professionals, has rendered traditional defense mechanisms inadequate. Traditional cybersecurity systems struggle to effectively counteract rapidly evolving cyber threats that increase in frequency, complexity, and severity [11,17].

© The Author(s), under exclusive license to Springer Nature Switzerland AG 2026
L. Martínez et al. (Eds.): IDEAL 2025, LNCS 16239, pp. 423–433, 2026.
https://doi.org/10.1007/978-3-032-10489-2_36

ACD has emerged as a promising research frontier with the aim of developing autonomous agents that monitor, detect, and respond to cyber threats in real-time. By combining human analysts with AI-powered defensive systems, ACD promises to continuously improve reaction time and adaptive strategies [6,11].

DRL has shown potential in creating autonomous agents to learn optimal defensive policies through trial-and-error interactions in a simulated environment. Previous work has demonstrated that DRL agents outperform traditional defense systems in intrusion detection, lateral movement containment, and service restoration [2,10,13].

LLMs are pretrained on vast corpora of text and technical documents, LLMs can reason over security inputs such as logs, alerts, and threat intelligence, and output plausible defensive actions. LLMs can articulate natural language justifications for their decisions. Early research suggests that LLM agents can adapt to threats using prior knowledge, explain their reasoning that enables human oversight and compliance auditing, and operate effectively with limited task-specific data through prompt engineering [3,9].

No prior study has conducted a controlled side-by-side evaluation of these paradigms under identical threat scenarios [9]. Our work addresses this gap through a comparative study of DRL and LLM defender agents in CybORG CAGE-2 environment, a simulation featuring multi-host networks, realistic services, and adaptive attackers. We contribute quantitative and qualitative evaluations of detection accuracy, mitigation speed, network uptime, cumulative reward, interpretability, and adaptability, as well as an adapter framework for seamless integration between the LLM agent and CybORG.

We develop two autonomous defenders: (i) a DRL agent based on PPO with ICM, and (ii) an LLM agent using prompt engineering to guide decision-making and explanation. Both are deployed in identical scenarios within CybORG CAGE-2, facing a red team that varies its attack strategies between runs. The evaluation spans quantitative performance metrics and qualitative assessments of defensive decisions.

The rest of the paper is structured as follows. Section 2 reviews relevant work in DRL and LLM based autonomous defense. Section 3 presents the experimental design, agent architectures, and evaluation methodology. Section 4 presents results, followed by an in-depth discussion in Sect. 5 of the observed strengths, weaknesses, and practical implications. Section 6 concludes with lessons learned and directions for future research toward reliable and adaptive ACD systems.

2   Background

ACD develops agents that select and execute defensive actions rapidly and reliably under partial observability and adversarial pressure. Research in ACD relies on simulation environments known as gyms that approximate enterprise networks with interacting blue defender, red attacker, and green user processes. Gyms expose diverse scenarios, controllable attacker behaviors, and standardized interfaces, allowing systematic exploration of state–action spaces, robustness, and reproducible evaluation.

CybORG provides a high-level API for hosts, services, observations, and actions, and serves as the underlying environment for the TTCP's CAGE series, which standardizes scenarios and scoring for reproducible ACD benchmarking [16]. CAGE-2 is a single-agent benchmark that models a three-subnet enterprise with benign green activity, a red adversary that begins in the user subnet, and natural pivot paths through the enterprise subnet toward an operational server. Red team behavior is captured by rule-based agents that execute predefined strategies: the B-Line agent launches a fast and aggressive attack with minimal reconnaissance to achieve impact quickly, while the Meander agent advances more slowly, performing extensive reconnaissance before escalating attacks to avoid detection [16]. The defender counters these threats by choosing from a set of five actions: Monitor, Analyze, Decoy, Remove, and Restore. Submissions are ranked by cumulative reward across episodes, balancing adversary disruption with service availability under partial observability and noisy signals. CAGE-2 has become the standard single-agent cybersecurity gym for controlled studies [18]. Figure 1 shows the topology of the network.

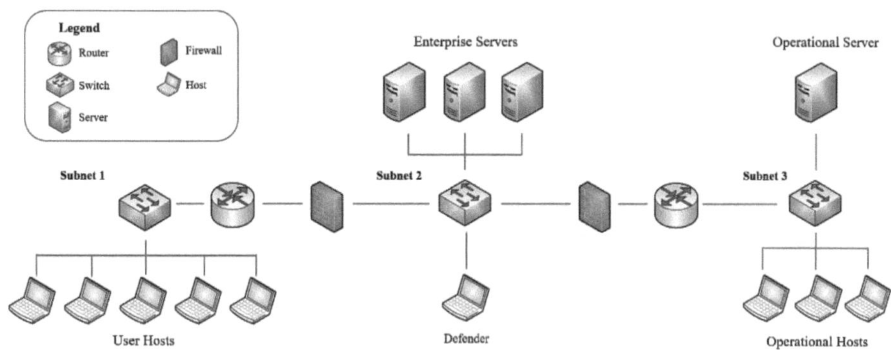

Fig. 1. Network topology of the CAGE-2 environment

DRL has been the dominant approach in CAGE-2. The top entries often use PPO variants, often with hierarchical control, to manage large action spaces and to specialize policies against different attacker styles [14,15]. Public write-ups describe controllers that route observations to specialist sub-policies, sometimes guided by lightweight bandit selection. Curiosity-driven exploration appears in several agents to mitigate sparse reward signals [12].

LLMs open a different path, they map security context, such as alerts, host summaries, and indicators, to high-level actions with a natural language explanation [9]. Recent studies push toward ACD settings and multiagent coordination in CAGE-4, arguing that LLMs can complement DRL with better generalization to novel events and auditability, while also noting failure modes such as hallucination, prompt brittleness, and need for guardrails [4,7].

Table 1. DRL hyperparameters.

Parameter	Value
Learning Rate (LR)α	0.0005
Gamma γ	0.99
Entropy coef	0.001
Clip	0.5
VF coef	1.0
ICM α	0.001
ICM β	0.2
ICM η	1.0
Train steps	100M

3   Methodology

3.1   DRL Agent

Our first approach is the DRL agent trained with PPO and augmented by the ICM. We selected PPO as the learning algorithm due to its stability and strong performance in complex DRL environments [14]. We use ICM to encourage exploration and give intrinsic rewards when the agent encounters a stage that is difficult to predict [8]. ICM helps the agent learn in a sparse-reward setting by prioritizing interesting or novel state transitions. In the CAGE-2 scenario, many events may not immediately yield extrinsic rewards, so the ICM bonus motivates the agent to investigate new hosts and actions, improving the coverage of attack patterns. We trained a single generalist blue agent instead of creating specialized agents that are overfitted to a specific red agent attack pattern. We created a wrapper environment that randomly picks a red agent every episode reset, it is trained to face any red agent [5]. This ensures the Blue agent experiences a variety of attack patterns and learns robust responses rather than overfitting to one red agent. We have reported the hyperparameters for the DRL agent in Table 1.

3.2   LLM Agent

We use LLMs to orchestrate blue agent actions using natural language. We prompt the LLMs to choose actions based on textual descriptions of the environment state. This models the defense scenario as a decision-making problem that the LLM can solve using its large knowledge base, reasoning abilities, and prompt engineering.

We developed a framework that converts the raw vector observations from the environment into a human-readable format. Specifically, we designed an adapter function that translates the bit-vector representation of host status and alerts into a tabular description of the network state. The complete state of all hosts is presented in a table format with the columns: Hostname, Activity, and Compromised. The description is then inserted into the prompt. This representation allows the LLM to interpret the network situation in semantic terms, closely mimicking the workflow of a human analyst [9].

We made another adapter that maps the LLMs output to an action that the CybORG environment understands, and we enforced a strict output schema within the prompt. We instructed the model to output its decision as a JSON object with two keys: *{"action": "<Action Target:Host>", "reason": "<Justification>"}*. The action field is the combination of both the defensive action and the target parameter of the host. We then pass action into our action parser which extracts the action and target parameter. We created a lookup table that maps the combination of the action and target to the action index the environment expects. We handled invalid or unparsable outputs by defaulting the action to a fallback action *Monitor*. This made sure that we had no runtime errors, and we also log invalid decisions for future observation.

We created prompts to review the network and assign it a role as an expert cybersecurity analyst that rules, examples, and, in some prompts, optimal strategies to defend the network. The LLM directly outputs its chosen action and rationale. This setup tests the LLM's ability to generalize and reason about states on the fly using the domain knowledge provided. The prompts are provided in our GitHub repository[1]. Table 2 summarizes the comparative performance of the different prompting strategies.

Zero-Shot Prompt: We design a prompt that provides high-level instructions and context, including the format of observations and actions, the list of possible actions with their descriptions, and the rules of the environment, such as network layout and attacker progression. We also give it domain knowledge, such as critical subnets and action descriptions, this allows the model to make informed decisions. We do not employ any strategy on the agent in this prompt, we just assign it a role.

Strategy Prompt: This prompt has all the high-level information and role instructions that the previous prompt has, and we also decided to enhance the agent with a strategy on how to approach defensive actions optimally. We observed that it is optimal to start off with decoys as soon as possible and to prioritize removes over restores depending on the situation or restores over removes depending on the status of the environment. In the CAGE-2 Challenge, the top submission on the leaderboard used the greedy decoy strategy that influenced some of our strategies [8].

Chain-of-Thought (CoT) Prompt: We added more explicit reasoning instructions, CoT prompting, such as encouraging the model to consider the threat indicators in the observation and the consequences of each action before deciding. The model was still required to output only the final action and a concise justification, as per the prompt guidelines, rather than a full reasoning trace. We found that providing the strategic context, such as the environment rules and

---

[1] https://github.com/HassanChowdhry/cage-challenge-2-LLM.

goals, and clearly instructing the output format was sufficient for the LLM to make sensible choices.

We also aimed to fine-tune the temperature to ensure optimal output from LLM inference. To explore this trade-off, we tested a range of temperature values while keeping other settings fixed. The results shown in the table below help identify the most effective settings for different models and prompt types, allowing us to achieve the best possible results.

Table 2. Results from prompts evaluated over 30 timesteps, compared against the red agents.

Model	Red Agent	Zero-Shot	Strategy	CoT
Gemini-2.5-flash-lite	Meander	−20.3	−11.6	−6.8
	B-Line	−130.8	−19.8	−2.8
Gemini-2.5-flash	Meander	−9.4	−10.9	−11.0
	B-Line	−7.6	−12.6	−10.3
Gemini-2.5-pro	Meander	−7.2	−7.2	−8.8
	B-Line	−7.8	−10.9	−7.6

4 Evaluation

We evaluated DRL and LLM agents in the CAGE-2 Challenge scenario. The CAGE-2 scenario involves a Blue defender tasked with defending an enterprise network against a Red attacker agent while maintaining normal services for legitimate users. The Red agent begins each episode with a foothold in the user host and attempts to escalate throughout the network's zones to impact a critical operational server. The Blue agent must take actions to detect and mitigate intrusion without disrupting the system [8].

To quantify the performance of an agent, the CAGE-2 environment provides a reward signal to the Blue defender agent after each action that reflects the state of the attack and the server. The reward function is defined as follows [8]:

$$R_{\text{Blue}} = -0.1\,(\text{host}_{Red} - 1) \; - \; \text{server}_{Red} \; - \; \text{Blue}_{Restore} \; - \; 10\,(\text{Red}_{impact}) \quad (1)$$

This formula penalizes the defender for each aspect of the attacker's progress. The Blue agent receives a small negative reward for each new host or server the attacker compromises and for each Restore action, and a negative reward if the attacker succeeds in disrupting the mission. The best possible reward in an episode is 0. Values closer to zero indicate stronger defense, which means that the Blue agent minimized the attacker's impact and avoided expensive responses [8].

We compare two Blue agents using different approaches: a DRL agent and an LLM agent. We deployed both agents against Red attacker agents B-Line and Meander. We ran each Blue agent against each Red Agent under episodes of timesteps: 50, 100 steps. This allows us to examine performance in both initial encounters and extended attack scenarios. We repeat each scenario multiple times for the DRL agent to account for stochasticity and measure average performance with confidence, as for the LLM agent due to the high cost of inference, we will limit the number of episodes. After each episode, we record the cumulative reward and sequence of actions taken by both DRL and Blue Agent.

The primary metric of CAGE-2 defender performance is the cumulative reward achieved by the Blue agent during an episode. Tables 3–4 report the results across 50, and 100 timesteps. The DRL agent keeps the rewards closer to zero, which indicates reduced attacker impact and fewer costly restores. At 100 steps, there is a clear divergence, DRL stabilizes its trajectory, while LLM degrades heavily, with several runs falling below -100 in CoT. These findings show that the DRL agent delivers stronger robustness under extended engagements, while LLM defenders can match performance briefly but lack long-term stability.

Table 3. 50-step cumulative reward for DRL and LLM (Gemini-2.5-pro) Agents.

Model	Red Agent	Temperature									
		0.1	0.2	0.3	0.4	0.5	0.6	0.7	0.8	0.9	1.0
Zero-Shot	Meander	-27.7	-26.2	-30.0	-72.8	-23.7	-35.9	-16.2	-98.3	-51.9	-21.0
	B-Line	-62.4	-14.9	-24.3	-19.9	-18.3	-22.6	-14.9	-12.8	-54.3	-20.6
Strategy	Meander	-24.0	-59.7	-39.3	-53.1	-40.2	-25.9	-43.8	-35.3	-24.7	-18.2
	B-Line	-22.9	-22.6	-22.3	-27.1	-21.1	-24.3	-18.3	-23.0	-21.9	-20.3
CoT	Meander	-29.9	-28.3	-43.1	-14.2	-70.7	-43.0	-25.9	-39.4	-41.4	-36.8
	B-Line	-17.6	-12.8	-21.7	-17.0	-14.2	-53.8	-17.6	-20.0	-18.3	-16.9
DRL	Meander	—				-15.1	—				
	B-Line	—				-4.7	—				

We analyze the distribution of actions taken by the Blue agents over 100 timesteps against the Meander attacker (Fig. 2). The DRL agent exhibited a strong preference for Decoy, allocating roughly three-quarters of its actions to deception, while using Analyze and Restore more, and rarely relied on Remove or Monitor. In LLMs, the Zero-Shot prompt favored Decoy (37.2%) and Restore (29.4%), defaulting to deception and recovery as its main defense. The Strategy prompt also emphasized Decoy and Restore, consistent with the instructions we provided. The CoT prompt diverged from this pattern, prioritizing Remove (38.9%), Analyze (25%), and Decoy (26.7%). These results indicate that LLM agents adapt their tactical preferences according to the prompt style: Zero-Shot and Strategy converge on decoys and restores, whereas CoT emphasizes removal and analysis.

Table 4. 100-step cumulative reward for DRL and LLM (Gemini-2.5-pro) Agents.

Model	Red Agent	Temperature									
		0.1	0.2	0.3	0.4	0.5	0.6	0.7	0.8	0.9	1.0
Zero-Shot	Meander	-77.7	-71.2	-75.0	-127.1	-47.7	-117.5	-36.2	-175.4	-80.9	-49.2
	B-Line	-142.4	-30.6	-49.3	-37.9	-40.3	-40.6	-36.5	-33.9	-129.3	-38.9
Strategy	Meander	-55.0	-176.8	-68.2	-90.8	-73.2	-63.9	-78.6	-95.7	-50.7	-43.1
	B-Line	-50.2	-47.6	-49.3	-57.1	-46.6	-47.2	-39.3	-45.9	-46.4	-49.0
CoT	Meander	-87.9	-252.2	-112.2	-53.2	-106.4	-78.9	-137.6	-115.4	-128.5	-117.5
	B-Line	-36.9	-31.9	-61.7	-39.0	-33.6	-100.7	-32.9	-32.0	-34.3	-31.9
DRL	Meander	—				-24.2	—				
	B-Line	—				-9.7	—				

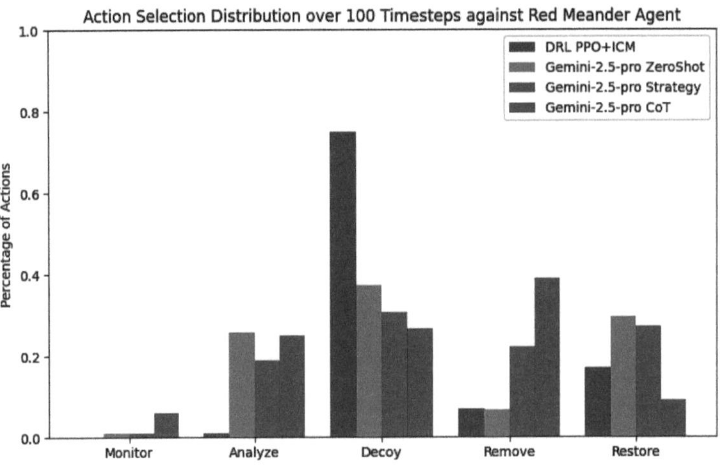

Fig. 2. Action Selection distributions for DRL and LLM agents at their best-performing temperatures.

A practical advantage of the LLM defender is that it produces a natural-language rationale alongside each action. We evaluated two properties of these outputs: (i) *Format Adherence*, the share of steps whose actions pass strict schema validation, and (ii) *Justification Relevance*, a rubric-based score checking entity grounding, indicator alignment, action–state coherence, absence of hallucination, and specificity. In our early experiments the model occasionally wrapped its JSON output inside fenced code blocks. We applied a preprocessing step to strip any leading or trailing fences prior to any validation. After this adjustment, we did not observe parsing errors and achieved a perfect 100% format adherence across all prompts and temperatures. This ensures that every LLM response could be reliably interpreted and executed by the environment. The justification relevance results are summarized in Table 5.

Table 5. LLM reasoning summaries by prompt style for Gemini-2.5-pro.

Prompt	Strategy	Key actions	Trade-offs	Lessons
Zero-Shot	Proactive hardening plus aggressive remediation; contain attacker to User/Enterprise; protect Operational subnet.	Shift to Restore-first after Remove proved ineffective; deploy decoys during quiet periods; prioritize Enterprise compromises.	Remove vs. Restore; reactive vs. proactive tension; early reliance on Remove allowed escalation.	Be decisive (Restore for serious threats); use downtime to expand decoys; prevention tools reduce reinfection cycles.
Strategy	"Protect the crown jewels" in three phases: early decoys, aggressive remediation, opportunistic hardening.	Immediate Restore on Enterprise; tiered user-host loop (Remove → Analyze → Restore); reactive hardening on probed hosts.	Prioritize critical assets; action economy and remediation cost; accept churn on User subnet to protect Enter-prise/Operational.	Remove often insufficient on users; Restore-first may be cheaper overall; diversify decoys earlier on weak-point servers.
CoT	Adaptive defense-in-depth: proactive decoys plus prioritized remediation for high-value assets (Defender, OpServer0).	Early/broad decoys; immediate Restore on Defender/-OpServer0; escalation from Remove → Analyze → Restore on Enterprise.	Trade-off between rapid containment and persistence eradication; Defender host frequently targeted → action sink.	Escalate to Restore immediately on critical assets; avoid reactive loops; protect the defender infrastructure first.

By examining these metrics, we can gain a more comprehensive understanding of each agent's capabilities. They help ensure that our evaluation covers both the average case performance and the reliability, safety, and practicality of the agents in ACD settings.

5   Discussion

These results highlight clear trade-offs between DRL and LLM defenders. In terms of raw performance, the DRL agent achieved higher cumulative rewards and limited attacker impact across all horizons. Its stability comes from explicit training to maximize the reward signal under both red agents. In contrast, LLM agents lacked reward optimization and instead relied on prompt-guided reasoning. The LLM agents demonstrated strong interpretability. Each decision was paired with a justification, making their reasoning transparent and suitable for human oversight, which is an important advantage over the DRL policy.

We observe several limitations in the LLM agent. The LLM agent exhibited instability in extended runs, often experiencing catastrophic failures when evaluated over 100 timesteps, indicating its inability to maintain stable defensive behavior over longer engagements. The LLM's decision-making incurred significant latency and computation cost, each LLM action required an inference from a large model, which introduces noticeable delays. This overhead is costly in real-time defense. The cost proposes another issue as in most cases you would need private LLMs as agents and would need significant compute power. The DRL agent does not have any of these issues and immediately produces actions.

An interesting result is that the two approaches show complementary strengths. The DRL agent excelled at fast and optimized threat mitigation under the specific condition in which it was trained. The LLM agent is much more transparent and has reasoning and is potentially more flexible. We propose a hybrid strategy in which a fast DRL policy handles routine, high-frequency actions. An escalation gate, implemented through anomaly detection using simple heuristics such as imminent compromise of the operational server, repeated failed removes, identifies high-impact states. When triggered, an LLM would interpret anomalies, recommend, or authorize costly actions such as Restore, and produce concise natural-language rationales for audit. An operator interface surfaces the LLM explanations and key telemetry for oversight while preserving automatic control by the DRL policy.

6   Conclusion

We presented a comparative study of DRL and LLM defender agents in the CAGE-2 scenario, evaluating their performance against multiple red strategies. DRL agents consistently achieved higher cumulative rewards, while LLM agents offered greater interpretability and adaptability, although constrained by prompt sensitivity, cost, and latency.

Despite these limitations, LLMs remain a promising direction for transparent and flexible defense. Recent research suggests that specialized small language models (SLMs) may provide comparable reasoning at reduced cost and latency [1]. Future research should expand evaluation beyond cumulative reward to include broader metrics such as mean detection and remediation time, number of successful compromises, and operational uptime, offering a more comprehensive view of defender effectiveness. In addition, expanding the evaluation beyond Gemini to include a wider set of commercial and open source LLMs will be relevant to improve generality and identify model-specific trade-offs. Finally, testing resilience against previously unseen attack strategies and exploring more in-depth hybrid strategies that combine DRL efficiency with LLM transparency will be essential steps toward autonomous defenders that are reliable, explainable, and capable of real-world deployment.

References

1. Belcak, P., et al.: Small language models are the future of agentic AI. arXiv preprint arXiv:2506.02153 (2025)
2. Cao, Y., Liu, K., Lin, Y., Wang, L., Xia, Y.: Deep-reinforcement-learning-based self-evolving moving target defense approach against unknown attacks. IEEE Internet Things J. **11**(20), 33027–33039 (2024). https://doi.org/10.1109/JIOT.2024.3423022
3. Castro, S.R., Campbell, R., Lau, N., Villalobos, O., Duan, J., Cardenas, A.A.: Large language models are autonomous cyber defenders. arXiv preprint arXiv:2505.04843 (2025)
4. Contractor, F., Li, L., Mallah, R.A.: Learning to communicate in multi-agent reinforcement learning for autonomous cyber defence. arXiv preprint arXiv:2507.14658 (2025)
5. Dutta, A., Chatterjee, S., Bhattacharya, A., Halappanavar, M.: Deep reinforcement learning for cyber system defense under dynamic adversarial uncertainties. arXiv preprint arXiv:2302.01595 (2023)
6. Hicks, C., Mavroudis, V.: Autonomous cyber defence: beyond games? (2024). https://doi.org/10.5281/zenodo.10974183
7. Kiely, M., et al.: Exploring the efficacy of multi-agent reinforcement learning for autonomous cyber defence: a CAGE challenge 4 perspective. In: Proceedings of the AAAI Conference on Artificial Intelligence, vol. 39, pp. 28907–28913 (2025). https://doi.org/10.1609/aaai.v39i28.35158, https://ojs.aaai.org/index.php/AAAI/article/view/35158
8. Kiely, M., Bowman, D., Standen, M., Moir, C.: On autonomous agents in a cyber defence environment. arXiv preprint arXiv:2309.07388 (2023)
9. Loevenich, J., Adler, E., Hürten, T., Lopes, R.R.F.: Design and evaluation of an autonomous cyber defence agent using DRL and an augmented LLM. Comput. Netw. **262**(C) (2025). https://doi.org/10.1016/j.comnet.2025.111162
10. Nguyen, T.T., Reddi, V.J.: Deep reinforcement learning for cyber security. IEEE Trans. Neural Networks Learn. Syst. **34**(8), 3779–3795 (2023). https://doi.org/10.1109/TNNLS.2021.3121870
11. Oesch, S., et al.: The path to autonomous cyberdefense. IEEE Secur. Privacy **23**(1), 38–46 (2025). https://doi.org/10.1109/MSEC.2024.3427640
12. Pathak, D., Agrawal, P., Efros, A.A., Darrell, T.: Curiosity-driven exploration by self-supervised prediction. In: ICML (2017)
13. Ren, K., Zeng, Y., Cao, Z., et al.: ID-RDRL: a deep reinforcement learning-based feature selection intrusion detection model. Sci. Rep. **12**(1), 15370 (2022). https://doi.org/10.1038/s41598-022-19366-3
14. Schulman, J., Wolski, F., Dhariwal, P., Radford, A., Klimov, O.: Proximal policy optimization algorithms. arXiv preprint arXiv:1707.06347 (2017)
15. Singh, A.V., et al.: Hierarchical multi-agent reinforcement learning for cyber network defense. arXiv preprint arXiv:2410.17351 (2024)
16. Standen, M., Lucas, M., Bowman, D., Richer, T.J., Kim, J., Marriott, D.: Cyborg: a gym for the development of autonomous cyber agents. arXiv preprint arXiv:2108.09118 (2021)
17. Vyas, S., Hannay, J., Bolton, A., Burnap, P.P.: Automated cyber defence: a review. arXiv preprint arXiv:2303.04926 (2023)
18. Wolk, M., et al.: Beyond cage: investigating generalization of learned autonomous network defense policies. arXiv preprint arXiv:2211.15557 (2022)

Special Session on Hybrid Intelligent Decision-Making: Integrating AI and MCDM for Real-World Applications

Inverse Reinforcement Learning for Reward Function in Traffic Signal Control

Do Thai Giang[1,2], Truong Cong Doan[1], and Phan Duy Hung[2(✉)] iD

[1] International School, Vietnam National University, Hanoi, Vietnam
{24078012,tcdoan}@vnu.edu.vn, giangdt26@fe.edu.vn
[2] FPT University, Hanoi, Vietnam
hungpd2@fe.edu.vn

Abstract. This research examines the utilization of Inverse Reinforcement Learning and Deep Q-Networks in traffic signal management to enhance urban mobility through the automation of reward function design. The study illustrates an 81.35% reduction in vehicle waiting time and a 60.15% decrease in queue length, achieved by training an Inverse Reinforcement Learning model with expert trajectory data and incorporating it into a Reinforcement Learning framework. Although throughput slightly decreased 25.11%, this trade-off enabled significant gains in congestion reduction. In comparison to the original Deep Q-Network methodology, the Inverse Reinforcement Learning controller realizes a 17.77% enhancement in waiting time and a 5.37% reduction in queue length, however throughput experiences a minor decline of 3.12%. The findings highlight the sensitivity to reward function weights, with the Inverse Reinforcement Learning controller attaining optimal equilibrium across performance measures, markedly enhancing urban traffic conditions. This study demonstrates the potential of Inverse Reinforcement Learning in optimizing traffic signals by deriving reward functions from expert tactics.

Keywords: Inverse Reinforcement Learning · Traffic Signal Control · Reward Function · Reinforcement Learning · Urban Mobility

1 Introduction

Traffic congestion has emerged as a significant challenge in urban areas globally, impacting millions daily [1]. As urban populations expand and vehicle ownership rises, road networks are failing to meet demand. In prominent urban centers such as New York, London, Beijing, and Mumbai, traffic congestion has become a commonplace occurrence, resulting in considerable delays and irritation for commuters [2]. Research indicates that traffic congestion not only consumes time but also incurs significant economic and environmental repercussions. Reports indicate that billions of dollars are forfeited annually due to diminished productivity, heightened fuel consumption, and vehicle maintenance expenses [3].

In addition to the inconvenience of extended commutes, traffic congestion significantly affects the environment. Vehicles idling in traffic emit substantial quantities of

L. Martínez et al. (Eds.): IDEAL 2025, LNCS 16239, pp. 437–448, 2026.
https://doi.org/10.1007/978-3-032-10489-2_37

carbon dioxide (CO_2) and other detrimental pollutants, exacerbating air pollution and climate change [4]. In densely populated urban areas, air quality frequently deteriorates below acceptable health standards, heightening the risk of respiratory ailments and other health issues [5]. Moreover, fuel inefficiency resulting from stop-and-go traffic exacerbates the economic strain on people as well as governments [6].

As road infrastructure development fails to keep up with increasing urban populations, enhancing traffic management has emerged as a crucial priority [7]. Numerous cities have sought to mitigate congestion by implementing strategies such as road expansion, instituting congestion fees, and enhancing public transportation systems. Nevertheless, these solutions alone frequently prove insufficient, highlighting the growing necessity for more intelligent and adaptive traffic control strategies [8].

This study's primary contributions introduce an innovative amalgamation of Deep Q-Networks (DQN) and Maximum Entropy Inverse Reinforcement Learning (MaxEnt IRL) to improve traffic signal management. In contrast to conventional reinforcement learning methods that depend on manually designed reward functions, our framework develops adaptive, expert-informed reward structures that more accurately represent real-world traffic management goals. Utilizing Inverse Reinforcement Learning (IRL), the system autonomously deduces optimal reward weights, thereby obviating the necessity for extensive manual calibration and enhancing generalization across varying traffic conditions. This study emphasizes the significance of reward function sensitivity in reinforcement learning applications. The study compares handcrafted and IRL-derived reward structures, illustrating that learned rewards surpass manually defined metrics, resulting in more effective and scalable traffic signal control strategies.

2 Literature Review

2.1 Reinforcement Learning in Traffic Signal Control

The utilization of Reinforcement Learning (RL) in traffic signal management has garnered considerable attention in recent years. Reinforcement learning methods allow traffic controllers to acquire optimal policies via interaction with the environment, without necessitating explicit models of traffic dynamics [9]. Research utilizing Q-learning and deep reinforcement learning methodologies has shown a reduction in delays of 15–30% relative to fixed-time controllers across diverse traffic conditions [7].

Notwithstanding these advancements, a core challenge in reinforcement learning based traffic signal control persists in the formulation of suitable reward functions. Most current methodologies employ manually designed reward signals derived from queue length, waiting time, or throughput [8]. Manually crafted reward functions frequently inadequately represent the intricacies of traffic optimization goals, resulting in suboptimal policies when implemented in real-world settings.

2.2 Inverse Reinforcement Learning for Traffic Control

IRL overcomes these limitations by deriving reward functions from expert demonstrations instead of manually defining them. Initially formalized by Ng and Russell [10], IRL deduces the intrinsic reward function that accounts for the observed behavior of experts. In the realm of traffic signal management, IRL presents numerous significant benefits. IRL can derive reward functions from human traffic engineering control patterns or from established high-performance adaptive systems [11]. By means of demonstration, IRL can implicitly encapsulate intricate trade-offs among competing objectives (e.g., throughput, fairness, emission reduction) that are challenging to articulate explicitly [12]. Moreover, learned reward functions may exhibit superior generalization across various intersections and traffic conditions compared to manually crafted rewards [13].

Recent research by Wu et al. [14] revealed that reward functions derived from IRL resulted in a 12% enhancement in average vehicle delay relative to traditional Reinforcement Learning methods utilizing manually crafted rewards. Likewise, Bhattacharyya et al. [15] demonstrated that IRL can proficiently acquire traffic control policies that align more closely with the preferences of human experts, thereby balancing efficiency and fairness.

2.3 Challenges in IRL for Traffic Signal Control

Implementing IRL in traffic signal management poses numerous challenges. The high-dimensional state space of traffic networks and the challenge of acquiring consistent expert demonstrations under varying conditions continue to pose substantial obstacles. Moreover, the computational complexity of the majority of IRL algorithms has constrained their utilization in extensive traffic networks.

Recent advancements in deep inverse reinforcement learning and maximum entropy inverse reinforcement learning methodologies [16] have started to tackle these challenges, facilitating more efficient learning from suboptimal demonstrations. The incorporation of these methods into practical traffic signal control systems continues to be a dynamic research domain with significant potential for innovation.

MaxEnt IRL, as presented by Ziebart et al. (2008) [11], enhances conventional IRL by integrating a probabilistic framework that guarantees learned policies do not excessively restrict agent behavior. This method posits that expert demonstrations adhere to a probability distribution that maximizes entropy while being congruent with observed trajectories. The principle of maximum entropy guarantees that, among all potential distributions adhering to expert constraints, the least biased distribution is selected, resulting in more resilient and flexible reward functions.

A recent paper presented at the IJCAI 2023 conference [17] proposes a solution utilizing an InitLight network founded on an adversarial architecture. Nevertheless, our survey indicates that no research has utilized the integration of MaxEnt IRL with sophisticated reinforcement learning algorithms like DQN to address the traffic light control issue. This research gap is significant, as the integration of MaxEnt IRL with DQN can leverage the strengths of both methodologies. MaxEnt IRL facilitates the precise learning of the reward function from expert data, whereas DQN capitalizes on its capacity to develop effective policies within the expansive and intricate state space of urban traffic. This study aims to develop an integrated approach utilizing MaxEnt IRL and DQN to learn the reward function and enhance the traffic light control policy, thereby overcoming the limitations of current methods and improving traffic management efficiency in dynamic urban settings.

## 3	Methodology

This study integrates Deep Reinforcement Learning (DRL) and Inverse Reinforcement Learning (IRL) to optimize traffic signal control in a simulated urban environment. A DQN agent is first trained in SUMO using a handcrafted reward function based on key metrics like waiting time, queue length, CO_2 emissions, and throughput.

Expert trajectories collected from the DQN agent are then used in MaxEnt IRL to infer a more adaptive, data-driven reward function. This learned reward replaces the original, and the agent is retrained to improve policy performance under dynamic traffic conditions. The model undergoes multiple training iterations and is evaluated across various traffic scenarios. Its performance is compared with both fixed-time control and manually tuned RL, demonstrating the effectiveness of IRL in enhancing urban traffic management. The proposed model architecture is shown in Fig. 1 below.

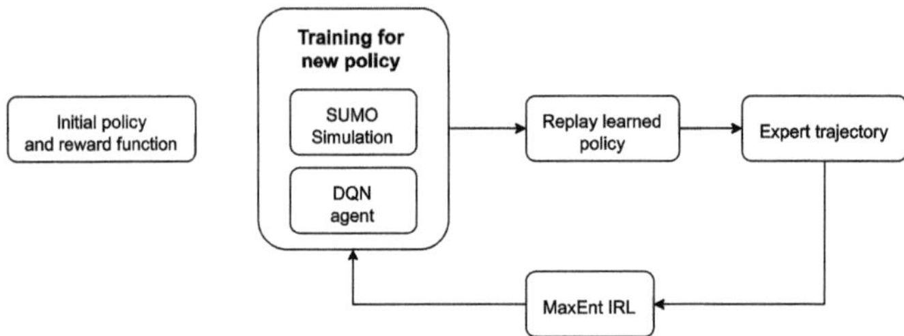

Fig. 1. Proposed method

The framework begins with a baseline policy, which can either be randomly initialized or pre-trained using a handcrafted reward function. This policy governs the initial behavior of the agent during simulation.

### 3.1	Deep Q-Network Algorithm

Deep Q-Network is an extension of traditional Q-learning, leveraging deep neural networks to approximate the action-value function $Q(s, a)$ in high-dimensional state spaces. This algorithm enables reinforcement learning agents to make complex decisions without requiring an explicit model of the environment, making it particularly suitable for traffic signal control optimization.

The core idea behind Q-learning is to learn the optimal action-value function $Q(s_t, a_t)^*$ that estimates the expected cumulative reward for selecting action a in state s and following the optimal policy afterward. The update rule for Q-learning is:

$$Q(s_t, a_t) = Q(s_t, a_t) + \alpha\left[r_t + \gamma \max_{a'} Q(s_{t+1}, a') - Q(s_t, a_t)\right] \tag{1}$$

where:

- s_t, a_t are state and action at time step t

- α is the step size
- r_t is the immediate reward received after taking action a_t
- γ is the discount factor that determines the importance of future rewards
- $\max_{a'} Q(s_{t+1}, a')$ represents the best possible future reward from state s_{t+1}

In the context of traffic signal control, states represent real-time traffic conditions such as vehicle density and speed, while actions correspond to selecting appropriate signal phases and durations. The reward function balances competing objectives such as minimizing congestion, reducing waiting time, optimizing throughput, and lowering emissions. DQN enables the system to adapt dynamically to real-world traffic patterns, learning optimal signal control strategies through repeated interaction with the simulation environment.

3.2 Maximum Entropy Inverse Reinforcement Learning

IRL seeks to infer the underlying reward function that drives expert behavior, enabling autonomous agents to learn optimal policies without explicit reward design. MaxEnt IRL extends traditional IRL by incorporating a probabilistic framework that ensures learned policies do not overly constrain agent behavior. This approach assumes that expert demonstrations follow a probability distribution that maximizes entropy while remaining consistent with observed trajectories. Mathematical Formulation In standard IRL, the goal is to recover the reward function $R(s)$, given expert demonstrations τ_E consisting of state-action sequences $(s_1, a_1, s_2, a_2, \ldots, s_t)$. MaxEnt IRL builds upon this by introducing a probability distribution over trajectories, ensuring that the agent explores multiple possible behaviors while favoring those demonstrated by the expert. The probability distribution over trajectories is defined as:

$$P(\tau) = \frac{e^{R(\tau)}}{Z} \tag{2}$$

where:

- τ is the trajectory $(s_1, a_1, s_2, a_2, \ldots, s_t)$
- $R_{(\tau)}$ is the cumulative reward of trajectory τ
- Z is the partition function, ensuring a valid probability distribution: $Z = \sum_{\tau} e^{R(\tau)}$

The maximum entropy principle ensures that among all possible distributions satisfying expert constraints, the most unbiased distribution is chosen. The reward function is typically modeled as a linear combination of feature expectations, leading to:

$$R(s) = w^T \phi(s) \tag{3}$$

where:

- w represents weight parameters learned during optimization,
- $\phi(s)$ is a feature vector extracted from state s, capturing relevant environmental properties.

MaxEnt IRL iteratively refines w by minimizing the difference between expert feature expectations $E\tau_E[\phi_s]$ and policy-derived feature expectations $E\pi[\phi_s]$. The gradient update follows:

$$\frac{\partial L}{\partial w} = E\tau_E[\phi_s] - E\pi[\phi_s] \tag{4}$$

where L is the likelihood function representing alignment between inferred and expert behaviors. Regularization techniques, such as L2 regularization, are often employed to ensure stability.

MaxEnt IRL is particularly useful in traffic management scenarios where reward function design is challenging due to the complexity of optimizing multiple competing objectives. By observing expert-like traffic patterns, MaxEnt IRL learns a reward function that balances waiting time reduction, queue length minimization, emissions control, and throughput maximization. This learned function enables reinforcement learning agents to adapt dynamically to real-world variations in traffic conditions, outperforming hand-crafted reward functions by offering greater generalization across different intersection configurations.

3.3 Environment

This study utilizes a simulation environment based on SUMO (Simulation of Urban Mobility), an open-source traffic simulator extensively utilized in urban transportation research. SUMO facilitates the microscopic simulation of individual vehicle dynamics, traffic signal management, and interactive behaviors within a roadway system. A four-way intersection was created for our implementation, featuring three lanes per approach and a flexible traffic signal control system that can dynamically modify signal phases according to real-time traffic circumstances.

The environment replicates various traffic patterns by incorporating randomized vehicle routing and several vehicle categories, such as passenger vehicles and buses. Traffic flow is generated probabilistically, enabling cars to adopt diverse paths, including straight travels, left turns, and right turns at varying frequencies. The simulation operates for 3600 time steps (equal to one hour) per episode, facilitating adequate data accumulation for reinforcement learning enhancement. To augment realism, SUMO incorporates emissions modeling, wait length assessments, and vehicle speed monitoring, facilitating a thorough comparison of traffic efficiency across various signal management schemes. Additionally, the Traffic Control Interface (TraCI) facilitates real-time interaction between reinforcement learning agents and the simulation, thereby ensuring adaptive decision-making in dynamic contexts.

3.4 State Space

The state space in this study is set up to capture important traffic conditions at the intersection, which helps reinforcement learning agents make good decisions. There are two main parts to the state representation include:

- Velocity Matrix is presented as a corresponding 12×12 velocity grid, tracking the speed of vehicles within the same spatial regions also show in Fig. 2. This allows the reinforcement learning agent to assess not only the presence of vehicles but also their movement dynamics, aiding in traffic signal optimization.
- Traffic Signal Phase Representation: A one-hot encoded vector representing the current active phase of the traffic light system. This encoding ensures that the agent is aware of the ongoing signal phase, enabling informed decisions on the next phase transition.
- A one-hot encoded vector that shows what phase of the traffic light system is currently active. This encoding makes sure that the agent knows what the current signal phase is, so it can make smart choices about when to move to the next phase.

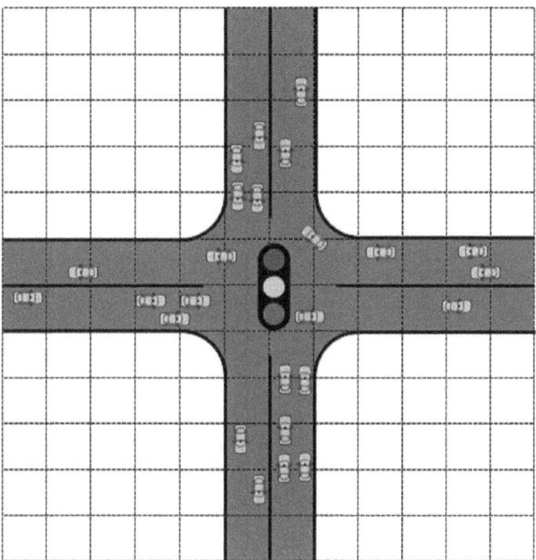

Fig. 2. Grid 12×12 of spatial and velocity

The state space also includes several important traffic metrics, such as queue length, waiting time, CO_2 emissions, average vehicle speed, and throughput. These give a complete picture of how well an intersection is working. These features let the reinforcement learning agent make smart changes that improve urban mobility while reducing traffic and environmental damage.

3.5 Action Space

The agent selects actions from a discrete set of signal phase-duration pairs to adaptively control traffic lights. Each action includes:

- Phase selection: One of six options covering straight and left-turn movements for both North-South and East-West directions, including combined movements.
- Duration selection: A green light duration of 15, 20, 30, or 40 s, chosen based on current traffic conditions.

This setup allows dynamic and efficient signal control. Yellow light transitions are computed automatically based on speed and queue length to ensure safe phase changes.

3.6 Initial Reward Function

The reward function is a critical component of reinforcement learning for traffic signal control, as it directly influences the agent's decision-making process. In this study, an initial handcrafted reward function was designed to incorporate multiple traffic efficiency objectives, ensuring balanced optimization of intersection performance.

The initial reward function is formulated as:

$$R = -\alpha \cdot CO_2 - \beta \cdot Q - \gamma \cdot W + \delta \cdot S + \epsilon \cdot T \tag{5}$$

where:

- CO_2 represents emissions (grams)
- Q represents queue length (number of vehicles)
- W represents waiting time (seconds)
- S represents average vehicle speed (m/s)
- T represents throughput (number of vehicles passing)

The weight coefficients $(\alpha, \beta, \gamma, \delta, \epsilon)$ are empirically tuned to balance competing objectives, ensuring that the traffic control agent optimizes flow efficiency while minimizing environmental impact and congestion. This handcrafted function serves as the foundation for initial reinforcement learning, providing structured feedback to guide policy development before transitioning to a learned reward function using IRL.

4 Experiment

The experimental assessment of the proposed traffic signal control framework was performed in a microscopic traffic simulation environment utilizing SUMO (Simulation of Urban Mobility). The experiments aimed to evaluate the efficacy of Deep Q-Network (DQN) reinforcement learning and Maximum Entropy Inverse Reinforcement Learning (MaxEnt IRL) in enhancing signal control strategies.

4.1 Data Collection

To effectively train reinforcement learning models, traffic data is gathered using SUMO in a simulated intersection where vehicles adhere to probabilistic routes exhibiting varied movement patterns. Each episode spans 3600 time steps, emulating one hour of urban traffic involving various vehicle types.

A proficient DQN agent engages with the environment by modifying signal phases and durations. The dataset encompasses state-action pairs, comprising vehicle density, velocity, queue length, waiting time, throughput, and CO_2 emissions, providing insights into traffic conditions and control effects.

Expert trajectories are documented for MaxEnt IRL, capturing state transitions and control actions to deduce a reward function from observed behavior. To uphold data quality, episodes characterized by significant delays or low throughput are excluded, thereby guaranteeing superior expert demonstrations for the acquisition of optimal policies.

4.2 Model Training

The training process includes two stages combining DQN and MaxEnt IRL for traffic signal optimization.

Stage 1 - DQN Training: A Deep Q-Network is trained using a handcrafted reward function in the SUMO environment. The agent observes traffic states (density, velocity, queue length, signal phase) and selects actions (phase and duration) to improve key metrics. Training parameters include a learning rate of $\alpha = 0.001$, discount factor $\gamma = 0.95$, and experience replay with 920 episodes. The result is a baseline control policy. Stage 2 - MaxEnt IRL: Expert trajectories from the trained DQN are used to infer a reward function via MaxEnt IRL, which balances competing objectives using entropy-based optimization. The learned reward replaces the handcrafted one, and the RL agent is retrained to adapt more effectively to dynamic traffic conditions.

This two-stage framework enhances generalization and minimizes manual reward tuning, leading to improved policy performance across various scenarios.

5 Result Analysis

The proposed traffic signal control framework's evaluation shows that it makes intersections work much better, especially by reducing traffic jams and making it easier for people to get around. We compared three control methods: a traditional fixed-time controller, a Deep Q-Network based controller with a custom reward function, and an IRL enhanced controller that changes the reward structure based on expert demonstrations.

The Fig. 3 shows that the IRL-based controller made traffic flow much better. Compared to the fixed-time method, it cut down on vehicle wait times by 81.35% and queue lengths by 60.15%. This made delays at the intersection much shorter for vehicles. Also, throughput went down by 25.11%, which shows that there is a trade-off between reducing congestion and increasing flow volume.

When we compared the IRL-based method to the original DQN model, we found even more ways to make it better. The learned reward function worked better than the manually made ones and also cut down on waiting time by 17.77% and made the queue shorter by 5.37%. Throughput went down by 3.12%, but the changes made the traffic management plan more balanced. It improves the flow of traffic at intersections without putting too much emphasis on the number of vehicles.

The IRL-based controller helped lower carbon emissions by 22.04% in addition to reducing congestion. This was because traffic flowed more smoothly and there were fewer stop-and-go patterns, which usually use more fuel. The IRL model also greatly improved signal phase transitions, cutting the number of phase changes by 32.78%. This made operations more stable and predictable, which reduced driver confusion and made intersections work better.

Fixed-time controllers prioritize throughput but often lead to congestion with longer delays and queues. In contrast, reinforcement learning models adaptively balance waiting time, queue length, and flow. The IRL-based model further refines this trade-off by aligning control strategies with real-world traffic dynamics rather than fixed rules. Although throughput slightly decreased compared to fixed-time scheduling, the IRL

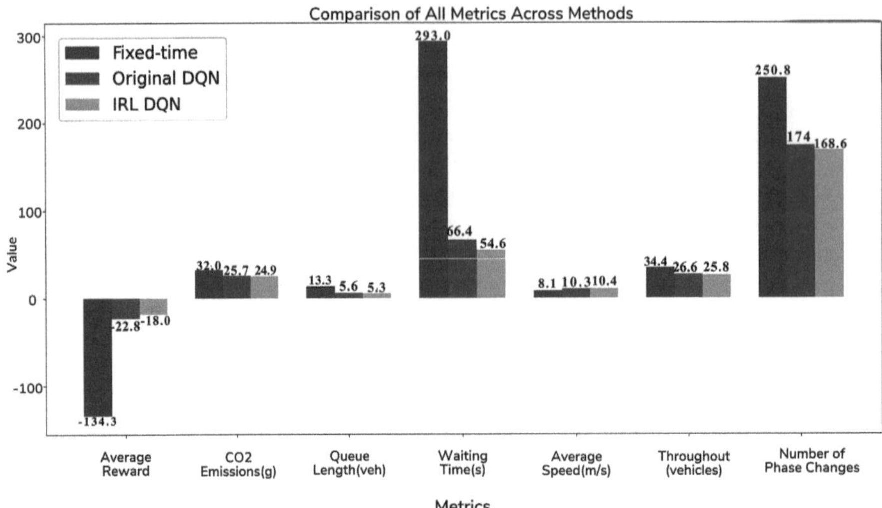

Fig. 3. Comparative evaluation of fixed-time, DQN, and IRL-based controllers on key metrics

model improved signal stability, reduced delays, and optimized queue management through adaptive phase timing and fewer unnecessary transitions.

The suggested IRL-enhanced controller exhibits both performance improvements and practical computational efficiency. Upon completion of training, the policy inference duration for each decision step is roughly 5 ms on CPU AMD Ryzen 7 8745H, rendering it appropriate for near-real-time implementation in urban traffic settings. This low-latency inference guarantees prompt signal modifications without causing operational delays, even in fluctuating traffic scenarios. The offline training phase, especially the MaxEnt IRL component, is computationally demanding due to its iterative reward optimization; nevertheless, this expense is spent only once and does not impact live performance. The system achieves a compromise between decision quality and computational practicality for real-world applications.

These findings confirm the effectiveness of learning-based traffic control, particularly when enhanced by IRL. A data-driven, adaptive approach proves beneficial for urban mobility. Future work should extend this framework to multi-intersection networks and real-time deployment.

6 Conclusion and Perspective

This study shows that using Deep Q-Networks and Maximum Entropy Inverse Reinforcement Learning together can improve the control of traffic signals in cities. The proposed framework learned reward functions from expert behavior, which greatly reduced congestion and wait times while keeping throughput stable. It did better than both fixed-time and manually tuned reinforcement learning controllers. The results show how important it is to design the reward function in reinforcement learning. By replacing static, manually created goals with dynamic, data-driven signals, IRL helps the system better adapt to different traffic situations.

There are still some problems to solve, even though the results look good. The current model only works for one intersection and doesn't take into account things like how many people are walking, how the weather changes, or traffic accidents that happen in real time. In addition, the high cost of IRL computing makes it hard to use on a large scale or in real time. In the future, we will work on making this framework work with networks that have more than one intersection, adding real-time sensor data, and making it work with more than one type of traffic, like pedestrians and public transportation. Also, making learned policies easier to understand and see through will be important for them to be used in real-world traffic management systems.

Acknowledgments. This research is funded by the International School, Vietnam National University, Hanoi (VNU-IS). It also receives funding from FPT University, Hanoi, Vietnam under grant number DHFPT/2025/17.

References

1. Schrank, D., Eisele, B., Lomax, T., et al.: Urban mobility report 2019 (2019)
2. Pozzoni, M., Ceccarelli, G., Gorrini, A., Manenti, L., Sanfilippo, L.: TomTom data applications for the assessment of tactical urbanism interventions: the case of Bologna. Sustainability **15**(17), 12716 (2023)
3. INRIX. Scorecard — inrix.com. https://inrix.com/scorecard/. Accessed 20 Apr 2025
4. Zhang, K., Batterman, S., Dion, F.: Vehicle emissions in congestion: comparison of work zone, rush hour and free-flow conditions. Atmos. Environ. **45**(11), 1929–1939 (2011)
5. World Health Organization, et al.: WHO global air quality guidelines: particulate matter (PM2. 5 and PM10), ozone, nitrogen dioxide, sulfur dioxide and carbon monoxide. World Health Organization (2021)
6. Barth, M., Boriboonsomsin, K.: Real-world carbon dioxide impacts of traffic congestion. Transp. Res. Rec. **2058**(1), 163–171 (2008)
7. Wei, H., et al.: Presslight: learning max pressure control to coordinate traffic signals in arterial network. In: Proceedings of the 25th ACM SIGKDD International Conference on Knowledge Discovery & Data Mining (KDD '19), pp. 1290–1298. Association for Computing Machinery, New York, NY, USA (2019)
8. Genders, W., Razavi, S.: Using a deep reinforcement learning agent for traffic signal control. arXiv:1611.01142 (2016)
9. El-Tantawy, S., Abdulhai, B., Abdelgawad, H.: Multiagent reinforcement learning for integrated network of adaptive traffic signal controllers (MARLIN-ATSC): methodology and large-scale application on downtown Toronto. IEEE Trans. Intell. Transp. Syst. **14**(3), 1140–1150 (2013)
10. Ng, Y.A., Russell, S.: Algorithms for inverse reinforcement learning. In: ICML, vol. 1, p. 2 (2000)
11. Ziebart, B.D., Maas, A., Bagnell, J.A., Dey, A.K.: Maximum entropy inverse reinforcement learning. In: Proceedings of the 23rd National Conference on Artificial Intelligence - Volume 3 (AAAI'08), pp. 1433–1438. AAAI Press (2008)
12. Arora, S., Doshi, P.: A survey of inverse reinforcement learning: challenges, methods and progress. Artif. Intell. **297**, 103500 (2021)
13. Abbeel, P., Ng, A.Y.: Apprenticeship learning via inverse reinforcement learning. In: Proceedings of the Twenty-First International Conference on Machine Learning, p. 1 (2004)

14. Wu, C., Kreidieh, A.R., Parvate, K., et al.: Flow: a modular learning framework for mixed autonomy traffic. IEEE Trans. Rob. **38**(2), 1270–1286 (2021)
15. Bhattacharyya, R.P., Phillips, D.J., Liu, C., et al.: Simulating emergent properties of human driving behavior using multi-agent reward augmented imitation learning. In: Proceedings of the International Conference on Robotics and Automation (ICRA), pp. 789–795. IEEE (2019)
16. Fu, J., Luo, K., Levine, S.: Learning robust rewards with adversarial inverse reinforcement learning. arXiv:1710.11248 (2017)
17. Ye, Y., Zhou, Y., Ding, J., et al.: Initlight: initial model generation for traffic signal control using adversarial inverse reinforcement learning. In: Proceedings of the Thirty Second International Joint Conference on Artificial Intelligence (IJCAI-23), pp. 4949–4958. International Joint Conferences on Artificial Intelligence Organization (2023)

Statistical–Fuzzy Hybrid Framework for Handling Uncertainty in Medical Data Processing

Zulmary Carolina Nieto Sánchez[1]([✉]) and Antonio José Bravo Valero[2]

[1] Corporación Universitaria Minuto de Dios CR, San José de Cúcuta, Colombia
zulmary.nieto@uniminuto.edu
[2] Universidad Simón Bolívar, Facultad de Ciencias Básicas y Biomédicas, Centro de Crecimiento Empresarial MACONDOLAB, Cúcuta, Colombia
antonio.bravo@unisimon.edu.co

Abstract. Reducing uncertainty and imprecision in medical data is essential to improve the quality of diagnoses and treatments. This paper presents a statistical—fuzzy hybrid framework that integrates information–theoretic metrics and classical statistical indicators with fuzzy logic techniques to address these challenges. The framework combines geometric mean filtering—as a computational operator—with generalized compensation operators and evaluates uncertainty reduction using mutual information, clairvoyance value (VoC), and traditional error indicators to enhance data clarity and minimize uncertainty. Validation was performed using synthetic and real–world datasets, including electrocardiographic (ECG) signals and computed tomography (CT) scans. The results showed a significant decrease in the VoC metric, indicating improved data quality and reduced uncertainty. In the ECG recordings, the VoC reduction was particularly notable, improving signal quality for subsequent analysis. Comparison with previous studies demonstrates that the proposed framework offers substantial improvements in uncertainty reduction. Practical implications include more accurate diagnoses and the potential for personalized treatments. The limitations of the study include the need for greater validation in diverse clinical datasets. In conclusion, the hybrid framework effectively improves the quality of medical data and contributes to emerging frontiers in medical data analysis through the integration of statistical methods, fuzzy logic, and uncertainty metrics to support optimized clinical decision-making.

Keywords: Uncertainty handling · Statistical methods · Fuzzy logic · Hybrid frameworks · Medical data processing

1 Introduction

The systematic reduction of uncertainty and imprecision in healthcare is a key factor in achieving significant advances in diagnostic processes, the development

© The Author(s), under exclusive license to Springer Nature Switzerland AG 2026
L. Martínez et al. (Eds.): IDEAL 2025, LNCS 16239, pp. 449–460, 2026.
https://doi.org/10.1007/978-3-032-10489-2_38

and application of treatments, and daily clinical practice [6]. Uncertainty in medical data can arise from multiple sources, including technical or human errors during data collection, variability inherent in human diversity, and limitations within the scientific and analytical models used to process, represent, and interpret clinical phenomena [9]. Adequate handling and mitigation of these sources of uncertainty are essential to ensure that the results obtained from clinical evaluations are reliable and valid–indispensable conditions for safeguarding patient safety and ensuring the effectiveness of medical interventions [1]. Systematically recognizing, measuring, and minimizing errors and ambiguities in diagnostic and therapeutic processes provides healthcare professionals with a stronger and more accurate foundation to make sound clinical decisions [11]. This consequently has a direct positive impact on clinical outcomes, contributing to the overall well–being of patients and raising quality standards in medical care.

In this context, the integration of conventional statistical approaches with fuzzy logic techniques represents an innovative and promising alternative to address the complexity and high variability of clinical data [29]. Classical statistical methods are robust tools for quantitative analysis, facilitating the detection of trends, the formulation of hypotheses, and the inference of causal relationships between variables [32]. However, these methods typically assume that data are accurate, complete, and free from ambiguity—an assumption rarely met in clinical practice, where data are often fragmented, error–prone, or inconsistent [13]. Fuzzy logic, by contrast, provides a framework for adequately managing ambiguity, subjectivity, and intermediate degrees of certainty [20]. This theoretical framework uses degrees of membership to represent phenomena that cannot be defined in absolute terms [35], providing a human–like interpretability that enables more realistic and flexible representations of uncertain medical data in cases where clinical categories are not clearly delineated [36].

A clear example of statistical and fuzzy logic integration is found in the computational analysis of medical images [17], where gradient–based operators and support vector machines are combined within segmentation pipelines to improve anatomical delineation. This approach enhances feature extraction and region identification, supporting the detection of clinical anomalies in complex imaging data [27]. Such integration is especially valuable in diagnosing complex diseases, including certain cancers [5] and cardiac conditions [33], where accurate image interpretation is essential for prognosis and treatment. Methodological fusion not only improves analytical accuracy but also strengthens clinicians' confidence in the results, ultimately contributing to better patient care [19].

Hybrid frameworks that combine traditional statistical approaches with fuzzy logic can significantly enhance the personalization of medical treatments [12]. By addressing inherent uncertainty [4] and imprecision in individual clinical data [26], these approaches enable more precise adjustments to treatment plans, tailored to each patient's characteristics. This approach is particularly valuable in chronic diseases, where continuous follow–up and adaptive therapeutic responses are essential. In precision medicine–grounded in genomic, physiological, and behavioral data–models that incorporate multiple layers of uncertainty are indispensable. Clinical knowledge in such contexts cannot rely on rigid categories but requires nuanced interpretation, making these frameworks key tools for improving medical care effectiveness [30].

The integration of traditional statistical approaches with fuzzy logic principles represents a significant advance in reducing uncertainty and imprecision that affect the interpretation of health data. This methodological combination generates hybrid frameworks that enhance the quality of analysis and strengthen healthcare professionals' ability to make more informed clinical decisions tailored to each patient. The ability to personalize treatments based on a more accurate interpretation of available information improves patient safety and increases the efficiency of healthcare systems [28]. As digital technologies and large-scale data platforms continue to evolve, these combined approaches are likely to be adopted across various medical disciplines, transforming the analysis and utilization of clinical information. Along these lines, the present study seeks to explore the emerging frontiers of medical data analytics by integrating statistical methods, fuzzy logic, and uncertainty indicators to optimize clinical decision–making.

2   Method

2.1   Study Design

This study was designed to develop and validate a hybrid framework that integrates statistical methods and fuzzy logic techniques to reduce uncertainty and imprecision in medical datasets. In this study, we focus specifically on two major categories of uncertainty: epistemic and ontological. Epistemic uncertainty arises from limitations in knowledge or data, including measurement noise, missing values, or incomplete models. Ontological uncertainty, in contrast, refers to the intrinsic variability and complexity of medical phenomena, such as inter-patient differences or ambiguous clinical boundaries. While other taxonomies describe uncertainty in terms of vagueness, ambiguity, or incompleteness, these perspectives can be subsumed under the epistemic—ontological distinction adopted here, which is particularly suited to medical data analysis. The research process was structured into several phases: (1) theoretical framework development, (2) computational implementation, (3) simulation and data collection, and (4) empirical validation. Each phase was carried out sequentially, ensuring that the outcomes of each stage informed and enhanced the subsequent phases. The validation process was conducted using both synthetic datasets and real–world medical records to assess the framework's effectiveness across diverse contexts and conditions.

2.2   Hybrid Framework Components

The hybrid framework developed in this study combines advanced statistical techniques with fuzzy logic to address uncertainty and imprecision in medical data. The main components of the framework include:

- **Geometric mean filter**: This operator is applied to reduce noise in medical datasets while preserving the integrity of meaningful patterns [15]. Beyond its conventional use in image and signal smoothing, the geometric mean filter has shown superior performance in handling both epistemic and ontological uncertainty in clinical data [26]. In medical imaging contexts—particularly CT and ECG signals—it minimizes the impact of noise and outliers without distorting anatomical or physiological structures, making it valuable for preserving diagnostic relevance. The filter was identified as the most stable among computational methods, maintaining consistent results even under simulated imperfections such as Poisson noise and artifacts. Although the geometric mean filter is grounded in a statistical measure of central tendency, in this study it is considered a computational smoothing tool.
- **Fuzzy connectors**: These are mathematical operators used to combine fuzzy membership values, extending classical concepts of intersection (AND) and union (OR) into a fuzzy framework [2]. The framework employs T–norms and T–conorms, which model fuzzy logical relationships under uncertainty [14]. T–norms generalize the logical intersection, supporting scenarios where data must meet all conditions (conservative approach), while T–conorms generalize the logical union, favoring inclusive interpretations. Common examples include the minimum and product (T–norms) or maximum and probabilistic sum (T–conorms). These connectors enable the construction of flexible mathematical relationships that adapt to epistemic and ontological uncertainty in the data, supporting more nuanced modeling of dependencies between processed and original datasets [26].
- **Generalized compensation operators**: These operators dynamically integrate processed and original data by adjusting the contribution of each, based on the hidden dependencies and shared information between them [10]. Unlike classical operators, generalized compensation functions balance between the strictness of T–norms and the inclusiveness of T–conorms through a compensation parameter that controls this trade–off [23]. Formally, these operators are expressed as a geometric combination of T-norm and T-conorm values, or as weighted aggregations using functions such as the geometric mean and identity functions [26]. This adaptive integration helps reduce uncertainty by modeling nonlinear relationships and compensating for distortions or inconsistencies in the data. Such operators are especially useful in contexts where preserving both global structure and individual variability is critical, as in the processing of noisy or heterogeneous medical datasets.
- **Statistical–fuzzy integration and adaptive compensation**: This stage articulates the output of the geometric mean filtering and fuzzy connectors through generalized compensation operators, generating a transformed dataset where statistical smoothing and fuzzy logic relationships are balanced adaptively. This cross–layer integration preserves meaningful dependencies while mitigating uncertainty and distortion, creating a robust foundation for subsequent uncertainty quantification and clinical analysis.
- **Mutual information metrics and clairvoyance value**: These analytical tools provide a quantitative assessment of uncertainty and imprecision in

complex datasets. Mutual information (MI) quantifies the amount of shared information between variables, capturing both linear and nonlinear dependencies [7]. Clairvoyance value (VoC), inspired by Howard's value of information theory [16], estimates the potential reduction in uncertainty if perfect information were available. In this framework, VoC is computed as the difference between the mutual information of the raw data (MI_{raw}) and that of the data processed through cross–layer integration previously described ($MI_{\text{processed}}$), as shown in the following expression:

$$\text{VoC} = MI_{\text{raw}} - MI_{\text{processed}} \tag{1}$$

This formulation quantifies the loss of information caused by noise and distortions, facilitating the classification of uncertainty into different levels and supporting the identification of hidden dependencies within the data. This approach aligns with broader methodological perspectives that emphasize the integration of uncertainty estimation and decision-making processes when analyzing structured or spatially distributed data [21], which in this study are extended to the context of medical datasets. In this sense, the formulation is the stage where formal statistical methods are applied to evaluate uncertainty reduction and pattern preservation.

2.3 Data Sources and Simulation

Two types of data were used to validate the framework: synthetic and real–world datasets. Synthetic data were generated through simulations designed to replicate essential characteristics of medical data, incorporating elements such as noise, artifacts, and the variability inherent in clinical records [34]. For electrocardiographic (ECG) signals, synthetic recordings were created using specific functions that simulate PQRST waves, with a sampling rate 360 Hz and a duration of 30 min per channel. These noise–free simulations allowed the framework's performance to be evaluated under ideal and controlled conditions, focusing the analysis exclusively on the signal structure.

The real–world data were obtained from well–established medical databases, such as the MIT–BIH Arrhythmia Database, which contains ECG recordings with detailed clinical annotations and various sources of uncertainty, including noise, artifacts, and variability in the interpretation of cardiac events [24]. This database enabled the assessment of the framework's applicability in practical scenarios using real data. Additionally, six synthetic datasets of computed tomography (CT) images were incorporated, designed to simulate different types and levels of imperfections–such as Poisson noise, streak artifacts, stair–step artifacts, and hybrid combinations–allowing for an analysis of the framework's response to varying degrees of distortion [34].

These data sources, both synthetic and real, enabled not only the validation of the framework in terms of its ability to detect nonlinear dependencies and quantify uncertainty, but also the establishment of reference ranges for interpreting the clairvoyance value (VoC) as an indicator of data quality and reliability.

2.4   Analytical Procedure and Validation

The analytical procedure was carried out in several stages:

- **Data preprocessing**: Both synthetic and real datasets were normalized to the $[0, 1]$ range to ensure consistency in processing. Geometric mean filtering was applied to reduce initial noise.
- **Application of fuzzy connectors**: T–norms and T–conorms were used to combine the processed and original data, modeling the fuzzy relationships between them.
- **Implementation of compensation operators**: Generalized compensation operators were applied to dynamically adjust the relative contributions of the original and processed data. This integration generated a new transformed dataset characterized by a significant reduction in uncertainty.
- **Evaluation of results**: The transformed data were evaluated using mutual information metrics and the clairvoyance value, quantifying the reduction in uncertainty and imprecision achieved by the framework. The results were compared with the original data to validate the framework's effectiveness.

Empirical validation was conducted through experiments on both real and synthetic datasets, evaluating the framework's ability to improve data quality by reducing noise and imprecision. The results demonstrated a significant decrease in uncertainty in most cases, confirming the robustness and applicability of the framework in medical contexts.

3   Results

Empirical validation of the proposed hybrid framework was carried out using both synthetic ECG recordings and real signals from the MIT-BIH Arrhythmia Database. The results are presented in Table 1, which shows the VoC metric before and after applying the generalized compensation operator.

The results show a significant reduction in uncertainty and imprecision in most recordings after applying the generalized compensation operator. For example, in record 100, channel 1, the VoC metric decreased from 1.099750 to 0.101779, indicating a substantial improvement in signal quality. This reduction indicates that noise and artifacts were removed, allowing clinically relevant ECG components to be more clearly delineated. In practical terms, a lower VoC value means a more reliable basis for detecting arrhythmias and reduces the risk of misinterpretation in clinical decision–making. The variability, including minimal reductions, reflects differences in initial noise levels: when raw signals are already of high quality, the framework produces smaller VoC changes while preserving fidelity and robustness.

Table 2 presents the results of the empirical assessment conducted on various synthetic CT datasets. Each dataset is identified by a code (DB1–DB6) and accompanied by a brief description of its characteristics, indicating whether it

Table 1. VoC results for ECG data with generalized compensation operator.

Record	Channel	VoC (Before)	VoC (After)
Synthetic	1	0.485598	0.475731
100	1	1.099750	0.101779
	2	1.146997	0.106152
101	1	0.900972	0.083383
	2	1.808339	1.564025
201	1	1.022111	0.884019
	2	1.216675	1.052297
205	1	0.993588	0.859351
	2	1.342678	1.161276

Table 2. VoC of CT data when applying generalized compensation operator.

Dataset	Description	VoC (Before)	VoC (After)
DB1: Ground Truth	Reference dataset free of imperfections.	0.093063	0.091172
DB2: Poisson	Dataset contaminated with Poisson noise.	2.147220	1.857122
DB3: Stair-Step	Dataset with simulated stair-step artifacts.	1.078029	0.106449
DB4: Streak	Dataset with simulated streak artifacts.	1.238907	0.114658
DB5: Artifacts	Dataset containing both stair-step and streak artifacts.	1.321857	0.122335
DB6: Hybrid	Hybrid dataset with Poisson noise and both artifacts.	2.194666	1.980512

is free of imperfections, contaminated with noise, or contains specific artifacts such as stair–step or streak patterns, or a combination of both.

The results demonstrate a significant reduction in VoC metric values after processing, indicating a decrease in uncertainty and imprecision. In particular, datasets containing specific artifacts, such as stair–step (DB3) and streak (DB4), exhibit notable reductions, highlighting the framework's ability to address specific variabilities in the data. In DB1, the reference dataset free of imperfections, the VoC change is minimal. This outcome is expected, since the data already exhibit optimal quality; the small variation simply confirms that the framework preserves the integrity of noise–free images without introducing distortions.

4 Discussion

The results demonstrate that the proposed hybrid framework is effective in reducing uncertainty and imprecision in medical data. For instance, in record 100, channel 1, the reduction of VoC from 1.099750 to 0.101779 reflects a marked improvement in the accuracy of the processed signals. These findings confirm that the combination of the geometric mean filter and generalized compensation operators preserves meaningful patterns while mitigating noise.

Cases with marginal reductions in VoC, such as high–quality ECG records or the DB1 ground–truth CT dataset, are expected when the raw data contain little or no noise. In these situations, the framework does not generate large improvements because there is little uncertainty to remove; instead, it confirms its adaptive behavior by preserving the fidelity of already clean signals and images. The practical implication is that the method enhances data when imperfections are present and safeguards information integrity when data are ideal, ensuring consistent performance across different quality contexts.

In recent years, several studies have explored the integration of fuzzy logic with machine learning models and statistical techniques to address the uncertainty inherent in medical data. A comprehensive review of fuzzy deep learning frameworks applied to uncertain medical datasets concludes that these approaches significantly enhance diagnostic accuracy in contexts characterized by ambiguity and incomplete information. Although this work highlights the effectiveness of fuzzy neural networks, it focuses primarily on predictive modeling, without delving into the structural preprocessing of data–as the present study does through the use of geometric mean filtering and generalized compensation operators [36].

Similarly, fuzzy morphological operators have proven effective for image enhancement and denoising, although they lack explicit uncertainty reduction and quantitative metrics like the VoC, a key contribution of this framework [3]. In the field of neuroimaging, fuzzy attention mechanisms integrated into deep neural networks have been proposed to enhance brain tumor classification. These approaches focus on detection and segmentation tasks using artificial intelligence, achieving notable improvements in diagnostic accuracy. However, their application is limited to a specific pathology and does not incorporate data preprocessing using classical operators, such as those employed in the framework, which could enhance the robustness of these models against noise and variability in the data [8].

In a complementary line of research, fuzzy logic has been integrated with classical segmentation techniques such as region growing and the watershed algorithm. Although this model demonstrates good results in tumor classification, it focuses specifically on brain structures and does not consider the construction of synthetic datasets or controlled evaluation in simulated scenarios–key aspects addressed in our study to assess the framework's generalization capacity [25].

Finally, a real–time volume filtering method for CT images based on fuzzy logic and accelerated by GPU has been developed to improve the diagnostic quality of images acquired with reduced radiation doses. While this approach shares the motivation of enhancing diagnostic reliability through fuzzy prepro-cessing, it does not incorporate dynamic compensation strategies or quantitative uncertainty assessment tools such as the VoC, which add an additional analytical dimension to the proposed framework [31].

The practical implications of these results are significant for the healthcare field. The ability to reduce uncertainty and imprecision in medical data can enhance the accuracy of diagnoses and the effectiveness of treatments. In the analysis of ECG signals, improved accuracy facilitates the early detection of arrhythmias and other cardiac abnormalities, thereby improving clinical out-comes for patients. In CT images, the reduction of artifacts and noise leads to more accurate and reliable diagnoses, which is essential for the detection and treatment of diseases such as cancer. The application of this hybrid framework in clinical practice can thus contribute to more precise and personalized medical care.

The results demonstrate the robustness of the proposed framework, sup-ported by approximately 5.85 million samples from eight clinical ECG recordings and one synthetic signal. While these well-known datasets provide a solid basis for evaluating uncertainty reduction, they represent a specific context. Broader validation across diverse populations, acquisition devices, and clinical condi-tions is needed to strengthen generalizability. Both substantial and marginal VoC reductions are consistent with the framework's adaptive behavior, which enhances noisy signals while preserving already reliable ones. Future work should also address processing speed, scalability for real-time use, and the inclusion of objective quality metrics, as well as integration with machine learning pipelines for automated diagnosis.

5   Conclusions

The study demonstrated that integrating statistical methods with fuzzy logic techniques is effective in reducing uncertainty and imprecision in medical data. The combination of the geometric mean filter and generalized compensation operators significantly improved data quality in both ECG signal recordings and CT datasets. The results showed a remarkable decrease in VoC metrics, validating the framework's ability to mitigate noise and preserve the integrity of meaningful patterns within the data.

The findings of this study have important implications for clinical practice and health research. The ability to reduce uncertainty and imprecision in med-ical data can enhance the accuracy of diagnoses and the effectiveness of treat-ments. In ECG signal analysis, improved accuracy facilitates the early detec-tion of arrhythmias and other cardiac abnormalities, leading to better clinical outcomes for patients. In CT image processing, the reduction of artifacts and noise contributes to more accurate and reliable diagnoses, which is critical for the detection and treatment of diseases such as cancer. The application of this hybrid framework, positioned at the emerging frontiers of medical data analysis, holds the potential to advance more precise and personalized medical care.

In order to maximize the impact and applicability of the proposed hybrid framework, and to contribute to the continuous improvement of accuracy and reliability in medical data analysis, several future research directions are suggested. First, the framework should be validated on more diverse medical datasets to confirm its generalizability and robustness across different contexts and clinical conditions. Second, it is recommended to investigate the optimization of the generalized compensation parameters using machine learning algorithms, aiming to further enhance the framework's capacity to address uncertainty and imprecision in complex scenarios. Furthermore, the scope of application could be expanded by exploring its applicability in other health domains, such as bioinformatics and big data analytics. Another promising avenue for future research is to integrate the proposed approach with deep learning [22] and extreme learning [18] techniques, enabling the analysis of unstructured data, including medical images and genomic sequences. Finally, it is recommended to develop a computational platform integrating the proposed techniques, enabling real–time processing of large–scale datasets in distributed or cloud environments and providing accessible tools for healthcare professionals and researchers.

References

1. Alizadehsani, R., et al.: Handling of uncertainty in medical data using machine learning and probability theory techniques: a review of 30 years (1991–2020). Ann. Oper. Res. 1–42 (2021). https://doi.org/10.1007/s10479-021-04006-2
2. Alsina, C., Trillas, E., Valverde, L.: On some logical connectives for fuzzy sets theory. J. Math. Anal. Appl. **93**(1), 15–26 (1983). https://doi.org/10.1016/0022-247X(83)90216-0
3. Anitha, J., Kalaiarasu, M.: MRI brain tumor segmentation with intuitionist possibilistic fuzzy clustering and morphological operations. Comput. Syst. Sci. Eng. **43**(1), 363–379 (2022). https://doi.org/10.32604/csse.2022.022402
4. Bao, Y., Velni, J.M., Shahbakhti, M.: Epistemic uncertainty quantification in state-space LPV model identification using Bayesian neural networks. IEEE Control Syst. Lett. **5**(2), 719–724 (2021). https://doi.org/10.1109/LCSYS.2020.3005429
5. Chacón, G., et al.: Computational assessment of stomach tumor volume from multi–slice computerized tomography images in presence of type 2 cancer. F1000Research **7**(1098) (2018). https://doi.org/10.12688/f1000research.14491.2
6. Chakraborty, S., Raut, R.D., Rofin, T., Chakraborty, S.: A comprehensive and systematic review of multi-criteria decision-making methods and applications in healthcare. Healthcare Anal. **4**, 100232 (2023). https://doi.org/10.1016/j.health.2023.100232
7. Cover, T.M., Thomas, J.A.: Elements of Information Theory, 2nd edn. Wiley-Interscience (2012)
8. de Campos Souza, P.V.: Fuzzy neural networks and neuro-fuzzy networks: a review the main techniques and applications used in the literature. Appl. Soft Comput. **92**, 106275 (2020). https://doi.org/10.1016/j.asoc.2020.106275
9. Doyle, E.E., et al.: Where does scientific uncertainty come from, and from whom? Mapping perspectives of natural hazards science advice. Int. J. Disaster Risk Reduct. **96**, 103948 (2023). https://doi.org/10.1016/j.ijdrr.2023.103948

10. Dubois, D., Prade, H.: Fuzzy sets and probability: misunderstandings, bridges and gaps. In: [Proceedings 1993] Second IEEE International Conference on Fuzzy Systems, vol. 2, pp. 1059–1068 (1993). https://doi.org/10.1109/FUZZY.1993.327367
11. Eachempati, P., Büchter, R.B., KS, K.K., Hanks, S., Martin, J., Nasser, M.: Developing an integrated multilevel model of uncertainty in health care: a qualitative systematic review and thematic synthesis. BMJ Global Health **7**, e008113 (2022). https://doi.org/10.1136/bmjgh-2021-008113
12. El-Ibrahimi, A., et al.: Fuzzy based system for coronary artery disease prediction using subtractive clustering and risk factors data. Intell.-Based Med. **11**, 100208 (2025). https://doi.org/10.1016/j.ibmed.2025.100208
13. Feder, S.L.: Data quality in electronic health records research: quality domains and assessment methods. West. J. Nurs. Res. **40**(5), 753–766 (2018). https://doi.org/10.1177/0193945916689084
14. Gaxiola, F., Melin, P., Valdez, F., Castillo, O., Castro, J.R.: Comparison of T–norms and S–norms for interval type–2 fuzzy numbers in weight adjustment for neural networks. Information **8**(3) (2017). https://doi.org/10.3390/info8030114
15. Hiary, H., Zaghloul, R., Al-Adwan, A., Al-Zoubi, M.B.: Image contrast enhancement using geometric mean filter. SIViP **11**(5), 833–840 (2016). https://doi.org/10.1007/s11760-016-1029-8
16. Howard, R.A.: Information value theory. IEEE Trans. Syst. Sci. Cybern. **2**(1), 22–26 (1966). https://doi.org/10.1109/TSSC.1966.300074
17. Huérfano, Y., Vera, M., Mar, A., Bravo, A.: Integrating a gradient–based difference operator with machine learning techniques in right heart segmentation. J. Phys: Conf. Ser. **1160**, 012003 (2019). https://doi.org/10.1088/1742-6596/1160/1/012003
18. Huérfano-Maldonado, Y., Mora, M., Vilches, K., Hernández-García, R., Gutiérrez, R., Vera, M.: A comprehensive review of extreme learning machine on medical imaging. Neurocomputing **556**, 126618 (2023). https://doi.org/10.1016/j.neucom.2023.126618
19. Kadry, S., Mahajan, S. (eds.): Data Science in the Medical Field. Academic Press (2025). https://doi.org/10.1016/C2023-0-01347-4
20. Kobrinskii, B.: Fuzzy and reflection in the construction of a medical expert system. J. Softw. Eng. Appl. **13**, 15–23 (2020). https://doi.org/10.4236/jsea.2020.132002
21. Lark, R., Chagumaira, C., Milne, A.: Decisions, uncertainty and spatial information. Spat. Stat. **50**, 100619 (2022). https://doi.org/10.1016/j.spasta.2022.100619
22. Litjens, G., et al.: A survey on deep learning in medical image analysis. Med. Image Anal. **42**, 60–88 (2017). https://doi.org/10.1016/j.media.2017.07.005
23. Mizumoto, M.: Pictorial representations of fuzzy connectives, part II: cases of compensatory operators and self-dual operators. Fuzzy Sets Syst. **32**(1), 45–79 (1989). https://doi.org/10.1016/0165-0114(89)90087-0
24. Moody, G.B., Mark, R.G.: The impact of the MIT-BIH arrhythmia database. IEEE Eng. Med. Biol. Mag. **20**(3), 45–50 (2001). https://doi.org/10.1109/51.932724
25. Narasimham, N., KeshavKumar, K.: Fuzzy logic–based system for brain tumour detection and classification. arXiv:abs/2401.14414 (2024). https://doi.org/10.48550/arXiv.2401.14414
26. Nieto, Z., Bravo, A.: Exploring computational methods in the statistical analysis of imprecise medical data: between epistemology and ontology. Salud, Ciencia y Tecnología **4**, 1341 (2024). https://doi.org/10.56294/saludcyt20241341
27. Sarker, I.H.: Data science and analytics: an overview from data-driven smart computing, decision-making and applications perspective. SN Comput. Sci. **2**(5), 1–22 (2021). https://doi.org/10.1007/s42979-021-00765-8

28. Sinha, R.: The role and impact of new technologies on healthcare systems. Discover Health Syst. **3**, 96 (2024). https://doi.org/10.1007/s44250-024-00163-w
29. Stoumpos, A.I., Kitsios, F., Talias, M.A.: Digital transformation in healthcare: technology acceptance and its applications. Int. J. Environ. Res. Public Health **20**(4) (2023). https://doi.org/10.3390/ijerph20043407
30. Sutton, R.T., Pincock, D., Baumgart, D.C., et al.: An overview of clinical decision support systems: benefits, risks, and strategies for success. NPJ Digit. Med. **3**, 17 (2020). https://doi.org/10.1038/s41746-020-0221-y
31. Tendero, C., Pérez, M., Arnal, J., Vidal, V., Blanco, E.: A GPU-accelerated fuzzy method for real-time CT volume filtering. PLoS ONE **20**(1), 1–27 (2025). https://doi.org/10.1371/journal.pone.0316354
32. Valbuena, O., Vera, M., Ramirez, G., Barrientos, R., Mojica, D.: Statistical techniques for digital pre-processing of computed tomography medical images: a current review. Displays **85**, 102835 (2024). https://doi.org/10.1016/j.displa.2024.102835
33. Vera, M., Bravo, A., Medina, R.: Improving ventricle detection in 3–D cardiac multislice computerized tomography images. In: Richard, P., Braz, J. (eds.) VISIGRAPP 2010. CCIS, vol. 229, pp. 170–183. Springer, Heidelberg (2011). https://doi.org/10.1007/978-3-642-25382-9_12
34. Vera, M., Bravo, A., Medina, R.: Description and use of three-dimensional numerical phantoms of cardiac computed tomography images. Data **7**(8) (2022). https://doi.org/10.3390/data7080115
35. Zadeh, L.: Fuzzy sets. Inf. Control **8**(3), 338–353 (1965). https://doi.org/10.1016/S0019-9958(65)90241-X
36. Zheng, Y., Xu, Z., Wu, T., Yi, Z.: A systematic survey of fuzzy deep learning for uncertain medical data. Artif. Intell. Rev. **57**, 230 (2024). https://doi.org/10.1007/s10462-024-10871-7

Application of Machine Learning Methods for Classification of Appeals in the Field of Housing and Communal Inspection

Denis Skvortsov(✉) ⓘ, Andrey Lemanov ⓘ, and Dmitriy Fedorov ⓘ

ITMO University, Kronverksky Prospekt 49, 197101 St. Petersburg, Russia
den.skvortsoff@gmail.com

Abstract. This paper explores the use of machine learning and natural language processing (NLP) techniques for the automatic classification of text-based appeals in the field of housing and communal inspection (HCI). The proposed approach includes a data processing module and a set of classification models. An experimental study was conducted comparing traditional classification algorithms such as Naive Bayes, Support Vector Machines, Decision Trees, K-Nearest Neighbors (KNN), FastText, and transformer-based models using BERT. The approach—combining preprocessed data with sampling based on class distribution in the training set and leveraging top predictions from multiple models—demonstrated effective performance based on evaluation metrics including accuracy, top-k accuracy, F1-score, and ROC-AUC. This method significantly reduces the workload on human operators and improves service quality in the HCI sector.

Keywords: Machine learning · Natural language processing · Text classification · Transformer models · BERT · Appeals classification · Housing and communal inspection

1 Introduction

1.1 Problem Statement

In the field of Housing and Communal Inspection (HCI), operators deal with a large volume of text-based appeals from citizens, which need to be classified by topic for further processing. Manual classification of such appeals is time-consuming, labor-intensive, and prone to human error, making the process inefficient [1]. This highlights the need to automate the process using machine learning and natural language processing (NLP) techniques.

HCI is one of the most socially important sectors, where the quality and speed of appeal processing directly affect citizen satisfaction and the effectiveness of public utility services. However, despite the potential for automation, many processes in this domain are still performed manually. This results in delays, classification errors, and, consequently, a decline in service quality.

L. Martínez et al. (Eds.): IDEAL 2025, LNCS 16239, pp. 461–470, 2026.
https://doi.org/10.1007/978-3-032-10489-2_39

A major challenge is the inconsistency, noisiness, and weak structure of the texts submitted by users [2]. Appeals may address a wide range of topics—from emergency situations and technical issues to subjects completely unrelated to HCI, such as threats or misdirected messages. These appeals are often written in informal language and may contain typos, abbreviations, and slang, which complicates automated processing.

The goal of this work is to apply a machine learning and NLP-based approach for the automatic classification of citizen appeals submitted to HCI, taking into account class imbalance and informal language. This approach aims to reduce the time required to process appeals, automate their categorization, and improve the operational efficiency of services responsible for housing and communal inspection.

1.2 Objectives

- Develop a classification model capable of effectively processing text-based appeals by considering their semantics, context, and the specific characteristics of the HCI domain.
- Compare and evaluate a range of classification approaches in order to identify the most suitable method for handling appeals in the HCI context.
- Assess model performance using appropriate evaluation metrics, with particular attention to the challenges of multi-class classification, informal and unstructured text, and imbalanced data.

Machine learning, natural language processing (NLP), and data preprocessing methods must be incorporated into the solution, as modern NLP technologies—particularly transformers—are capable of analyzing text while accounting for context, semantics, and stylistic nuances [3]. Therefore, the research problem lies in developing and evaluating a comprehensive methodology that combines existing ML, NLP, and data preprocessing techniques to classify citizen appeals in the HCI domain, while addressing challenges such as class imbalance, noisy data, and informal language.

2 Research

2.1 Literature Review

Modern text processing algorithms, initially designed for machine translation, classification, and semantic analysis [4, 5], require domain adaptation for classifying appeals in the housing and communal infrastructure (HCI) context.

Classical approaches include decision trees [5], k-nearest neighbors (KNN) with optimization enhancements [6], Naive Bayes [7, 8]—effective for smaller datasets— and Support Vector Machines (SVM) [8], which remain widely used in classification tasks. A comprehensive overview of these methods, vectorization techniques (Word2Vec, FastText), and evaluation metrics (Accuracy, F1, AUC) is given in [9].

Transformer-based deep learning models, especially BERT, dominate in modern NLP. RuBERT-Base [10], adapted for Russian, performs well in classification and entity recognition. RuRoBERTa extends this with fine-tuning on Russian corpora. FastText [11] offers fast classification and good performance on morphologically rich texts through subword tokenization.

Real-world applications include GIS HCI [12], "Dobrodel" [13], and Just AI [14], though most still rely on rule-based routing or generic classification schemes. Notably, [15] used BERT for hotline complaint classification, with preprocessing significantly improving results.

While these systems automate parts of appeal handling, they are not optimized for HCI-specific inputs such as informal, noisy, or domain-specific data, which limits their effectiveness. Thus, adapting modern models to the HCI domain remains essential for robust, accurate classification.

2.2 Research Materials

An anonymized dataset obtained from the Housing and Communal Inspection (HCI), comprising 4,772 citizen appeals, was analyzed to extract key structural and statistical characteristics. The dataset includes 50 thematic categories covering a diverse range of issues. On average, each appeal contains approximately 1,000 characters, which significantly complicates manual processing. In addition, the presence of informal language, slang, and typographical errors, combined with class imbalance and category overlap, adds to the complexity of automatic classification.

To construct a representative and sufficiently balanced dataset for training and evaluation, 35 relevant categories were selected based on their semantic clarity and data volume. While all selected categories were used in the experiments, Table 1 provides a brief overview of top 10 representative topics, chosen for illustrative purposes due to space constraints and their prevalence in the dataset.

Experts from the Housing and Communal Inspection identified the most important aspects of the data: the text of appeals and their classification topics. Data selection was further refined to focus on the most relevant categories for the study and to reduce noise during analysis. The dataset was standardized by grouping appeals into topics and sorting them by their relative share in the database. This filtering removed categories that were underrepresented and unsuitable for classification, allowing the study to concentrate on the most significant types of cases based on their dominance in percentage terms, as shown in Table 1.

The data were provided in a format easily convertible for supervised learning without additional labeling, but required thorough cleaning and structuring to reduce noise. This step is especially critical for practical applications and analysis without relying on advanced NLP or deep learning techniques.

These data can support more advanced analyses in the future, such as building predictive models, generating responses to citizen appeals, and identifying trends in the housing and communal sector. For example, it will be possible to analyze appeal dynamics by category (Table 1), detect seasonal fluctuations, or pinpoint the most problematic areas. This will not only improve service quality but also optimize management processes in the sector.

In summary, this study produced a representative data sample (Table 1) suitable for further analysis and decision-making in housing and communal services.

Table 1. Sample from the HCI Database

Category No.	Number of Records	Topic Name	Share of Total Records (%)
0	833	Disagreement with the response from the management company / HOA / HBC (common property maintenance)	17,46%
1	471	Maintenance of stairwells and courtyard areas	9,87%
2	358	Management companies and housing maintenance/repair contracts	7,50%
3	309	Disagreement with the response from the management company / HOA / HBC	6,48%
4	285	Consumer rights regarding incorrect calculation of utility service costs	5,97%
5	284	Interruptions in heat supply	5,95%
6	276	Information disclosure	5,78%
7	269	Holding general meetings of homeowners	5,64%
8	241	Engineering infrastructure	5,05%
9	169	Disagreement with the response from the management company / HOA / HBC (information disclosure)	3,54%
…	…	…	…
49	1	Complaints about failure to enforce court decisions	0,02%

3 Practical Use

3.1 Model Selection and Configuration

To evaluate the effectiveness of each machine learning method described earlier, classification accuracy was measured for different types of predictions—Top-1, Top-3, and Top-5 [16]. Top-k accuracy measures the fraction of cases where the true category appears among the model's k highest-probability predictions ($k = 1, 3, 5$). The weighted F1-score accounts for class imbalance by weighting per-class F1-scores according to the number of samples in each class. Additionally, macro-averaged ROC-AUC was computed to assess overall ranking performance across all classes.

The results demonstrated that fine-tuning the RuRoBERTa model on the HCI domain using a transfer learning approach [17] achieved the highest performance in the Top-1 category (see Table 2). This can be attributed to RuRoBERTa's optimization for the Russian language and its ability to capture complex contextual relationships within the text.

The RuBERT-Base model also showed strong results, particularly in Top-3 and Top-5 scenarios, benefiting from its bidirectional context understanding. Meanwhile, FastText performed well on tasks involving shorter contexts and was particularly effective when morphological features were important. However, it lagged behind transformer-based models in handling long and semantically rich texts (Table 2).

These findings highlight that model selection should be guided by the specific task, data size, and requirements for accuracy and interpretability. For more complex natural language processing tasks involving Russian-language texts such as nuanced citizen appeals in the HCI domain, transformer-based models like RuRoBERTa and RuBERT-Base are preferable. Conversely, FastText may be more appropriate for lightweight tasks where morphological features are key and computational efficiency is a priority.

All fine-tuning and model training were performed in a cloud-based environment using Google Colaboratory. Deep learning models were trained on an NVIDIA T4 GPU (12 GB) (hereafter referred to as GPU), while classical machine learning algorithms were executed on a standard 8-core CPU (hereafter referred to as CPU), as detailed in Table 2.

The solution builds upon the implementation provided in [8], with development carried out using the following libraries:

- PyTorch and Transformers for neural network modeling
- Scikit-learn (sklearn) for classical ML algorithms and evaluation
- Pandas and numpy for data preprocessing and manipulation

Table 2. Comparison of the models' results

Model Name	Accuracy (Top-1)	Accuracy (Top-3)	Accuracy (Top-5)	F1	ROC-AUC	Training Time	Used Memory
RuBert-Base	67,33%	85,76%	91,52%	64,02%	84,45%	16 min 7 epochs	12 GB CPU
RuRoberta	70,99%	89,63%	94,45%	69,52%	85,92%	28 min 4 epochs	8.5 GB CPU
FastText	16,02%	26,70%	34,03%	10,86%	61,30%	2 min 50 epochs	<4 GB CPU
Naive Bayes	39,69%	66,70%	72,88%	36,34%	65,22%	2 min	<4 GB CPU
Decision Tree	46,07%	53,09%	53,61%	42,35%	66,85%	11 min	<4 GB CPU
Support Vector	64,61%	81,99%	88,38%	63,15%	83,87%	16 min	<4 GB CPU
KNN	45,24%	71,83%	80,21%	44,15%	-	2 min	<4 GB CPU

Additionally, a comparative evaluation was conducted between the two best-performing models, RuRoBERTa and RuBERT-Base, using the Top-5 Accuracy metric across individual categories (see Fig. 1). The results clearly demonstrate the performance gap between the models: RuBERT-Base fails to produce correct predictions (Top-5 Accuracy = 0) for 10 categories in which RuRoBERTa achieves non-zero accuracy. This highlights the superior generalization capabilities of RuRoBERTa in handling diverse and imbalanced categories within the HCI domain.

Furthermore, RuRoBERTa is able to distinguish undefined categories (a group formed from the unused 18 categories) in a noticeable manner, which RuBERT-Base struggles to capture. Overall, this demonstrates the effectiveness of RuRoBERTa as a tool for general classification, since it provides a broader and more reliable spectrum of results across multiple categories, whereas RuBERT-Base surpasses it only in a limited number of cases.

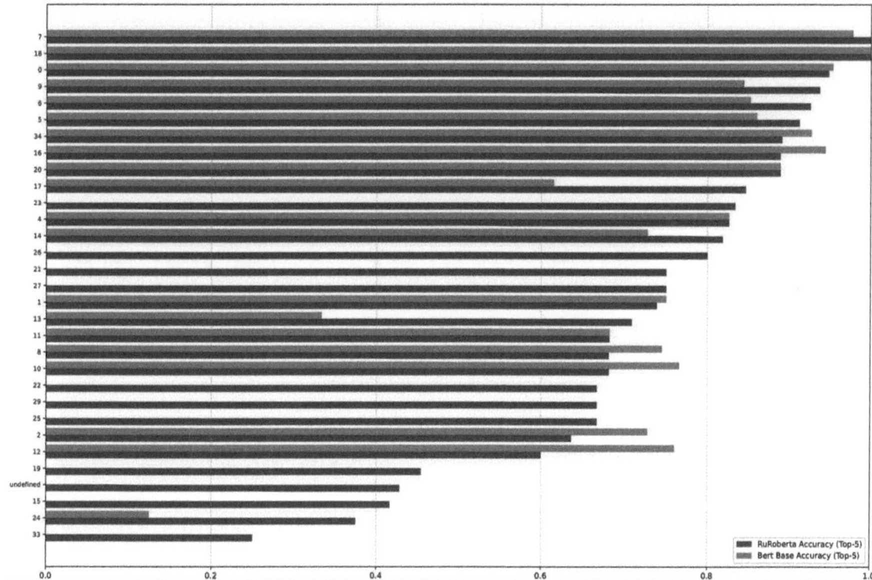

Fig. 1. Comparison of Transformer Models' Results by Category.

3.2 Research Results

Fine-tuning the RuRoBERTa model on the HCI domain using a transfer learning app-roach [17] achieved the highest performance in the Top-1 category (see Table 2). This can be attributed to RuRoBERTa's optimization for the Russian language and its ability to capture complex contextual relationships within the text.

The results of by-topic models' comparison (see Fig. 1) reveal clearer distinctions between the two transformer models. RuRoBERTa successfully predicts all proposed categories, whereas RuBERT-Base produces no results in 10 out of 30 cases (33%) where RuRoBERTa gives a positive prediction. And Top-5 Accuracy is zero for both models in only three of the tested topics. Examples of RuRoBERTa's performance on individual queries are presented in Table 3, where the intended categories appear consistently as the most probable ones (Probability column).

Traditional machine learning methods such as Naive Bayes, decision trees, and sup-port vector machines (SVM) showed substantially lower accuracy. Among them, SVM achieved the best Top-1 accuracy (64.61%) but still lagged far behind transformer mod-els. Naive Bayes and decision trees also underperformed, while FastText demonstrated the lowest effectiveness with only 16.02% Top-1 accuracy due to its simplified text rep-resentation. However, its speed training on CPU makes it suitable for tasks with highly formalized data.

In summary, BERT-based models, particularly RuRoBERTa, proved to be the most appropriate solution for processing appeals in the HCI domain. They offer high accuracy and robustness in handling noisy and complex textual data when sufficient computing resources are available. Classical models, while more lightweight, cannot match transformer performance. Among them, SVM offers the best balance between interpretability, training time, and accuracy.

Table 3. Example of the appeal and its classification

Category Number	Category Name	Sample Query	Predicted Categories' Numbers (RuRoberta)	Confidence (RuRoberta Probability) respectively
0	Disagreement with the response from the management company …	#### в #### я направил в управляющую организацию #### запрос по теме Проблемы в подвале. ##### ### заявка перешла в статус "Исполнена", но мне поступил неудовлетворительный ответ. (Напишите, почему вас не устраивает ответ - например, ответ направлен с нарушением сроков, не проведены мероприятия по устранению проблемы или не устраивает качество проведенных работ). В связи с этим прошу принять меры, направленные на устранение нарушений моих прав и решения обозначенной мною проблемы! добрый день! сегодня произошло четверное затопление подвода с Кладовками жильцов! Прошу срочно принять меры ответственности к УК, пока собственники массово не стали обращаться в администрацию Губернатора.	0; 3; 14	41.37%; 29.52%; 8.66%

3.3 Application Diagram

The sequence diagram (see Fig. 2) shows the full cycle of processing citizen appeals in the housing and communal sector using an automated classification system based on the RuRoBERTa transformer. A complaint form is filled out on the site, sent to the server for pre-processing and model prediction, and the 5 most probable topics are returned to the user for visualization. For context, a typical appeal contains ~2000 characters, which a human reads in about one minute. As a rough estimate of the system's capabilities, RuRoBERTa can process 15 such texts in 3–5 s, yielding an efficiency gain of approximately 300 times compared to manual review.

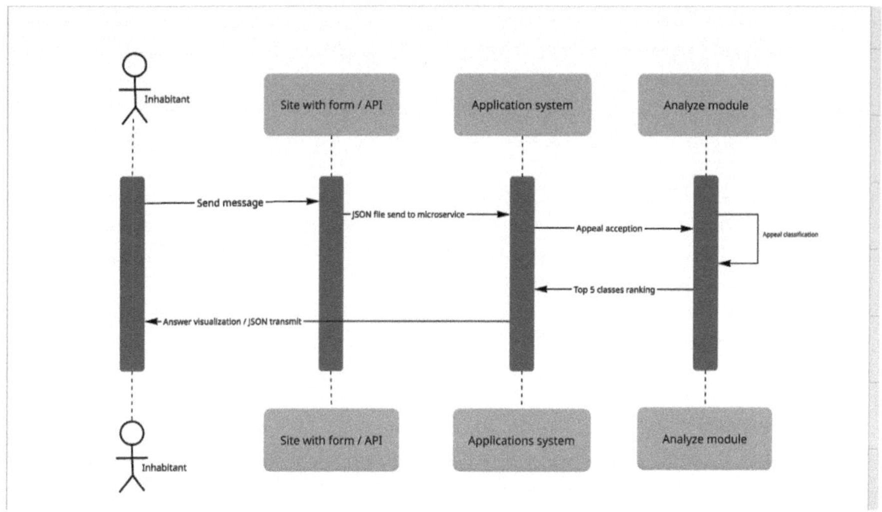

Fig. 2. Sequence diagram of receiving, analysing and classifying of appeals.

4 Conclusion

This study evaluated a range of algorithms and optimization methods for classifying citizen appeals and confirmed the clear advantage of transformer-based models. RuRoBERTa (Sberbank) achieved the highest accuracy and was selected for practical implementation, while classical methods, though faster and more lightweight, remained less effective.

The proposed approach significantly reduces operator workload in housing services and improves classification accuracy. Transformer models combined with NLP techniques proved effective, even under limited resource conditions.

The scientific contribution lies in adapting transformer models (RuBERT-Base, RuRoBERTa) to the Russian-language HCI domain—addressing challenges such as informal style, class imbalance, limited labels, and domain-specific nuances.

Future work includes extending the model to all 50 categories and integrating it into public digital platforms. The system has been tested on real-world data and is open-source: https://github.com/LISA-ITMO/Communal-Services-ML.

References

1. Chang, W.-C., Yu, H.-F., Zhong, K., Yang, Y., Dhillon, I.: Taming pretrained transformers for extreme multi-label text classification. arXiv:1905.02331 (2019). https://doi.org/10.48550/arXiv.1905.02331. Accessed 10 Mar 2025
2. Cheng, Z., et al.: A debiased nearest neighbors framework for multi-label text classification. arXiv:2408.03202 (2024). https://doi.org/10.48550/arXiv.2408.03202. Accessed 10 Mar 2025
3. Gochhait, S.: Comparative analysis of machine and deep learning techniques for text classification with emphasis on data preprocessing. Symbiosis Institute of Digital and Telecom Management (2023). Accessed 10 Mar 2025

4. Lynn, T., Aji, A.F., Wong, D.F., Wang, L.: A paradigm shift: the future of machine translation lies with large language models. arXiv:2305.01181 (2023). https://doi.org/10.48550/arXiv.2305.01181. Accessed 10 Mar 2025

5. Jijo, B.T., Abdulazeez, A.M.: Classification based on decision tree algorithm for machine learning. Technical College of Informatics, Duhok Polytechnic University, Iraq (2022). Accessed 10 Mar 2025

6. Li, B., Yu, S., Lu, Q.: An improved k-nearest neighbor algorithm for text categorization. Department of Computer Science and Technology (2003). Accessed 10 Mar 2025

7. Raschka, S.: Naive Bayes and text classification. arXiv:1410.5329 (2014). https://doi.org/10.48550/arXiv.1410.5329. Accessed 10 Mar 2025

8. Batta, V.M.B.: Text classification in natural language. Int. J. Adv. Res. Sci. Comput. Technol. https://doi.org/10.48175/IJARSCT-17645. Accessed 01 Apr 2025

9. Kowsari, K., Jafari Meimandi, K., Heidarysafa, M., Mendu, S., Barnes, L.E., Brown, D.E.: Text classification algorithms: a survey. arXiv:1904.08067 (2019). https://doi.org/10.48550/arXiv.1904.08067. Accessed 10 Mar 2025

10. Zmitrovich, D., et al.: A family of pretrained transformer language models for Russian. arXiv:2309.10931 (2023). https://doi.org/10.48550/arXiv.2309.10931. Accessed 10 Mar 2025

11. Bojanowski, P., Grave, E., Joulin, A., Mikolov, T.: Enriching word vectors with subword information. arXiv:1607.04606 (2016). https://doi.org/10.48550/arXiv.1607.04606. Accessed 10 Mar 2025

12. GIS HCS. https://dom.gosuslugi.ru. Accessed 21 May 2025

13. Government of the Moscow Region: "Dobrodel" portal – citizen request processing system with AI elements. https://dobrodel.mosreg.ru. Accessed 21 May 2025

14. Just AI: Application of generative AI in corporate environments. https://just-ai.com. Accessed 21 May 2025

15. Bunova, E., Serova, V.: Methodology for classifying appeals to the presidential hotline. Bull. South Ural State Univ. Comput. Technol. Autom. Radioelectron. **22**, 29–40 (2022). https://doi.org/10.14529/ctcr220203. Accessed 21 May 2025

16. Messalas, A., Makris, C., Kanellopoulos, Y.: Model-agnostic interpretability with Shapley values. In: 10th International Conference on Information, Intelligence, Systems and Applications (IISA 2019). IEEE (2019). https://doi.org/10.1109/IISA.2019.8900669. Accessed 10 Mar 2025

17. Do, C.B., Ng, A.Y.: Transfer learning for text classification. https://ai.stanford.edu/~chuongdo/papers/transfer.pdf. Accessed 01 Apr 2025

Towards Bitcoin Trend Prediction: A Machine Learning Approach Using Blockchain-Derived Data

Arthur G. Bubolz[1]([✉]), Marcos C. Freitas[1], Giancarlo Lucca[2] , Rafael A. Berri[1] , Eduardo N. Borges[1] , and Bruno L. Dalmazo[1]

[1] Federal University of Rio Grande, Rio Grande, Brazil
{arthurgomesbubolz,marcos.freitas,rafaelberri,eduardoborges,
dalmazo}@furg.br
[2] Catholic University of Pelotas (UCPel), Pelotas, Brazil
giancarlo.lucca@ucpel.edu.br

Abstract. As cryptocurrencies redefine global finance, accurate trend forecasting becomes essential for data-driven decision-making. This work proposes an approach to forecasting trends in the Bitcoin market using on-chain data and supervised machine learning. Instead of predicting exact price values, the focus is on identifying market direction, which may benefit not only investors, but also data scientists and risk analysts involved in financial forecasting. Data preprocessing included labeling with moving averages, normalization, and dimensionality reduction using t-SNE. Models were evaluated using TimeSeriesSplit, with precision as the primary metric. The results showed consistent performance across the classifiers, with average precision ranging from 65% to 70%, highlighting the potential of trend-based forecasting in highly volatile markets.

Keywords: Bitcoin · Blockchain · On-chain Data · Machine Learning · Trend Prediction

1 Introduction

Bitcoin has established itself as one of the most relevant and disruptive assets in the global financial market, driven by its decentralization proposal and growing adoption in different economic contexts [1]. Its volatility and atypical behavior compared to traditional assets make cryptocurrency an object of interest for investors, analysts, and researchers, especially with regard to predicting future market movements.

Most studies focused on Bitcoin prediction focus on estimating exact values [9], which, despite being technically challenging, is not always the most useful approach from a practical point of view. Decision-making in volatile markets benefits more from models capable of identifying trends than absolute price prediction. However, there is still a gap in the literature on trend-based approaches, especially those that use on-chain data as a main source [13], dealing with the problem with a perspective of a classification task, and not a regression one, considering that there is a broad base of studies demonstrating the effectiveness of machine learning methods applied to classification in this context [3].

Given this scenario, this work proposes a predictive model based on supervised machine learning to classify the future direction of the Bitcoin market. Instead of focusing on predicting value, the objective is to indicate whether there is an upward or downward trend, using metrics extracted directly from the blockchain as input. The model is complemented with techniques such as normalization by temporal windows and reduction of dimensionality with t-SNE.

The main contributions include: (i) the use of on-chain, public, and transparent data, as a basis for the predictive model; (ii) applying a labeling process based on moving averages to capture trends; (iii) the choice of the precision metric as an evaluation criterion, as it penalizes false positives, which imply greater financial risk; and (iv) the use of temporal cross-validation (TimeSeriesSplit), which respects the sequential structure of the data.

This study is organized as follows: Sect. 2 presents the theoretical foundations of Bitcoin, blockchain, on-chain data, and machine learning algorithms. Section 3 describes the adopted methodological proposal, including the construction of the data set and the feature engineering procedures. Section 4 details the experiments carried out and analyzes the results obtained. Finally, Sect. 5 presents conclusions and directions for future work.

2   Theoretical Foundation

This section presents the fundamental concepts that underlie the development of the study, enabling the understanding of essential concepts to contextualize the application of the methods and techniques used, as well as for the interpretation of the results obtained.

2.1   Bitcoin and Financial Decentralization

Bitcoin represents a major milestone in financial decentralization. Introduced in 2008 by the anonymous Satoshi Nakamoto, it proposed a digital currency operating without intermediaries [10]. Emerging during the global financial crisis, Bitcoin challenged the fragility of centralized systems and encouraged the pursuit of more secure and autonomous alternatives.

Currently, Bitcoin is widely adopted and functions as both a decentralized digital asset and a speculative instrument. Its high volatility drives active trading on exchanges and generates abundant financial data, such as opening/closing prices, highs and lows, trading volume, and market capitalization, valuable inputs for market behavior analysis.

The public and auditable nature of the blockchain reinforces the decentralization of Bitcoin, allowing participants to access data, validate blocks, and execute transactions independently. In doing so, Bitcoin introduced new standards for transparency and laid the groundwork for alternative forms of economic interaction and data-driven market analysis.

2.2 Blockchain

Blockchain is a distributed ledger in which transactions are recorded in a secure, transparent, and immutable way. Its structure consists of linked data blocks, each containing transaction records and a unique hash that connects to the previous block, ensuring data integrity [5].

A core feature of blockchain is decentralization. Instead of relying on centralized entities, the network is maintained by multiple distributed nodes that validate transactions through consensus mechanisms. Bitcoin uses Proof of Work (PoW), in which nodes solve complex mathematical problems to validate new blocks, making manipulation costly and impractical.

Although first popularized by cryptocurrencies, blockchain technology has broader applications, including smart contracts, logistics, and identity verification. Its combination of security, decentralization, and auditability makes it a valuable foundation for building trust in digital systems and a rich source of on-chain data for market analysis.

2.3 On-Chain Data

On-chain metrics are indicators extracted directly from the blockchain, offering insight into the network's internal activity and usage patterns. These metrics enhance crypto market analysis by providing public, transparent data [14].

Among them, Weight Mean reflects block space usage efficiency, while Difficulty Mean tracks mining complexity, adjusting to maintain block production rates. Reward Mean captures average miner earnings per block, including fees and newly issued coins. Total Transactions, Inputs, and Outputs help assess activity and capital flow across the network.

Witness Data (via SegWit) improves transaction efficiency and security, and Total Fees indicates network demand and congestion, rising when space is limited and transaction volume is high.

Together, these metrics help identify buying/selling pressure, adoption cycles, and liquidity shifts, providing valuable inputs for data-driven forecasting models.

2.4 Machine Learning Models

Machine learning enables the construction of models that learn from historical data by identifying patterns between input variables and corresponding outputs. In supervised learning, algorithms are trained on labeled data to make predictions or classifications on new, unseen data.

This study focuses on supervised classification algorithms, chosen for their robustness and applicability to multivariate problems. Among linear models, Logistic Regression [16] is widely used for binary and multiclass tasks, especially when data shows linear separability. K-Nearest Neighbors (KNN) [6] bases classification on data proximity, making it intuitive and effective when local structure matters.

Support Vector Classifier (SVC) [15] identifies optimal hyperplanes between classes, even in high-dimensional spaces. Naive Bayes [2], though based on variable independence assumptions, offers computational efficiency and solid performance in various contexts.

Ensemble models like Random Forest and Gradient Boosting [4] aggregate weak learners, the former through independent tree voting, the latter by sequentially correcting previous errors. Advanced variants like XGBoost and LightGBM offer high accuracy and scalability on complex datasets.

The Multilayer Perceptron (MLP) [11], a type of neural network inspired by biological neurons, excels at capturing non-linear relationships, particularly in data with limited linear correlation.

Combining these diverse algorithms enables a broad evaluation of model behavior and predictive capacity, allowing comparisons in terms of performance, generalization, and suitability for financial time series analysis.

2.5 Dimensionality Reduction with t-SNE

Dimensionality reduction is a key technique in data science for representing complex, high-dimensional data in fewer dimensions while preserving important relationships.

Among these methods, t-Distributed Stochastic Neighbor Embedding (t-SNE) is notable for projecting high-dimensional data into two or three dimensions, facilitating human visualization and interpretation. Proposed by Van der Maaten and Hinton [8], t-SNE excels at preserving local structures, maintaining proximity among similar points while separating dissimilar ones.

It converts distances into neighborhood probabilities in both original and reduced spaces and minimizes their divergence via a cost function based on Kullback-Leibler divergence [7]. This ensures that data clusters or separations in the original space are preserved after reduction.

Though often used for exploratory visualization, t-SNE also reveals hidden groupings and behavioral patterns, aiding understanding before applying predictive models.

2.6 Cross-Validation and TimeSeriesSplit

Cross-validation estimates the generalization ability of machine learning models, preventing overfitting by evaluating performance on unseen data [12].

However, for sequential data like time series, random splits can disrupt temporal order and bias results. To address this, TimeSeriesSplit respects chronology by creating progressive train-test splits where training data always precedes test data.

This method simulates real-world forecasting, ensuring the model trains only on past data and tests on future data, avoiding "peeking" ahead.

TimeSeriesSplit is particularly relevant for financial applications, such as predicting crypto market trends, where past behavior informs future outcomes.

Using this approach provides a realistic and reliable evaluation of model predictive power.

2.7 Confusion Matrix and Precision

The confusion matrix is a fundamental tool for evaluating supervised classification models, allowing visualization of both correct predictions and error types. It organizes outcomes into four categories: true positives (correct positive predictions), false positives (incorrect positive predictions), true negatives (correct negative predictions), and false negatives (missed positive cases). This detailed structure enables a comprehensive performance analysis (Fig. 1).

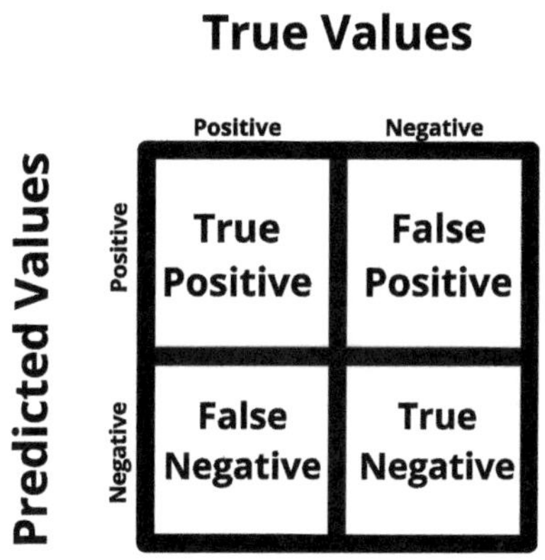

Fig. 1. Illustrative diagram of the confusion matrix.

From the confusion matrix, various metrics can be derived. Among them, precision is critical in contexts where false positives carry a significant cost.

Precision is defined as the ratio of true positives to all positive predictions made:

$$\text{Precision} = \frac{\text{TP}}{\text{TP} + \text{FP}}$$

This metric answers the question: 'Of all the times the model predicted the positive class, how many did it actually get right?' Thus, a model with high accuracy commits few false positives; that is, it tends to be more reliable when it indicates the occurrence of a relevant class.

We chose to use the precision metric due to its ability to penalize incorrect predictions of the positive class, in this case, false positives. In a scenario where the model classifies the data binary as rise (1) or fall (0), a false positive represents a prediction of appreciation when, in reality, a devaluation occurs. This type of error is particularly critical as it can lead to decisions that result in direct financial loss.

On the other hand, a false negative, when the model does not predict an increase that will occur, only represents the loss of a profit opportunity and not necessarily an actual loss. Given this, we seek to prioritize the accuracy of the model when indicating an increase, ensuring that such predictions are more reliable and aligned with capital preservation. The binary structure of classes and their operational foundation will be explored in more detail in Section XX.

3   Proposal

This section describes the steps followed in the development of the study, clearly and objectively explaining how the data were collected, which techniques and tools were used, and the criteria adopted for building, evaluating, and validating the proposed models.

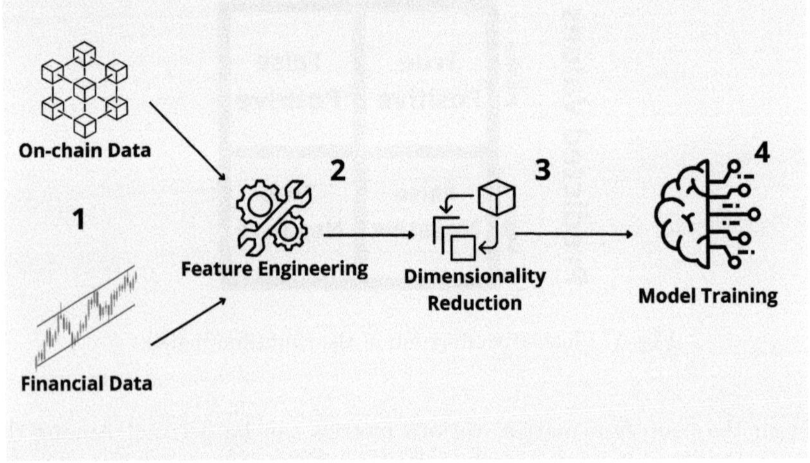

Fig. 2. Pipeline of Model Construction.

In the first step (1 in Fig. 2), the on-chain and financial data are extracted, which provides a detailed view of transactions and activities on the blockchain aligned with market behavior, in step two (2 in Fig. 2), the process of feature engineering is used to develop the indicators that represent the different moments of the market, which are used as input variables in the model, in addition to the target variable, step three (3) involves the dimensionality reduction of the dataset, which helps the model to train more efficiently.

With all this normalized data, in step four 4 in Fig. 2, the models are trained and evaluated through cross-validation, allowing the performance of each approach to be estimated with greater reliability, in addition to choosing the best model.

The main idea it's to propose a model that aims to automatically classify market moments, based on previously defined manual labeling, considering data analysis from an annual perspective. In other words, labeling was initially carried out manually, based on observation of data throughout the year, and the model now aims to automate this process, reproducing the identification of market moments without the need for human intervention.

3.1 Dataset Construction

The first step consisted of collecting historical Bitcoin price data, obtained through the CoinMarketCap platform, covering the period from July 14, 2010 to June 7, 2025. To make this process more efficient, the data was downloaded separately, year by year. Then, using the Google Colab environment together with the Pandas library, all files were grouped into a single dataset, organizing them in ascending chronological order and saving them in a new CSV file.

With the consolidated dataset, it was possible to create candlestick-type charts, which allow a more intuitive visualization of price changes over time. These charts were essential for identifying recurring market patterns and behaviors in certain periods.

Based on the visual analysis of the graphs, temporal intervals with similar or particularly striking behavior in the variation of Bitcoin prices were identified and delimited. To achieve this, a systematic approach was adopted: the graphs were plotted separately, year by year, from the consolidated dataset. To facilitate the visual interpretation of price trends, continuous lines were drawn over the most representative points of each interval, highlighting the predominant upward, downward, or stable movements.

This additional graphical feature allowed for a clearer visualization of price direction over time and made it possible to identify patterns of behavior with greater confidence. From this visual inspection, periods that had a well-defined direction (ascending, descending, or lateral) were selected. Based on this, an auxiliary file was created containing, for each interval, the start date, the end date, and the corresponding label.

Subsequently, on-chain metrics obtained through the blockchain platform were incorporated into the study. This information, described previously, was integrated into the Bitcoin price data set through an inner join operation, using the date of the records as the common key. This procedure allowed the temporal alignment of market data and data extracted from the blockchain, enabling more robust and contextualized analyses based on the combination of these two complementary sources of information.

3.2 Feature Engineering

The next step involved creating a new column in the dataset, using the Pandas library, called "label", designed to classify the behavior of Bitcoin's price within specific ranges. We established a simple coding for these labels: 0 representing a price drop, 1 indicating stability or trading range, and 2 corresponding to a price increase.

Finally, a Python script was developed to automate the data labeling process. The strategy adopted consisted of going through the main dataset and assigning, for each line, the corresponding value in the label column, according to the time intervals previously defined in an auxiliary file. This file contained, in each line, a start date, an end date, and the label associated with the price behavior in that period (0 for fall, 1 for stability, and 2 for rise).

Initially, the main database was loaded using the Pandas library, and the date column (timestamp) was converted to the datetime type in order to allow more accurate temporal comparisons. Then the label column was created and initialized with the default value of zero. The script then went through, line by line, the auxiliary file containing the previously defined intervals, interpreting the start and end dates, as well as the label associated with each period. With this information, the dataset records that corresponded to each time interval were updated with the respective label value, thus completing the historical data classification process.

This approach made it possible to automatically label thousands of records based on observed historical patterns, making it possible to observe the results of each class over time associated with its price in the figure (above or below, depending on where it is located), ensuring consistency and agility to the process. This process was essential to ensure that the learning algorithms could correctly associate market behavior patterns with each prediction class, increasing the accuracy of estimates regarding the future movement of Bitcoin's price.

To enable the application of supervised classification algorithms, it was necessary to structure a target variable that reflected, in a simplified way, the future behavior of the asset. The adopted strategy was based on the moving average of fixed windows throughout the price time series.

Initially, the data was segmented into consecutive blocks of 30 records. For each block, the average of the values in the Close column was calculated, representing the average price for that interval. This average was then compared with the average of the 30 immediately subsequent records. If the future average was higher than the current average, the block was classified as 1, representing an expectation of growth. Otherwise, the value 0 was assigned, indicating a downward trend.

This binary approach allowed labeling data based on a logic of comparison between adjacent windows, facilitating the training of predictive models and maintaining alignment with the sequential nature of the problem.

3.3   Normalization and Data Preparation

To ensure greater consistency in the models' input and facilitate their learning capacity, a process of normalization of numerical data was carried out using the Min-Max Scaling technique, with values scaled to the interval between 0 and 1. This procedure was applied within annual windows (30 days as a base), considering that the labeling of the target variable was also done in a segmented manner by periods. In this way, normalization respects the temporal structure of the data, preserving a more contextualized and macroeconomic view of the patterns observed over time. Categorical variables, in turn, were transformed using One-Hot Encoding, ensuring their incorporation into the model without introducing artificial ordinal relationships between the values.

Secondly, aiming to reduce the complexity of the sample space and facilitate visualization and separation between classes, a dimensionality reduction technique was applied using t-SNE. This procedure generated two new variables representing projections in a two-dimensional space, which approximately maintain the relationships of proximity and divergence between the points originally present in multiple dimensions. This contributed not only to the performance of the models but also to a clearer and more interpretable discretization of the regions associated with each class.

From the two columns generated by the application of t-SNE, it was possible to create two columns that don't present a significant linear correlation (Pearson correlation: -0.002933), contributing to the mitigation of serial autocorrelation. This process allowed the reduction of the dimensionality of the data, maintaining its expressiveness, but without the excessive inclusion of columns, especially those that are strongly correlated with each other.

4   Prediction Tests and Data Comparison

To evaluate the performance of the prediction models applied to Bitcoin records, experiments were carried out focusing on supervised classifiers, considering the sequential nature of the data as a time series with strong dependence between the records, dealing with a validation approach that preserved the temporal order via the TimeSeriesSplit method with 5 divisions $n_splits = 5$ for the cross-validation process, ensuring that the data used for training is chronologically positioned before the test data, respecting the temporal structure of the problem and avoiding leakage of information from the future to the past.

The models were evaluated using precision as the main metric. This choice is justified by the asymmetric nature of the risk involved: it is more harmful to the investor if the model indicates an appreciation (indicating a moment of purchase) and the price subsequently falls, resulting in a real loss, than the opposite, where the model indicates a fall and the price rises, representing only a loss of opportunity. Thus, precision allows capturing this bias, favoring hits in situations where appreciation is expected and penalizing false positives in this sense.

During the experiments, different classification algorithms were tested, including Random Forest, Support Vector Machines (SVM), Gradient Boosting and logistic regression models. Each was trained on defined time windows and subsequently evaluated based on test data from each TimeSeriesSplit split. The code used to carry out these tests maintained a modular structure, ensuring that the pre-processing, training and evaluation pipeline was reproducible and consistent across the different algorithms tested (Fig. 3).

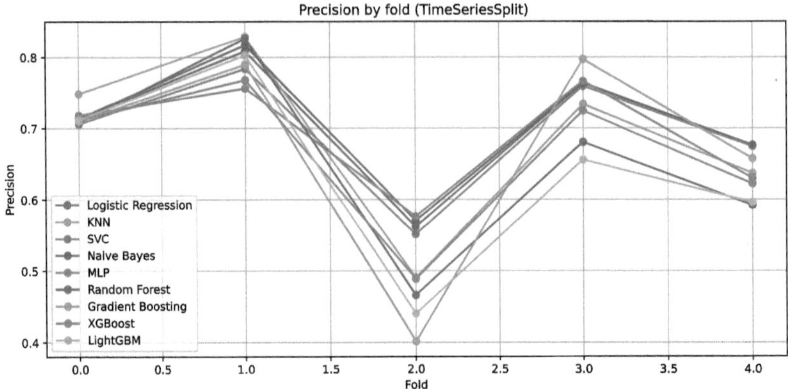

Fig. 3. Cross-validation results by fold.

In the figure presented, the performance of each algorithm is observed throughout the different divisions of cross-validation. Although some models achieved slightly higher scores in certain folds, the general behavior between them was quite similar. In general, when a given fold presented satisfactory results for a model, the other algorithms also tended to perform well in that same division. Likewise, in folds where performance was lower, this downward trend was observed across all models. This pattern suggests that the difficulty

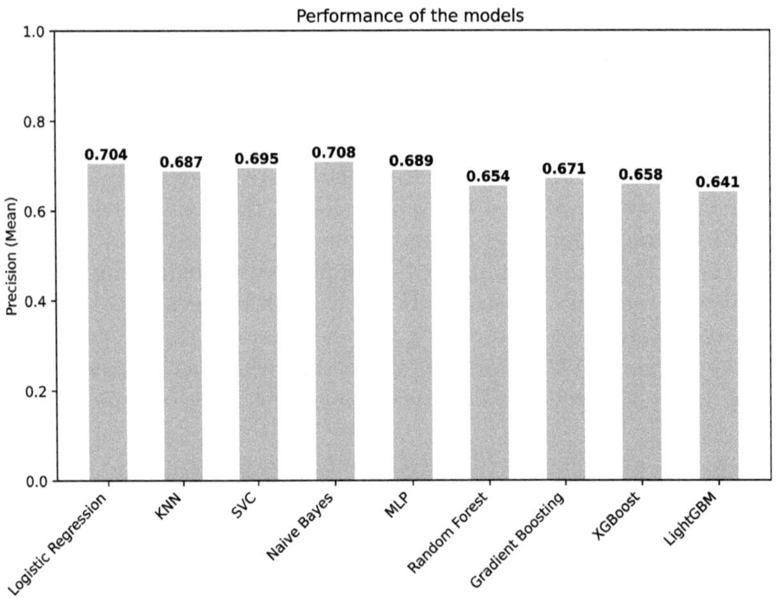

Fig. 4. Mean of Precision by model.

or ease of prediction in certain parts of the time series is an influential factor common to all tested classifiers (Fig. 4).

Based on the arithmetic average precision for each algorithm tested via cross-validation, the linear and simpler algorithms, such as Logistic Regression and Naive Bayes, consistently performed better throughout the tests. However, among the models that operate with non-linear decision functions, the Support Vector Classifier (SVC) stood out, achieving results comparable in quality to the aforementioned linear models. This indicates that, although simpler models adapted well to the data structure, the SVC demonstrated a similar capacity for generalization even in the face of the greater complexity involved in its operation.

5   Conclusions

The experiments carried out in this study showed promising results for Bitcoin trend prediction using supervised machine learning models, with average precision ranging from 65% to 70% across different algorithms and cross-validation folds. These outcomes are relevant given the inherent volatility of the crypto market.

A key methodological strength lies in the use of TimeSeriesSplit, which preserved the chronological order of data and prevented information leakage from future to past, enhancing the reliability of the evaluation. Additionally, the choice of precision as the primary metric proved appropriate, as it aligns with real financial risks, especially prioritizing accuracy in predicting upward movements to avoid misleading buy signals.

Overall, the findings suggest that it is possible to capture meaningful behavioral patterns in Bitcoin time series. Although not a definitive decision-making system, the models offer valuable insights and open the way for future research involving feature refinement, hyperparameter tuning, and the application of more advanced architectures such as recurrent neural networks. For evaluation and transparency of results, we maintain all source codes publicly available in a git repository[1].

References

1. Baur, D.G., Hong, K., Lee, A.D.: Bitcoin: medium of exchange or speculative assets? J. Int. Finan. Markets. Inst. Money **54**, 177–189 (2018)
2. Bayes, T.: Naive Bayes classifier. Article Sources and Contributors, pp. 1–9 (1968)
3. Bubolz, A.G., et al.: Analysis of bitcoin trends through the integration of on-chain financial indicators and machine learning. In: Gervasi, O., et al. (eds.) Computational Science and Its Applications – ICCSA 2025. pp. 35–50. Springer, Cham (2025). https://doi.org/10.1007/978-3-031-96997-3_3
4. Callens, A., Morichon, D., Abadie, S., Delpey, M., Liquet, B.: Using random forest and gradient boosting trees to improve wave forecast at a specific location. Appl. Ocean Res. **104**, 102339 (2020)

---

[1] https://github.com/arthur-bubolz/implementation_codes_ideal_2025.git.

5. Crosby, M., Pattanayak, P., Verma, S., Kalyanaraman, V., et al.: Blockchain technology: beyond bitcoin. Appl. innovation **2**(6–10), 71 (2016)
6. Guo, G., Wang, H., Bell, D., Bi, Y., Greer, K.: KNN model-based approach in classification. In: Meersman, R., Tari, Z., Schmidt, D.C. (eds.) OTM 2003. LNCS, vol. 2888, pp. 986–996. Springer, Heidelberg (2003). https://doi.org/10.1007/978-3-540-39964-3_62
7. Kullback, S.: Kullback-leibler divergence. Technical report (1951)
8. Maaten, L.V.D., Hinton, G.: Visualizing data using t-SNE. J. Mach. Learn. Res. **9**, 2579–2605 (2008)
9. McNally, S., Roche, J., Caton, S.: Predicting the price of bitcoin using machine learning. In: 2018 26th Euromicro International Conference on Parallel, Distributed and Network-Based Processing (PDP), pp. 339–343. IEEE (2018)
10. Nakamoto, S.: Bitcoin: a peer-to-peer electronic cash system (2008)
11. Popescu, M.C., Balas, V.E., Perescu-Popescu, L., Mastorakis, N.: Multilayer perceptron and neural networks. WSEAS Trans. Circuits Syst. **8**(7), 579–588 (2009)
12. Refaeilzadeh, P., Tang, L., Liu, H.: Cross-validation. In: Liu, L., Özsu, M.T. (eds.) Encyclopedia of Database Systems, pp. 532–538. Springer, Boston (2009). https://doi.org/10.1007/978-0-387-39940-9_565
13. Sharma, A.: On-chain analysis and cryptocurrency price forecasting using on-chain metrics (2023)
14. Stober, A., Sandner, P.: Using on-chain and market metrics to analyze the value of crypto assets. Technical report, FSBC Working Paper. Frankfurt am Main: Frankfurt School of Finance (2020)
15. Suthaharan, S.: Support vector machine. In: Machine Learning Models and Algorithms for Big Data Classification. ISIS, vol. 36, pp. 207–235. Springer, Boston (2016). https://doi.org/10.1007/978-1-4899-7641-3_9
16. Worster, A., Fan, J., Ismaila, A.: Understanding linear and logistic regression analyses. Can. J. Emergency Med. **9**(2), 111–113 (2007)

FC Assistant. Enhancing Squad Selection in EA SPORTS FC Ultimate Team with an Intelligent Decision Support System

Juan Carlos Vargas-Camacho, Francisco José Quesada-Real[✉],
and Álvaro Labella

Computer Science Department, University of Jaén, Jaén, Spain
`jcvc0006@red.ujaen.es`, `{fqreal,alabella}@ujaen.es`

Abstract. Squad selection in EA SPORTS FC Ultimate Team (FCUT) poses a complex multi-objective problem where users must balance technical performance, team chemistry, and budgetary constraints. The dynamic nature of the in-game transfer market and the vast number of available player items make this task cognitively demanding, highlighting the need for intelligent decision support tools. This work presents a modular Decision Support System (DSS) that formulates squad building as a constrained multi-criteria optimization problem. Player evaluation is performed using the Technique for Order of Preference by Similarity to Ideal Solution (TOPSIS), which aggregates heterogeneous attributes into interpretable scores based on user-defined weights. To construct complete squads under formation and budget constraints, we employ an Ant Colony Optimization algorithm enhanced with local search, using TOPSIS scores as heuristic guidance and combining them with team chemistry in a composite objective function. The system integrates these components within a scalable web-based architecture and provides explainable recommendations through an interactive assistant. An illustrative example demonstrates that the proposed approach effectively generates competitive squads that align with user preferences, offering a flexible and transparent solution for complex team-building scenarios in dynamic digital environments.

Keywords: Multi-criteria decision-making · Optimization · FC Ultimate Team · Decision support system

1 Introduction

Team selection in modern sports simulation games has evolved into a sophisticated exercise in strategic thinking and resource optimization. EA SPORTS FCUT exemplifies this trend, challenging players to build and manage a squad that balances performance, budget, team chemistry, and tactical fit. With

L. Martínez et al. (Eds.): IDEAL 2025, LNCS 16239, pp. 483–495, 2026.
https://doi.org/10.1007/978-3-032-10489-2_41

thousands of player items, fluctuating in-game markets, and constant content updates, assembling an optimal squad has become a cognitively demanding and time-intensive task [6, 12].

Players often feel decision fatigue when navigating large player databases and unclear trade-offs. Community forums frequently highlight common frustrations: How to choose the best player for a role under budget constraints? How to balance team chemistry while integrating diverse leagues and nationalities? While third-party tools have emerged to support filtering or price tracking, most remain rule-based, static, and lack strategic personalization or decision transparency [6].

To address such challenges, Artificial Intelligence (AI) has been increasingly adopted in digital games to improve personalization, support complex decisions, and enhance user engagement [2, 10]. AI-driven companions and recommendation systems now leverage reinforcement learning, clustering, or deep learning to adapt to individual playing styles and preferences. However, these systems often fall short in explainability, limiting their usefulness in environments where users need justifiable recommendations aligned with domain constraints.

Multi-Criteria Decision-Making (MCDM) techniques offer a complementary and rigorous framework for modeling trade-offs among competing objectives [5]. An MCDM problem typically involves evaluating a finite set of alternatives according to multiple, and often conflicting, criteria—each of which may have different units, priorities, or levels of importance to the decision maker. These methods enable systematic comparison and prioritization of alternatives by aggregating individual criteria into a global assessment that reflects user-defined preferences. MCDM has been successfully applied in complex domains such as healthcare, energy, education, and sustainability [1, 9, 11], where decisions must account for heterogeneous and context-dependent priorities.

This paper bridges both perspectives by proposing an intelligent assistant for FCUT that integrates MCDM principles with AI-based optimization and user modeling. Unlike existing tools, our assistant enables players to express strategic priorities and constraints explicitly, processes real-time player data, and generates explainable squad recommendations tailored to evolving in-game contexts.

The main contributions of this paper are as follows:

- We formalize the squad selection process in FCUT as a multi-criteria decision-making problem, incorporating tactical, economic, and compatibility factors.
- We design and implement a modular intelligent assistant that combines MCDM techniques with AI-driven data acquisition, user preference modeling, and optimization algorithms.
- We describe the system architecture in detail, emphasizing component interaction and extensibility.
- We illustrate the system's functionality and user impact through a realistic case study and example queries.

This work aims to bridge the gap between theoretical decision models and practical gaming assistance, contributing to the design of intelligent, explainable, and user-aligned support tools for digital environments characterized by complexity, uncertainty, and personalization demands.

The remainder paper is structured as follows: Sect. 2 reviews some relevant concepts related to the proposal. Section 3 formalizes the problem to be solved by the proposed system. Afterward, in Sect. 4, the developed system is described. To show the utility and performance of the system, Sect. 5 introduces an example of use of the DSS. Later on, Sect. 6 discusses the main advantages and limitation of the tool. Finally, Sect. 7 draws some conclusions.

2   Background

This section introduces the gameplay environment in which the squad selection task takes place, and presents the theoretical foundations that support the proposed modeling and optimization approach.

2.1   FC Ultimate Team

FCUT[1] is a game mode in which users build a personalized squad of players represented as collectible cards. Each player card corresponds to a real-world footballer and includes various attributes such as pace, shooting, passing, physicality, and an overall rating (see Fig. 1). These attributes are used by the game engine to simulate player behavior during matches.

Fig. 1. Card with the attributes of a player (Extracted from Futbin). (Source: www. futbin.com, accessed July 2025.)

Player cards are categorized by league, club, and nationality, and belong to specific positions (e.g., goalkeeper, center back, striker). In addition to individual performance metrics, squad construction in FCUT must consider the

---

[1] https://www.ea.com/es/games/ea-sports-fc/fc-25/features/ultimate-team.

concept of *team chemistry*, a gameplay mechanic that reflects how well players work together based on shared affiliations (league and club) and nationalities. High chemistry improves in-game responsiveness, while low chemistry can impair player effectiveness regardless of technical ratings.

Cards are acquired through packs or the in-game transfer market, and each has a dynamic market value that fluctuates over time. Users face budget constraints and must make trade-offs between cost, performance, and chemistry. Building a competitive squad thus requires balancing multiple interdependent factors, making the decision process both strategic and complex.

This structure turns squad building into a multi-faceted problem involving heterogeneous attributes, hard formation rules, and gameplay-specific synergies, making it well-suited for formal modeling through MCDM and constrained optimization techniques.

2.2 MCDM and the TOPSIS Method

MCDM refers to a family of methods designed to support decisions involving multiple, often conflicting, criteria [5]. MCDM explicitly accounts for trade-offs between attributes, enabling decision makers to structure preferences, compare alternatives, and identify the most suitable option according to their priorities. MCDM techniques have been widely applied across domains such as healthcare, supply chain management, and sustainability [1], and their use is increasingly relevant in digital environments where users face complex, preference-driven decisions.

The squad-building process in FCUT naturally aligns with the structure of an MCDM problem. Each player item represents an alternative that must be evaluated across several criteria—including technical attributes, market value, positional fit, and chemistry contribution—with different users exhibiting distinct tactical objectives and constraints. Modeling this task within an MCDM framework provides structured decision support, enhances personalization, and increases transparency in the recommendation process.

Among the various MCDM methods, the TOPSIS is particularly notable for its simplicity, computational efficiency, and intuitive rationale [4]. The core idea behind TOPSIS is that the best alternative is the one simultaneously closest to the ideal solution and farthest from the anti-ideal. Alternatives are ranked based on their relative distances to these reference points, resulting in a score that reflects their overall desirability.

TOPSIS is especially well-suited to scenarios such as FCUT, where alternatives are characterized by numerical attributes on heterogeneous scales, and decision makers may wish to assign weights that emphasize specific priorities (e.g., favoring pace over strength for wingers).

2.3 Optimization Under Constraints

Many decision-making problems involve selecting a subset of options that jointly satisfy a set of structural, budgetary, or logical constraints. These are typically modeled as constrained combinatorial optimization problems, where the objective is to maximize (or minimize) a utility function defined over combinations of elements, subject to feasibility conditions. In such problems, solution space is exponentially large, and exact optimization methods may become computationally intractable as the number of variables increases.

To cope with this complexity, metaheuristic algorithms have emerged as effective tools for finding near-optimal solutions in reasonable time. Among them, *Ant Colony Optimization* (ACO) [3] is a population-based approach inspired by the foraging behavior of real ants. ACO algorithms simulate a set of artificial agents (ants) that incrementally construct solutions by selecting components based on a combination of pheromone intensity (reflecting experience) and heuristic desirability (representing domain knowledge).

Formally, ACO operates over a construction graph where nodes represent components of partial solutions, and transitions are governed by a probabilistic rule that balances exploitation and exploration. Pheromone trails are updated iteratively based on the quality of the solutions found, reinforcing promising regions of the search space. Its advantages include flexibility in encoding constraints, the ability to incorporate heuristic guidance, and natural support for parallelization. These properties make ACO particularly well-suited for complex decision problems involving heterogeneous criteria, large search spaces, and dynamic constraints, as the problem we are dealing with.

3 Problem Formalization

Squad selection in FCUT involves choosing a set of players that collectively maximize team effectiveness while satisfying user-defined priorities and a variety of structural and economic constraints. This section formalizes the underlying decision problem.

3.1 Definition

Let $\mathcal{P} = \{p_1, p_2, \ldots, p_N\}$ denote the set of all available players in the game, and let $\mathcal{S} \subset \mathcal{P}$ represent a candidate squad composed of K players (typically $K = 11$ for starting squads, optionally extended to 15–23 to include substitutes).

Each player p_i is described by a vector of attributes over a set $\mathcal{A}$, which may include technical characteristics (e.g., pace, passing, strength), economic indicators (e.g., market value), tactical properties (e.g., work rate, position), and gameplay-related metrics (e.g., chemistry potential). These attributes can be numeric, categorical, or derived from combinations of card metadata.

The squad selection task can be formulated as a multi-objective combinatorial optimization problem:

$$\mathcal{S}^* = \arg\max_{\mathcal{S} \subset \mathcal{P}} [f_1(\mathcal{S}), f_2(\mathcal{S}), \ldots, f_m(\mathcal{S})] \tag{1}$$

subject to:

- $|\mathcal{S}| = K$,
- $H_j(\mathcal{S}) = \text{true}$ for all hard constraints H_j,
- $S_k(\mathcal{S}) \leq \varepsilon_k$ for all soft constraints S_k.

Hard constraints (H_j) define the feasibility of a squad and must be strictly satisfied. Typical hard constraints include:

- Squad size must be exactly K.
- Each position in the selected formation must be filled by exactly one player.
- Total cost of the squad must not exceed the user-defined budget.

Soft constraints S_k, in contrast, encode user preferences or desirable properties that may be violated within tolerable bounds. Examples include:

- Favoring players from a specific league or nationality.
- Avoiding players with low stamina or weak-foot ratings.
- Maximizing chemistry links between specific pairs of players.

Soft constraints are integrated into the objective function via weighted penalties or preference scores, and can be adjusted dynamically to reflect user-defined priorities.

3.2   Optimization Objective

Building upon the abstract problem structure defined in Sect. 3, this section introduces the concrete optimization objective used in our system. While the formal model allows for an arbitrary number of evaluation criteria and constraints, we adopt a practical formulation that combines team chemistry with individual player scores derived from TOPSIS. The resulting function provides a unified, normalized metric that guides the search for high-quality squads under realistic gameplay constraints.

Each player is first evaluated using the TOPSIS method, producing a normalized score that reflects their overall desirability based on selected attributes. These scores serve as heuristics to guide the optimization process. The final quality of a squad is then assessed by aggregating the average TOPSIS score of all players with the normalized total chemistry value, as follows:

$$f_{\text{objective}} = \max\left(\frac{1}{2} \left(\frac{\text{chemistry}}{33} + \frac{1}{n} \sum_{i=1}^{n} \text{TOPSIS}_i \right) \right) \tag{2}$$

where n is the number of selected players, and the chemistry score is normalized by its maximum value of 33.

To solve this optimization problem, we employ an ACO algorithm with local search refinement. Each artificial ant constructs a full squad by selecting one player per position, guided by pheromone trails and TOPSIS-based heuristics. The generated solution is then improved by a local search that explores neighboring squads through controlled substitutions. This hybrid strategy allows the system to efficiently explore the solution space while adapting to user-defined priorities and game constraints.

4   System Architecture

The system follows a modular architecture composed of independent subsystems designed for maintainability, scalability, and seamless integration of new functionalities (see Fig. 2).

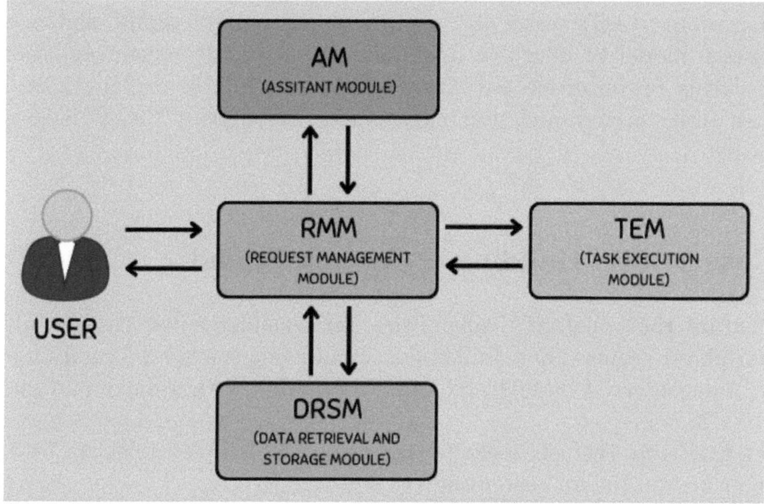

Fig. 2. System architecture.

Although inspired by the classical structure of a DSS, the proposed architecture extends the traditional model by incorporating a virtual assistant module and deploying the platform as a web-accessible service to provide interactive and explainable recommendations. The main modules are described below:

- *Data Retrieval and Storage Module (DRSM)*. This module implements the data subsystem of the DSS. It collects and maintains up-to-date information on player items—including attributes, chemistry metadata, and market values—through automated web scraping. The data are stored in a distributed database to ensure scalability, consistency, and availability, providing a reliable foundation for decision-making.
- *Request Management Module (RMM)*. Acting as the communication interface between users and the system, this module manages incoming queries, coordinates task execution, and returns results through a web interface. It includes a front-end server for interaction, a back-end server for orchestration, and a Redis layer for asynchronous processing and low-latency response handling, ensuring efficient performance under concurrent load.
- *Task Execution Module (TEM)*. This module forms the core of the model subsystem. It processes user preferences, applies TOPSIS to rank players based on multiple criteria, and runs the Ant Colony Optimization with Stagnation

Avoidance (ACOS) algorithm to generate viable squad configurations [7]. The optimization tasks are parallelized across worker servers and supported by a lightweight local database optimized for fast access to relevant subsets of players. Its modular design facilitates the future integration of alternative metaheuristics.

– *Assistant Module (AM)*. This module enhances the system by providing intelligent interaction and explainability. It integrates a Retrieval-Augmented Generation (RAG) server [8], a curated document corpus, and a natural language model to interpret user queries and justify recommendations. By translating optimization outputs into human-readable explanations, it promotes transparency and user engagement throughout the decision-making process.

5   Case Study: Building a Custom Squad

To illustrate the assistant's capabilities, we consider a use case in which an amateur player requests help building a squad using the 4-4-2 formation (1 goalkeeper, 4 defenders, 4 midfielders, and 2 forwards), with a maximum budget of 20,000 FC25 coins.

As a first step, the user seeks to understand which playstyles are best suited to this formation before generating the squad.

> **User Query**
>
> What playstyles are effective when using the 4-4-2 formation?

In response to this query, the assistant leverages the LLM in combination with a domain-specific RAG system based on FC 25 Ultimate Team information to provide an answer evaluating the suitability of using this formation.

> **Assistant Response**
>
> The 4-4-2 formation in FCUT is highly versatile and supports several playstyles depending on player attributes. The most effective approaches are high pressing, counterattack, and direct play. Choosing the optimal style depends on your squad's pace, stamina, and technical skills.

The assistant also provides information on the different playstyles previously mentioned, outlining the key positions required for each style and suggesting suitable player archetypes accordingly. Additionally, it highlights performance-enhancing factors, notes certain limitations, and offers further recommendations.

> **Recommended playstyles for 4-4-2**

- *Direct Play*: Best with fast, agile wingers and clinical forwards. Requires overlapping full-backs to support wide attacks.
- *Counterattack*: Effective if your wide and central attackers are quick and good at positioning. Midfielders should have strong long passing to launch transitions.
- *High Pressing*: Demands stamina and aggression from strikers and midfielders. Full-backs must be fast and recover quickly.
- *Possession Play* (less common): Requires midfielders with vision and short passing, and forwards who can combine well (e.g., target man + mobile striker).

Key Position Roles by Playstyle

Position	Direct / Counterattack	Possession
ST	Mobile + Finisher	Mobile + Target Man
LM/RM	Fast dribbler	Creative passer
CM	Box-to-box + Defensive mid	Two playmakers
LB/RB	Attacking full-backs	Balanced or conservative

Performance Factors

- CM pace and stamina to sustain pressing and transitions.
- Winger attributes determine vertical vs. combinational play.
- Full-back-midfield coordination ensures defensive balance when attacking.
- Striker pairing should be complementary (e.g., pace + positioning).

Limitations
Due to the lack of information about your current squad, recommendations are based on general performance profiles

Additional Advice

- Start with direct or counterattack if you have fast wingers.
- Combine a mobile striker with a poacher or target man.
- Let full-backs join the attack only if they have sufficient pace and stamina.
- Use a high defensive line for pressing; a medium block for counterattack.
- Switch to possession play if your wingers are creative rather than fast.

After defining the tactical approach, the user requests a squad compatible with the *counterattack* playstyle, constrained by a 20,000 FC25 coin budget (see Fig. 3).

Squad selection was based on the relative importance of player attributes, prioritizing those best suited to the specified playstyle. For example, in a coun-

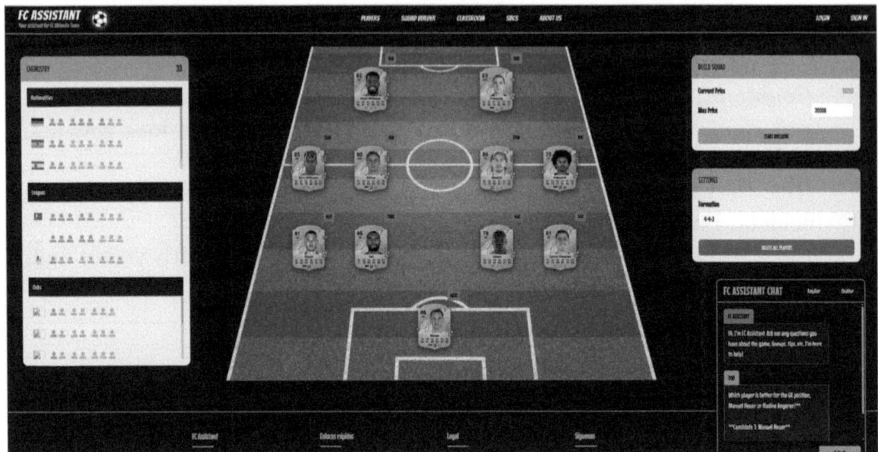

Fig. 3. Automatically generated 4-4-2 squad for counterattack strategy (20,000-coin budget).

terattack strategy, wingers and forwards must be highly pace-driven; in this case, players such as Nico Williams and Adeyemi were selected as wingers, along with Iñaki Williams and Freigang as forwards, all known for their speed.

Player selection was further optimized using an ant colony optimization algorithm, which considers the weighted criteria discussed earlier while adhering to the predefined budget. Additionally, the algorithm aims to maximize team chemistry. As a result, the final squad includes six German players and five from the Spanish league, some of whom play for the same club—further enhancing team chemistry through shared affiliations.

The assistant also provides a justification for attribute prioritization per position, as shown in Table 1.

6   Discussion

The proposed system demonstrates several strengths. By framing squad selection as a constrained multi-criteria optimization problem, it provides a structured and transparent decision process for a task traditionally driven by heuristics or trial-and-error. The combination of TOPSIS for player evaluation with an ACO algorithm enables the integration of heterogeneous attributes, budgetary constraints, and chemistry requirements into a unified framework. The modular architecture further enhances maintainability and scalability, supporting dynamic data updates and user customization.

Another key advantage is the interpretability of the recommendations. The use of TOPSIS as a heuristic layer allows the system to justify player choices based on weighted criteria, increasing user trust and facilitating interactive

Table 1. Tactical priorities by position for 4-4-2 counterattack strategy

Position	Key Attributes	Tactical Role
GK	Positioning (90), Reflexes (90), Sweeper Keeper (80)	Reliable in 1v1s, launches counterattacks with accurate distribution.
CB	Interceptions (91), Defensive Awareness (90), Standing Tackle (85), Ball Control (85)	Disrupt attacks without losing shape, initiate buildup via short passes.
LB/RB	Sprint Speed (90), Stamina (85), Crossing (80), Attacking Full-Back (85)	Support wide transitions with overlapping runs; recover quickly on defense.
CM	Vision (91), Short Passing (90), Stamina (85), Box-to-Box (85)	Link defense and attack; enable rapid transitions and spatial coverage.
LM/RM	Sprint Speed (90), Dribbling (85), Crossing (80), Winger (85)	Break defensive lines, deliver crosses, create width and imbalance.
ST-9 (Poacher)	Finishing (91), Positioning (90), Shot Power (85), Advanced Forward (85)	Primary goal threat; exploits space behind the defense.
ST-10 (False 9)	Finishing (90), Vision (70), False 9 (85)	Drops deep to create space and drag defenders, supports link-up play.

adjustments. The integration of a virtual assistant also lowers the entry barrier for non-technical users, aligning the tool with the principles of explainable decision support.

Despite its strengths, the system presents certain limitations. The quality of its recommendations relies heavily on the accuracy and freshness of player data, making efficient scraping and update mechanisms critical. The current optimization model uses fixed weights and a predefined formation, limiting its adaptability to evolving strategies. Although the hybrid TOPSIS–ACO approach offers a good trade-off between exploration and efficiency, it does not guarantee global optimality and may require parameter tuning across different scenarios. Incorporating an intelligent agent capable of retrieving up-to-date FCUT information and dynamically enriching the document corpus could address data currency issues and enhance system responsiveness.

7   Conclusions

This work introduced a modular DSS for automated squad selection in FCUT, formulated as a constrained multi-criteria optimization problem. The approach combines TOPSIS for player evaluation with an ACO algorithm enhanced by local search to generate squads that balance chemistry, attributes, and budget under user-defined preferences. The architecture integrates data acquisition, multi-criteria analysis, and metaheuristic optimization within a scalable web-based platform. An illustrative example demonstrates the effectiveness and transparency of the system.

Future work will expand the range of criteria, incorporate advanced tactical metrics, and explore dynamic user modeling. The methodology also shows potential for broader applications in constrained team formation and resource allocation, highlighting the value of combining MCDM techniques with bio-inspired optimization in interactive DSS.

Disclosure of Interests. The authors have no competing interests to declare that are relevant to the content of this article.

References

1. Aruldoss, M., Lakshmi, T.M., Venkatesan, V.P.: A survey on multi criteria decision making methods and its applications. Am. J. Inf. Syst. **1**(1), 31–43 (2013)
2. Chan, L., Hogaboam, L., Cao, R.: Artificial intelligence in video games and esports. In: Applied Artificial Intelligence in Business: Concepts and Cases, pp. 335–352. Springer, Cham (2022). https://doi.org/10.1007/978-3-031-05740-3_22
3. Dorigo, M., Birattari, M., Stutzle, T.: Ant colony optimization. IEEE Comput. Intell. Mag. **1**(4), 28–39 (2007)
4. Hwang, C.L., Yoon, K.: Methods for multiple attribute decision making. In: Multiple Attribute Decision Making: Methods and Applications a State-of-the-Art Survey, pp. 58–191. Springer, Cham (1981). https://doi.org/10.1007/978-3-642-48318-9_3
5. Ishizaka, A., Nemery, P.: Multi-criteria Decision Analysis: Methods and Software. Wiley (2013)
6. Kaminski, L., Becker, M.: AI applications in fantasy sports and ultimate team games: a systematic review. ACM Comput. Surv. **55**(8), 1–36 (2022)
7. Laalaoui, Y., Drias, H., Bouridah, A., Ahmed, R.: Ant colony system with stagnation avoidance for the scheduling of real-time tasks. In: 2009 IEEE Symposium on Computational Intelligence in Scheduling, pp. 1–6. IEEE (2009)
8. Lewis, P., et al.: Retrieval-augmented generation for knowledge-intensive NLP tasks. Adv. Neural. Inf. Process. Syst. **33**, 9459–9474 (2020)
9. Malik, D.A.A., Yusof, Y., Na'im Ku Khalif, K.M.: A view of MCDM application in education. J. Phys. Conf. Ser. **1988**, 012063 (2021)
10. Miyake, Y.: Current status of applying artificial intelligence in digital games. In: Nakatsu, R., Rauterberg, M., Ciancarini, P. (eds.) Handbook of Digital Games and Entertainment Technologies, pp. 253–292. Springer, Singapore (2017). https://doi.org/10.1007/978-981-4560-50-4_70

11. Quesada-Real, F.J., Moya-Pérez, F., Rodriguez-Garcia, M., Dutta, B.: A transparent and ecologically sustainable DLT-based approach for tendering processes. J. Univ. Comput. Sci. (JUCS) **31**(3) (2025)
12. Vignesh, M., Ramanujam, N.: FIFA ultimate team: economic incentives and behavioral challenges in competitive squad building. Int. J. Game-Based Learn. **11**(2), 45–59 (2021)

Intelligent Multicriteria Classification of ETFs

Pavel Álvarez[1]([✉]) [iD], Manuel Muñoz-Palma[2] [iD], María Bernal[3] [iD],
Tanya García-Gastélum[4] [iD], and Pavel López-Parra[1] [iD]

[1] Universidad Autónoma de Occidente, Culiacán, México
{pavel.alvarez,pavel.lopez}@uadeo.mx
[2] Universidad de Sonora, Hermosillo, México
manuel.munoz@unison.mx
[3] Universidad Tecnológica de Culiacán, Culiacán, México
[4] Universidad Autónoma de Sinaloa, Culiacán, México
tanya.garcia@uas.edu.mx

Abstract. The paper addresses the classification of Exchange-Traded Funds
(ETFs) to propose a set of profitable ETFs for portfolio selection. An intelligent
classification tool has been applied to infer the weights of the decision criteria from
holistic preference information from the decision-maker. It classifies the ETFs in
three risk categories. The resulting classification is not sensitive to minor modifications of the weight values due to the exploration of values of decision variables
that match the decision makers' preferences.

Keywords: intelligent classification · genetic algorithm · multicriteria sorting

1 Introduction

Exchange-Traded Funds (ETFs) have gained significant attention since 2020, as reported
by the Investment Company Institute [1], with a total of \$5.4 trillion in assets in 2020.
ETFs, compared to mutual funds (MFs) and closed-end funds (CEFs), have shown less
growth; however, understanding this difference makes them an interesting alternative
investment vehicle. In [2] examines the complementary effect between ETFs and MFs,
and the substitution effect between ETFs and CEFs.

ETFs are a relatively safe and time-efficient decision [3] They provide investors with
exposure to the underlying indexes and enable them to trade at a high frequency with
lower costs [2].

ETFs can be used as portfolio investments, similar to assets on the stock exchange
market. They can be evaluated to select those with the most profit (more risk) or the
lowest profit (less risk), depending on the investor's profile. However, selecting the best
based on the investor's profile is challenging because the subjective information from
the expert or the investor needs to be added to the model that selects the best ETFs and
emits a recommendation.

For the problem of selecting or classifying elements regarding the decision-makers'
preferences, different models are applied, such as model-driven or data-driven. Particularly, those approaches that use intelligent tools can be neural networks, deep learning,

L. Martínez et al. (Eds.): IDEAL 2025, LNCS 16239, pp. 496–502, 2026.
https://doi.org/10.1007/978-3-032-10489-2_42

or genetic algorithms, and others. However, in the research community, approaches that consider the investor's profile or adjacent information (consumer, expert, or decision-maker) aim to support the classification (selection) process, thereby helping customers and providing a satisfying experience.

1.1 Intelligent Tool Using Customers' Preferences

The study of AI is crucial for improving it and understanding its implications when it fails to meet consumer expectations. Those implications are found in [4], where failures driven by AI classification lead to consumer regret over using AI, influencing subsequent behavioral outcomes, such as subscription and service quitting. The AI classification failures are inevitable [5, 6] as the subjectivity is inherent to the classification experience.

In [4] is observed by information recollected with experts on AI technologies and marketing AI requires the information and point of view customer subjectivity. In this sense, to knowledge the effects of AI on consumers' self-identification and self-expression.

In [4], information is gathered from experts in AI technologies and marketing, high-lighting the need for customer-centric insights. In this sense, to understand the effects of AI on consumers' self-identification and self-expression.

A text classification for the digital transformation of library services is proposed by [7], developing an improved version of K-nearest neighbor. It improved the accuracy of sports, tourism, and education categories.

An improved version of K-means clustering and class mean distance is used to characterize and extract text information with a vector space model [8]. They classified English text resources and provided support for readers to obtain the required information quickly.

In [9], an intelligent deep learning tool for the detection and classification of surface defects of porcelain insulators is developed. The study demonstrates the achievement of efficient and automated defect detection in actual industrial applications, providing a strong guarantee for the safe operation of the power system.

There is a need to include preference information and/or self-expression of customers (decision-makers) on prediction models to generate a recommendation that satisfies them. It will help customers identify the proposed solution.

The present work aims to classify ETFs in three risk categories to support analysts in the selection of ETFs to form an investment portfolio. The classification is performed by an intelligent procedure that generates a recommendation of weights regarding the analyst's preferences (customer's preferences) about the relative importance of decision criteria and uses them in a multicriteria sorting method to classify the ETFs in three risk categories.

The rest of the paper is organized as follows. Section 2 describes the intelligent classification procedure. An empirical application of the intelligent tool is shown in Sect. 3, describing the outcomes of the procedure and the classified ETFs. Section 4 shows the conclusion of the paper.

2 Methodology

The intelligent procedure is designed in two phases. The first phase involves defining the intelligent criteria weights. The second phase involves classifying ETFs into a risk category.

2.1 Genetic Algorithm for Inferring Weights

Phase (1): It requires the expert/decision-maker/customer to provide some adjacent information about the decision-criteria (ratio indicators of the ETFs). In the rest of the document, we will use "expert" to refer to the person who defines this preference information. The information corresponds to the order of decision criteria regarding the relative importance of criteria. That information serves as input for a genetic algorithm that will infer thousands of weight vectors matching the expert's preference. The weight values must match the relative order of the criteria. The most important criterion should be the highest weight value. The least important criterion should be the lowest weight value. The large number of weight vectors will be used to generate a stable weight vector that can express robust parameters and, therefore, a robust classification.

A criterion g_j is an element of the set of decision criteria $g_j \in G = \{g_1, g_2, ..., g_n\}$, $j = \{1, 2, ..., n\}$. For each criterion corresponds a weigh value, the weight vector W is the set of values inferred by the genetic algorithm where an individual is represented as follow.

1	2	3	4	...	n
0.2	0.3	0.1	0.2	...	0.1
$g1$	$g2$	$g3$	$g4$	...	gn

Fig. 1. Genotype of the individual for the inferring weight vector problem.

The genotype of Fig. 1 shows the n values representing the weights of the n decision criteria. For criterion $g1$, a weight of 0.2 is assigned, and for criterion n, a value of 0.1 is assigned.

In the current phase, it is considered that the expert is willing to provide information about the relative importance of criteria in order of importance, from the least important to the most important. It is translated to a criteria comparison; if criterion j is more important than criterion i ($g_j \succ g_i$), the corresponding weight value is higher ($w_j > w_i$).

The inferring model of the genetic algorithm is defined in Eq. 1.

$$Max|w|$$
$$\text{s.a.}$$
$$\sum_{i=1}^{n} w_j = 1$$

$$w_j \geq w_i \ \textit{iff} \ g_j \succ g_i$$
$$w_i \leq \sum_{\substack{i=1 \\ i \neq j}}^{n} w_j, i = 1, 2, \ldots, n$$
$$w_i \geq 0$$

(1)

2.2 Multicriteria Sorting Method

The MR-sort method is applied to sort the ETFs in a risk category. The Eq. (2) is an outranking relation. The ETF is an alternative a_i; to assign it to the category C_h, the condition of Eq. (2) must be fulfilled.

$$\sum_{\substack{j \in N: \\ g_j(a_i) \geq b_{h-1,j}}} w_j \geq \lambda \ and \sum_{\substack{j \in N: \\ g_j(a_i) \geq b_{h,j}}} w_j < \lambda \tag{2}$$

w_j is the weight of the criterion j, λ is a cut level representing the coalition of criteria has reached a majority condition.

The categories are represented by limit profiles in this method. The lower profile $(b_{h-1,})$ and the upper profile (b_h). Each profile is a vector of attributes values $(b_{h,1}, \ldots, b_{h,j}, \ldots b_n)$. In Eq. (2), the value of each criterion j of alternative a $(g_j(a_i))$ is compared with the respective attribute of the profile $b_{h,j}$. In this sense the coalition of the criteria supporting $g_j(a_i) \geq b_{h-1,j}$ is higher than cut threshold λ means that a_i should be assigned to C_h, if and only if, the majority of criteria do not support a_i should be assigned to C_{h+1}.

3 Classification of ETFs

In Table 1 are listed 20 ETFs from different stock exchanges around the world. The index ratio information is gathered from statistical reports from 2024. For the present analysis, the analyst defined three risk categories: high risk, medium risk, and low risk. Because the investor is interested in significant profit, he agrees to be considered an aggressive investor. In this sense, he is looking at high-risk ETFs to invest.

The application of the intelligent procedure was carried out as follows. In phase (1), the analyst was required to order the decision criteria in ascending order. The least

Table 1. List of Exchange-Traded Funds

No.	Exchange-Traded Funds	No.	Exchange-Traded Funds
1	AMLP	11	AlphaDEX
2	QQQ	12	EZA
3	EWZ	13	FLLA
4	EZU	14	FTXL
5	EPI	15	GXG
6	HEDJ	16	IDX
7	KRE	17	TUR
8	EWJ	18	EWW
9	FXI	19	FBZ
10	EWT	20	SOXX

important is first, and the most important is on the bottom. The analyst ordered the decision criteria as follows: $g_5 \succ g_4 \succ g_2 \succ g_3 \succ g_8 \succ g_6 \approx g_1 \approx g_7$.

The analyst identifies g_5 as the most important criterion, while the group of three criteria ($g_6 \approx g_1 \approx g_7$) is considered the least important. The genetic algorithm parameters were defined with the following values.

Population size: 40
Number of generations: 5000
Crossover index: 0.9
Mutation index: 0.9

The genetic algorithm inferred 31987 weight vectors that match the analyst's preference order, and the sum of the eight criteria values is one for each vector. An arithmetic mean was computed for each criterion; the final weight vector is presented in Table 2. The highest weight value (0.2070) is defined as the most important criterion (g_5); the lowest value (0.0758) is defined as the lowest group of criteria ($g_6 \approx g_1 \approx g_7$). Table 2 also shows the defined limit profiles for each category. We can see that Category 1 (High risk) has b2 a lower limit profile. The same profile (b2) is the upper limit profile of Category 2 (Medium risk), and its lower profile is b1. At the same time, b1 is the upper profile of Category 3 (Low risk).

At Phase (2) of the intelligent procedure, the MR-sort multicriteria method was applied, using the defined weight from Phase (1) and the limit profiles from Table 2. Table 3 shows the resulting classification of each ETF in its respective category. It is shown that ETFs assigned to the high-risk category are AMLP, EWT, AlphaDEX, FLLA, GXG, EWW, SOXX. Those ETFs from the high-risk category are what the investor is looking for. He wants high profit, even if they are risky.

A brief discussion about the applied methodology is provided. The traditional approach considers mainly the profit of the funds or assets. It is well known that historical data can positively affect funds or assets. However, from the perspective of an expert and

Table 2. Parameters and profiles of risk categories

Label	Weight	b1	b2
g1	0.07580495	−9.41	8.84
g2	0.15455825	0.92	1.05
g3	0.12826048	−0.15	1.01
g4	0.18070575	20	23
g5	0.20703047	0.0312	0.0388
g6	0.07586602	2.33	127.86
g7	0.07580495	2.34	62.02
g8	0.10196915	1	2

Table 3. Assignation of ETFs to risk categories

Category	EFTs	Category	EFTs	Category	EFTs
High risk	AMLP	Medium risk	QQQ	Low risk	EPI
	EWT		EWZ		HEDJ
	AlphaDEX		EZU		EWJ
	FLLA		KRE		IDX
	GXG		FXI		
	EWW		EZA		
	SOXX		FTXL		
			TUR		
			FBZ		

the investor's profile, other objectives related to the funds can be considered. As the type of funds (different sectors) an investor may be interested in, they may be interested in specific companies that include sustainable activities, such as ecological and sustainable production, efficient use of resources, and others.

In this sense, the application of an intelligent multicriteria decision-making approach allows the expert team and the investor the ability to evaluate funds or assets from their respective points of view, considering how the portfolio will be selected in terms of profit, investor preferences, and expert experience.

4   Conclusion

The intelligent classification procedure generated thousands of weight vectors that were used to propose a median weight vector. The resulting weight vectors were inferred considering the relative importance of the criteria expressed by the analyst. The final weight vector is a median vector that presents certain robustness because it is constructed with variation of decision variables explored in a restricted area where the preferences of the analyst are considered. The weight vector used in the MR-sort method generates a robust solution where the result is not changed significantly by some smooth modification of some of the weight values.

The result can be used to second EFTs analysis, where a portfolio is constructed with the selected category for the investor with an aggressive profile. Here, the optimization of the portfolio is needed to identify the amount of money that should be invested in each selected EFT.

The research is ongoing; more steps need to be added to the procedure to ensure the robustness of the parameters and the result. Some of those steps are: 1) classify ETFs with inferred weights of each vector to use those where the result is almost the same. 2) Make a descriptive analysis with the classification of each vector. For further research, it would be interesting to test other multicriteria sorting methods.

References

1. ICI: Investment Company Fact Book, 61st edn. ICI Research Staff and Publications, Washington, DC (2021)
2. Tang, L., Tan, E.K.M., Low, R.: Complements or substitutes? The effect of ETFs on other managed funds. Int. Rev. Financ. Anal. **95**, 103414 (2024)
3. Elton, E.J., Gruber, M.J., Busse, J.A.: Are investors rational? Choices among index funds. In: Investments and Portfolio Performance, pp. 145–172 (2011)
4. Rita Gonçalves, A., Costa Pinto, D., Gonzalez-Jimenez, H., Dalmoro, M., Mattila, A.S.: Me, myself, and my AI: how artificial intelligence classification failures threaten consumers' self-expression. J. Bus. Res. **186**, 114974 (2025)
5. Puntoni, S., Reczek, R.W., Giesler, M., Botti, S.: Consumers and artificial intelligence: an experiential perspective. J. Mark. **85**, 131–151 (2020)
6. Yalcin, G., Punton, S.i.: How AI affects our sense of self. Hav. Bus. Rev. **9**, 1–12 (2023)
7. Wang, C., You, F., Wang, Y.: Text intelligent classification model based on improved KNN algorithm in library service transformation. Syst. Soft Comput. **7** (2025)
8. Xu, Q.: Application of an intelligent English text classification model with improved KNN algorithm in the context of big data in libraries. Syst. Soft Comput. **7**, 200186 (2025)
9. Tan, L., Hu, Y.: Accurate detection and classification of surface defects in electric porcelain insulators based on deep learning intelligent algorithms. Procedia Comput. Sci. **261**, 582–588 (2025)

Causal Inference and Predictive Modeling for Evaluating the Effectiveness of Online Marketing Campaigns

Antoine Marmonier[1]($\boxtimes$) (iD) and Carlos Monserrat Aranda[2] (iD)

[1] École Nationale Supérieure des Mines de Saint-Étienne, Saint-Étienne, France
`antoine.marmonier@etu.emse.fr`
[2] Universitat Politècnica de València, València, Spain

Abstract. In the age of digital marketing, firms must allocate advertising budgets efficiently by identifying which campaign strategies truly drive return on investment (ROI). This paper proposes a data-driven framework combining causal inference techniques with predictive modeling to assess and forecast the impact of campaign characteristics on marketing ROI.

We employ different causal methods: Matching, Linear Regression, Inverse Probability Weighting (IPW), and Causal Forests to estimate the Average Treatment Effect on the Treated (ATT) of key features such as campaign type. These methods allow us to account for selection bias and reveal heterogeneous effects across subpopulations.

Our findings confirm the negative impact of certain campaign types on ROI while highlighting the consistent positive effects of others. We further integrate these causal insights into a predictive model using SHAP (SHapley Additive exPlanations) values to quantify the contribution of each feature to the expected ROI of a given campaign.

This research provides a robust, interpretable methodology to guide marketers in optimizing campaign design, tailoring strategies to specific audiences, and forecasting ROI with greater confidence.

Keywords: Causal Inference · Predictive Modeling · Marketing Analytics

1 Introduction

In the context of growing economic constraints, companies are seeking to optimize every dimension of their activity, including marketing strategies[1] The digitalization of marketing has made it easier to deploy large-scale campaigns across multiple platforms, yet it has also introduced new challenges in measuring their true effectiveness. While performance metrics such as click-through

---

[1] McKinsey article on marketing cutting costs: https://www.mckinsey.com/capab ilities/growth-marketing-and-sales/our-insights/beyond-belt-tightening-how-market ing-can-drive-resiliency-during-uncertain-times.

© The Author(s), under exclusive license to Springer Nature Switzerland AG 2026
L. Martínez et al. (Eds.): IDEAL 2025, LNCS 16239, pp. 503–515, 2026.
https://doi.org/10.1007/978-3-032-10489-2_43

rate or impressions are readily available, they do not capture the causal impact of a campaign on the company's bottom line [19]. For companies with limited marketing budgets, identifying which campaigns truly drive Return on Investment is not only strategic but also essential for survival. However, identifying the true causal impact of a marketing campaign on ROI remains a complex task due to the absence of randomized experiments and the presence of confounding factors that influence both campaign exposure and outcomes.

To address these challenges, this paper proposes a data-driven framework that combines causal inference methods with predictive modeling to identify the key drivers of campaign efficiency and estimate their individual impact on ROI. To achieve this, we propose the use of counterfactual (Level 3 of Causal Inference) [6], where recent studies have demonstrated that machine learning techniques can successfully approximate unobservable outcomes and enable robust causal reasoning. Then, we aim to provide evidence that the use of this tool will enable companies, especially those with budget constraints, to focus on the most effective marketing levers.

Unlike randomized controlled trials, real-world marketing data is observational by nature and prone to selection bias [16]. Traditional performance analyses may therefore produce misleading conclusions. To overcome this, we use causal inference techniques to estimate the treatment effects across various campaign types. These insights are subsequently integrated into a machine learning model to forecast ROI, combining interpretability and predictive accuracy to support more informed and impactful marketing decisions.

Our contributions are threefold:(i) we demonstrate how machine learning methods can enhance causal inference by applying advanced techniques such as Matching, Inverse Probability Weighting, and Causal Forests to real-world marketing data; (ii) we integrate machine learning techniques, notably Random Forests and SHAP values, to interpret the contribution of each campaign feature to the predicted ROI, thereby enhancing the model's transparency and robustness to noise; (iii) we propose a predictive framework that combines outcome modeling with causal insights using a doubly robust estimator, offering a practical tool for campaign design and budget optimization under uncertainty.

2   Related Work

The question of how to estimate the true causal impact of marketing campaigns using observational data has attracted attention across various fields, including econometrics, marketing science, and machine learning. Traditional approaches, most notably linear regression, have been widely used due to their simplicity. However, they often fail to account for unobserved confounding and non-random treatment assignment, leading to biased or unreliable estimations of campaign effectiveness [12].

Early empirical studies addressed this challenge by employing randomized hold-out designs, either at the individual or geographic level, to estimate incremental return on ad spend (IROAS). Paired geo-tests using optimal matching techniques substantially reduced variance, particularly in contexts where only a small number of geographic units were available [3]. Given the high cost of always-on experimentation, more recent contributions have proposed meta-learning frameworks trained on multiple small-scale randomized controlled trials (RCTs) to predict the incremental lift of untested campaigns. For example, a Predictive Incrementality by Experimentation (PIE) framework leverages hundreds of Facebook campaigns and consistently outperforms traditional methods such as propensity score matching and double machine learning [8]. Beyond the limitations of classical models, robust causal inference techniques have gained prominence. Among them, Inverse Probability Weighting (IPW) stands out as a powerful method for adjusting observational data, by creating pseudo-populations in which treatment assignment is independent of observed covariates [10]. This approach has become a cornerstone in modern causal analysis.

In parallel, recent work has increasingly focused on uncovering heterogeneous treatment effects, acknowledging that campaign performance often varies across segments. In this context, tree-based algorithms, particularly Causal Forests [18], have shown a remarkable ability to model non-linear interactions and estimate individual-level treatment effects, an essential feature for marketing, where personalization is key. This line of research has been further developed through the introduction of Orthogonal Random Forests [15], which are well-suited to high-dimensional settings, as well as the doubly robust framework [4]. These methodological advances have laid the groundwork for numerous applied studies, including recent work on the analysis of email creatives [5].

Meanwhile, structural challenges such as endogeneity and multicollinearity in aggregate channel spending have renewed interest in Marketing-Mix Modeling (MMM). A recent advancement, CausalMMM, uses variational inference and Granger causality constraints to learn shop-specific causal graphs and mitigate bias in traditional regression-based models [7]. Additionally, decision-focused causal learning now enables direct optimization of budget allocation, yielding substantial real-world gains [11].

At the intersection of causal inference and model interpretability, SHAP values [13] offer a consistent and theoretically grounded method to quantify each variable's contribution in complex models. While widely adopted in machine learning, their integration into causal pipelines remains relatively underexplored.

Despite increasing interest in combining causal inference and predictive modeling, few studies offer a unified framework capable of both estimating treatment effects and forecasting outcomes. Our approach combines Matching, IPW, and Causal Forests with SHAP-based interpretation and predictive modeling to jointly estimate treatment effects and forecast ROI, offering a comprehensive tool for both evaluation and campaign planning.

3   Materials and Methods

This section details the different processes we used to compute estimated causal effects on Return on Investment (Eq. 1), our measure of effectiveness in this study. Usually, it is defined as:

$$\text{ROI} = \frac{\text{Net Returns}}{\text{Investment Costs}} \tag{1}$$

We implemented the methodology using Python, following best practices on how to apply causal inference in real-world tech industry datasets using Python tools (especially the 'econml' package) and machine learning approximation [6].

The database[2] we use for this analysis contains 200,000 marketing campaigns, each described by a variety of features that define their key characteristics.

Each campaign in the dataset is uniquely identified by a Campaign_ID and linked to a specific company. The analysis focuses on several key variables, including Target_Audience (Men 18–25, Women 18–25, All ages, etc.), Channel_Used (YouTube, Instagram, Email, Website, Google Ads, Facebook), Acquisition_Cost (i.e., the amount spent on the campaign), and ROI, which serves as our main effectiveness metric. Additional attributes include Location, Duration, Language, and the campaign's Start_Date.

As with any causal inference approach based on observational data, certain identification assumptions must be satisfied—most notably unconfoundedness, overlap (positivity), and consistency. These assumptions underpin all methods used in this study and are assumed to hold throughout the analysis unless stated otherwise.

In addition, we assume that the reported ROI reflects the isolated performance of each campaign, without interference from concurrent campaigns. Since all campaign types and channels are delivered online, it is reasonable to expect that conversions can be accurately attributed to specific campaigns.

In this context, since it is impossible to observe both potential outcomes for a given unit, we rely on the Average Treatment Effect on the Treated [10], which focuses exclusively on treated units. It estimates how much better (or worse) they performed compared to how they would have performed without the treatment.

$$\text{ATT} = \mathbb{E}[Y(1) - Y(0) \mid T = 1] \tag{2}$$

Although Y(0) is not directly observable, we use machine learning models to estimate the counterfactual outcome by relying on comparable campaigns as proxies. This allows us to estimate what would have been the outcome of the treated unit under another campaign type. In this context, our objective is not to evaluate the effect of one specific alternative, but rather to determine whether the chosen treatment performs better on average compared to other available options. This is precisely what the ATT (Eq. 2) captures, making it a particularly relevant and powerful metric for causal analysis in marketing.

3.1 Matching

Matching methods have been widely used for causal inference and reviewed extensively in the literature [17]. In this study, we consider that the treatment

---

² Online database: https://www.kaggle.com/datasets/manishabhatt22/marketing-campaign-performance-dataset.

T depends on which variable we want to see the causal effect on ROI. Yet, to respect the matching technique that is used to have comparable campaigns, we have to condition on different variables. We usually have a feature that can be considered as the treatment, but in this case, the treatment is a combination of features. This is essential to make sure that we have comparable campaigns.

In the database we consider only one continuous variable which is the Acquisition cost, so in order to do the matching we had to set a criteria. We choose a 10% margin in difference between two different campaigns. The Duration in this study will be considered as a categorical variable as it is only taking three levels categorized as: Short (15 days), Medium (30–45 days) and Long (60 days).

Due to the strict filtering criteria applied to identify comparable campaigns, the analysis suffers from high variance and sometimes lacks sufficient results. To increase the number of observations without introducing significant bias, plausible assumptions were made—notably grouping cities with similar population sizes (e.g., Houston and Chicago) into a new variable called CityGroup. The final filtering process was then based on language, CityGroup, month, company, campaign duration, channel used, target audience, customer segment, and a 10% threshold on acquisition cost differences.

Then we stand a Treatment T as treated (T = 1), if the campaign has the specific campaign type that we are analysing or not treated (T = 0) otherwise, considering all the previous filters we set up.

For the matching, the ATT is obtained by computing the mean of the difference in ROI for comparable campaigns (Eq. 3):

$$\text{ATT} = \frac{1}{N} \sum_{n=1}^{N} \mathbb{E}[Y_{1n} - Y_{0n} \mid T = 1] \tag{3}$$

- Y_{1n} is the observed outcome (ROI) for the treated unit;
- Y_{0n} is the counterfactual outcome for the comparable unit that has not been treated (i.e., a campaign with a different type);
- $T = 1$ denotes treated units only;
- N is the total number of treated units.

Using the matching method helps reduce bias from unobserved confounding and omitted variables by conditioning on a rich set of covariates. The positivity assumption is reasonably met, as treatment groups show sufficient overlap in covariate distributions. While not all bias can be eliminated, this is less critical from a managerial perspective, as decision-makers typically focus on actionable variables.

However, matching provides only a local estimate of the ROI, as it filters on nearly all campaign attributes. This results in wide confidence intervals due to the limited number of comparable campaigns—especially when applying strict criteria like the 10% threshold on Acquisition Cost. As a consequence, we cannot draw strong conclusions about the causal impact of campaign types using matching alone.

3.2 Linear Regression

Due to the sample size limitations of matching, we adopt linear regression as a model-based alternative that leverages the full dataset to estimate average treatment effects while adjusting for covariates.

Obviously, because $Y(0)$ is unobservable, we need additional tools to compute it. One of the most widely used tools in this context is linear regression, which allows us to estimate the effect of campaign type while controlling for potential confounding variables. We therefore apply this method to isolate the impact of the treatment variable by holding all other covariates constant, providing a more global and interpretable estimate of the causal effect.

As we wanted to estimate the causal effect with minimal assumptions, as in the matching approach. Linear regression offers another way to assess the impact of each variable on ROI but remains vulnerable to omitted variable bias and residual confounding. To mitigate these issues, we applied some regularization techniques inspired by the approach in [9].

Given the limitations of matching in terms of sample size, we next turn to linear regression, which enables broader generalization.

3.3 Inverse Probability Weighting

While linear regression offers a simple approach to estimate treatment effects, it relies on strong functional form assumptions and remains sensitive to residual confounding. To mitigate these biases and move closer to identifying true causal effects, we adopt Inverse Probability Weighting, a method specifically designed to balance covariates across treatment groups. Originally developed for binary treatments, IPW has since been extended to accommodate multi-valued treatments [20], making it particularly suited for our multi-campaign context.

The core idea behind IPW is to approximate a randomized experiment by reweighting the data to create a pseudo-population where the assignment of campaign types is independent of observed covariates. This adjustment helps remove confounding bias and enhances the credibility of causal estimates derived from observational data.

Basically, in this method each treated observation receives a weight of $\frac{1}{e(X)}$, and each control observation receives a weight of $\frac{1}{1-e(X)}$, where $e(X)$ is the estimated propensity score or the conditional probability of receiving the treatment. The definition of the propensity score is $e(X) = P(T = 1 \mid X)$, it allows you to avoid every possible backdoor path that could include a biais in your model by conditioning directly on the propensity score.

The propensity scores were computed with a logistical regression. We respect the positivity assumption. Otherwise, we would have some definition problems on weights as they are inversely proportional to the propensity score.

However, after assessing the balance of covariates before and after weighting, we observed that global balance was misleadingly good due to the dilution effect of heterogeneous language-specific distributions. By filtering the dataset to include only English-language campaigns, we ensured greater overlap and comparability across treatment and control units. This filtering substantially improved the quality of the propensity score model, resulting in better estimations. In this way, we computed the ATT with IPW for each language and then we averaged these results to obtain an estimation of the ATT for each campaign type. To evaluate these results, we have been using a bootstrap procedure with 200 resamples, allowing us to construct confidence intervals and compute p-values for statistical inference.

3.4   Causal Forest

Yet, IPW assumes that treatment effects are homogeneous across the population, which is rarely the case in marketing. So, to capture potential heterogeneity and uncover more granular insights, we employed Causal forests [18]. This method offers an effective way to estimate individual-level treatment effects. It is a non-parametric machine learning method designed to estimate heterogeneous treatment effects, meaning how the effect of a treatment varies across different subpopulations defined by covariates such as target audience, company, location, or acquisition cost. This approach allows us to go beyond average effects and uncover nuanced insights about which strategies work best for whom.

The Heterogeneous Treatment Effect (HTE) refers to the idea that the causal effect of a treatment is not constant across all individuals, but rather varies depending on their characteristics. In the context of Causal Forests, the model is explicitly designed to estimate the Conditional Average Treatment Effect (CATE) for each observation Xi, defined as in Eq. 4:

$$\mathrm{CATE} = \tau(x_i) = \mathbb{E}[Y(1) - Y(0) \mid x = x_i] \tag{4}$$

This quantity represents the individualized treatment effect. In other words, it is how much we expect the treatment to change the outcome for a specific unit with features Xi. This approach is more realistic and interpretable for marketing campaigns. Causal forests give the HTE, yet we want to have an overview of which type is more efficient. To have this information we have to compute the mean of the HTE for each subgroup to have the average ATT (Eq. 5) for one campaign type :

$$\mathrm{ATT}_{\mathrm{Type}} = \mathbb{E}[Y(1) - Y(0) \mid T = 1, \mathrm{Campaign\_Type} = \mathrm{Type}] \tag{5}$$

This approach leverages the fact that treatment effects were estimated while accounting for heterogeneity in each subgroup, before summarizing the results through averaging to obtain an interpretable overall effect per group.

To estimate the conditional treatment effects, we implemented a Causal Forest model using the 'CausalForestDML' estimator with a Random Forest Regressor as the base learner for the outcome model. The use of Random Forests enhances robustness by reducing model misspecification and naturally handling high-dimensional interactions. In addition, the Causal Forest framework provides direct access to statistical inference through confidence intervals and p-values, allowing us to assess the significance of the estimated effects in a principled way, as demonstrated in applied causal literature [1,18].

3.5 ROI Prediction

After estimating the treatment effects using a causal forest, our objective is to forecast the expected ROI of a campaign based on its specific characteristics. To improve interpretability and understand which variables influence ROI predictions, we compute SHAP values [13] on a random forest model trained to predict ROI from campaign features. SHAP values allow us to decompose the prediction into additive contributions of each input variable [14]. However, SHAP values are not used directly in the ROI prediction process. They serve solely as an interpretability tool for the predictive model, helping us identify which features contribute positively or negatively to ROI estimates.

SHAP values can be sensitive to correlated features, so we checked the data for multicollinearity. As expected, categorical variables exhibit internal correlation (only one level can be selected per variable). Additionally, we observed that Acquisition_Cost is correlated with several campaign features, which could potentially distort SHAP attribution.

We compute SHAP importances from a model excluding Acquisition_Cost to avoid the mechanical attribution caused by ROI's cost term, yielding a clearer ranking of non-budget levers. Budget impact is evaluated separately by varying Acquisition_Cost in prediction (other covariates fixed).

The ROI itself is predicted using a random forest, and the causal impact of the campaign type is estimated separately via a causal forest. To obtain a robust ROI prediction that adjusts for selection bias, we apply a doubly robust estimator [2] that combines both models. This allows us to combine both the outcome model and the treatment model and ensures reliable predictions as long as at least one of the two models, the propensity score model or the causal effect model, is correctly specified.

The adjusted ROI prediction is given by the Eq. 6:

$$ROI_{\mathrm{DR}} = ROI_{\mathrm{mean}} + (T - \hat{e}) \cdot \hat{\tau} \tag{6}$$

- ROI_{mean}: is the average ROI predicted by the model regardless of treatment;
- $\hat{\tau}$: is the estimated causal effect of the tested campaign type, obtained using the Causal Forest method.;
- $\hat{e}$: is the propensity score, i.e., the estimated probability of receiving the treatment (campaign type);
- ROI_{DR}: is the doubly robust estimate of ROI, that integrates both predicted outcomes and treatment assignment correction.

This approach helps refine predictions by correcting for possible bias due to selection effects, while leveraging the causal effect insights previously computed.

4   Results

The matching method is interesting but it's not optimizing computations. It is really hard to find two very same campaigns with all the filters we considered to have no bias. Even if the source database is huge, it always leads to non-significant results or even no comparison at all.

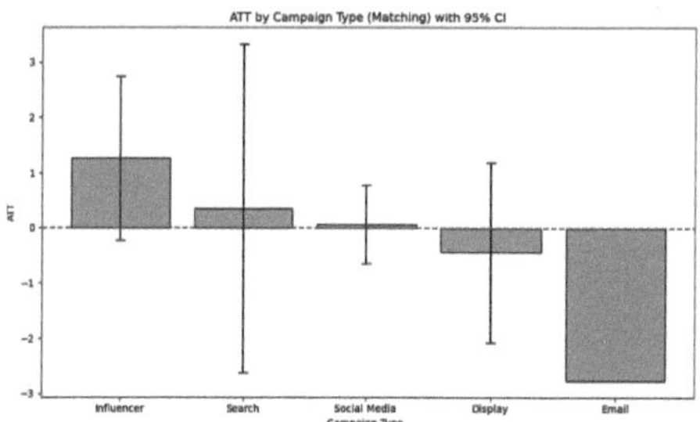

Fig. 1. ATT per campaign with matching method. No confidence interval is reported for the Email campaign type because, under our matching criteria, only a single Email campaign was retained, which precludes confidence interval estimation.

In this Fig. 1, we estimate the ATT for "Innovates industries" campaigns that took place in Houston or Chicago (as those cities are considered comparable), on the segment Women 24–35, targeting fashionistas, using instagram as a channel to promote this campaign and which is an english speaking campaign. To overcome this problem of very local analysis, we looked at the results obtained by linear regression method. Identically, we still cannot conclude about a specific campaign type that improves the ROI. The only Campaign_Type where we have a result is Email where we can assume that it has a significantly negative impact on ROI as Fig. 2 shows.

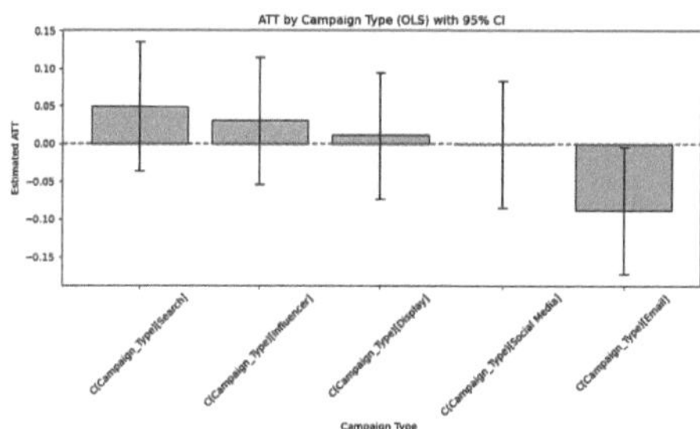

Fig. 2. ATT per campaign with linear regression method.

However, we cannot clearly have assumptions on which Campaign_Type is really impacting the ROI. This may be due to confounding biases that are still not considered inside the database.

That's why we used an Inverse Probability Weighting method, and it gave us significant results of the impact of certain campaign types with wide confidence intervals as shown in Fig. 3. This method removes confounding bias and gives us a clearer idea of the estimation of the Average Treatment Effect on the Treated for each campaign type.

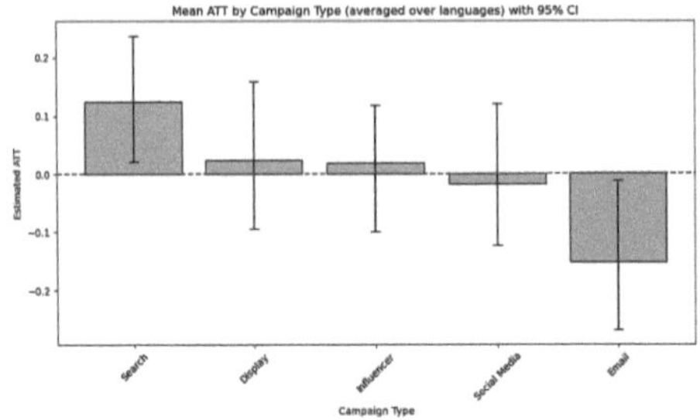

Fig. 3. ATT per campaign with IPW method.

To go even further, causal forest analysis allows us to obtain narrower confidence intervals than in the previous method. Here, we clearly have some insightful results almost for every campaign type (Fig. 4). By capturing non-linear interactions and subgroup-specific responses, this method provides stronger evidence for prioritizing certain campaign types in future marketing strategies.

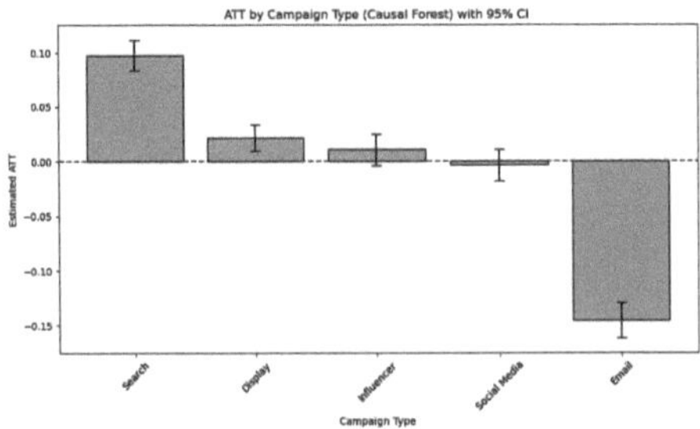

Fig. 4. ATT per campaign with causal forest method.

The second objective of this study was to estimate the ROI for each campaign type and all other things being equal by making predictions.

The results presented in Table 1 demonstrate that, based on our doubly robust predictive framework, the influencer campaign shows the highest predicted ROI (6.21), confirming its relative advantage in this scenario.

Table 1. ROI predictions for the specific campaign using Instagram, targeting Women aged 25–34 within the "Outdoor Adventurers" segment, conducted in English and located in Miami—the advertiser is Innovate Industries. The campaign is scheduled to launch in April for a duration of 30 days, with an investment cost of $13,500.

Campaign_Type	ROI_{DR}	ROI_{pred}	$\hat{\tau}$	$\hat{e}$
Influencer	6.2108	6.0418	0.2304	0.2664
Search	6.0538	6.0418	0.0142	0.1577
Display	6.0302	6.0418	−0.0160	0.2766
Social Media	5.9871	6.0418	−0.0654	0.1645
Email	5.9567	6.0418	−0.0987	0.1382

To evaluate prediction performance, we compare the RMSE (Root Mean Squared Error) to the standard deviation of the observed ROI. An RMSE of 0.856, with a standard deviation of 1.734, suggests a strong predictive model (RMSE/SD = 0.4953). As the ratio between the standard deviation and RMSE is less than 0.5, we can consider that the prediction is correct because this indicates that the model captures meaningful variation in ROI beyond random noise.

To enhance the interpretability of our predictions, we conducted a SHAP value analysis [13](as shown in Fig. 5), that decomposes the predicted ROI into additive contributions of each input variable. The analysis reveals that some features, such as "Month 04," which is included in the predicted campaign configuration, contributed positively to the predicted ROI. This indicates that launching the campaign in April aligns well with historically successful campaigns.

However, other strong positive predictors like "Language French" were not selected in this campaign, and their absence reduced the ROI potential. According to the model, French-language campaigns have historically yielded higher returns. Based on past data, targeting Women 25–34 shows a negative SHAP contribution, lowering the prediction.

This contrast highlights the importance of feature selection: aligning campaign attributes with historically high-performing variables can be a strategic lever to enhance effectiveness, assuming alignment with business constraints and market context.

5   Conclusion and Future Work

Identifying the causal drivers of Return on Investment (ROI) in digital marketing remains a complex yet essential challenge. Through this study, we combined traditional causal inference methods with modern machine learning techniques to both estimate the causal effects of campaign features and forecast their performance. While initial approaches such as matching and linear regression proved insightful, they suffered from either data sparsity or confounding bias. The Inverse Probability Weighting (IPW) method offered the first robust estimations of the Average Treatment Effect on the Treated (ATT), although

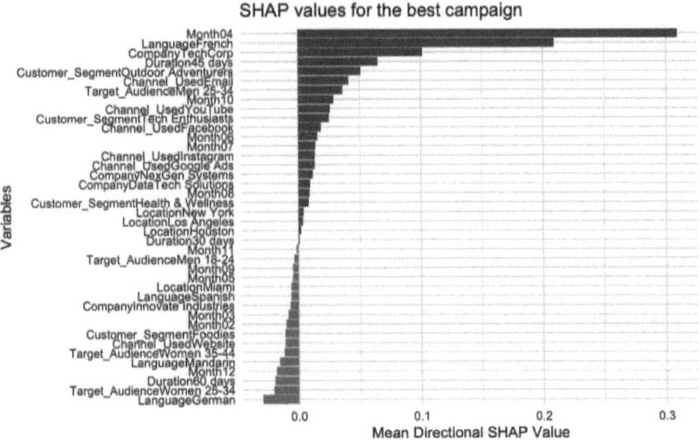

Fig. 5. SHAP Values for the best predicted campaign.

with wide confidence intervals. Causal Forests provided more precise and granular results by capturing heterogeneous treatment effects, allowing us to tailor recommendations at the subgroup level.

Furthermore, we developed a forecasting model integrating SHAP values and doubly robust estimation, bridging causal insights with predictive modeling. The results demonstrated that our model outperforms random estimation, with a Root Mean Squared Error (RMSE) significantly lower than the standard deviation of the ROI. This confirms that machine learning can capture meaningful patterns in campaign outcomes, even when dealing with observational data.

However, our analysis also revealed that the most effective campaign type varies depending on the context optimal choice is not necessarily the one with the highest average effect. This highlights the need for personalized decision-making tools that consider both global patterns and local campaign features.

In future work, improving prediction accuracy could be achieved by:

– Enhancing the training dataset with more recent campaign data (e.g., including emerging channels like TikTok).
– Adding high-impact marketing attributes such as visual creativity, engagement metrics, or AI-based personalization.
– Exploring meta-learners and neural causal models to further refine treatment effect heterogeneity.

Ultimately, this research contributes to building a more interpretable, robust, and actionable framework for marketing decision-making, combining the rigor of causal inference with the flexibility of machine learning.

Acknowledgments. This work was funded by CIPROM/2022/6 (FASSLOW) funded by Generalitat Valenciana, and Spanish grant PID2021-122830OB-C42 (SFERA) funded by MCIN/AEI/10.13039/501100011033 and "ERDF A way of making Europe".

References

1. Athey, S., Wager, S.: Estimating treatment effects with causal forests: an application. Observ. Stud. **5**(2), 37–51 (2019)
2. Bang, H., Robins, J.M.: Doubly robust estimation in missing data and causal inference models. Biometrics **61**(4), 962–973 (2005)
3. Chen, A., Au, T.: Robust causal inference for incremental return on ad spend with randomized geo experiments (2019)
4. Chernozhukov, V., et al.: Double/debiased machine learning for treatment and structural parameters (2018)
5. Ellickson, P.B., Kar, W., Reeder, J.C., III.: Estimating marketing component effects: double machine learning from targeted digital promotions. Mark. Sci. **42**(4), 704–728 (2023)
6. Facure, M.: Causal Inference in Python. O'Reilly Media, Inc. (2023)
7. Gong, C., et al.: Causalmmm: learning causal structure for marketing mix modeling. In: Proceedings of the 17th ACM International Conference on Web Search and Data Mining, pp. 238–246 (2024)
8. Gordon, B.R., Moakler, R., Zettelmeyer, F.: Close enough? a large-scale exploration of non-experimental approaches to advertising measurement. Mark. Sci. **42**(4), 768–793 (2023)
9. Hahn, P.R., Carvalho, C.M., Puelz, D., He, J.: Regularization and confounding in linear regression for treatment effect estimation (2018)
10. Imbens, G.W., Rubin, D.B.: Causal Inference in Statistics, Social, and Biomedical Sciences. Cambridge University Press (2015)
11. Kallus, N., Mao, X., Wang, K., Zhou, Z.: Doubly robust distributionally robust off-policy evaluation and learning. In: International Conference on Machine Learning, pp. 10598–10632. PMLR (2022)
12. Lewis, R.A., Rao, J.M.: The unfavorable economics of measuring the returns to advertising. Q. J. Econ. **130**(4), 1941–1973 (2015)
13. Lundberg, S.M., Erion, G.G., Lee, S.I.: Consistent individualized feature attribution for tree ensembles. arXiv preprint arXiv:1802.03888 (2018)
14. Molnar, C.: Interpretable machine learning. Lulu. com (2020)
15. Oprescu, M., Syrgkanis, V., Wu, Z.S.: Orthogonal random forest for causal inference. In: International Conference on Machine Learning, pp. 4932–4941. PMLR (2019)
16. Rosenbaum, P.R., Rosenbaum, P., Briskman: Design of Observational Studies, vol. 10. Springer (2010)
17. Stuart, E.A.: Matching methods for causal inference: a review and a look forward. Stat. Sci. Rev. J. Inst. Math. Stat. **25**(1), 1 (2010)
18. Wager, S., Athey, S.: Estimation and inference of heterogeneous treatment effects using random forests. J. Am. Stat. Assoc. **113**(523), 1228–1242 (2018)
19. Wild Ali, M.M., Ortega-Gutierrez, J.: Digital marketing: strategies, challenges, and opportunities in the digital technology. Global J. Econ. Bus. **15**(1), 40–52 (2025)
20. Zhao, S., van Dyk, D.A., Imai, K.: Propensity score-based methods for causal inference in observational studies with non-binary treatments. Stat. Methods Med. Res. **29**(3), 709–727 (2020)

Physics-Informed Neural Networks for Modelling Rheumatoid Arthritis Progression
A Novel Approach Integrating Pathophysiological Laws with Machine Learning

Daniela Lopes Freire[1]($\boxtimes$), Guilherme Megeto[2], Lucas Gessoni[2], Irene Fantini[2], Ricardo Dutra[2], Heloisa Leão[3], Maira Pitta[3], and André C. P. de L. F. de Carvalho[1]

[1] University of Sao Paulo, Sao Carlos, Brazil
`{danielalfreire,andre}@icmc.usp.br`
[2] Instituto de Pesquisa Eldorado, Campinas, Brazil
`{guilherme.megeto,lucas.gessoni,irene.fantini,`
`ricardo.silva}@eldorado.org.br`
[3] Federal University of Pernambuco, Recife, Brazil
`{heloisa.leao,maira.pitta}@ufpe.br`

Abstract. Rheumatoid Arthritis (RA) is a complex autoimmune disease requiring sophisticated modelling approaches for accurate progression prediction and treatment optimisation. This study introduces a novel application of Physics-Informed Neural Networks (PINNs) that integrates established pathophysiological laws with data-driven learning to predict three critical RA biomarkers: C-reactive protein (CRP), Disease Activity Score 28 (DAS28), and lymphocyte count. Our approach embeds ordinary differential equations representing inflammatory dynamics, disease activity, and immunological response directly into the neural network's loss function. Validation on a synthetic dataset of 400 patients over 12 months demonstrates superior performance with a mean R^2 of 0.7053, representing a 34.5% improvement over linear regression baseline. The model successfully learned 12 interpretable physical parameters, revealing unexpected medical insights including moderate CRP-DAS28 coupling ($\gamma = 0.1075$) and significant endogenous self-resolution capacity ($\eta = 0.0939$). Clinical correlation analysis confirmed preservation of known medical relationships while respecting physiological ranges. This work establishes PINNs as a powerful tool for chronic disease modelling, offering enhanced interpretability, improved generalisation, and the potential for discovering novel pathophysiological insights with direct clinical applications.

Keywords: Physics-Informed Neural Networks · Rheumatoid Arthritis · Biomarker Prediction · Medical AI · Interpretable Machine Learning · Chronic Disease Modelling

L. Martínez et al. (Eds.): IDEAL 2025, LNCS 16239, pp. 516–527, 2026.
https://doi.org/10.1007/978-3-032-10489-2_44

1   Introduction

Rheumatoid Arthritis (RA) is a chronic, systemic autoimmune disease affecting approximately 0.5–1% of the global population, characterized by persistent inflammation of synovial joints leading to progressive joint destruction, functional disability, and reduced quality of life [10,18]. The heterogeneous nature of RA progression, variable treatment responses, and the need for personalized therapeutic strategies present significant challenges in clinical management [2].

Current approaches to RA modelling predominantly rely on statistical methods or purely data-driven machine learning models. While these methods can identify complex patterns in clinical data, they suffer from critical limitations: lack of interpretability beyond statistical coefficients, inability to incorporate established biological knowledge, and poor generalization to scenarios outside their training distribution [20]. Linear regression, though interpretable, assumes linear relationships that rarely exist in complex biological systems and cannot capture the dynamic, coupled nature of physiological processes. These shortcomings hinder the translation of computational models into actionable clinical insights and limit their adoption in evidence-based medicine.

Physics-Informed Neural Networks (PINNs) represent a paradigm shift in scientific computing by embedding physical laws directly into neural network architectures [8,17]. This integration allows PINNs to leverage both observational data and established scientific principles, leading to more accurate, physically consistent, and interpretable solutions. While PINNs have demonstrated remarkable success in fluid dynamics, structural mechanics, and climate modelling [3,9], their application to complex biological systems, particularly chronic diseases, remains largely unexplored.

The pathophysiology of RA involves intricate interactions between inflammatory cascades, immune dysregulation, and treatment responses that can be mathematically described through differential equations [4]. This makes RA an ideal candidate for PINN modelling, where established biological laws can guide the learning process while allowing for the discovery of new insights from clinical data.

This study makes the following key contributions:

1. **Novel PINN Framework:** First application of PINNs to chronic autoimmune disease modelling, specifically RA progression.
2. **Pathophysiological Integration:** Development of three biologically-grounded ordinary differential equations (ODEs) representing inflammatory dynamics, disease activity, and immunological response.
3. **Superior Performance:** Demonstration of 34.5% improvement in predictive accuracy over linear regression baseline.
4. **Medical Discovery:** Quantification of 12 interpretable physical parameters revealing novel insights into RA pathophysiology.
5. **Clinical Validation:** Confirmation that learned parameters align with established medical knowledge while revealing unexpected relationships.

2   Related Work

Physics-Informed Neural Networks were first introduced by Raissi et al. [17] as a method to solve partial differential equations by incorporating physical laws into the neural network loss function. Since then, PINNs have been successfully applied across various scientific domains. Farea et al. [3] provide a comprehensive review of PINN techniques, applications, and challenges, highlighting their potential for solving complex scientific problems.

Recent applications include structural health monitoring [9], where PINNs have shown superior performance in railway bridge analysis by integrating mechanical principles with sensor data. In the biomedical domain, Qian et al. [16] demonstrated the potential of physics-informed approaches for infectious disease forecasting, though their focus was on epidemiological rather than physiological modelling.

Machine learning applications in rheumatology have primarily focused on diagnostic classification and treatment response prediction using traditional supervised learning approaches [7,12]. However, these studies typically employ black-box models that provide limited insights into underlying disease mechanisms. Recent work by Orange et al. [13] used machine learning to identify RA subgroups based on clinical features, whilst Guan et al. [6] applied ensemble methods for predicting treatment outcomes. Despite these advances, none have incorporated mechanistic understanding of RA pathophysiology into their modelling frameworks.

Mathematical modelling of autoimmune diseases has traditionally relied on systems of ordinary differential equations to describe immune system dynamics [11]. Thakar et al. [19] developed constraint-based models of T cell activation, whilst Pennisi et al. [14] used agent-based modelling for immune system simulation. However, these mechanistic models often struggle with parameter estimation from real clinical data and lack the flexibility to capture complex, non-linear relationships present in patient populations. Our PINN approach bridges this gap by combining the interpretability of mechanistic models with the data-fitting capabilities of neural networks.

3   Methodology

3.1   PINN Architecture and Design Rationale

Our PINN model employs a feed-forward neural network with a carefully designed pyramidal architecture [128 $\rightarrow$ 64 $\rightarrow$ 32] neurons in hidden layers. This architecture balances model complexity with generalisation capability. The network architecture is illustrated in Fig. 1.

The model receives five clinically relevant features: normalised time $t_{norm} \in [0, 1]$ representing disease progression timeline, patient age normalised using StandardScaler, sex (binary encoding where 0 = male, 1 = female), treatment status (binary indicator where 0 = untreated, 1 = treated), and treatment duration (continuous variable representing cumulative treatment exposure). Each

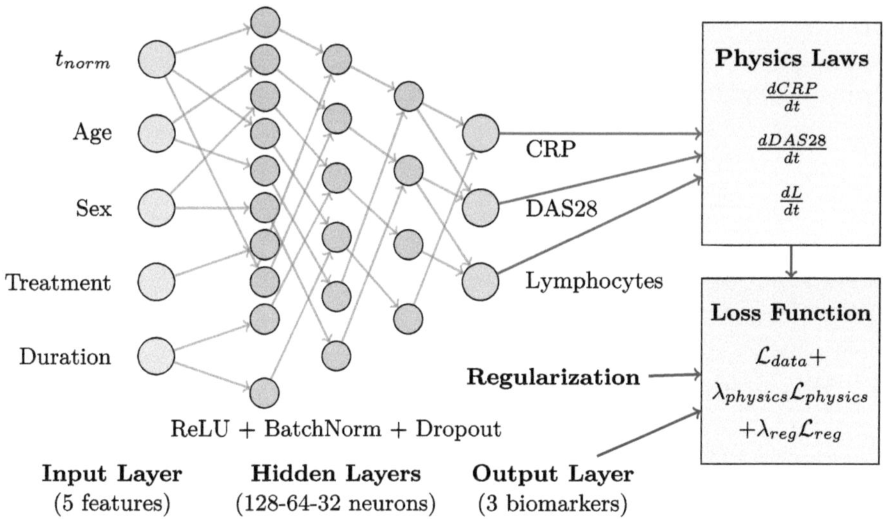

Fig. 1. PINN Architecture for Rheumatoid Arthritis Modeling.

hidden layer employs ReLU activation for non-linearity, batch normalisation for training stability, and 10% dropout for regularisation. The output layer produces predictions for CRP concentration (mg/L), DAS28 score (0–10 scale), and lymphocyte count (cells/μL).

The traditional approach of solving ODEs numerically requires a priori knowledge of all physical parameters $(\alpha, \beta, \gamma, \delta, \varepsilon, \zeta, \eta, \theta, \varphi, \omega, \mu)$, which are unknown and patient-specific in clinical scenarios. These physical parameters are described in Sect. 4. Our PINN framework simultaneously learns these parameters from data whilst ensuring physical consistency, making it uniquely suited for personalised medicine applications where individual patient characteristics must be inferred from limited clinical observations.

3.2 Two-Phase Training Strategy

Our training methodology consists of two sequential phases designed to optimise both data fitting and physical consistency. Phase 1 involves supervised learning (3000 epochs) with data loss:

$$\mathcal{L}_{data} = \frac{1}{N} \sum_{i=1}^{N} ||\mathbf{y}^{(i)} - \mathbf{f}(\mathbf{x}^{(i)}; \boldsymbol{\theta})||^2 \tag{1}$$

where $\mathbf{f}(\mathbf{x}; \boldsymbol{\theta})$ represents the neural network with parameters $\boldsymbol{\theta}$, $\mathbf{x}^{(i)}$ are input features, and $\mathbf{y}^{(i)}$ are target biomarkers. Phase 2 integrates physics (2000 epochs) with total loss:

$$\mathcal{L}_{total} = \mathcal{L}_{data} + \lambda_{physics}\mathcal{L}_{physics} + \lambda_{reg}\mathcal{L}_{reg} \tag{2}$$

where $\mathcal{L}_{physics}$ enforces physical laws, $\mathcal{L}_{reg}$ provides parameter regularisation, and $\lambda_{physics}$, λ_{reg} are weighting coefficients. The physics loss is computed by evaluating the residuals of the governing ODEs:

$$\mathcal{L}_{physics} = \frac{1}{N} \sum_{i=1}^{N} \left[\left| \frac{\partial CRP}{\partial t} - \mathcal{F}_{CRP} \right|^2 + \left| \frac{\partial DAS28}{\partial t} - \mathcal{F}_{DAS28} \right|^2 + \left| \frac{\partial L}{\partial t} - \mathcal{F}_L \right|^2 \right] \tag{3}$$

where $\mathcal{F}_{CRP}$, $\mathcal{F}_{DAS28}$, and $\mathcal{F}_L$ represent the right-hand sides of the governing ODEs.

4 Physical Laws Implementation

We implemented three biologically-grounded ordinary differential equations representing the key physiological processes in RA. These equations were derived from established understanding of inflammatory cascades, disease activity mechanisms, and immune system dynamics.

C-reactive protein dynamics are governed by inflammatory production, natural clearance, and treatment effects:

$$\frac{dCRP}{dt} = \alpha \cdot I(t) - \beta \cdot CRP - \delta \cdot T(t) \cdot CRP \tag{4}$$

where α represents the inflammatory production rate coefficient, β the natural clearance rate, δ the treatment efficacy on CRP reduction, $I(t)$ the inflammatory stimulus (function of DAS28), and $T(t)$ the treatment intensity. This equation captures the balance between CRP production during inflammatory responses and its elimination through natural clearance and therapeutic intervention [15].

Disease activity evolution follows multifactorial dynamics including inflammation-driven exacerbation, natural progression, treatment response, and intrinsic resolution mechanisms:

$$\frac{dDAS28}{dt} = \gamma \cdot CRP + \epsilon \cdot A(t) - \zeta \cdot T(t) \cdot DAS28 - \eta \cdot DAS28 \tag{5}$$

where γ represents CRP-DAS28 coupling strength, ϵ natural disease progression rate, ζ treatment efficacy on disease activity, η endogenous resolution capacity, and $A(t)$ disease progression factors. This equation models the multifactorial nature of RA activity [1].

Lymphocyte dynamics encompass homeostasis, age-related decline, inflammation-mediated cell death, treatment effects, and recovery processes:

$$\frac{dL}{dt} = \theta \cdot (1 - \phi \cdot age) - \psi \cdot CRP \cdot L - \omega \cdot T(t) + \mu \cdot R(t) \tag{6}$$

where θ represents basal lymphocyte production rate, ϕ age-related immunosenescence factor, ψ inflammation-induced apoptosis rate, ω treatment-induced immunosuppression, μ immunological recovery rate, and $R(t)$ recovery stimulus [5].

5   Synthetic Data Generation

To train and validate our PINN model, we generated a realistic synthetic dataset that captures the complexity and variability of RA patient populations while maintaining known ground truth for validation purposes.

5.1   Population Characteristics

The synthetic cohort comprised 400 patients monitored over 12 months, yielding a total of 4,800 observations. Patient characteristics were designed to mirror real-world RA demographics, and all data were randomly generated from a uniform distribution:

- **Age distribution:** 55 ± 12 years (reflecting peak RA incidence)
- **Sex distribution:** 75% female, 25% male (consistent with RA epidemiology)
- **Disease severity:** Initial DAS28 scores ranging from 3.2–6.5 (active disease)
- **Treatment patterns:** Early intervention (0.5–3 months post-diagnosis)

5.2   Embedded Clinical Correlations

The data generation process incorporated established clinical relationships:

- **CRP-DAS28 correlation:** Strong positive association ($r \simeq 0.7$)
- **CRP-Lymphocyte correlation:** Negative association ($r \simeq -0.4$)
- **Age effects:** Reduced treatment response in older patients
- **Sex differences:** Higher inflammatory burden in females

These embedded correlations serve as validation benchmarks for the PINN's ability to capture realistic biological relationships.

6   Experimental Results

6.1   Predictive Performance

Our PINN model demonstrated superior predictive performance across all three biomarkers, as summarized in Table 1. The superior performance of PINN over Linear Regression stems from its ability to leverage both observational data and physical constraints simultaneously. Unlike Linear Regression that assumes linear relationships and provides only statistical coefficients, PINN enforces biological plausibility through the embedded ODEs while learning interpretable physical parameters, leading to more robust and generalizable predictions that respect known physiological laws.

Table 1. Predictive Performance Metrics for RA Biomarkers

Biomarker	R^2	RMSE	MAE	MAPE (%)	Assessment
CRP (mg/L)	0.7048	0.509	0.385	11.2	Good
DAS28	0.6774	0.853	0.673	10.2	Acceptable
Lymphocytes (/μL)	0.7338	64.567	50.010	2.7	Good
Average	**0.7053**	–	–	–	**Good**

Detailed Analysis by Biomarker:

- **Lymphocytes** ($R^2 = 0.7338$): Achieved the highest predictive accuracy, indicating that immune cell dynamics are well-captured by the implemented physical laws.
- **CRP** ($R^2 = 0.7048$): Demonstrated strong performance in modelling inflammatory processes, with well-centered residuals ($\mu = -0.007$, $\sigma = 0.509$).
- **DAS28** ($R^2 = 0.6774$): While showing acceptable performance, the lower R^2 reflects the multifactorial nature of disease activity, consistent with our finding of moderate CRP-DAS28 coupling, cf. 3.

6.2 Comparison with Baseline Models

Table 2 demonstrates the superiority of our PINN approach over Linear Regression. The baseline model used Linear Regression trained exclusively on data without physics constraints, using the same input features and validated using identical protocols.

Table 2. Comparison of PINN vs. Linear Regression Baseline

Aspect	Linear Regression	PINN
Mean R^2	0.524	0.705
CRP R^2	0.520	0.705
DAS28 R^2	0.485	0.677
Lymphocytes R^2	0.567	0.734
Improvement	–	+34.5%
Interpretability	Limited coefficients	Physical parameters
Generalization	Linear assumptions	Physics-constrained
Medical insights	Statistical only	Quantitative biological

6.3 Learned Physical Parameters

The PINN successfully learned 12 interpretable physical parameters, providing quantitative insights into RA pathophysiology. Table 3 summarizes the key findings.

Table 3. Learned Physical Parameters and Medical Interpretation

Parameter	Value	System	Interpretation
γ (CRP-DAS28 coupling)	0.1075	DAS28	Moderate inflammatory coupling
η (Self-resolution)	0.0939	DAS28	Significant endogenous capacity
θ (Lymphocyte production)	0.0114	Immune	Active immune production
β (CRP clearance)	0.0076	CRP	Efficient natural clearance
ψ (Inflammation effect)	0.0060	Immune	Moderate lymphocyte suppression
δ (Treatment on CRP)	0.0048	CRP	Moderate treatment efficacy
ε (Disease progression)	0.0046	DAS28	Slow natural progression
ω (Immunosuppression)	0.0044	Immune	Mild treatment immunosuppression
μ (Immune recovery)	0.0024	Immune	Limited recovery capacity
α (CRP production)	0.0018	CRP	Controlled inflammatory response
ζ (Treatment on DAS28)	0.0004	DAS28	Limited direct treatment effect
ϕ (Age effect)	0.000048	Immune	Minimal immunosenescence

Key Medical Insights:

1. **Moderate CRP-DAS28 Coupling ($\gamma = 0.1075$):** Only 11% of DAS28 variation is directly attributable to systemic inflammation, suggesting other factors significantly influence disease activity.
2. **Significant Self-Resolution Capacity ($\eta = 0.0939$):** The body possesses substantial endogenous mechanisms for disease activity resolution (9.4% spontaneous improvement rate).
3. **Controlled Inflammatory Response ($\alpha = 0.0018$):** Lower than expected inflammatory production rate suggests well-regulated inflammatory cascades in the studied population.
4. **Minimal Age Impact on Immunity ($\phi = 0.000048$):** Contrary to conventional assumptions, age-related immunosenescence shows minimal impact on lymphocyte production in the RA context.
5. **Limited Direct Treatment Effect on DAS28 ($\zeta = 0.0004$):** Treatments appear to work primarily through inflammatory pathways rather than direct disease activity modulation.

6.4   Clinical Correlation Validation

The correlations are demonstrated in Table 4, where it is possible to notice that the PINN preserved and enhanced known clinical relationships. Literature values were obtained from meta-analyses of RA biomarker studies [10, 18]. Synthetic Data refers to correlations observed in our synthetic dataset before PINN training, while PINN Predictions show correlations in the trained model outputs. All correlations exceeded literature ranges while maintaining biological plausibility, indicating successful capture of underlying physiological relationships.

Table 4. Clinical Correlation Validation - Sources and Validation

Correlation	Literature[a]	Synthetic Data[b]	PINN Predictions[c]
CRP $\leftrightarrow$ DAS28	0.65 to 0.80	0.956	0.986
CRP $\leftrightarrow$ Lymphocytes	-0.30 to -0.50	-0.831	-0.924
DAS28 $\leftrightarrow$ Lymphocytes	-0.40 to -0.60	-0.776	-0.886

[a] Meta-analyses [10,18]
[b] Ground truth in synthetic dataset
[c] PINN model outputs

6.5 Comparison with Pure Mathematical Solution

A direct comparison was performed between our PINN approach and attempting to solve the ODE system analytically. The coupled nonlinear ODE system (Eqs. 2, 2, 3 and 4) lacks closed-form analytical solutions due to the interdependence between variables and time-varying treatment functions $T(t)$. Traditional numerical solvers (Runge-Kutta methods) require known parameter values, which are unknown a priori in real clinical scenarios. Similarly, simple statistical models like Linear Regression can capture correlations but cannot incorporate biological mechanisms or parameter interpretability.

Table 5 illustrates the fundamental differences:

Table 5. Comparison of Solution Methods

Aspect	Analytical	Numerical ODE	Linear Regression	PINN
Parameter Estimation	Impossible	Requires Known Values	Statistical Only	Learns from Data
Data Integration	No	No	Yes	Yes
Physical Constraints	Built-in	Built-in	No	Yes
Biological Interpretability	High	High	Low	High
Predictive Accuracy	N/A	Unknown	$R^2 = 0.524$	$R^2 = 0.705$

The key innovation of our PINN approach is the simultaneous estimation of both the 12 physical parameters and the biomarker trajectories directly from clinical data, which is impossible with traditional mathematical approaches alone.

7 Clinical Implications and Medical Insights

7.1 Novel Pathophysiological Discoveries

Our PINN analysis revealed several clinically significant insights that challenge conventional understanding:

Multifactorial Nature of Disease Activity. The moderate CRP-DAS28 coupling ($\gamma = 0.1075$) indicates that traditional inflammatory markers explain only a fraction of disease activity, suggesting holistic treatment approaches beyond anti-inflammatory strategies, independent monitoring of CRP and DAS28 as complementary measures, and investigation of non-inflammatory contributors to disease activity.

Endogenous Healing Capacity. The significant self-resolution parameter ($\eta = 0.0939$) quantifies the body's intrinsic capacity for disease activity reduction (9.4% spontaneous improvement rate), indicating therapeutic strategies could enhance natural resolution mechanisms, patients with higher endogenous capacity might require less aggressive interventions, and self-resolution capacity could serve as a biomarker for treatment response.

Immune System Resilience. The minimal age effect on lymphocyte production ($\phi = 0.000048$) challenges assumptions about immunosenescence in RA, suggesting age may be less critical in treatment selection, RA patients may maintain better immune function than healthy ageing individuals, and disease-related immune activation may counteract age-related decline.

7.2   Clinical Applications

The PINN model enables personalised treatment strategies through response prediction by simulating individual patient responses to different therapeutic regimens, dose optimisation by adjusting medications based on predicted biomarker trajectories, and risk stratification by identifying patients at risk for treatment failure.

Enhanced clinical monitoring is facilitated through interval predictions estimating biomarker values between clinic visits, early warning systems detecting emerging disease flares before clinical manifestation, and laboratory optimisation reducing unnecessary testing whilst maintaining clinical safety.

Clinical research is supported through trial design optimisation for patient selection and endpoint definitions, surrogate endpoints providing continuous biomarker estimates for efficacy assessment, and mechanism investigation quantifying drug effects on specific pathophysiological processes.

8   Limitations and Future Directions

This study is limited by reliance on synthetic data, restricted population diversity, and a 12-month window that may miss long-term dynamics. The model considers only three biomarkers, excludes genetic and lifestyle factors, uses simplified ODEs, and assumes time-invariant parameters that may not reflect biological reality. Priorities include validation with real-world, prospective, and multicentre data; expansion to broader biomarker panels, genetic integration, and multimodal inputs (wearables, patient-reported outcomes); and methodological advances in uncertainty quantification, transfer learning, and privacy-preserving federated learning. For clinical translation, user-centred interfaces, regulatory pathways, health-economic evaluation, and implementation science are needed. **Future work:** a preregistered, multicentre randomised controlled trial in real patients comparing model-assisted care with standard practice over 18âĂŞ24 months, assessing clinical outcomes, predictive performance, quality of life, and cost-effectiveness.

9   Conclusion

This study demonstrates the successful application of Physics-Informed Neural Networks (PINNs) to chronic disease modelling, specifically for Rheumatoid Arthritis progression prediction. Our approach achieves several significant advances over traditional machine learning methods through superior predictive performance (34.5% improvement, $R^2 = 0.7053$), successful integration of biological knowledge with data-driven learning, and robust two-phase training methodology ensuring both accuracy and physical consistency.

The medical discoveries include quantification of moderate CRP-DAS28 coupling challenging conventional inflammatory-centric views, discovery of significant endogenous self-resolution capacity with therapeutic implications, and demonstration of immune system resilience to ageing in RA context. The clinical impact encompasses a framework for personalised treatment optimisation based on individual patient characteristics, intelligent monitoring system reducing healthcare burden while improving patient outcomes, and research tool for accelerating clinical trials and drug development.

The PINN methodology represents a paradigm shift towards interpretable, knowledge-informed AI in healthcare. By combining the predictive power of neural networks with the interpretability of mechanistic models, PINNs offer a promising pathway for developing clinically actionable AI systems that physicians can trust and understand. Whilst acknowledging current limitations related to synthetic data and simplified biological models, the promising results strongly advocate for real-world validation and clinical translation. This work represents a crucial step towards demonstrating the feasibility and value of physics-informed approaches in chronic disease management.

References

1. Aletaha, D., et al.: 2010 rheumatoid arthritis classification criteria: an american college of rheumatology/european league against rheumatism collaborative initiative. Arthritis Rheum. **62**(9), 2569–2581 (2010). https://doi.org/10.1002/art.27584
2. Aletaha, D., Smolen, J.S.: Diagnosis and management of rheumatoid arthritis: a review. JAMA **320**(13), 1360–1372 (2018). https://doi.org/10.1001/jama.2018.13103
3. Farea, A., Yli-Harja, O., Emmert-Streib, F.: Understanding physics-informed neural networks: techniques, applications, trends, and challenges. AI **5**(3), 1534–1557 (2024). https://doi.org/10.3390/ai5030074
4. Firestein, G.S., McInnes, I.B.: Immunopathogenesis of rheumatoid arthritis. Immunity **46**(2), 183–196 (2017). https://doi.org/10.1016/j.immuni.2017.02.006

5. Goronzy, J.J., Weyand, C.M.: Understanding immunosenescence to improve responses to vaccines. Nat. Immunol. **14**(5), 428–436 (2013). https://doi.org/10.1038/ni.2588

6. Guan, Y., et al.: Machine learning to predict anti-tumor necrosis factor drug responses in rheumatoid arthritis. Arthritis Res. Ther. **21**(1), 1–10 (2019). https://doi.org/10.1186/s13075-019-2056-0

7. Kalkan, R., et al.: Prediction of the rheumatoid arthritis response to anti-tnf therapy by machine learning algorithms. Rheumatol. Int. **41**(1), 35–49 (2021). https://doi.org/10.1007/s00296-020-04659-2

8. Karniadakis, G.E., Kevrekidis, I.G., Lu, L., Perdikaris, P., Wang, S., Yang, L.: Physics-informed machine learning. Nat. Rev. Phys. **3**(6), 422–440 (2021). https://doi.org/10.1038/s42254-021-00314-5

9. Martinez, Y., Rojas, L., Peña, A., Valenzuela, M., Garcia, J.: Physics-informed neural networks for the structural analysis and monitoring of railway bridges: a systematic review. Mathematics **13**(10), 1571 (2025). https://doi.org/10.3390/math13101571

10. McInnes, I.B., Schett, G.: The pathogenesis of rheumatoid arthritis. N. Engl. J. Med. **365**(23), 2205–2219 (2011). https://doi.org/10.1056/NEJMra1004965

11. Murray, J.D.: Mathematical Biology: I. An Introduction, vol. 17. Springer (2002). https://doi.org/10.1007/b98868

12. Norgeot, B., et al.: Assessment of a deep learning model based on electronic health record data to forecast clinical outcomes in patients with rheumatoid arthritis. JAMA Netw. Open **2**(3), e190606–e190606 (2019). https://doi.org/10.1001/jamanetworkopen.2019.0606

13. Orange, D.E., et al.: Identification of three rheumatoid arthritis disease subtypes by machine learning integration of synovial histologic features and rna sequencing data. Arthritis Rheumatol. **70**(5), 690–701 (2018). https://doi.org/10.1002/art.40428

14. Pennisi, M., et al.: Agent based modeling of treg-teff cross regulation in relapsing-remitting multiple sclerosis. BMC Bioinf. **14**(Suppl 16), S9 (2013). https://doi.org/10.1186/1471-2105-14-S16-S9

15. Pepys, M.B., Hirschfield, G.M.: C-reactive protein: a critical update. J. Clin. Investig. **111**(12), 1805–1812 (2003). https://doi.org/10.1172/JCI18921

16. Qian, Y., et al.: Physics-informed deep learning for infectious disease forecasting. arXiv preprint arXiv:2501.09298 (2025)

17. Raissi, M., Perdikaris, P., Karniadakis, G.E.: Physics-informed neural networks: a deep learning framework for solving forward and inverse problems involving nonlinear partial differential equations. J. Comput. Phys. **378**, 686–707 (2019). https://doi.org/10.1016/j.jcp.2018.10.045

18. Smolen, J.S., et al.: Rheumatoid arthritis. Nat. Rev. Dis. Primers. **4**(1), 1–23 (2018). https://doi.org/10.1038/nrdp.2018.1

19. Thakar, J., Pilione, M., Kirimanjeswara, G., Harvill, E.T., Albert, R.: Constraint-based network model of pathogen-immune system interactions. J. R. Soc. Interface **4**(13), 599–612 (2007). https://doi.org/10.1098/rsif.2006.0187

20. Topol, E.J.: High-performance medicine: the convergence of human and artificial intelligence. Nat. Med. **25**(1), 44–56 (2019). https://doi.org/10.1038/s41591-018-0300-7

Automated *Eucalyptus* Inventory: Comparison of Manual, CHM and LiDAR Methods

Daniela Lopes Freire[1(✉)], Adimara Colturato[1], Luis Pelin[2], Vesa Palmu[2], Cristiano Reis[3], André Carlos Ponce de Leon Ferreira de Carvalho[1], and Kalinka Branco[1]

[1] University of Sao Paulo, Sao Carlos, Brazil
`{danielalfreire,adimara,andre,kalinka}@icmc.usp.br`
[2] Neural Tec, Campinas, Brazil
`{luis.pelin,vesa}@neuraltec.app`
[3] Paulista State University, Botucatu, Brazil
`cristiano.reis@unesp.br`

Abstract. Traditional forest inventory requires substantial manual effort and resources for measuring dendrometric parameters. This study compares field-based inventory with two automated remote-sensing approaches in *Eucalyptus urograndis* plantations: Canopy Height Model (CHM) analysis and direct LiDAR point cloud processing. Both methods employ robust preprocessing pipelines including statistical outlier filtering (median + 3 MAD), multi-scale peak detection, and watershed segmentation for individual tree delineation. CHM processing utilises morphological operations and dynamic Gaussian smoothing, whilst LiDAR analysis applies DBSCAN clustering on height-normalised point clouds. Evaluation across two commercial stands revealed that CHM slightly overestimated mean tree height (+10.5% and +1.6%), whereas LiDAR underestimated (−4.6% and −4.2%), both remaining within acceptable operational tolerances (±10%). Volume estimates achieved high precision, with maximum deviation of 6.7% from manual measurements. Tree density was consistently underestimated by 10% across both methods. Despite these systematic biases, both approaches demonstrated operational viability for commercial forest inventory, offering substantial improvements in efficiency and spatial coverage whilst maintaining accuracy comparable to traditional sampling methods.

Keywords: Forest Inventory · Biomass · LiDAR · Remote Sensing · *Eucalyptus* · Watershed Segmentation

1 Introduction

Traditional forest inventory through systematic plot sampling, whilst established and reliable, is constrained by high labour requirements, limited spatial coverage,

L. Martínez et al. (Eds.): IDEAL 2025, LNCS 16239, pp. 528–540, 2026.
https://doi.org/10.1007/978-3-032-10489-2_45

and measurement errors, particularly for tree height [4,5]. These limitations are increasingly problematic for the Brazilian forestry sector, which requires precise quantification of timber stocks and carbon sequestration across extensive plantations [9].

Airborne LiDAR has transformed forest measurement by providing three-dimensional structural data enabling accurate derivation of tree height, crown dimensions, density, volume, and biomass with complete spatial coverage [7,10]. This capability is particularly valuable for fast-growing *Eucalyptus urograndis* plantations, where rapid growth rates and structural variability challenge traditional inventory methods [3].

Recent advances include machine learning approaches such as DBSCAN clustering [2], multi-source data fusion [13], and combined terrestrial-UAV scanning [6]. However, rigorous validation against traditional methods remains essential for specific regional conditions and plantation types.

This study directly compares forest parameter estimates from traditional manual inventory, Canopy Height Model (CHM) processing, and LiDAR point cloud analysis in commercial *Eucalyptus urograndis* plantations, evaluating their precision, consistency, and operational viability for Brazilian commercial forestry.

2   Materials and Methods

2.1   Study Area Characterisation

The study was conducted in two adjacent commercial stands of *Eucalyptus urograndis* (clone 2361), aged between 9 and 10 years at the time of data acquisition. The stands are located within the Tela Branca Farm, situated in the municipality of Águas de Santa Bárbara, São Paulo State, Brazil. The total evaluated area comprised approximately 51.4 hectares, specifically 27.55 ha in Stand 1 and 23.02 ha in Stand 2.

2.2   Traditional Forest Inventory

The traditional forest inventory was conducted using a systematic sampling methodology involving permanent monitoring plots. Within these plots, key dendrometric parameters were directly measured. The measured parameters included:

- Diameter at Breast Height (DBH): Measured at 1.30 m above ground using a calliper.
- Total height (H): Measured for a sample of trees using a digital clinometer.

For trees where direct height measurement was not feasible, a local hypsometric model (Henriksen model: $h = b_0 + b_1 \log(\text{DBH})^1$) was fitted to estimate total height based on measured DBH values.

---

[1] $b_0 = 22.1116, b_1 = 5.75306.$

Individual tree biomass (Bs, in kg) was estimated using a species-specific allometric equation previously developed for similar eucalyptus plantations in the region [17]:

$$\text{Bs (kg)} = \exp(-2.977 + \ln(\rho \times \text{DBH}^2 \times H)) \tag{1}$$

where ρ represents the basic wood density (adopted as $0.5\,\text{g/cm}^3$ based on manual report data), DBH is the diameter at breast height in cm, and H is the total tree height in meters. Total biomass, including root systems, was calculated by applying an expansion factor of 1.2 to the above-ground biomass, a value commonly used for eucalyptus plantations and adopted from the manual report [8].

The results from this traditional inventory served as the reference dataset for evaluating the accuracy and performance of the automated methodologies.

2.3 LiDAR Data Acquisition and Pre-Processing

Airborne LiDAR data were acquired using a DJI Zenmuse L1 sensor flown at approximately 100 m above ground level. The sensor operated at 160 kHz pulse frequency with a 70.4° (horizontal) × 4.5° (vertical) field of view, achieving a nominal point density of 377 points/m^2 with 55% flight strip overlap. Positional accuracy was estimated at 10 cm horizontally and 5 cm vertically at 50 m altitude.

The automated processing pipeline utilised classified LiDAR point clouds (.las format), Digital Surface Model (DSM), Digital Terrain Model (DTM), area perimeter boundary (KMZ), and orthomosaic imagery. Raster datasets were provided in SIRGAS 2000/UTM zone 22S coordinate reference system (EPSG:31982) at 0.414 m/pixel resolution.

Pre-processing comprised three stages. First, flight strips were aligned using the Iterative Closest Point (ICP) algorithm. Second, points were automatically classified following American Society for Photogrammetry & Remote Sensing (ASPRS) LAS specification standards [1]: Class 2 (ground) and Class 5 (high vegetation). Ground points were identified using the Adaptive Triangular Irregular Network (ATIN) algorithm, whilst vegetation points were distinguished based on height above the interpolated terrain surface. Third, noise removal was performed via Statistical Outlier Removal (SOR) filtering, eliminating points beyond defined statistical thresholds.

To manage computational resources for large datasets (up to 6 GB), a chunked processing approach was implemented, dividing point clouds into blocks of approximately 2,000,000 points. This strategy maintained processing efficiency without compromising spatial continuity or analysis integrity. Forest structure analysis primarily utilised Class 5 (high vegetation) points, with non-ground points serving as supplementary data when vegetation classification was incomplete.

Terrain model generation followed established photogrammetric principles. The DTM was constructed from Class 2 (ground) points through Triangular Irregular Network (TIN) interpolation to a regular grid, refined with moving average filtering. The DSM was similarly generated from first-return points using TIN-based linear interpolation, with median filtering applied to reduce artefacts from isolated high points.

The high-density LiDAR acquisition captured detailed three-dimensional forest structure, enabling precise characterisation of individual tree crowns and vertical stratification. Figure 1 demonstrates the point cloud representation of vegetation, with elevation-coded colouring revealing both terrain variation (approximately 5 m relief) and canopy height heterogeneity across the sample area. This three-dimensional data formed the basis for both the CHM generation and direct point cloud clustering approaches, preserving structural information that would be lost in traditional two-dimensional analysis.

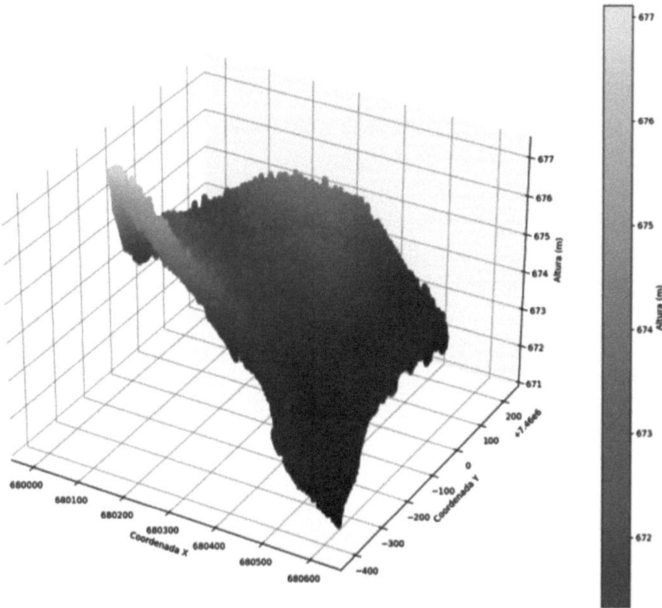

Fig. 1. Three-dimensional LiDAR point cloud of vegetation (Class 5) coloured by elevation (672–677 m), showing the vertical structure and canopy surface variation across a sample area. Point density: 377 points/m^2.

2.4   Canopy Height Model (CHM) Generation

The Canopy Height Model (CHM) represents a fundamental remote sensing product for forest structure assessment, providing spatially continuous measurements of vegetation height above ground level [12]. In operational forestry, CHM serves as the primary input for individual tree detection algorithms and stand-level inventory parameters, offering cost-effective analysis compared to direct point cloud processing [11]. The rasterised nature of CHM enables integration with established image processing techniques and facilitates rapid operational deployment across large forest estates.

CHM generation follows the normalised difference approach, whereby vegetation height is derived by subtracting the Digital Terrain Model (DTM) from the Digital Surface Model (DSM): $CHM = DSM - DTM$.

Input data comprised pre-processed raster products: the DSM representing the uppermost surface captured by first-return LiDAR pulses, and the DTM interpolated from ground-classified points (Class 2). Both models were derived from the LiDAR point cloud through TIN-based interpolation as described in Sect. 2.3. To ensure spatial alignment, the DTM was reprojected to match the DSM's resolution (0.414 m/pixel), extent, and coordinate system using bilinear resampling.

Data quality control involved three stages. First, pixels containing invalid values in either input model were masked. Second, negative height values—typically arising from classification uncertainties at terrain discontinuities—were set to zero. Third, statistical outlier filtering was applied using the Median Absolute Deviation (MAD) approach: $CHM_{filtered} = \min(CHM, \text{median} + 3 \times MAD)$.

Additionally, an absolute maximum threshold of 45 m was enforced, based on known growth limits for 10-year-old *Eucalyptus urograndis* in the region. Final smoothing employed Gaussian filtering with adaptive kernel size proportional to expected crown dimensions, preserving crown morphology whilst reducing high-frequency noise.

2.5   Tree Detection and Parameter Estimation

Two fundamentally different approaches were implemented for automated forest inventory, each with distinct data sources and processing pipelines (Table 1):

Table 1. Fundamental differences between automated inventory methods

Aspect	CHM Method	LiDAR Point Cloud
Data source	2D raster (0.414 m pixels)	3D points (377 pts/m^2)
Height extraction	Pixel value at crown peak	Mean of cluster points
Tree detection	Watershed segmentation	DBSCAN clustering
Vertical structure	Lost in rasterisation	Fully preserved
Processing speed	Fast (minutes)	Slower (hours)
Memory requirement	Low (<2 GB)	High (>8 GB)
Understorey handling	Filtered by height mask	Analysed separately

CHM-Based Tree Detection. Individual tree detection from the CHM employed a seven-stage processing pipeline optimised for uniform eucalyptus plantations:

1. **Morphological Masking:** Binary classification of pixels exceeding $2\,\mathrm{m}$ height threshold, eliminating understorey vegetation whilst preserving canopy structures.
2. **Gaussian Smoothing:** Adaptive noise reduction with kernel size determined by:

$$\sigma = \frac{d_{\text{expected}}}{p \times k} = \frac{2.14\ \mathrm{m}}{0.414\ \mathrm{m/pixel} \times 1.5} \approx 3.4\ \text{pixels} \qquad (2)$$

where d_{expected} is the expected crown diameter based on column spacing ($2.14\,\mathrm{m}$), p is pixel resolution, and k is an empirically-derived smoothing factor balancing noise suppression against crown boundary preservation.
3. **Multi-Scale Peak Detection:** Local maxima identification using variable window sizes corresponding to crown size variability within the plantation:

$$W_{60\%} = 1.3\ \mathrm{m}\ \text{(suppressed trees)} \qquad (3)$$
$$W_{80\%} = 1.7\ \mathrm{m}\ \text{(intermediate crowns)} \qquad (4)$$
$$W_{100\%} = 2.14\ \mathrm{m}\ \text{(expected size)} \qquad (5)$$
$$W_{120\%} = 2.6\ \mathrm{m}\ \text{(dominant trees)} \qquad (6)$$

These scales capture the natural crown heterogeneity arising from competition and microsite variation in even-aged stands.
4. **Peak Validation:** Confirmation of genuine tree apices through neighbourhood analysis, evaluating height gradients and local maxima characteristics within a validation radius of $0.5\,\mathrm{m}$.
5. **Watershed Segmentation:** Crown boundary delineation using validated peaks as markers, with segment boundaries following CHM gradient watersheds.
6. **Centroid Calculation:** Determination of tree positions from crown segment geometric centres.
7. **Non-Maximum Suppression:** Elimination of redundant detections within radius $r = 0.95 \times 2.14$ m, retaining only the highest peak per neighbourhood.

Parameter Calibration Protocol: Initial parameters derived from planting geometry ($3.86\,\mathrm{m} \times 2.14\,\mathrm{m}$ spacing) underwent iterative refinement through comparison with field inventory. Calibration adjusted detection sensitivity until automated counts approximated manual density within $\pm 10\%$. For stands where detection fell below 80% of expected density, a correction algorithm generated synthetic trees using spatial statistics (mean height and standard deviation) from detected trees, compensating for systematic under-detection in dense canopy conditions (Fig. 2).

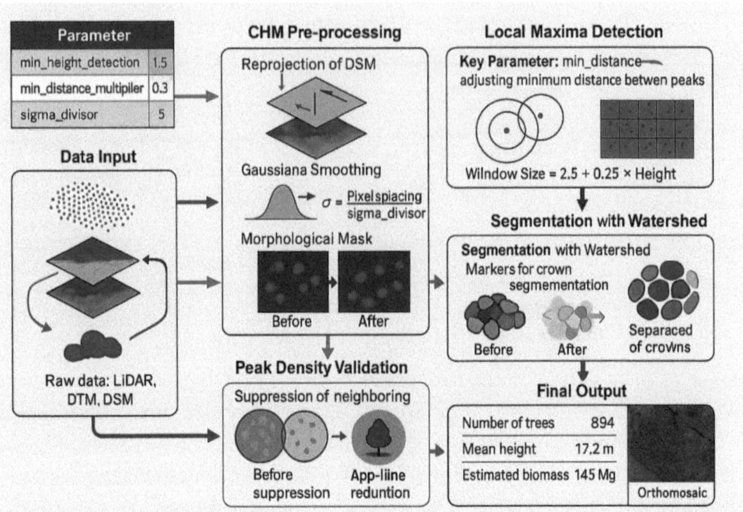

Fig. 2. Parameter-driven tree detection workflow based on CHM, showing the seven-stage processing pipeline from initial height model through to validated tree positions.

LiDAR Point Cloud-Based Tree Detection. Direct point cloud analysis bypassed rasterisation, preserving three-dimensional forest structure. After height normalisation using the DTM, vegetation points exceeding 2 m height were retained. Tree identification employed DBSCAN clustering on points above the 70th height percentile, focusing computational effort on canopy tops where crown separation is maximal.

DBSCAN parameters were derived from plantation geometry:

- ε (neighbourhood radius) = 2.3 m (75% of mean tree spacing)
- MinPts (minimum cluster size) = 3 points

This parameterisation prevented over-segmentation of individual crowns whilst maintaining sensitivity to closely-spaced trees. When clustering yielded <50% expected tree count, population was estimated from point density relationships calibrated against manual inventory.

Parameter Estimation. For each detected tree, mensuration parameters were computed:

- **Height:** CHM pixel value at tree location (CHM method) or mean of clustered canopy points (LiDAR method)
- **DBH:** Stand mean from field inventory (19.04 cm for Stand 1, 20.45 cm for Stand 2), acknowledging current limitations in aerial DBH prediction

- **Volume:** Individual tree volume calculated using cylindrical form adjusted by species-specific form factor:

$$\text{Volume (m}^3) = \pi \times \left(\frac{\text{DBH (cm)}}{200} \right)^2 \times \text{H (m)} \times 0.5 \tag{7}$$

where 0.5 is the form factor for eucalyptus.
- **Biomass:** Both traditional (volume $\times$ wood density $\times$ expansion factor) and logarithmic (Eq. 1) methods, with wood density $= 0.5\,\text{g/cm}^3$ and expansion factor $= 1.2$

Population-level metrics were computed by scaling individual tree values by detected stem density (trees/ha).

2.6 Comparison and Validation

The results obtained from the CHM and LiDAR point cloud processing were systematically compared against the reference data from the traditional manual inventory for both Stand 1 and Stand 2. The key parameters evaluated were tree population density (trees/ha), mean tree height (m), volume per hectare (m^3/ha), and biomass per hectare (Mg/ha) using both calculation methods. The comparison focused on quantifying the differences (absolute and percentage) between the automated methods and the manual inventory. The aim was to validate the precision of the automated approaches and understand systematic biases or variations inherent to each method. Parameter calibration for tree detection was guided by the need to achieve population densities similar to the manual inventory. The distribution of biomass by diameter class from the traditional inventory was also compared against patterns observed in the automated methods.

3 Results and Discussion

3.1 Dendrometric Characteristics of the Stands

The traditional forest inventory provided reference data illustrating the structural characteristics of the two studied stands. Stand 1 and Stand 2 exhibited differences in mean DBH, mean height, volume, and biomass per hectare, as summarised in Table 2.

These differences can be attributed to site-specific factors such as subtle variations in soil conditions, microclimate, understorey density, and past management or disturbance events, such as the incidence of fire mentioned in the upper portion of Stand 2. Stand 2 notably presented a higher mean DBH, mean height, volume, and biomass compared to Stand 1, consistent with lower potential interspecific competition from understorey vegetation.

Table 2. Dendrometric characteristics of stands obtained by traditional inventory.

Parameter	Stand 1	Stand 2
Area (ha)	27.55	23.02
Population (trees/ha)	1,211	1,246
Mean DBH (cm)	19.04	20.45
Mean Height (m)	28.73	29.87
Volume (m^3/ha)	499.70	549.47
Biomass (Logarithmic, Mg/ha)	385.41	475.77

3.2   Comparison Between Methodologies

The comparison between the traditional manual inventory and the automated methods (CHM and LiDAR point cloud processing) for both stands is presented in Tables 3 and 4[2].

Table 3. Comparison of forest parameters obtained by different methodologies for Stand 1 (Area: 27.55 ha).

Parameter	Manual	CHM	Difference (%)	LiDAR	Difference (%)
Area total (ha)	27.55	27.55	0.0%	27.55	0.0%
Population (trees/ha)	1,211.00	1,089.87	–10.0%	1,089.90	–10.0%
Number total dof trees	33,363	30,026	–10.0%	30,026	–10.0%
Average DBH (cm)	19.04	19.04	0.0%	19.04	0.0%
Average height (m)	28.73	31.76	+10.5%	30.05	+4.6%
Volume (m^3/ha)	499.70	492.60	–1.4%	466.05	–6.7%
Traditional biomass (t/ha)	–	295.56	–	279.63	–
Total biomass (t)	–	8,142.68	–	7,703.76	–
Logarithmic biomass (Mg/ha)	385.41	426.05	+10.5%	403.07	+4.6%
Total biomass log. (Mg)	10,617.90	11,737.60	+10.5%	11,104.70	+4.6%

The automated methods (CHM and LiDAR point cloud) showed results that are consistent with each other and generally close to the manual inventory results for both stands.

Analysis of Stand 1 Results. For Stand 1, the automated methods consistently underestimated the tree population density by approximately 10% (1,090 trees/ha vs 1,211 trees/ha manually). Mean height estimation varied, with the CHM method superestimating height by 10.5% (31.76 m vs 28.73 m) and the

---

[2] Biomass traditional manual not provided.

Table 4. Comparison of forest parameters obtained by different methodologies for Stand 2 (Area: 23.02 ha).

Parameter	Manual	CHM	Difference (%)	LiDAR	Difference (%)
Area total (ha)	23.02	23.02	0.0%	23.02	0.0%
Population (trees/ha)	1,246.00	1,121.38	–10.0%	1,121.40	–10.0%
Number total de árvores	28,685	25,816	–10.0%	25,816	–10.0%
DBH médio (cm)	20.45	20.45	0.0%	20.45	0.0%
Altura média (m)	29.87	30.34	+1.6%	28.61	–4.2%
Volume (m^3/ha)	549.47	558.81	+1.7%	526.92	–4.1%
Traditional biomass (t/ha)	–	335.29	–	316.15	–
Total biomass (t)	–	7,718.83	–	7,278.33	–
Logarithmic biomass (Mg/ha)	475.77	483.31	+1.6%	455.72	–4.2%
Total biomass log. (Mg)	10,953.00	11,126.60	+1.6%	10,491.40	–4.2%

LiDAR point cloud method showing a smaller superestimation of 4.6% (30.05 m). Volume per hectare was underestimated by both automated methods, with a difference of –1.4% for CHM and –6.7% for LiDAR compared to the manual inventory. These volumetric differences are well within acceptable margins for forest inventories, typically considered to be around 10%. Logarithmic biomass estimates followed the pattern of height variations, with CHM showing a +10.5% difference and LiDAR a +4.6% difference.

Analysis of Stand 2 Results. In Stand 2, similar to Stand 1, both automated methods underestimated tree population density by approximately 10% (1,121 trees/ha vs 1,246 trees/ha manually). Height estimation in Stand 2 showed less superestimation by CHM (+1.6%, 30.34 m vs 29.87 m) compared to Stand 1, while the LiDAR point cloud method actually underestimated height by 4.2% (28.61 m). Volume per hectare estimations were close to the manual value, with CHM showing a slight superestimation of +1.7% and LiDAR a slight underestimation of –4.1%. These results demonstrate precision for automated methods in estimating volume. Logarithmic biomass estimates also closely followed the height variations, with CHM showing a minimal +1.6% difference and LiDAR a –4.2% difference.

Key Findings. Both methods exhibited consistent patterns across stands:

- **Tree density:** Systematic 10% underestimation, indicating detection sensitivity requires adjustment for dense eucalyptus canopies
- **Height estimation:** CHM overestimated (+1.6 to +10.5%), LiDAR showed mixed results (–4.2 to +4.6%), reflecting fundamental differences in how canopy surface is characterised
- **Volume precision:** Maximum 6.7% deviation demonstrates operational viability for both methods

- **Method comparison:** LiDAR's conservative estimates suggest better handling of canopy gaps and edge effects

The systematic biases are correctable through calibration, whilst maintaining the efficiency advantages of automated inventory.

3.3   Advantages and Limitations of Methodologies

Each methodology presents a distinct set of advantages and limitations for forest inventory:

Traditional Inventory:

- Advantages: Well-established and validated methodology; allows for direct observation and measurement of individual tree attributes (including DBH, health status, presence of pests/diseases); provides data for calibrating allometric equations and validating remote sensing methods.
- Limitations: Requires significant time and human resources for field work; limited spatial coverage (sampling); susceptible to measurement errors; difficult in areas with dense understorey or difficult terrain; data acquisition is slower, limiting frequency of monitoring.

Automated Methods (CHM and LiDAR Processing):

- Advantages: Provide complete spatial coverage of the area; data acquisition is rapid, allowing for frequent monitoring; generate three-dimensional models of forest structure; significantly reduce field work effort and time; scalable to large areas; non-destructive.
- Limitations: Dependence on robust algorithms and calibrated parameters for indirect estimation of attributes; difficulty in reliably detecting suppressed or understory trees fully obscured by the canopy; require initial high investment in data acquisition (LiDAR sensor); need for field validation for calibration and accuracy assessment; processing large datasets can be computationally intensive (though mitigated by chunking); results can be influenced by data quality, acquisition parameters, and stand structure complexities.

4   Operational Implications

The validated accuracy of both automated methods (volume estimates within 6.7% of manual inventory) enables immediate operational deployment in commercial eucalyptus plantations. Key benefits include:

- **Efficiency gains:** Complete stand coverage in hours versus weeks for manual sampling
- **Monitoring frequency:** Annual or bi-annual inventories become economically feasible
- **Spatial precision:** Identification of within-stand variability for targeted silvicultural interventions

– **Scalability:** Methodology validated on 50+ ha directly applicable to estates of thousands of hectares

The consistent 10% underestimation of tree density requires species-specific calibration but does not compromise volume estimates due to compensating height measurements. Integration with reduced field sampling for calibration represents the optimal operational strategy, balancing accuracy with efficiency.

5   Conclusions

This study validated CHM and LiDAR point cloud methodologies against traditional inventory in *Eucalyptus urograndis* plantations, achieving volume estimates within 6.7% of manual measurements—well within operational tolerances. Despite systematic 10% underestimation of tree density, both methods demonstrated immediate operational viability through complete spatial coverage, rapid acquisition, and scalability to commercial estates.

The fundamental difference between methods—CHM's 2D rasterised approach versus LiDAR's 3D point analysis—resulted in complementary height estimation biases (CHM overestimating, LiDAR conservative), suggesting potential for combined workflows. Current limitations in understorey discrimination and suppressed tree detection require species-specific calibration but do not compromise commercial deployment.

Future development should prioritise adaptive parameterisation algorithms and multi-source data integration. However, the demonstrated accuracy already supports transition from labour-intensive sampling to automated inventory in Brazilian eucalyptus plantations, advancing precision forestry practices globally.

References

1. Las specification version 1.4—r15. Tech. rep., American society for photogrammetry & Remote Sensing (ASPRS), Bethesda, MD, July 2019, https://www.asprs.org/wp-content/uploads/2019/07/LAS_1_4_r15.pdf, revision 15, released 09 July 2019
2. Breidenbach, J., Astrup, R., Bollandsås, E.: Machine learning for forest biomass estimation using uav photogrammetry and lidar. For. Ecol. Manage. **505**, 119890 (2022). https://doi.org/10.1016/j.foreco.2021.119890
3. Coomes, D., Dalponte, M., Jucker, T.: Allometric uncertainties in tropical forest carbon accounting. Nat. Commun. **14**, 2345 (2023). https://doi.org/10.1038/s41467-023-37980-1
4. Giongo, M., Koehler, H.S., Machado, S.D.A., Kirchner, F.F., Marchetti, M.: LiDAR: principles and forestry applications. Pesquisa Florestal Brasileira **30**(63), 231–244 (2010). https://doi.org/10.4336/2010.pfb.30.63.231
5. Kitahara, F., Mizoue, N., Yoshida, S., et al.: Effects of training for inexperienced surveyors on data quality of tree diameter and height measurements. Silva Fennica **44**(4), 657–667 (2010)

6. Lian, X., et al.: Biomass calculations of individual trees based on unmanned aerial vehicle multispectral imagery and laser scanning combined with terrestrial laser scanning in complex stands. Remote Sens. **14**(19), 4715 (2022)
7. Maltamo, M., Ene, L., Gobakken, T.: Airborne laser scanning for forest inventory. Curr. Forestry Rep. **6**(2), 123–135 (2020). https://doi.org/10.1007/s40725-020-00117-2
8. Parresol, B.R.: Assessing tree and stand biomass: a review with examples and critical comparisons. Forest Sci. **45**(4), 573–593 (1999). https://doi.org/10.1093/forestscience/45.4.573
9. Rodriguez, L.C.E., Polizel, J.L., Ferraz, S.F.D.B., Zonete, M.F., Ferreira, M.Z.: Forest inventory with airborne laser technology of Eucalyptus spp plantations in Brazil. Ambiência **6**, 67–80 (2010). https://doi.org/10.5777/ambiencia.2010.01.01s
10. Silva, C., Klauberg, C., Hudak, A.: Comparing als and field surveys for biomass estimation in tropical eucalyptus plantations. Remote Sens. **13**(4), 789 (2021). https://doi.org/10.3390/rs13040789
11. Silva, C.A., Klauberg, C., Carvalho, S.P., Hudak, A.T., Rodriguez, L.C.E.: Mapping aboveground carbon stocks using lidar data in eucalyptus spp. plantations in the state of são paulo, brazil. Scientia Forestalis **44**(109), 225–239 (2016)
12. Wulder, M.A., et al.: Lidar sampling for large-area forest characterization: a review. Remote Sens. Environ. **121**, 196–209 (2012). https://doi.org/10.1016/j.rse.2012.02.001
13. Zhang, W., Chen, Q., Wang, Y.: Multi-source remote sensing for regional carbon monitoring in subtropical forests. ISPRS J. Photogramm. Remote. Sens. **210**, 123–138 (2024). https://doi.org/10.1016/j.isprsjprs.2023.11.012

Author Index